AF846046

€95
'18
Technik

Multi-Disciplinary Engineering for Cyber-Physical Production Systems

Stefan Biffl • Arndt Lüder • Detlef Gerhard
Editors

Multi-Disciplinary Engineering for Cyber-Physical Production Systems

Data Models and Software Solutions for Handling Complex Engineering Projects

Springer

Mängelexemplar

Editors
Stefan Biffl
Institute of Software Technology
and Interactive Systems
Technische Universität Wien
Wien, Austria

Arndt Lüder
Institute of Ergonomics, Manufacturing
Systems and Automation (IAF)
Otto von Guericke University Magdeburg
Magdeburg, Germany

Detlef Gerhard
Institute of Engineering Design
and Logistics Engineering
Technische Universität Wien
Wien, Austria

ISBN 978-3-319-56344-2 ISBN 978-3-319-56345-9 (eBook)
DOI 10.1007/978-3-319-56345-9

Library of Congress Control Number: 2017940637

© Springer International Publishing AG 2017
This work is subject to copyright. All rights are reserved by the Publisher, whether the whole or part of the material is concerned, specifically the rights of translation, reprinting, reuse of illustrations, recitation, broadcasting, reproduction on microfilms or in any other physical way, and transmission or information storage and retrieval, electronic adaptation, computer software, or by similar or dissimilar methodology now known or hereafter developed.
The use of general descriptive names, registered names, trademarks, service marks, etc. in this publication does not imply, even in the absence of a specific statement, that such names are exempt from the relevant protective laws and regulations and therefore free for general use.
The publisher, the authors and the editors are safe to assume that the advice and information in this book are believed to be true and accurate at the date of publication. Neither the publisher nor the authors or the editors give a warranty, express or implied, with respect to the material contained herein or for any errors or omissions that may have been made. The publisher remains neutral with regard to jurisdictional claims in published maps and institutional affiliations.

Printed on acid-free paper

This Springer imprint is published by Springer Nature
The registered company is Springer International Publishing AG
The registered company address is: Gewerbestrasse 11, 6330 Cham, Switzerland

Foreword

Being university professor implies the attempt to provide young engineers with the required knowledge enabling them to successfully work within a field of science, in my case the field of mechanical engineering. This knowledge shall be sufficient to also cope with challenges that will come up in the next few years.

Following this line of thought, the professional life of mechanical engineers, and in my case, product engineers, has strongly changed during the last 20 years. Within the field of product engineering, the increasing capabilities of information processing have resulted in two main trends.

First, the new capabilities of information processing enable radically improved or even new engineering methodologies. Examples for improved methodologies are more detailed analysis methodologies based on finite element methods or improved simulation methodologies, now also applying improved physics simulations. Examples of new methodologies are the development of advanced creativity techniques, optimization-based problem solution strategies, for example, exploiting swarm intelligence or genetic algorithms, or even new product prototype realization methodologies, such as 3D printing.

Second, the product itself can become more intelligent and, thereby, provide advanced product features, such as advanced user interaction for product customization, or product-related services, such as self-maintenance or self-adaptation.

All these new methodologies and technologies are based on advanced application of information processing. Thus, information creation, management, and use are key results, and also challenges, in the professional life of an engineer. Thus, student capabilities shall be trained to apply these improved or new methodologies and technologies. In addition, students shall be enabled to adopt upcoming concepts, methods, and technologies in their work environment efficiently and successfully.

To make this challenge more complicated also in product engineering, engineers will not work in isolation. Product engineers work in collaborations, in changing groups of engineers, who together aim at solving an engineering problem. Product engineers have to share knowledge with/from different engineering disciplines to enable the appropriate use of this knowledge.

analytics in manifold ways increase the capabilities and efficiency of CPPS. CPPS-based Product Service Systems open new business opportunities.

Wien, Austria
February 2017

Stefan Biffl
Detlef Gerhard
Arndt Lüder

Contents

1 Introduction to the Multi-Disciplinary Engineering for Cyber-Physical Production Systems 1
Stefan Biffl, Detlef Gerhard, and Arndt Lüder

Part I Product and Systems Design

2 Product and Systems Engineering/CA* Tool Chains 27
Kristin Paetzold

3 Cyber-Physical Product-Service Systems 63
Stefan Wiesner and Klaus-Dieter Thoben

4 Product Lifecycle Management Challenges of CPPS 89
Detlef Gerhard

Part II Production System Engineering

5 Fundamentals of Artifact Reuse in CPPS 113
Arndt Lüder, Nicole Schmidt, Kristofer Hell, Hannes Röpke, and Jacek Zawisza

6 Identification of Artifacts in Life Cycle Phases of CPPS 139
Arndt Lüder, Nicole Schmidt, Kristofer Hell, Hannes Röpke, and Jacek Zawisza

7 Description Means for Information Artifacts Throughout the Life Cycle of CPPS 169
Arndt Lüder, Nicole Schmidt, Kristofer Hell, Hannes Röpke, and Jacek Zawisza

8 Engineering of Next Generation Cyber-Physical Automation System Architectures 185
Matthias Foehr, Jan Vollmar, Ambra Calà, Paulo Leitão, Stamatis Karnouskos, and Armando Walter Colombo

9 Engineering Workflow and Software Tool Chains of Automated Production Systems 207
Anton Strahilov and Holger Hämmerle

10 Standardized Information Exchange Within Production System Engineering 235
Arndt Lüder, Nicole Schmidt, and Rainer Drath

Part III Information Modeling and Integration

11 Model-Driven Systems Engineering: Principles and Application in the CPPS Domain 261
Luca Berardinelli, Alexandra Mazak, Oliver Alt, Manuel Wimmer, and Gerti Kappel

12 Semantic Web Technologies for Data Integration in Multi-Disciplinary Engineering 301
Marta Sabou, Fajar J. Ekaputra, and Stefan Biffl

13 Patterns for Self-Adaptation in Cyber-Physical Systems 331
Angelika Musil, Juergen Musil, Danny Weyns, Tomas Bures, Henry Muccini, and Mohammad Sharaf

14 Service-Oriented Architectures for Interoperability in Industrial Enterprises 369
Ahmed Ismail and Wolfgang Kastner

15 A Deterministic Product Ramp-up Process: How to Integrate a Multi-Disciplinary Knowledge Base 399
Roland Willmann and Wolfgang Kastner

16 Towards Model Quality Assurance for Multi-Disciplinary Engineering 433
Dietmar Winkler, Manuel Wimmer, Luca Berardinelli, and Stefan Biffl

17 Conclusions and Outlook on Research for Multi-Disciplinary Engineering for Cyber-Physical Production Systems 459
Stefan Biffl, Detlef Gerhard, and Arndt Lüder

Index 469

List of Contributors

Oliver Alt Lieber Lieber GmbH, Vienna, Austria

Luca Berardinelli Business Informatics Group, Technische Universität Wien, Wien, Austria

Stefan Biffl Technische Universität Wien, Wien, Austria

Tomas Bures Department of Distributed and Dependable Systems, Charles University Prague, Prague, Czechia

Ambra Calà Otto-v.-Guericke University, Magdeburg, Germany

Siemens AG Corporate Technology, Erlangen, Germany

Armando Walter Colombo University of Applied Sciences Emden/Leer, Emden, Germany

Rainer Drath ABB Research, Ladenburg, Germany

Fajar J. Ekaputra Technische Universität Wien, Wien, Austria

Matthias Foehr Siemens AG Corporate Technology, Erlangen, Germany

Detlef Gerhard Technische Universität Wien, Wien, Austria

Holger Hämmerle EKS InTech, Weingarten, Germany

Kristofer Hell Volkswagen AG, Wolfsburg, Germany

Ahmed Ismail Institute of Computer Aided Automation, Technische Universität Wien, Wien, Austria

Gerti Kappel Business Informatics Group, Technische Universität Wien, Wien, Austria

Stamatis Karnouskos SAP, Walldorf, Germany

Wolfgang Kastner Institute of Computer Aided Automation, Technische Universität Wien, Wien, Austria

Paolo Leitão Polytechnic Institute of Bragança, Bragança, Portugal

Arndt Lüder Otto-v.-Guericke University/IAF, Magdeburg, Germany

Alexandra Mazak Business Informatics Group, Technische Universität Wien, Wien, Austria

Henry Muccini DISIM Department, University of L'Aquila, L'Aquila, Italy

Angelika Musil Institute of Software Technology and Interactive Systems, Technische Universität Wien, Wien, Austria

Jürgen Musil Institute of Software Technology and Interactive Systems, Technische Universität Wien, Wien, Austria

Kristin Paetzold UniBW München, Munich, Germany

Hannes Röpke Volkswagen AG, Wolfsburg, Germany

Marta Sabou Technische Universität Wien, Wien, Austria

Nicole Schmidt Otto-v.-Guericke University/IAF, Magdeburg, Germany

Mohammad Sharaf DISIM Department, University of L'Aquila, L'Aquila, Italy

Anton Strahilov EKS InTech, Weingarten, Germany

Klaus Dieter Thoben Universität Bremen/BIBA, Bremen, Germany

Jan Vollmar Siemens AG Corporate Technology, Erlangen, Germany

Danny Weyns Department of Computer Science, KU Leuven, Leuven, Belgium

Department of Computer Science, Linnaeus University, Växjö, Sweden

Stefan Wiesner BIBA—Bremer Institut für Produktion und Logistik GmbH at the University of Bremen, Bremen, Germany

Roland Willmann Carinthia University of Applied Sciences, Villach, Austria

Manuel Wimmer Business Informatics Group, Technische Universität Wien, Wien, Austria

Dietmar Winkler SBA Research gGmbH, Vienna, Austria

Technische Universität Wien, Wien, Austria

Jacek Zawisza Otto-v.-Guericke University/IAF, Magdeburg, Germany

Chapter 1
Introduction to the Multi-Disciplinary Engineering for Cyber-Physical Production Systems

Stefan Biffl, Detlef Gerhard, and Arndt Lüder

Abstract The *Internet of Things and Services* opens new perspectives for goods and value-added services in various industrial sectors. Engineering of industrial products and of industrial production systems is a multi-disciplinary, model- and data-driven engineering process, which involves engineers coming from several engineering disciplines. These engineering disciplines exploit a variety of engineering tools and information processing systems. This book discusses challenges and solutions for the required information processing and management capabilities within the context of multi-disciplinary engineering of production systems. The authors consider methods, architectures, and technologies applicable in use cases according to the viewpoints of product engineering and production system engineering, and regarding the triangle of (1) the product to be produced by (2) a production process executed on (3) a production system resource.

This chapter motivates the need for better approaches to *multi-disciplinary engineering* (MDE) for *cyber-physical production systems* (CPPS) and provides background information for non-experts to explain the interaction between production engineering, production systems engineering, and enabling contributions from informatics. Furthermore, the chapter introduces a set of research questions and provides an overview on the book structure, chapter contributions, and benefits to the target audiences.

Keywords Multi-disciplinary engineering • Cyber-physical production systems • Product lifecycle management • Make-to-order • Model-based systems engineering

---

S. Biffl (✉) • D. Gerhard
Technische Universität Wien, Wien, Austria
e-mail: Stefan.Biffl@tuwien.ac.at; Detlef.Gerhard@tuwien.ac.at

A. Lüder
Otto-von-Guericke Universität Magdeburg, Magdeburg, Germany
e-mail: Arndt.Lueder@ovgu.de

© Springer International Publishing AG 2017
S. Biffl et al. (eds.), *Multi-Disciplinary Engineering for Cyber-Physical Production Systems*, DOI 10.1007/978-3-319-56345-9_1

1.1 Motivation

Designing and developing smart products and systems comprising embedded systems and *Internet of Things* (IoT) technology—often referred to as *Cyber-Physical Systems* (CPS)—requires the extensive collaboration of several engineering disciplines. Product creation processes embrace engineering processes of the definition of products (or modules/systems) and of the required production system. This book essentially deals with the challenges of domain-spanning engineering processes of complex technical systems in the production area. This particular focus is mirrored in the term *Cyber-Physical Production System* (CPPS), which is also used in the title of this book. CPPS depicts the projection of the CPS concept to the production domain. Nonetheless, CPPS with emphasis on smart products, smart production, and product service systems have several links to CPS concepts focusing on other domains, e.g., "smart grid" in the energy domain and "smart mobility" in the mobility domain.

Within engineering processes for CPPS, several software solutions are used for different tasks in the sense of *Model-Based Systems Engineering* (MBSE). These engineering processes lead to many different but linked models, which have to be managed and maintained. To achieve this goal, typically several types of business information systems, e.g., *Product Data Management* (PDM), *Enterprise Resource Planning* (ERP), and *Manufacturing Execution Systems* (MES), form a company-specific *Product Lifecycle Management* (PLM) solution. The more complex engineering projects and associated models get, the more emphasis has to be put on interoperability and the ability to capture of the semantics of data in interfacing different systems.

This chapter introduces key characteristics of smart product design and production system engineering and derives requirements for informatics approaches that facilitate information modelling and data integration as foundation for multidisciplinary engineering of CPPS.

The world of manufactured products, industrial goods, and services with associated businesses is changing its face. On the one hand side, there is *demand pull*. Drivers for this effect are manifold. New technical solutions are one approach to solve existing problems of the twenty-first century—often referred to as mega challenges—on a global scale. Examples are global warming, fresh water or energy shortage, and population growth. Tackling these challenges often leads to concepts, which require an increase of cost or resource efficiency, while high quality standards have to be maintained. In consequence, the complex and interconnected challenges result in complex technical systems with advanced information technology required to make them "smart". Additionally, and sometimes in contrast to the stated global challenges, huge portions of the world are living in an unprecedented wealth. This also leads to steadily growing demands in terms of high-end consumer products, mobility and transport solutions, smart homes etc. Particularly, the demand for individualized products has increased and the lifecycle of products has shortened significantly.

In the early 1990s, car manufacturers had about 3–10 different models. 25 years later, they have a huge variety of models, crossovers, and derivatives easily exceeding 50–70 major variants. The development time of a car including production system has shrunk from 6 years to 2–3 years within the same time span. The sales lifecycle duration of a car was in some cases 30 years or more and is now about 8 years on average. In the consumer electronics industry, this effect is even stronger. Once a new smart phone is on the market, the predecessor does not sell any more, and the time span is not even 1 year. Sometimes, there is even an artificially generated customer demand, which cannot be explained by rational means, but western economies heavily rely on growth, and marketing experts do their best in generating demands.

These effects have a huge impact on industrial production. Production systems have to be quickly established in parallel to product development and furthermore, agile and flexible in order to be able to respond rapidly to changed production demands and variants. There is a strong demand to transform mass production to "lot-size-one" production while—at the same time—maintaining high quality and low production cost.

On the other hand—besides *demand pull*—there is a strong *technology push*. This has an impact on the products themselves but also on the production system. Besides the progress in production technologies enabling producers to exploit improved production processes, there is progress in automation and control technology based on information processing. Recent developments in PC-based technologies make it much cheaper to integrate intelligence into production system components enabling new control system architectures and new ways of control decision taking (Vogel-Heuser et al. 2013). For instance, condition monitoring of a machine tool or production system offers the option to perform preventive maintenance tasks and thereby reduce downtime or repair costs.

Together, these drivers lead to more complex production systems, see Fig. 1.1. This complexity has to be faced within both engineering and use of production systems as well as products produced within them. To do so, engineers have developed methods and tools like mechatronical engineering, agile programming, and plug-and-play of devices assisting them in dealing with system complexity and in dealing with the necessary quality of the engineering results.

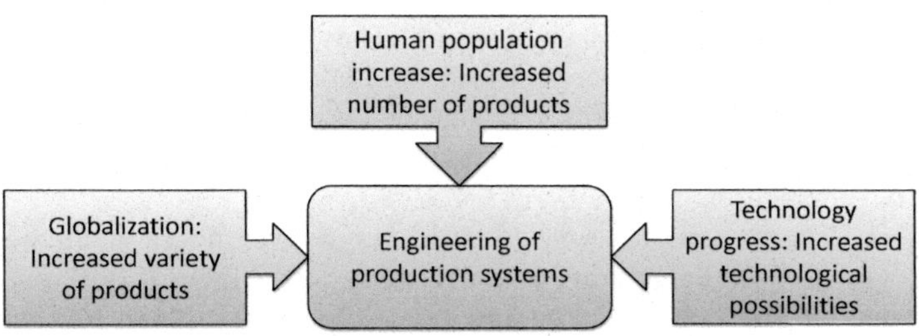

Fig. 1.1 Impact factors on production system engineering

One the one hand, we now face several engineering disciplines developed to enable the best possible engineering of a part of the overall production system. Initially in the 1950s, the engineering disciplines mainly were mechanical and electrical engineering, we now see on the one hand specialized disciplines emerging from the named two, such as multi-body simulation, computational fluid dynamics, tribology, or material sciences coming from mechanical engineering, and wiring, enclosure design, or communication system engineering coming from electrical engineering. In addition, we have seen emerge several new disciplines like control programming for *programmable logic controllers* (PLCs) and robots or optical system engineering (for laser-based welding). All of these disciplines have developed their special engineering methods, models, and terminologies applied within, and special tools to be used.

On the other hand, we see engineering process chains increasing in duration, complexity of the engineered technical system, and complexity of the required discipline-related skills, knowledge and activities. For example, engineering of a bodywork line for a car manufacturer contains around 25 engineering steps, which are executed by 25 different engineering tools. Well-known engineering activities are mechanical engineering design, electrical wiring design, and control programming. However, there are also less-known engineering steps, such as reachability analysis for welding points. All these engineering steps depend on each other's results. These dependencies form a tightly knit network. For example, the reachability analysis for welding points depends on the engineering design of the welding cell and the selection of a welding gun. In turn, the results of the reachability analysis for welding points have an impact on the engineering design of the welding cell and the welding gun selection.

Most of the engineering-discipline-specific tools have been enriched and detailed to engineering tools for the special engineering activities to be executed. Thereby, engineering-step-related dialects of the engineering methods, models and terminologies have emerged.

Engineering of production systems is conducted today in a multi-domain, multi-model, and multi-method environment, with a multitude of organizational, technical, and social dependencies. There are some initial works to analyze and optimize the raised complex engineering organizations, e.g., the VDI Guideline 3695 (VDI 3695 2009). However, the editors of this book are convinced that the improvement of production system engineering requires detailed knowledge about the boundary conditions of the engineering. These boundary conditions include possibilities of upcoming cyber-physical structures of production systems and (enforced by them) new possibilities of data and knowledge acquisition, integration, consistency evaluation, and management within collaborative multi-discipline and multi-model engineering.

CPPS is a very general term. In order to derive the research needs and challenges for CPPS engineering, it is necessary to distinguish different product types, production concepts, and production types. In the first place, four production concepts, reflecting the procedure during order processing, can be differentiated (Higgins et al. 1996): *Make-to-Stock* (MTS), or alternatively *Pick-to-Order* (PTO), reflects on production concepts for standard products without variants, which takes place independent of customer requests and orders, e.g. consumer electronics, hand machining tools, household appliances. *Assemble-to-Order* (ATO), or *Build-to-Order* (BTO), reflects on preproduction of standard products with manufacturer-specific variants irrespective of the order but connected to a customer-specific final production or assembly, e.g. cars and or personal computers. *Make-to-Order* (MTO) depicts production of standard products with customer-specific variants that are partly composed of pre-defined components and partly made up of only pre-designed components like gas turbines, airplanes, or kitchen furniture. Accordingly, new components are also created in this concept. Examples are complex machines for special products, machining tools or utility vehicles. With the *Engineer-to-Order* (ETO) concept, products according to customer specifications are produced, e.g. plant construction or shipbuilding. Because of specialization and considerable number of new components that have to be designed specific to an order, those products cannot be completely pre-engineered. A key characteristic of MTO and ETO production is the combination of existing standard parts with the new or adapted design of individual parts.

MTS products are typically produced in larger volumes (series or mass production) using a specialized production system. Those production systems have a special engineering process, which starts at a certain maturity level of the product. Largely they are optimized, often in a clocked flow production. Adaptability and reconfigurability are not main concerns in terms of linking product engineering with production system engineering. ETO products are in general fixed site fabrications with job shop pre-manufacturing of single parts or pre-assemblies. Since there are only single items or small batches to be produced, there is no special engineering of the production system, but individual workshop production on standard *numeric control* (NC) machining tools. Furthermore, intra-logistics and material handling is in general not automated. The trend towards individualized products with customer specific requirements is moving industry away from MTS and mass production towards ATO or MTO in order to meet customer demands. ATO and MTO are often considered as sub-classes of *Configure-to-Order* (CTO). Configuration of products is the essential part of the order process prior to manufacturing and assembly. Typically, the components of the product cannot be chosen independently, i.e., dependencies have to considered. However, in ATO production, the dependencies are rather simple in nature; components of the product are defined in detail and may be prefabricated in stock. In MTO production, the dependencies are more complex compared to ATO, components are manufactured as needed. This requires additional flexibility in the production process, particularly in terms of detailed production and material flow planning.

These types of production processes are mainly addressed with CPPS approaches. Additionally, a far greater collaboration of product engineering and production system engineering as well as integration of the respective IT systems is required. A high degree of flexibility for variant rich and customized products requires the adjustment product structures accordingly. This leads to higher efforts for product modularization and product line definition. A thoughtful product structure is the necessary basis of the often referred to products that control their own production. The essential task of optimization in production is to increase efficiency in terms of four each opposing target dimensions: variability, quality, speed, and economy. This applies especially for CPPS approaches.

1.2 Background

Within the prior section, production systems have been named as product of the ETO approach. In the following, the distinction between product and production system will be clarified.

Technical systems are often distinguished in product and production system. In (Stark 2015), a product is characterized as the reason a company exists for, i.e. it is created and applied within the company business making profit by selling the product. Products can be tangible like cars and cameras or intangible like a repair service for cars or a print service for photos. The combination of tangible products and associated services is referred to as *Product Service System* (PSS). In contrast, production systems are seen by the different authors in (El Maraghy 2009) as a means to create products by appropriate combination of production factors. Production factors exploited are among others materials, used work-in-progress, applied production resources (machines), and the human workers executing the activities. As easily visible, the same object, for example a bakery, can be regarded as product (by the bakery system integrator) and as a production system for cake production (by the bakery owner).

Nevertheless, there are strong dependencies between product and production system. On the on hand, the product requires a production system to be created. The production system defines boundary conditions to the properties of products possibly to be created within. On the other hand, products define requirements to the production system able to produce them. For example within production systems of optical components of cameras, dedicated cleanness conditions have to be fulfilled. Hence, within the engineering of a production system, requirements coming from the products to be created are relevant; within the engineering of the product the capabilities and boundary conditions of the production system need to be reflected, see Fig. 1.2.

Facing these dependencies, the engineering of products and production systems are interlinked and in some way equivalent. To understand this interlinking and equivalence, the term engineering needs to be understood. With respect to this book, the definition given by IEEE seems to be most appropriate. In IEEE (1941) engineering is defined as a process consisting of a sequence of activities that creatively apply scientific principles to design or develop structures, machines, apparatus, or manufacturing processes; all as respects of an intended function, economic and safe operation.

All engineers involved in an engineering project of a technical system, together with its necessary technical, economical, and management resources, shall be seen

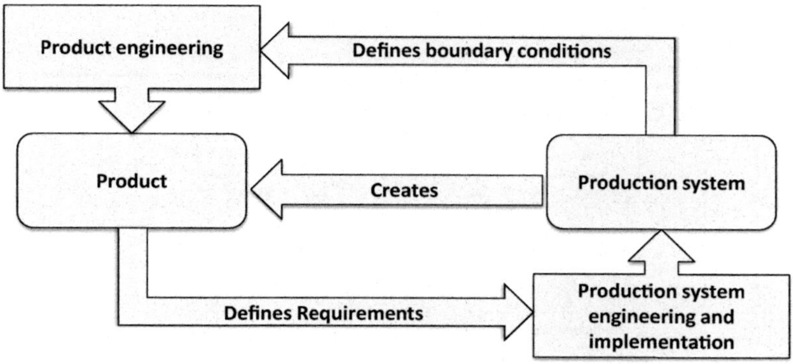

Fig. 1.2 Relations between product and production system based on (Biffl et al. 2016)

as engineering organization. The VDI 3695 Guideline "Plant engineering" (VDI 3695 2009) defines an engineering organization as a set of engineering firms or engineering subunits of a company (supplier, plant manufacturer, plant operator), which is involved in the engineering process of a technical system. This organization is involved in planning, realization, and commissioning of new technical systems and, if necessary, in upgrading, optimizing or modernizing existing technical systems. Note that an engineering organization is the execution environment and the executor of MDE.

Widening the picture of MDE engineering in the field of products and production systems, not only its dependencies buy also its life cycles, shall be reviewed. VDI/VDE (2014a) gives a good overview about these life cycles related to this chapter.

Figure 1.3 provides an overview of activities and their relations to the product and production system life cycle. The product life cycle contains engineering of a product as an individual entity to be sold. However, in this model, engineering of one product is not independent from engineering of the other products to be produced in the considered production system. Here, product line development covers the informed management of similarities and differences of products required to fulfil all relevant costumer needs and (in parallel) to not overburden the technological capabilities of the production system. Finally, a product discontinuation management belongs to each product (product family) as on the one hand customers require information about newer versions of their products to be replaced by new acquisitions and on the other hand technological progress shall be reflected within product lines. To each existing product type, the production system needs to be able to process production orders, which need to be generated, shipped and (possibly) maintained at customer sites.

The link between product and production system is the *production process* executing finally the product creation based on orders. The life cycle of a production system requires engineering of the production system before production execution. In addition, production system engineering requires the development of production

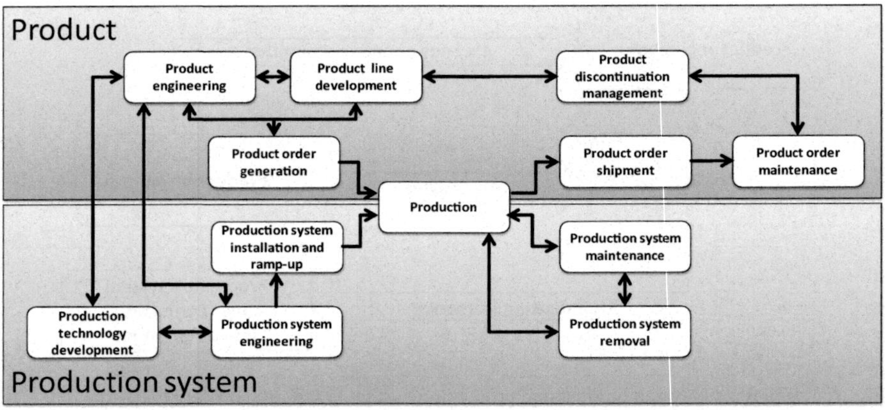

Fig. 1.3 Value-chain-oriented view on the product and production system life cycle based on (VDI/VDE 2014a) and (Biffl and Sabou 2016)

technologies to be applied within the production system. Based on the set of available production technologies (and the set of engineered products), the production system can be engineered and used in production. In addition, the production system life cycle contains production system maintenance activities as well as, in case of production system deterioration, production system removal activities.

Engineering of products and production systems involves several stakeholders. Obviously, these life cycles will add additional stakeholders relevant for the engineering of products and production systems, which especially will be responsible for the definition of the boundary conditions of intended function as well as economic and safe operation of products and production systems as it is intended in the definition of engineering. Figure 1.4 depicts the interactions between the stakeholders.

First, there is the *plant owner*. He is responsible for the economic success of the production system and, therefore, is involved in the definition of the product to be produced and the capabilities of the production system to produce the products.

The plant owner will instruct the *product engineer* with the engineering of the product as described above. He will collect all necessary boundary conditions related to the intended function of the product as well as its economic and safe operation from *potential customers* and *regulation bodies*. In addition, he collects technical boundary conditions related to the necessary production process from the *production system builder*.

In parallel, the plant owner will instruct the *production system builder* (often also named *plant integrator*) to set up a production system able to produce the intended set of products. Together the *production system engineer* and the production system builder will engineer, install, and ramp-up the production system. Therefore, *production system builder* and *engineer* will collect all necessary boundary conditions related to the intended function of the production system from

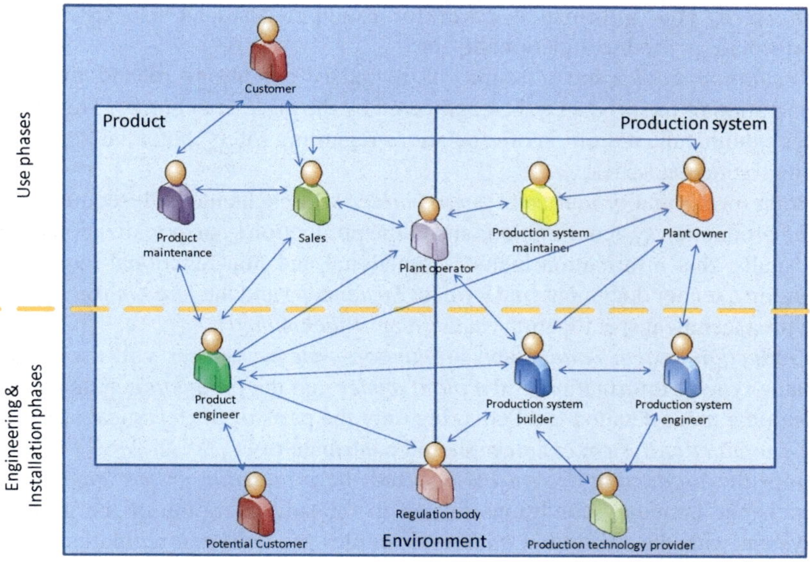

Fig. 1.4 Stakeholders in added-value chains related to industrial plant engineering based on (Biffl et al. 2016)

the *product engineer* and the *production technology provider*. Boundary conditions related to the economic and safe operation of the production system will come from *regulation bodies*, *plant operator*, and *plant owner*.

After the product and production system are engineered and (in case of the production system) set up, the production system can be used by the *plant operator* to produce products. The plant operator will get all necessary information how to produce the product from the *product engineer* and about how to use the production system from the *production system builder*. He will get product orders and their economic and technical boundary conditions, such as the due dates from *sales departments*. To ensure a long-lasting and safe operation of the production system, the *plant operator* interacts with the *production system maintainer*.

After a product has been produced, it is shipped by *sales* to the *customer* to be used. During this use phase of the product, the *customer* may interact with *sales* and *product maintenance* to ensure the economic and safe operation of the product.

Among these stakeholders, information relevant for the engineering of product and production system will be exchanged. The discussion of the complete flow of information goes far beyond the scope of this chapter. Some of the interaction flows will be considered in detail in later chapters of this book. Here, we will focus on discussing selected illustrative examples relevant for product and production system engineering.

- *Potential customers* are a source of information related to boundary conditions for the intended functions of the product to be engineered by the *Product engineer*. This information cover for example customer use cases, quality information, product functionality, etc.

- R*egulation bodies* are a source of boundary conditions related to the safe operation of the product to be engineered by the *product engineer*. This includes for example the definition of regulations regarding safety, potential hazards, and environmental issues.
- *Plant owner* and *production system builder* will exchange both requirements to the production system functions and to the production system realization process. Usually, this information includes functional and non-functional requirements within a tender document (in German: *Lastenheft*) and the *plant maker* will reply with a technical specification (in German: *Pflichtenheft*).
- *Production system builder* and *production system engineer* will exchange the same type of information as the *plant owner* and the *production system builder*, but on a more detailed level covering only the parts of the technical system that the *production system engineer* should contribute to.
- Both the *production system builder* and the *production system engineer* will exchange boundary conditions related to the safe operation of the production system with the *regulation bodies*. Examples are pollution regulations, energy consumption monitoring regulations, and human safety regulations.
- In addition, the *production system builder* and the *production system engineer* will exchange information related to possible functions of the production system and its usability in the production and/or production system setup, control, and maintenance. Among others, this covers manufacturing methods, devices required for the realisation of the manufacturing methods, control code used to control the manufacturing methods.

A dependency similar to the dependency between product and production system also exists between production system and production technologies. Within production system engineering and installation, the production system is set up based on the appropriate combination of production system components (Wagner et al. 2010). These components provide capabilities for production process execution (and in addition capabilities for its integration in the production system during installation, ramp-up, and maintenance) and can eventually be regarded as CPPS. These capabilities limit the possibilities within production system engineering and implementation. In the opposite direction, production system engineering requires special production technology capabilities to enable the creation of the intended products, which need to be reflected by production technology development. Thus, the production system in general cause requirements to further production technology development. These dependencies are depicted in Fig. 1.5.

The named dependencies between production systems and production system technologies can also be seen in a different light. Each production system component, which provides certain technological functions used within the production system, is itself a product of a company. These companies act as production technology providers and are interested in fulfilling the needs of their customers, the plant owners and production system builders, to the best extent possible.

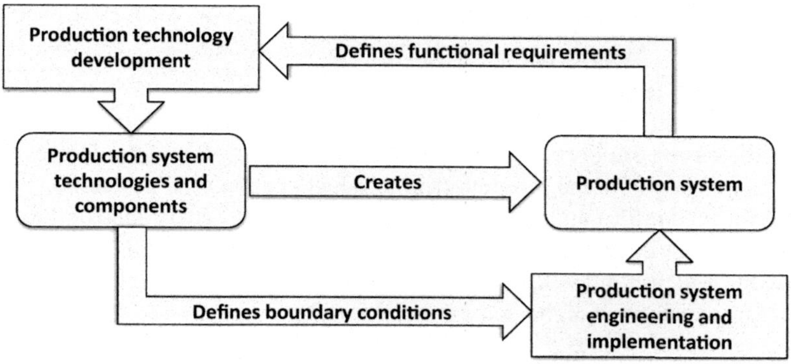

Fig. 1.5 Relations between production systems technologies and the production system based on (Biffl et al. 2016)

The sketched dependencies between product, production system, and production system technologies and components in front of the required efficient engineering involving multiple engineering disciplines is one of the sources of the newly intended comprehensive review and redesign of engineering processes like in the *Industrie 4.0* approach.

Within this initiative, companies and research institutions intend to apply technologies developed within information process and IT sciences for the implementation of flexibility and adaptability capabilities of production system resources and production processes. They are focusing on IoT and CPS (Kagermann et al. 2013). Key elements are (among others)

- Self-aware and self-adaptable production system components,
- The intelligent networking components to provide flexibility on system level using adaptation capabilities and plug-and-work capabilities, and
- The integrated exchange component information related to engineering and runtime phases along the production system life cycle.

As the *Industrie 4.0* component is a controlled part of a production system including manufacturing physics as well as control intelligence the *Industrie 4.0* component shall be considered as a *Cyber Physical Production System* (CPPS) (VDI/VDE 2014b) and shall be considered in the triangle of products, production processes, and resources (production system components). As indicated above, each product requires for its production the processes defined in its product engineering. These processes will be processed on a production system component. Each production system component will process sets of products and will be able to execute processes. Finally, each process is used for the production of products and can be executed by production system components (Pfrommer et al. 2013). Facing this fact, the production process is the lock stone within the roof architecture of the building integrating product, production system, and production system technology and components.

1.3 Research Questions

Looking on the described multi-disciplinary nature of the life cycles of products, production systems, and production system technologies and components, engineers require increasing support to ensure high quality work efficiently. However, this support requires additional research from product engineering, production systems engineering, and informatics communities. Seen from the background of the editors, three of the most interesting research fields related to the necessary support are the field of information modelling, the field of integrated information flows, and the field of key capabilities of the considered objects.

Modelling Within the life cycle of production systems, several information sets are created and applied. It is common sense that these information sets shall be represented by models and other means for description that are best applicable for the involved engineers and technical systems (hard- and software). In this field, the following research questions are of interest.

RQ M1: Modelling the structure and behavior of CPPS. *How can model-based methodologies be exploited to address the specific multi-disciplinary requirements for the representation of the structure and behavior of CPPS?* This question requires for example (a) the consideration of requirements for model-based engineering and model-based application of CPPS, (b) an investigation of usual CPPS architectures, and (c) the exploration of approaches for automating the multi-disciplinary engineering of CPPS.

RQ M2: Modelling in CPPS life cycle phases. *How can model-based methodologies support information creation and processing in the different life cycle phases of a CPPS?* Related to this question are methodologies (a) for the automation of engineering, commissioning, and use of CPPS, (b) for the application of CPPS by service providers or agents, as well as (c) for addressing the quality needs for models of CPPS.

Integrated Information Flows Supporting informed decisions by engineers requires that the relevant information is available when needed in the right quantity and quality independent of its source. This is valid for the life cycles of product, production system, and systems operation. From this need, we derive the following research questions.

RQ I1: Information integration in and across value chains. *Which methods and technologies support the integration on information within and across value chains of products, production systems, and production technologies? Are there benefits accessible from the exploitation of CPPS?* This question addresses for example (a) the links between product, production technology, and production systems engineering, (b) the horizontal and vertical integration within production systems and production value chains, and (c) the digital links between engineering and operation phases.

RQ I2: Quality assurance for information exchange. *Which methods and technologies support assuring the required information quality for information exchange?* This question includes (a) the analysis of typical requirements for the integration of engineering project data coming from heterogeneous data sources and typical requirements in a CPPS supply chain, e.g. concerning the ramp-up of a production system, the examination of multi-disciplinary knowledge integration and representation, as well as (b) the study of required information quality in different life cycle phases of CPPS.

RQ I3: Description of plug-and-play capabilities and interfaces for engineering and run time. *Are there specific aspects of information exchange related to the life cycle of CPPS?* It is assumed that relevant aspects will come from the consideration of typical requirements from product engineering and from production systems engineering on information modelling and integration in the multi-disciplinary engineering of CPPS. For example, the product engineering may provide a description of the production process (required for product creation) in a way enabling its automatic interpretation and execution in the production system. This is only possible with appropriate rich description means.

Key Capabilities of CPPS A CPPS will provide by its nature advanced capabilities like parameterizable function access and provision of state and health information related to necessary activities for its design and use along its life cycle phases, which usually requires cooperative involvement of all technical and informational parts of the CPPS. The support of multi-disciplinary work in this context will benefit from answers on the following questions.

RQ C1: Modelling of CPPS flexibility and self-adaptation capabilities. *How can model-based approaches improve the flexibility and self-adaptation of production systems? What are the roles of product, production technology, and production system models in this context?* This question includes (a) the consideration of typical requirements for flexibility and adaptability in software and in hardware systems, and (b) the analysis of methods and tools for closing the gap between product engineering and production system engineering as well as (c) the analysis of typical requirements for self-adaptation of CPPS.

RQ C2: Linking discipline-specific engineering views for flexible and self-adaptable CPPS. *How shall several disciplines in product and production system engineering be linked to support the engineering of flexible and self-adaptable CPPS?* Within this research question, the exchange of information between both engineering processes and their relation to the problem of cyber physical systems is relevant. Especially the digital shadow of products and production systems need to be considered (Table 1.1).

Table 1.1 Requirements and book chapter contributions: book chapter A addresses research question B strongly (X)

Book chapter A addressing RQ B		M1	M2	I1	I2	I3	C1	C2
Part I	**Product and Systems Design**							
Chap. 2	Product and Systems Engineering/CA* Tool Chains	X	X	X				
Chap. 3	Cyber-Physical Product-Service Systems		X		X			X
Chap. 4	Product Lifecycle Management Challenges of CPPS		X	X			X	X
Part II	**Production System Engineering**							
Chap. 5	Fundamentals of Artifact Reuse in CPPS	X	X					
Chap. 6	Identification of Artifacts in Life Cycle Phases of CPPS	X	X	X	X			
Chap. 7	Description Means for Information Artifacts Throughout the Life Cycle of CPPS	X		X	X			
Chap. 8	Engineering of Next Generation Cyber-Physical Automation System Architectures					X	X	
Chap. 9	Engineering Workflow and Software Tool Chains of Automated Production Systems			X	X	X		
Chap. 10	The Problem of Standardized Information Exchange within Production System Engineering			X	X			
Part III	**Information Modeling and Integration**							
Chap. 11	Model-Driven Systems Engineering: Principles and Application in the CPPS Domain	X	X	X	X			
Chap. 12	Semantic Web Technologies for Data Integration in Multi-Disciplinary Engineering		X	X	X			
Chap. 13	Patterns for Self-Adaptation in Cyber Physical Systems						X	
Chap. 14	Service Oriented Architecture Middleware for Vertical Integration in Industrial Enterprises			X				
Chap. 15	A Deterministic Product Ramp-Up Process—How to Integrate a Multi-Disciplinary Knowledge Base		X	X		X	X	
Chap. 16	Towards Model Quality Assurance for Multi-Disciplinary Engineering—Needs, Challenges, and Solution Concept in an AutomationML Context		X		X			

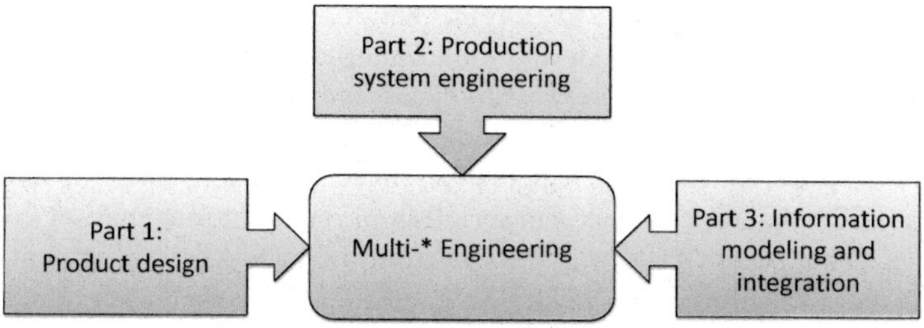

Fig. 1.6 General book structure

1.4 Book Structure

The aim of this book is to provide insight into the field of multi-* engineering, where * can stand for discipline, domain, and/or model. The book is written within the context of the upcoming next generation of production systems envisioned by research and development initiatives, such as *Industrie 4.0* in Germany, *Industrial Internet Consortium* in USA, *Factory of the Future* in France and UK, or *Made in China 2025* from China and will cover the engineering of industrial products and industrial production systems, with their dependencies named above.

The book in hand discusses topics including

- The multi-disciplinary and multi-model nature of engineering processes,
- Data integration needs along the various value adding chains,
- Dependencies between products, production processes, and production systems within engineering processes,
- Architectures of products and production systems enabling improved engineering processes, and
- Needs and approaches for information modelling and integration.

Therefore, the book is structured into three main parts, see Fig. 1.6, dedicated to product design, production system engineering, and information modelling and integration.

1.4.1 Part I: Product Design

Part I on *Product Design* discusses challenges of and approaches for designing and developing products with varying degrees of flexibility. These products provide added value to users and added complexity to production process and system engineers. An important part of engineering is the multi-disciplinary process creating information models for the evaluation of product concepts and for reuse in production systems engineering.

Chapter 2: Product and Systems Engineering/CA Tool Chains* discusses the specifics of engineering processes and the development of CPS from a mechanical engineering design point of view. Emphasis is put on *Model-Based Systems Engineering* (MBSE) methods and the required *software* tools to cope with existing challenges of different domains especially related to system analysis and system integration. The chapter contains a description of data and information flows from an organizational point of view as well as from the product development point of view. This includes information models (e.g., in SysML) as well as organization and tailoring of tools and tool chains.

Chapter 3: Cyber-Physical Product Service Systems is discussing the important topic of product related services, which are deeply integrated in the product development and use like the provision of a machining capability service useable by a producing company. It gives a definition of service-based product service systems (PSS) and unveils the state-of-the-art of CPS-based PSS with major research issues. The evolution from products to solutions and servitization is shown as well as the elements and life cycle of CPS-based PSS including hardware, software, and service elements with integration of product and service life cycles. Based on industrial use cases, this chapter also deals with challenges for engineering a CPS-based PSS in terms of complexity, end user involvement with distributed stakeholders, and involvement of multiple disciplines (e.g., mechanical engineering, information systems, and service science). This discussion of challenges leads to implications for designing engineering processes, particularly cross-domain requirements engineering and design, but also for designing servitized business models enabled by CPS (i.e., business models related to product services).

Chapter 4: Product Lifecycle Management Challenges of CPPS summarizes data and information management issues arising from the advanced use of *Model-Based Systems Engineering* (MBSE) methods that result from engineering processes of smart systems and individualized products with high complexity and variability. The chapter focuses on challenges of the life-cycle integration of products and the respective CPPS especially addressing the information exchange oriented possible dependencies between engineering, production, and use phases of products. Furthermore, data and information management problems coming from integration of the named life cycle phases of products and systems in terms of forward and backward information flows are addressed.

1.4.2 Part II: Production System Engineering

Part II on *Production System Engineering* discusses the design of flexible production systems, which can be adapted effectively and efficiently to provide a scope of production processes and address the challenges coming from products and production processes, which have advanced requirements on flexibility. Key topics are concepts, methods, and tools to deal with dependencies between production system model parts and discipline-specific sub models. An important part is the simulation and virtual commissioning of flexible production systems to reduce the risks coming from added flexibility.

Chapters 5, 6, and 7 build a common frame for the consideration of hierarchical and modular production system architectures and related information along their life cycle. These chapters provide a discussion of the question, which parts of a production system can be regarded as components within the hierarchy and which functionalities and information are assigned to them.

Chapter 5: Fundamentals of Artifact Reuse in CPPS discusses meaningful layers within the hierarchy of production system components and their life cycle. Based on a literature survey and practical experiences candidates for hierarchy layers are identified and their identification criteria are named. In addition, main life cycle phases of production systems are discussed. The thereby developed hierarchy serves as a foundation for the reusability and modularization of *Industrie 4.0* components.

Chapter 6: Identification of Artifacts in Life Cycle Phases of CPPS considers in detail the information sets relevant for a production system component along the life cycle of a production system. For each of the three main life cycle phases named in Chap. 5 relevant artifacts are identified, assigned to the different layers of the production system hierarchy, and discussed against main cases of information reuse within the life cycle of production systems. Thereby, it is intended to enable an identification of hierarchy layers based on relevant information sets.

Chapter 7: Description Means for Information Artifacts Throughout the Life Cycle of CPPS again takes up the artifacts and description means related to them in each of the three life cycle phases on each layer of the hierarchical production system structure as proposed in Chap. 5. These artifacts are clustered and generic artifact classes are derived from the fragmented information artifact landscape. Description means are assigned to the artifact classes, enabling a holisting information management and paving the way for future research on this topic.

Chapter 8: Engineering of Next Generation Cyber-Physical Automation System Architectures provides a summary of non-hierarchical control system architectures that could be applied in industrial automation domain as well as a review of their commonalities. The chapter aims to point out the differences between the traditional centralized and hierarchical architecture to the discussed architectures, which rely on decentralized decision-making and control. The chapter also explores the challenges and impacts that industries and engineers face in the process of adopting decentralized control architectures, analyzing the obstacles for industrial acceptance and the necessary new interdisciplinary engineering skills. In the end, the chapter gives an outlook of possible mitigation and migration activities required to implement decentralized control architectures.

Chapter 9: Engineering Workflow and Software Tool Chains of Automated Production Systems presents an overview of tool chains that are applied in the production system engineering process. The current workflow of production system engineering is described. In particular, three essential phases of the workflow are considered in detail, namely mechanical design, electrical design, and software design. With respect to those essential phases, tool chains are presented that are well established in industry and applied by practitioners. In addition, the tool chain of planning and simulating production processes is discussed. In this regard, various engineering data formats and information that is required as input or results as output by engineering tools is explained. One conclusion that can be derived from the described workflow is the necessity of a standardized data format to exchange engineering data along the entire production system engineering process. As a consequence the role of *AutomationML* as a potential standardized data format is addressed in this chapter and exemplarily presented for the case of virtual commissioning of a production system.

Chapter 10: The Problem of Standardized Information Exchange within Production System Engineering discusses the problem of appropriate structuring (syntax) and meaning (semantics) definition for a file based data exchange technology applicable within information exchange among life cycles, engineering disciplines, and engineering activities of information driven production systems. Based on a set of use cases challenges of the information exchange and application within information driven production systems have been highlighted. The use cases have been accompanied of current standardization activities undertaken to make the use cases possible. In addition, information exchange technologies will be discussed starting with requirements an information exchange technology has to fulfil in an information driven production system and discussing the fulfilment level of these requirements provided by different existing information exchange technologies.

As a special case of file-based information exchange *AutomationML* is considered. It is discussed how *AutomationML* deals with the standardization of syntax and semantics and how the five main challenges of the standardization of data exchange formats can be fulfilled.

1.4.3 Part III: Information Modeling and Integration

Part III on *Information Modeling and Integration* discusses an informatics view on concepts, methods, and software tools for data management in heterogeneous cyber-physical production-system-engineering environments. This part will discuss data models and software solutions exploiting Model-Based System Engineering, Semantic Web, and service-oriented approaches for handling engineering projects of typical size and complexity. Several chapters discuss alternative approaches for representing engineering knowledge as foundation for designing applications to improve the effectiveness and efficiency of engineering processes in the context of multi-disciplinary engineering or CPPS. As a result, the reader can make a better informed decision on which selection of engineering knowledge representation approaches is likely to be most appropriate in a given application context.

Chapter 11: Model-Driven Systems Engineering: Principles and Application in the CPPS Domain discusses advantages and current challenges towards the adoption of model-based approaches in *cyber-physical production system* (CPPS) engineering. In particular, the chapter discusses how modeling languages and model transformations are employed to support current system engineering processes and show their application for a *Pick-and-Place Unit* (PPU) production system.

This chapter follows the model-based software engineering approach, which sees models and their metamodels as the central artifacts for engineering and for automating engineering processes. Abstraction, a key concept of modeling, can become a challenge at integration points during the engineering process in a multi-disciplinary environment, as different stakeholders may choose abstractions that are hard to reconcile with the modeling choices of other stakeholders.

Chapter 12: Semantic Web Technologies for Data Integration in Multi-Disciplinary Engineering investigates how Semantic Web technologies can support multi-disciplinary engineering processes in CPPS engineering. The chapter discusses typical requirements for intelligent data integration and access in the context of CPPS engineering and shows how these can be addressed by Semantic Web technologies and tools. For this, we draw on our own experiences in building Semantic Web solutions for engineering environments as well as on a survey of other Semantic-Web-enabled engineering projects. This chapter summarizes material published in the Springer Book entitled "Semantic Web for Intelligent Engineering Applications" (2016).

This chapter follows the Semantic Web approach, which puts the focus on the representation and integration of linked engineering knowledge as foundation for intelligent engineering applications. The Semantic Web approach originated from the need to harness the heterogeneity of information representation on the Internet. Therefore, the Semantic Web inherently assumes a variety of information models as input to designing software application for automating business and engineering processes.

Chapter 13: Patterns for Self-Adaptation in Cyber-Physical Systems investigates existing studies of CPS with regard to self-adaptation mechanisms and models, applied across the technology stack. From this investigation, we derive recurring patterns and adaptation models, consolidating design knowledge on self-adaptation in CPS, in particular CPPS. The patterns and models can support future CPS designers with the realization and coordination of self-adaptation concerns. Finally, this chapter outlines a research agenda to advance self-adaptation and coordination in the domain of CPS.

Chapter 14: Service-Oriented Architecture Middleware for Vertical Integration in Industrial Enterprises focuses on the technological aspects involved in developing a service-oriented solution for vertical integration in a heterogeneous CPPS context. The chapter addresses the typical state of industrial enterprises and the core technologies currently available for the development of a *gateway service bus* (GSB). Therefore, the chapter will discuss aspects related to enterprise and network architectures, constraints and technologies to discern the challenges to vertical integration and suggest methods for integrating GSBs in enterprises. In addition, the chapter will discuss connectivity strategies and standards that may be used to coordinate the GSB and its services, and to integrate PPS to finally generate a holistic framework for the secure operation of CPPS-based industrial plants.

This chapter follows the *Service Oriented Architecture* (SOA) approach, which represents systems as service interfaces that allow flexibly designing application systems even if the technologies of the underlying services differ and the run-time availability of services changes.

Chapter 15: Deterministic Product Ramp-Up Processes—How to Integrate a Multi-Disciplinary Knowledge Base describes the involvement of a multi-disciplinary knowledge base in a production environment in order to address the challenge of knowledge distribution across product development, production engineering, and elements of the supply chain. The chapter highlights how production data has to be maintained and prepared for the automated support of ramp-up project planning. Through this improvement of planning quality based on reusing existing production knowledge, ramp-up projects can be improved towards deterministic ramp-up processes. This chapter provides an example application for the Semantic Web approach.

Chapter 16: Towards Model Quality Assurance for Multi-Disciplinary Engineering—Needs, Challenges, and Solution Concept in an AutomationML Context discusses how models and their quality play an important role in *multi-disciplinary engineering* (MDE) projects as inputs to and outputs of engineering processes. MDE projects include various disciplines, such as mechanical, electrical, and software engineering. These disciplines apply generic and domain-specific models in their engineering context. Important challenges include model synchronization (of often-heterogeneous input from various disciplines) and *model quality assurance* (MQA) that is covered insufficiently in current MDE practices. The chapter focuses on the needs and approaches for MQA in isolated disciplines as well as in MDE environments, where engineers from different disciplines have to collaborate. Further, the chapter includes related work on MDE and MQA and presents concepts and an initial evaluation of MQA approaches in the context of selected MDE processes.

1.5 Who Shall Read This Book?

This book will be of interest to several target groups: decision makers, product and production system engineering professionals, researchers, and students within the various fields of production system engineering and information processing related sciences. All of these groups will better understand the challenges and needs of engineering project stakeholders coming from the dependencies between products and production systems with increased variability.

Decision Makers, such as industrial managers, and business professionals are interested in a general point of view on how best to make use of the capabilities that products and production systems provide. These groups will take away from this book an up-to-date view on future production system capabilities, in particular, on the challenges of and approaches for designing and developing products with varying degrees of flexibility. The CPPS vision will bring added value to users and added complexity to production process and system engineers. This added value and complexity have to be harnessed by novel kinds of families of systems, such as Product Service Systems or Product Lifecycle Management systems.

To support structuring decision making in a CPPS context, the book will provide better understanding of the benefits and limitations of applicable methods, architectures, and technologies for selected use cases.

Regarding information modelling and integration, the book will highlight the heterogeneous nature of data needed in multi-disciplinary engineering for decision making and explain data integration needs along the various value adding chains. To address data representation and integration the book will support making better informed decisions on which engineering knowledge representation approaches are likely to be most appropriate in a given application context to provide the knowledge needed for making key decisions.

Beyond information modelling and integration, the book will provide inside in the needs of information generation, processing, and use along the life cycle of products and production systems, enabling decision makers to take more informed decisions related to the management and improvement of engineering and use processes within production system environments.

Finally, the book will give an overview on informatics approaches that provide strong contributions to decision making with intelligent information representation, integration, quality assurance, and access in the context of CPPS engineering, such as Model-Based System Engineering, Semantic Web, and service-oriented approaches for CPPS engineering.

Users of Production Systems will become aware of the challenge of knowledge distribution across product development, production engineering, and elements of the supply chain. They will get an overview on approaches to select and use relevant integrated knowledge with appropriate methods, based on case studies, such as *deterministic product ramp-up*.

Engineering Professionals, including engineers of products and of production systems, will become aware of the major challenges of and approaches for designing and developing products with varying degrees of flexibility. They will better understand the viewpoints of the different engineering disciplines involved in CPPS engineering, as well as the benefits and limitations of applicable methods, architectures, and technologies for selected use cases. A core topic is the need for data integration along the various value adding chains, in particular, needs and approaches for information modelling and integration coming from engineering processes of smart systems and individualized products with high complexity and variability.

Product engineers will get better insight into the capabilities of CPPS, so they can consider these capabilities for designing innovative products. They will come to better understand the multi-disciplinary process of creating information models for the evaluation of product concepts and for reuse in production systems engineering, which is essential to achieve the key benefits of CPPS engineering.

Production systems engineers coming from different disciplines, e.g., mechanical, electrical, and software engineering, will better understand architectures of products and production systems enabling improved engineering processes, which in turn is the foundation for improved creative interaction with product engineers, and for understand flexibility options better. They will appreciate approaches for better forward and backward information flow between engineering and operation phases as a foundation for focused improvement of engineering designs and optimizing systems operations with knowledge coming from engineering models.

Finally, both product engineers and production system engineers will get inside in the needs and challenges of the other set of engineers enabling the improvement of a mutual discussion of upcoming challenges within their interaction as well as enabling the reuse of engineering information on both sides.

Researchers from the fields of product engineering, industrial production systems engineering, and information modeling and integration, will benefit from better awareness on the challenges, needs, and approaches in the multi-disciplinary and multi-model engineering of CPPS.

Product engineering researchers can consider how to use the information around engineering to design better capabilities for product engineering processes. They will be introduced to information sources from production systems engineering, e.g., using the emerging standard *AutomationML*, and from operation, e.g., using the standard *OPC UA*, that can be used for improving the product design process.

Industrial production systems engineering researchers will get a better understanding of the challenges and requirements of multi-disciplinary engineering that will guide them in future research and development activities. They will get ideas on how to use the information available around engineering and operation to design better capabilities for CPPS engineering processes. They will become aware of alternatives to hierarchical control system architectures, their potential challenges and impacts on production systems engineering. They will get a better overview of selected tool chains that are evaluated in the production system engineering process towards virtual commissioning, *AutomationML* for data exchange and engineering knowledge accumulation, and selected mechanisms for the self-adaptation in cyber-physical systems. As a consequence, they will be able to make better informed decisions on which engineering knowledge representation approaches are likely to be most appropriate in a given application context.

IT researchers will be enabled to better understand the application domain of CPPS engineering to provide relevant information management methods as a foundation to address the dependencies between products, production processes, and production systems within engineering processes. They will be supported in making the decision on which engineering knowledge representation approaches are likely to be appropriate in a given application context, based on case studies that allow comparing the contributions of Model-Based System Engineering, Semantic

Web, and service-oriented approaches for CPPS engineering. They will get an overview on key IT capabilities for CPPS engineering, such as modeling languages and model transformations, as well as intelligent data integration and access in a heterogeneous CPPS environment, as a foundation for designing and evaluating informatics contributions to CPPS engineering.

Finally, *students* of various disciplines related to production and information processing systems can use this book as textbook to gain understanding of various architectures for information creation, processing, and use within the interrelated life cycles of products and production system. For example, they will find discussions about the interrelations of life cycles, the description of special life cycles, the description of production system hierarchies, and the description of a methodology for defect identification within engineering data.

Thus, students will especially benefit from the book during their final graduation activities finding detailed representation of the state of the art related to multi-domain model-driven engineering.

Acknowledgment This work was supported by the Christian Doppler Forschungsgesellschaft, the Federal Ministry of Science, Research and Economy, and the National Foundation for Research, Technology and Development in Austria.

References

Biffl, S., Sabou, M.: Introduction. In: Biffl, S., Sabou, M. (eds.) Semantic Web Technologies for Intelligent Engineering Applications. Springer, Berlin (2016)

Biffl, S., Lüder, A., Winkler, D.: Multi-disciplinary engineering for *Industrie 4.0* – semantic challenges and needs. In: Biffl, S., Sabou, M. (eds.) Semantic Web Technologies for Intelligent Engineering Applications. Springer, Berlin (2016)

El Maraghy, H.A.: Changeable and Reconfigurable Manufacturing Systems. Springer, London (2009)

Higgins, P., Le Roy, P., Tierney, L.: Manufacturing Planning and Control – Beyond MRP II. Springer, Berlin (1996)

IEEE: Engineers council for professional development. Science. **94**(2446), 456 (1941)

Kagermann, H., Wahlster, W., Helbig, J. (eds.): Umsetzungsempfehlungen für das Zukunftsprojekt *Industrie 4.0* – Deutschlands Zukunft als Industriestandort sichern. Forschungsunion Wirtschaft und Wissenschaft, Arbeitskreis *Industrie 4.0* (2013)

Pfrommer, J., Schleipen, M., Beyerer, J.: PPRS: production skills and their relation to product, process, and resource. In: 18th IEEE Conference on Emerging Technologies and Factory Automation (ETFA 2013), Cagliary, Italy, Proceedings, pp. 1–4, September 2013

Stark, J.: Product Lifecycle Management – Volume 1: 21st Century Paradigm for Product Realisation. Springer, New York (2015)

VDI Richtlinie 3695: Engineering von Anlagen – Evaluieren und optimieren des Engineerings. Beuth Verlag, Berlin (2009)

VDI/VDE:*Industrie 4.0* – Wertschöpfungsketten. VDI/VDE Gesellschaft Mess- und Automatisierungstechnik. Statusreport (2014a)

VDI/VDE-Gesellschaft Mess- und Automatisierungstechnik (GMA): Fachausschuss "*Industrie 4.0*": *Industrie 4.0* – Gegenstände, Entitäten, Komponenten. Status report. http://www.vdi.de/technik/fachthemen/mess-und-automatisierungstechnik/industrie-40/ (2014b). Last access Feb 2015

Vogel-Heuser, B., Diedrich, C., Broy, M.: Anforderungen an CPS aus Sicht der Automatisierungstechnik. Automatisierungstechnik. **61**(10), 669–676 (2013)

Wagner, T., Haußner, C., Elger, J., Löwen, U., Lüder, A.: Engineering Processes for Decentralized Factory Automation Systems, Factory Automation 22. In-Tech, Austria. http://www.intechopen.com/articles/show/title/engineering-processes-for-decentralized-factory-automation-systems

Part I
Product and Systems Design

Chapter 2
Product and Systems Engineering/CA* Tool Chains

Kristin Paetzold

Abstract For the development of interdisciplinary technical systems such as CPS, systemic approaches which stringently summarize the logic of development are currently available. These approaches are suitable to support the complexity of both the CPS as well as the related developmental processes. However, these development methods are relatively generic. An adaptation or a tailoring to specific conditions of both the products under consideration as well as the development of boundary conditions is absolutely necessary to use them effectively and efficiently. For the development of CPS also a variety of IT tools which effectively support the product development but only if they are well coordinated with the corresponding processes, are already available. If the interfaces are described sufficiently and comprehensively, and the data characteristics of the results of the various development activities are taken into account, media discontinuities can be reduced. The major challenge in the development of complex technical systems is the overall system analysis and the system integration. To this end, modern methods such as model-based engineering in general and model-based Systems Engineering in specific, provide powerful approaches that must be applied and adjusted for the purposes of the product and process characteristics. This adjustment process to product development and the integration of MBSE approaches into the IT-structures may be seen as the main challenges for the future.

Keywords Product development • IT-structure • Systems engineering • Data- and information flow • Model-based systems engineering

2.1 Introduction

Developing Cyber Physical Systems (CPS) whose functionality is caused by strong interactions between physical and computational components (Sztipanovits 2007), pose major challenges to the development and especially the design of development

K. Paetzold (✉)
University of German Federal Armed Forces (UniBW), Munich, Germany
e-mail: kristin.paetzold@unibw.de

© Springer International Publishing AG 2017

S. Biffl et al. (eds.), *Multi-Disciplinary Engineering for Cyber-Physical Production Systems*, DOI 10.1007/978-3-319-56345-9_2

processes. CPS include solution approaches from various engineering fields such as mechanics, electrical engineering, computer science, control engineering but also thermodynamics or materials engineering. The system behaviour can ultimately no longer be derived just as a sum of individual partial functions. Instead, synergies can be skimmed off by the diverse interactions of sub-functions. This product characteristic results in a number of challenges for both the process itself as well as the methods and tools used in the development process.

Processes in general as well as product development processes in particular can be understood as a series of interrelated activities that give rise to a valuable result for the company (Hammer 2001). For product development, these activities can be specified as that they include all operations, from the product idea to the start of production (as Ehrlenspiel and Meerkamm 2013). Development processes are also characterized by a certain uniqueness. It is not necessarily the aim of achieving an always equal result but rather finding a customized solution to specific customer requirements and operating conditions, which is characterized by a high level of functionality at an equivalent quality. This leads to a paradox: in a company is never expected the same exact result of a development process is never expected twice, which is associated with the fact that the development processes are different in each case. Nevertheless, not only for reasons of efficiency and effectiveness it requires clear procedures in the design of the development processes which are connected to a standardization of these. In addition, development processes are distinguished by a high degree of innovation and creativity (Kline 1995), which again the mentioned paradox supports.

Basically, product development processes can be characterized by the following four characteristics (see Fig. 2.1):

- Data and information about products in general and to CPS in particular arise only in the context of development. In order to still be able to work

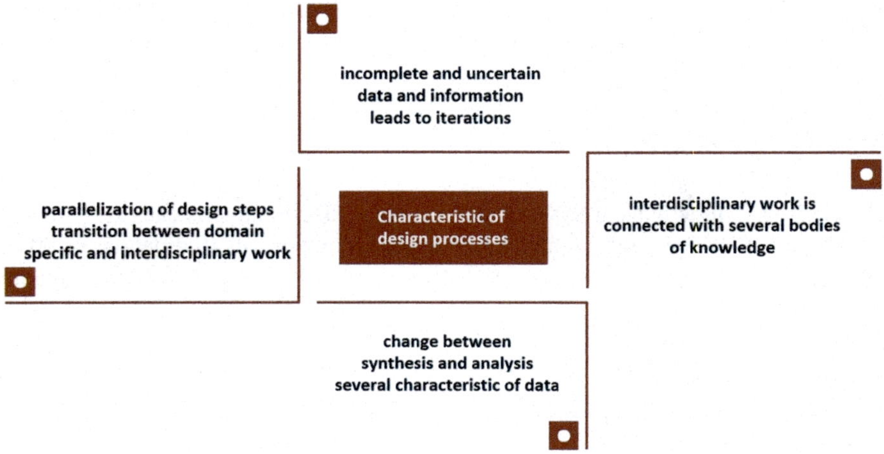

Fig. 2.1 Characteristics of the product development processes

result-oriented, it is common practice to make assumptions first which need to be concretized and evaluated in the later stages. Development processes are in accordance with the fact that there must be dealt with uncertain and incomplete data and information (Freisleben and Schabacker 2002) that become reality just in the course of development.

- The development of CPSs is highly interdisciplinary. Therefore, the different domains need to cooperate closely in all stages of the development. The challenge is that in the different domains, different models are used for the development of (Horvath and Gerritsen 2012), which in turn differ from the model approaches of system integration. The models describe the same technical system with different perspectives on it, what leads to a high variance in the information content of models.
- In terms of a concurrent engineering, developments in the individual departments run parallel. For the domain-specific development, data and information from other departments are required in general. This requires individual activities and tailored interface management to ensure that the data and information are available with sufficient quality at any stage of development.
- Development processes consist of a permanent exchange between analysis and synthesis. Data and information which are defined as part of the synthesis, need to be investigated and assessed by appropriate analytical steps regarding the fulfilment of requirements, which in turn may lead to corrections if there is the need. This is connected to the thought, that the characteristics of data for synthesis and analysis can be distinguished (Weber 2005).

The development processes of complex technical systems as CPSs thus require complex processes to secure the expectations in terms of functionality and quality. Such properties result in the fact that iterations are essential during the development. Concurrency of activities and the strong links between these individual activities through data and information exchange lead to pronounced nonlinearities in the development processes. These nonlinearities are only accessible through a detailed view of the data and information flows within the development but need to be taken into account in the design of processes.

In defining of development processes in companies, at least the corporate knowledge is manifested in the designed technical systems. The approach in the development is the result of an evolutionary process which does not only take the product of evolution into account but also effects the "Lessons Learned" or historical data and information. Reversed, development processes will be anchored in the definition of departments, team structures and responsibilities, which are in turn the basis for associate -processes such as release decisions or the actual project execution. Furthermore, emerge from the descriptions of the development process for CPS data and information requirements for each activity that must be ultimately provided and managed by the available IT infrastructure.

Process models for development are therefore not only used to represent the inherent logic in the development but also serve as basis for the design of the organizational and operational structure in the development departments and the design of the IT environment in enterprises. They also form the basis for associated processes such as risk management, change management, or the verification and validation management, which are often implemented as part of the division of labour as independent tasks of product development. Therefore, models for the description of the development process have to be considered at three levels also. Generic process models which reflect the logic of development, proved to be just little helpful for the practices of process design and optimization, because they are coarsely granular. It requires a context-specific refinement and adaptation of process models to the specific conditions in the company, to the market or the industry and, ultimately, to support the development in terms of workflow processes. In order to achieve this aim, the data and information flows are analysed within the development to link targeted activities with each other. This relatively fine-grained level of process description appears necessary to assess uncertainties due to incomplete and uncertain data, in order to increase the product maturity with the initiated activities and thus to avoid unnecessary iterations. Processes on such a fine-granular level are mainly characterized by detailed interface descriptions which have to take three aspects into account:

- Procedural interfaces result from the logical sequence of development steps which contribute to the increase of product maturity;
- Organizational interfaces define responsibilities for process steps or release mechanisms, and thereby support the quality assurance during the development;
- Formal interfaces link the IT tools which are used during the development in order to secure consistent data and information flows.

A closer look at these three levels of process description, the underlying models and the methods and approaches to each process support will be taken in the following subsection.

2.2 Generic Procedures for the Development of Interdisciplinary Products

Cyber Physical Systems are not only distinguished by the wide range of functions but also by the strong interlinkage between those functions. This distinction guarantees the variability in the response of the system in different states respectively its flexibility and robustness of the systems behavior. Such complex systems require adequate procedures in development. Systemic approaches will be used to make the strong dependences between systems design and process design explicit.

Haberfellner et al. (Haberfellner et al. 2015) recommend to organize the processes based on models for the system description, which map again both functional and structural aspects of the system being to designed (Fig. 2.2). Such holistic approach forms the basis of today's popular procedural models. System development itself is based on defined requirements of the overall system, which are

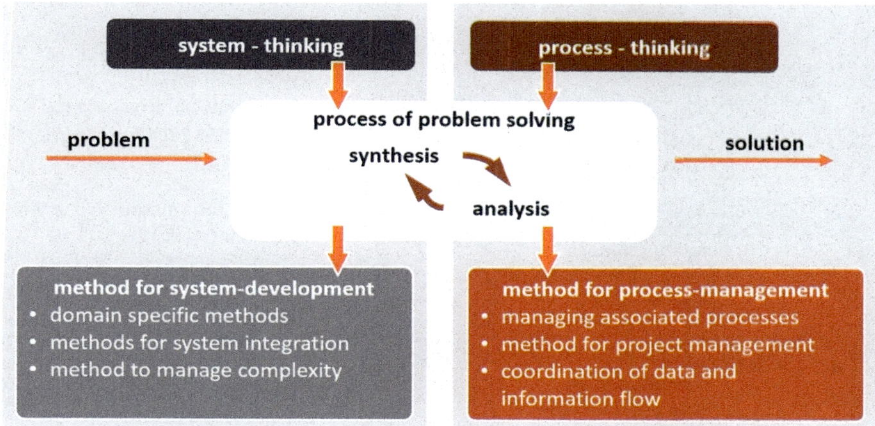

Fig. 2.2 Systemic thinking in the development of complex technical systems (according to Haberfellner et al. 2015)

successively decomposed to smaller units, where solution approaches are needed (from rough to detail). This approach in a phase structure (macro logic) with the basic steps planning, designing, and finishing of the results (e.g. Pahl and Beitz 2007). This phase structure in turn needs to be put in concrete terms, which depend on the system structure and the area of expertise, in which the development takes place.

Within these process models, the main task of the development is to seek for solutions and to check them with regard to the requested performance. This problem-solving process (micro logic) (Ehrlenspiel and Meerkamm 2013) manifests itself in a permanent alternation of synthesis and analysis and can be deduced from typical approaches to problem solving of individuals.

Individuals as developers need methodological support in the development from two perspectives: on the one hand, methods for system design and to control the system's complexity are needed and on the other hand, methods for process support and coordination of the development tasks are required. Both perspectives are explained briefly below.

2.2.1 Micro-logic in Development

The micro-logic in the development describes the operations at the level of concrete project work (Gausemeier et al. 2004) and supports the systematic processing for solving partial of problems within individual phases of the development process. The basis for this processes are generic procedures of psychology of thinking (Miller et al. 1973).

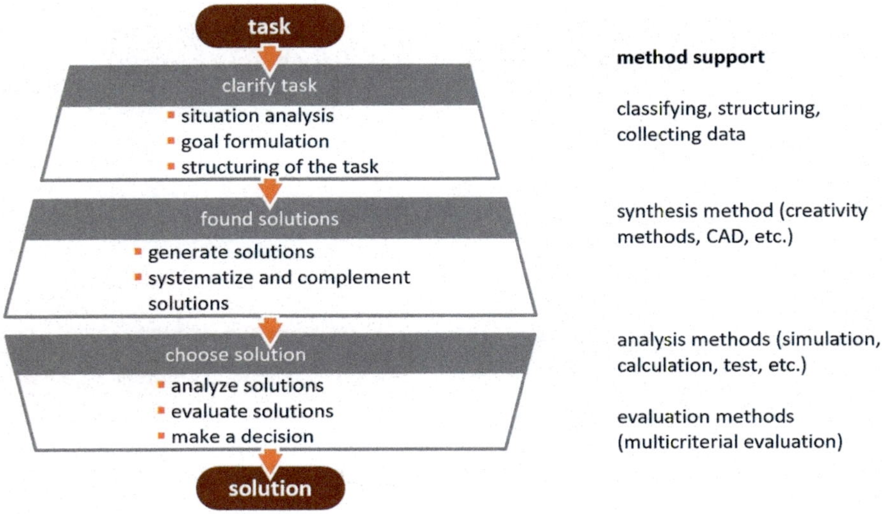

Fig. 2.3 Problem solving cycle according to Ehrlenspiel (Ehrlenspiel and Meerkamm 2013)

The problem-solving cycle (according to Ehrlenspiel and Meerkamm 2013) is shown in Fig. 2.3. After defining the roles, the broadest possible solution space is created within the framework of the synthesis which is then analysed and evaluated with respect to the achievement of objectives. Thus, the solution space is always limited. This constant interplay of analysis and synthesis is ultimately one of the reasons for iterations in the development process.

The product life cycle determines some implications for the coordination of data and information flows as well as the tool integration in the development. The individual steps can be associated with categories of methods as it is illustrated in Fig. 2.3. Special attention is given on the methods and tools for synthesis and analysis. The two process steps are interconnected via data and information flows (Fig. 2.4). It is crucial that for each of the two steps different categories of data must be captured.

For this purpose, two categories of data must be distinguished (Weber 2005):

- Characteristics which define the product; they are defined by the developer and thus serve as an adjusting screw to manipulate the properties.
- Features which describe the product behaviour; the properties cannot be influenced directly by the developer.

Exemplary for this mind set is the rough designing and dimensioning in the design. In order to meet the properties' conditions several components, for instance, part lengths or materials (characteristics) with respect to the strength or weight of the part have to be defined which must fulfil specified functions (features) (Weber 2005).

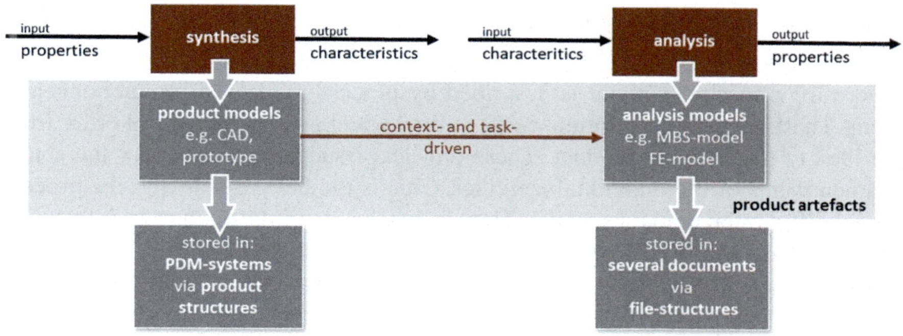

Fig. 2.4 Information flows in product development

Input values of the synthesis are properties that are requested by the technical system. Those values it is necessary to find corresponding characteristics. By analysing those values characteristics are examined to determine whether the property's performance is guaranteed or specified characteristics support the functional performance. Consequently, properties can be divided into target properties, from the customer requirements and actual properties which are the result of a development step.

Which of the various methods is suited best, is on the one hand determined by the objectives of the analysis and on the other hand by the product's maturity or rather by the developmental progress and its associated data quality respectively the uncertainty regarding the data (Reitmeier 2015). This implies the need to distinguish between product models as a result of the synthesis and analysis models. Result of synthetic steps are product models such as CAD models, prototypes or test setups. During the analysis it is necessary to transform the product models into analysis models in order to make them accessible for calculations, simulations and experiments. Examples of analysis models, such as FE models, Matlab Simulink models or test body illustrate the diversity which needs to be handled. Below both categories are summarised to product artefacts. Such a distinction is necessary to understand for instance breaks in the data and information flow.

The resulting problem solving cycle (Ehrlenspiel and Meerkamm 2013) as approach of process description on the problem-oriented level is therefore non-specific, which means that it is equally suitable for all disciplines. The problem-solving cycle is primarily used to provide situation-specific methods. Resulting instructions have descriptive character. It addresses the question of HOW the solution finding should be done (Lindemann 2007).

2.2.2 Process Models as Macro-logic in Development

Macro-logic in development is described by process models in product development. These process descriptions address the logical and chronological order from the idea to the finished product. These physical issues are relevant for the actual characteristics of the individual activities. This is the best reason why the process model differ quite domain specific. They follow a prescriptive nature and deal with the question of the WHAT within in the Process (Lindemann 2007).

Process models, as phase concepts, have a high degree of generality even within the domain. They require a sector-specific adaptation and concretisation for being used as process basis in terms of project management. It is their big advantage, that the development process is divided into manageable sections which partly can be found in the operating structures of the development departments in the company again. The structure is classified in two ways (Lindemann 2007):

- Logical: During development, an abstract situation will be concretized, whereby the data and information base progressively will be completed. Thus, the adaption for the methods which are used for synthesis and analysis is needed.
- Time: The order of the process steps and the details of the data base can therefore be used for project planning.

Process models describe generally how a predefined target system can be transferred to a more or less abstract level in a specific technical system (Negele 1998). Process models vary depending on the technical domain in which solutions for technical systems are being developed, because this solution finding and specification are very strongly influenced by the used physical effects and relationships. Nevertheless, independent of the considered domain, four phases can be identified at a very generic level (Fig. 2.5), which remind of (Pahl and Beitz 2007).

Planning Phase: This phase conduce the definition of the task and of the demands on the system to be developed. In addition to the customer or user requirements it is necessary to consider the affecting or limiting constraints from both perspectives, the company's internal situation as well as from the environment or the use out. The aim of this phase is to prepare all the development factors influencing, so that therefrom parameters for the development itself can be derived. Results besides the requirement specification is the definition of the system purpose and the expected system behaviour.

Concept phase: This phase concerns a system decomposition based on the system purpose and the expected system behaviour (Andreasen 1980). The related detailing is accompanied by a permanent exchange between function synthesis and synthesis principle. Function definition and refinement are always associated with the search for fundamental solutions, which can be derived from the use of physical effects. The choice of physical effects in turn influences the further breakdown into sub-functions (Ponn and Lindemann 2011). These partial solutions are assembled to form a working structure, which in turn defines the basic structure of technical systems.

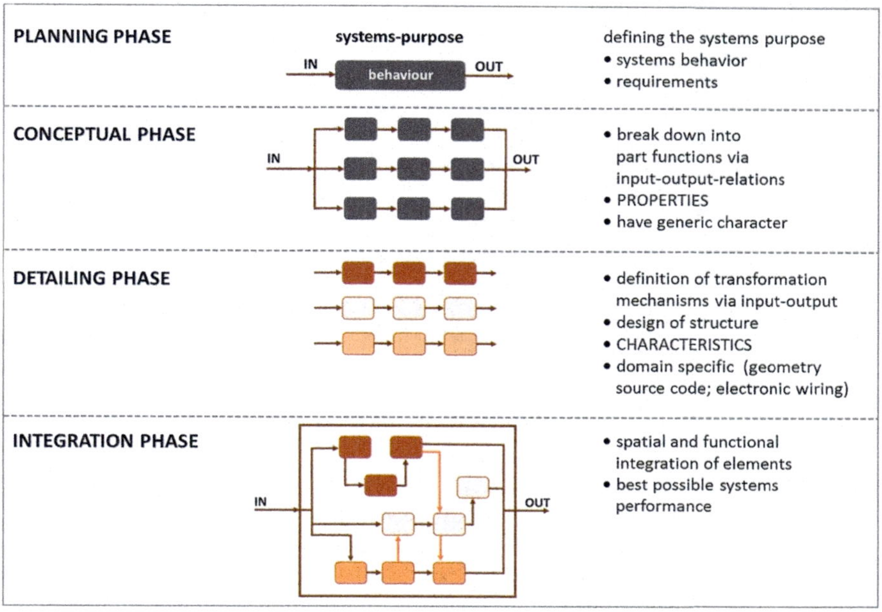

Fig. 2.5 Principle stages in product development

Design phase: The design phase is mainly influenced by the fact that the individual functions and fundamental solutions are specified and refined. Depending on the size and complexity of the technical system to be developed this design work is carried out in several different teams who then are specialized in the development of individual components.

Integration phase: Since today also in the domain-specific development, a high level on the division of labour can be observed, the integration of the results into an overall system is becoming increasingly important. In this context takes place both, a spatial and functional integration. Adaptions of the integration adjustments to the partial results or components are made such that the overall system then has the best performance. In and of itself it does not make sense to consider the system integration as downstream stage in development. Both in terms of efficiency as well as from a qualitative and functional point of view out a parallel consideration during development is strongly preferred. While this is quasi still immanent in the concept phase, it is challenging especially for further detailing.

Depending on the specific domains, different strengths of effort are required for each stage in development, which results from the physical relationships and the mathematical description associated. While the construction has to rely on the use of geometry and materials for continuous-time descriptions, in development of integrated circuits e.g. for the description of individual components, discrete event

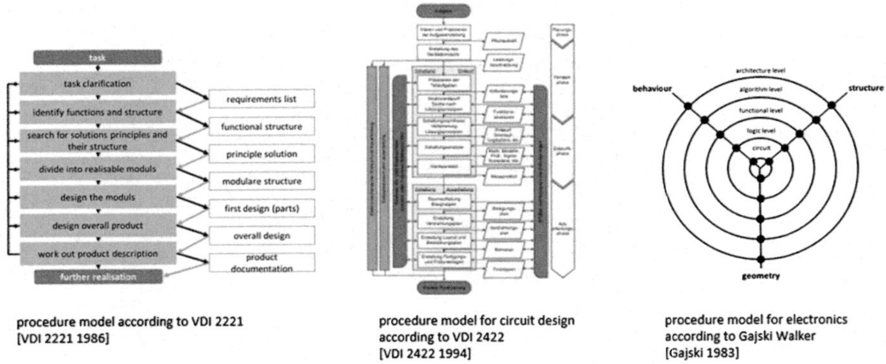

Fig. 2.6 Examples of domain-specific process models

approaches can be used. As a result, at a certain level the development is automated in circuit design, which is not feasible for mechanical design.

Differences in the specialized domains also arise because for function descriptions there is a cross-domain understanding, but the structure depths in the domains are very different (geometry in engineering, integrated circuits in electronics, source code in software development).

For these reasons, the process models in the specialized domains are correspondingly varied. Fig. 2.6 shows examples of procedures from the domain engineering, electrical engineering and software development. The differentiation of the domains is not only reflected in the structural form but also in individual activities and their designations. A detailed compilation of domain specific process models can be found in (Eigner et al. 2014).

Process models provide first only a sequential series of more or less detailed process steps, following the logic “from the abstract to the concrete”. In the development of this and especially as a result of adaptation to the specifics of different disciplines to the process models has been attempted to consider the general character of development processes. Correspondingly to the structure of the process models three main types can be differentiated. The simplest structure is a sequential approach as “logic-beam”. Partly, this structure is also called waterfall model. Since the need of iterations is well known, this was integrated into the waterfall model (Fig. 2.7a). Exemplary adaptations can be found in the VDI 2221 (VDI 2221 1986), the approach to circuit design according to VDI 2422 (VDI 2422 1994), generally waterfall model of software development (Boehm 1979) but also in the Y-chart to Gajsky-Walker, which takes place in the electronic development (Gajski 1983; Walker and Thomas 1985). In the latter, there are three logic-rays that reflect and merge the three typical ways of looking at an integrated circuit.

By the logic beam is folded over at the level of activities of the draft to a V-model (Fig. 2.7b), one can point out the importance of the property protection for development results. In terms of the function of protection and a high quality product

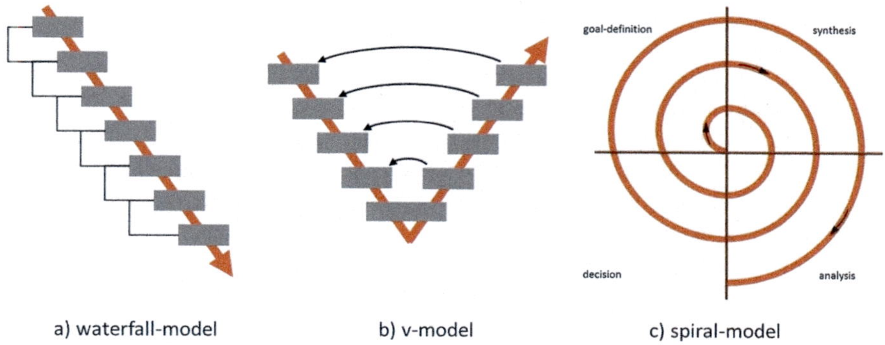

Fig. 2.7 Structures of process models

as by increasing demands in terms of system reliability, verification and validation of technical systems is becoming increasingly important. This type of visualization and implementation supports the way of thinking. V-model approaches can be found in software development (Forsber and Mooz 1991), but especially for the description of the development of complex technical systems as in mechatronics (VDI 2206 2004).

Another structural adjustment is the spiral model (Fig. 2.7c), which is more commonly used in software development (e.g. Boehm 1988). Ultimately, the logic-beam thereby will be "wound". This form of modelling illustrates not only the paradigm of successive refinement (NASA 1995). The introduction of the quadrants also addresses and integrates the logic of the problem solving cycle with the phases tasks careful synthesizing, analysing, evaluating and deciding that must be passed through to each stage of the procedure. This way of thinking are as the application of agile methods in software development as a basis. In this way of thinking is based e.g. the application of agile methods in software development.

2.2.3 Process Models for CPS as an Interdisciplinary Technical System

To explain the procedure for developing CPS as interdisciplinary technical systems, domain-specific approaches are not appropriate. But the logic illustrated with the four phases, needs to be considered due to the inherent systemic view. In considering such interdisciplinary products there are other aspects of importance:

- The complexity of CPS is not based solely on the number of elements (functions and components) but also on the cross-linking diversity. The structural characteristics of individual sub-elements are very heterogeneous, with the result that interfaces can be partly difficult to detect and to interpret. Emergent system behaviour of CPS can be the result.

- The expected from CPS behaviour is more diverse. This is not simply because the CPS must adequately respond to different environmental conditions, but also to its own system states. New business models, offering services in the context of the CPS and the requirement for an intensive consideration of the product life cycle mean that different stakeholders have different demands on the system behaviour or expect a specially specified behaviour from CPS.
- The division of labour for developing a CPS presents itself much more heterogeneous than in conventional technical systems. Many companies cannot hold the entire technological know-how for the solutions. In some industries, significant shifts in the leading competencies of the company can be observed. Core competencies are no longer alone in the mastery of specific technologies but in system integration. In addition to intensive cooperation with suppliers, components of the shelf (COTS) gain importance. This requires special attention to the problems associated with the product development processes, such as requirements management, risk management or the security management.

For the development of interdisciplinary technical systems like CPS the V-model has been proved as structural approach (Fig. 2.8). This reflects not only the four phases shown in the preceding chapter, but also explicitly address aspects such as property protection and modelling aspects. Thereby, not only the substantiation of the development results is discussed but there are also first indications for the description and analysis of data and information flows given.

For each phase now the areas of responsibility need to be expanded.

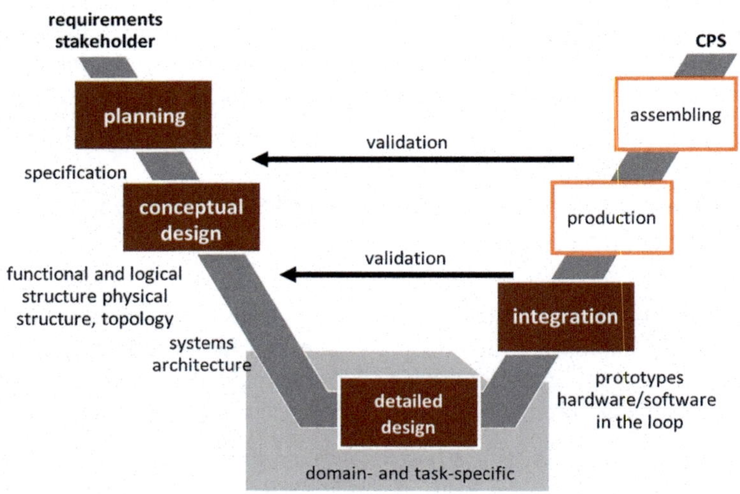

Fig. 2.8 V-model as a fundamental approach to CPS

In the *planning phase*, it is needed to analyse precisely the stakeholders of the CPS and their goals and expectations on the system behaviour. Since the CPS should adequately react to different situations, the definition of just one system response is not sufficient. Instead, it requires an intensive analysis of possible situations to specify the system behaviour. In addition, there remains a challenge, boundary conditions of the application environment, from the market and the competition, as well as from within the company to identify.

Characteristic for the *concept phase* is processing from rough to detail (Fig. 2.5). Under this phase it is necessary to substantiate the defined system behaviour by a functional analysis and to complement the logical order with function execution (operations). Structural analysis is the condition for a description of the physical architecture. In this context it is important to clarify how or what features are summarized in terms of subsystems or components. Hereby, considerations as the evolution of technical systems play a major role (Kossiakoff et al. 2010). Only in a few cases, radical new developments take place in product development. In general, it is necessary to consider incremental innovations. The physical architecture is then more or less given and serves as a basis for development. Product and variant management in the company are also influencing. This not only defines the product structure by the stored modularisation strategies. Finally, the physical architecture is characterized by the company's knowledge in development and by providing access to technologies. Linked to this is also the organizational assignment of development tasks to individual development teams in the specialized domains. Figure 2.9 summarizes these aspects of the system architecture together.

Result of the *design phase* is a system architecture, which describe structural description and system behaviour from different perspectives (Rechtlin and Maier 2000). The concept phase is of particular importance for development because it not only predefines the resulting CPS, but also establishes a system understanding,

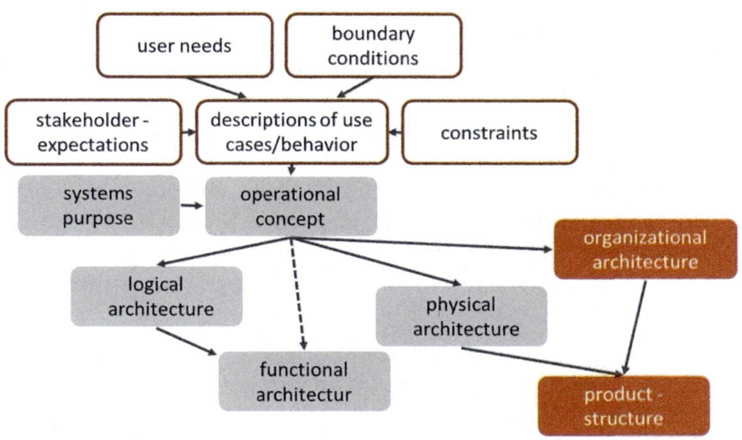

Fig. 2.9 Consideration to system architecture

to which the project partners must commit (Blanchard and Fabrycky 2012). Accordingly, the concept phase requires a highly interdisciplinary cooperation in which is assumed that the participants from different disciplines to develop a mutual understanding of the concerns of other specialized domains.

In contrast, the expertise and the specific knowledge of the individual domains are needed in the design phase. Since they summarize domain-specific knowledge, subject-specific process models are used to deal with the development tasks. In terms of the interdisciplinary nature of the entire development task, a special awareness of the overall system needs is necessary while the processing of subtasks. Decisions that were taken in one domain also influence the work and decisions of other development teams. A strong interface management is required, thus also in the domain-specific development phases, the overall objective won't get lost.

The phase of *system integration* provides companies especially in the development of CPS with special challenges. Individual solutions from different development teams, based on specifications, which only represents part of the complete the overall system requirements, are optimized in this way usually. It is in the nature of things that not all possible critical interfaces are considered to other subsystems. Often simply lack the domain-specific knowledge in detailed questions. In the phase of system integration, therefore much effort on the adaptation of individual solutions is put in terms of overall performance. This is associated with time and cost consuming iterations. It therefore makes sense to support the phase of system integration from the planning and design phase through methods and tools, as well as procedural and organizational interfaces by which interfaces between requirements, stakeholders, subsystems and functions between development teams and can be made visible. In addition, it requires methods to provide decision situations with sufficient data and information.

2.2.4 Systems Engineering as an Interdisciplinary Approach for Development of CPS

While in the previous chapter, the procedure from a more domain-specific point of view has been continuously broaden to draw conclusions for interdisciplinary product development, Systems Engineering provides an approach to support the theme from a cross-domain development perspective. Similar to the already shown systemic development approaches, Systems Engineering refers to a long history, (NASA 1995), especially for the development of large-scale systems (Chestnut 1967).

> *Systems Engineering (SE) is an interdisciplinary approach and means to enable the realization of successful systems. It focuses on defining costumer needs and required functionality early in the development cycle, documenting requirements, and then proceeding with design synthesis and system validation while considering the complete problem: operations, cost, schedule, performance, training and support, test, manufacturing, and disposal. SE*

considers both the business and the technical needs of all costumers with the goal of providing a quality product that meets the user needs. (INCOSE 2010, p. 6)

The definition shows that not only a focus on the stakeholders takes place, but also the entire product life cycle of the resulting system is made explicit in development. These limits of consideration are very broad in terms of description of a socio-technical system from the outset.

Also as part of SE approaches, development of technical systems is understood as top-down process with iterative character (Eisner 2008). The focus lies more on a total system approach and on how to find an optimal solution for complex tasks and problems (Hitchins 2007). Of course, in SE detailing phases in the domain-specific development are also needed. Here is less discussed how to proceed in detail, but more which tasks a system engineer has to fulfil, to be able to reasonably coordinate the results of these domain-specific phases to each other and to be able to integrate them into an overall system. The methods used in the context of the SE therefore primarily focus on the control and management of complexity (Kossiakoff et al. 2010), analysis and description of interactions between to developing systems and the environment as super-system, and between the sub-systems. Especially the latter is of course based on the expertise of different domains, but should be supported by the system engineer, from whom a broad technical understanding is expected (Blanchard and Fabrycky 2012).

Oliver et al. (1997) explain that SE approaches need to support two ways of looking at development:

SE management process describes the technical and organizational effort within the product lifecycle and thus focuses more on the typical tasks of project management. Target of management procedures is to provide information and to evaluate it in order to support the decision-making in terms of trade-offs between efficiency and costs (NASA 1995). It is necessary to consider restrictions on cost, time and potential risks, both in the development itself as well as in the evaluation of the performance of the technical system.

SE technical process includes all activities from the first request of requirements up to development through to verification and validation of the results. Both, procedures represented in literature as well as procedures known from practice, based on systemic approaches and mind sets that have been explained in detail in the previous chapter. As process model now commonly a V-model is used as basis (e.g. (Blanchard and Fabrycky 2012; Kossiakoff et al. 2010)).

Typical methods of SE focus less on the support of individual activities but more on the analysis and description of the system complexity. Therefore, it should be given particular mention to:

- Methods for modelling and simulation of complex systems (e.g. NASA 1995)
- Methods for analysing system contexts both at a functional and structural level; for this, graph-based approaches are used as well as network-based approaches (conclusion e.g. in (Parraguez 2015))

- Methods for the preparation and presentation of data and information flows in the technical system, taking into account constraints and stakeholders through approaches of model-based systems engineering (MBSE) (e.g. Delligatti 2013).

From the perspective of systems engineering, the emphasis in the description of data and information flows lies in summarizing the variety of data and information and in preparing a structure to firstly identify interfaces between sub-systems and the environment within the technical system. To this end, data and information from the synthesis of partial elements are grouped in such a way that dependencies between elements of the overall system can be identified.

Especially the extensions within the V-model for development of interdisciplinary products such as CPS give an indication what to look for in design but for the concrete process design and support, they are only partially suitable. Here, a significantly more granular fine-mapping of processes is required. Therefore, in the following chapter approaches are presented to substantiate and refine the process descriptions.

2.3 Concretisation of Process Descriptions in the Sense of a Workflow

Especially the extensions within the V-model for development of interdisciplinary products such as CPS give an indication what to look for in design but for the concrete process design and support, they are only partially suitable. Here, a significantly more granular fine-mapping of processes is required. Therefore, in the following chapter describes approaches to substantiate and refine the process descriptions.

Looking at the data and information flows within the context of process models, both, product models as a result of synthesis and analysis models shall be considered. This requires a detailed process description, which in turn forms the basis for the design of IT structures. It should focus on that system descriptions within the process are successively refined through the use of different methods and related tools. This leads not only to different versions within a product model but also contributes to a plurality of partly also very heterogeneous product models, which need to be managed and tackled. With increasing maturity of the technical system, the data quality improves and data uncertainties are gradually reduced.

Analysis models are based on product models and the associated level of maturity of the considered subsystem. In addition, analysis models depend in their detailing or their structure from the objectives which are pursued with the analysis in each case. Product models require a corresponding contextual preparation to make them accessible for the analysis.

Both model types, product models and analytical models contain data and information which describe each viewpoints or parts of the resulting CPS. For preparation of the product models for analysis and specifically for simulations, engineering-workbenches find use. The aim of this is, on the one hand to support the modelling for simulation by the prepared context information for pre-processing and assigned to the product model. On the other hand, the analysis results are pro-cessed in such a way that they are within the meaning of the decision support for the development process available in post-processing. Libraries for modelling, load cases, material values, etc. supplement this context information for analysis and can simultaneously be understood as knowledge repositories.

For collection and delivery of all product artefacts various IT tools are available. Product data mainly from construction will be deposited directly into product data management systems (PDM systems). Depending on the structure and format of these tools, this data can be linked to analysis results. Such processing and provision of data and information facilitates the provision of data for individual development steps and the development situation significantly, but also implies that the PDM systems need to be adjusted in terms of product lifecycle management to the development processes. In addition to PDM solutions for managing the documents, data filing systems in which the files themselves are stored, are used. This in turn is caused by the use of various tools in development. In sum it is necessary to support development processes with a diverse IT landscape. Their efficiency depends on how successfully typical development processes for the own organization can be mapped and finally adapted to the data and information flows.

Figure 2.10 gives an indication to the components of a holistic IT infrastructure for development. Besides PDM tools and engineering workbenches support file storage systems the structuring and management of data. These are linked to the for development necessary producer systems both for the synthesis (for example, CAD) as well as for the analysis (for example, FEA, MBS). Within these IT systems in turn, templates, libraries and assistance systems find use to support individual

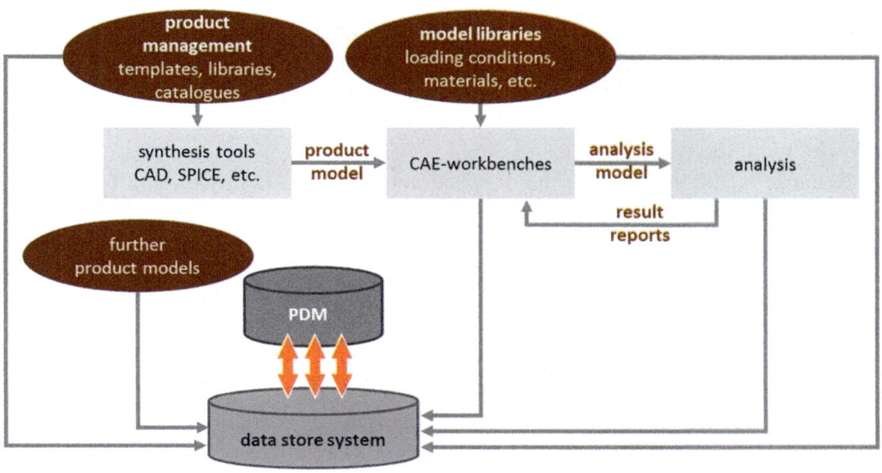

Fig. 2.10 Overview of components of an IT infrastructure for development

tasks through targeted provision of knowledge, but also to enable a holistic and overarching product management. Finally, it requires suitable data interfaces to ensure the exchange and further use of data within the development process. It quickly becomes clear that for a concrete definition of an IT infrastructure, a detailed treatment of the data and information flows is necessary, which cannot be guaranteed by the process models described in the previous chapter. The process models require a significant specification, for which the influencing factors are described below.

2.3.1 Adaptation of Development Processes to the Context

The refinement of the development process is not only required to substantiate to processes in the development and to support decision-making processes through targeted information delivery. This is done not only on the basis of interdisciplinary tasks in the development but also by the integration of the development in the company's organization. The link to production and assembly, sales and marketing, logistics, procurement etc. is especially required to consider all influences on development in terms of product lifecycle reasoning. As a result, so called swim lane models for process description can be found in the companies (exemplarily shown in Fig. 2.11). This also goes hand in hand, that for development necessary IT structures are linked to the IT tools of other company divisions. In this context should be referred specifically to the link to Enterprise Resource Planning systems (ERP systems).

The processes depicted with such Swim lane models are very company specific. They resemble each other but generally are not equal because they are influenced not only by industry, competition and market but also by corporate strategies. Here, for example, aspects such as competitive and technology strategy or typical customer patterns play a decisive role. Such process descriptions and therefore the support of developers are not only based on typical processes, but also can be explained to some extent from the company's history. Last but not least are manifested in the development processes and the associated IT structures the experience and knowledge of the company.

Because of the diversity in the expression of the factors influencing the process design, a generic oriented refinement makes little sense. Below shall therefore be briefly outlined, which factors are in what way to take into account in order to reflect the company's internal processes as a whole can. In turn, starting points and specifications for IT structures can be derived.

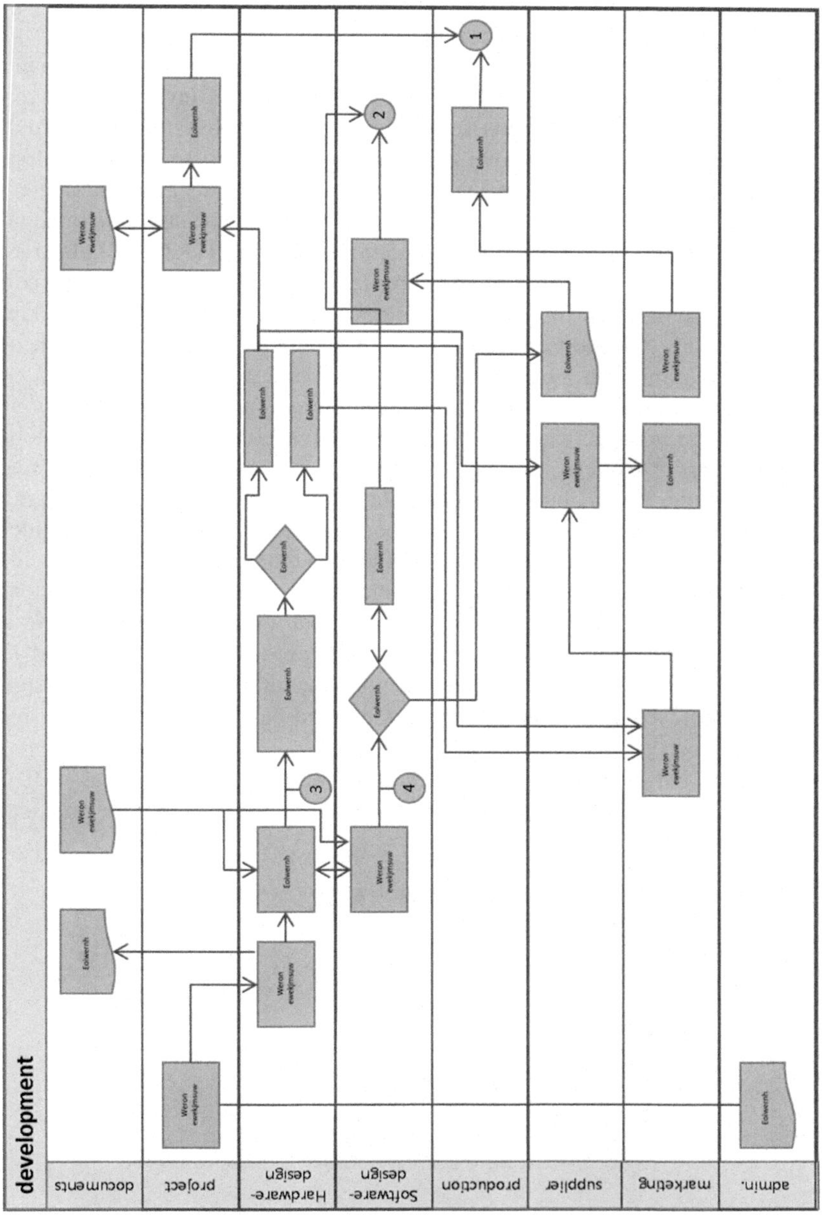

Fig. 2.11 Example of a development process as swim lane model in the company

2.3.2 *Identification of Context Factors for Adaption of Development Processes*

The adaptation of the generic process models is described in the literature for two reasons: on the one hand requires the pronounced interdisciplinarity especially for CPS the consideration of typical procedures of disciplines involved to obtain functional and high-quality solutions for components and subsystems (e.g. Pugh 1991; Gericke and Blessing 2011). On the other hand, the adjustment of development processes to the corporate context (industry, market, product group strategies) is necessary to explicitly include the resulting specific framework for development (e.g. Skalak et al. 1997; Meißner and Blessing 2004). Gericke et al. (2013) illustrate two types of customization options of development processes: the Augmenting and Tailoring. The Augmenting ultimately describes the extension of the procedure description not only by additional process steps but also by the integration of additional information such as guidelines, design practices, specific methods, or the like. Target of Tailoring is to carry out industry or company-specific adaptations of the development process, whereby it is mainly about to make the process steps explicit, which results from the product or the industry (Gericke and Moser 2012). In the aviation sector this includes, for example, admission procedures or hedging measures. For CPS are, for example, specific measures in relation to consider the data security or the human-machine interaction, required. The challenge for both forms of process adjustment is to completely grasp the development context and describe it in its characteristics. Hales and Gooch (2004) illustrate this in a framework to classify those factors influencing the product development and to identify (Fig. 2.12). Gericke et al. (2013) use this structuring approach and assign more detailed context factors that have been identified based on a comprehensive literature review.

The resulting list of over 230 contextual factors (in Gericke et al. 2013) provides basic evidence, what to look for in the concrete detailing, but for practical

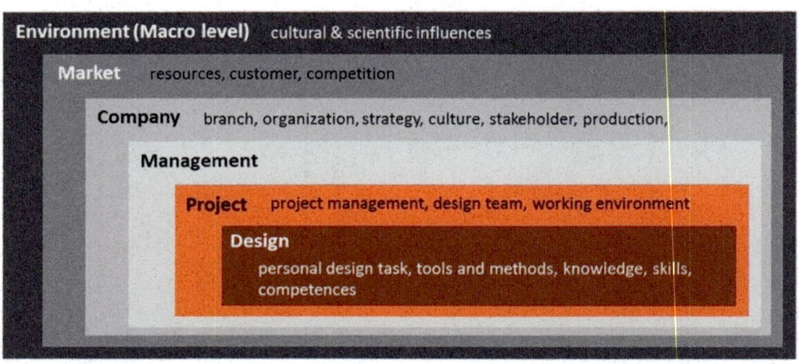

Fig. 2.12 Influencing factors to product development according to (Hales and Gooch 2004) improved in (Gericke et al. 2013)

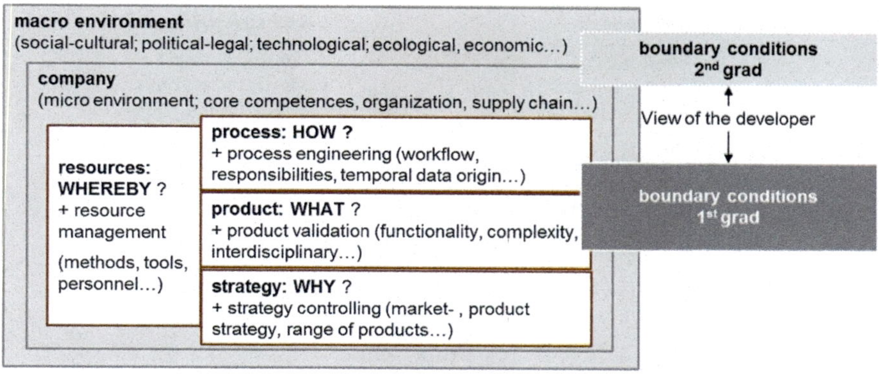

Fig. 2.13 Detection of the contextual factors for describing the data and information flows (Reitmeier 2015)

application, this list is a bit bulky. Also not included in the rather linear view is the fact, that between the contextual factors a number of dependencies are observed.

Reitmeier proposes to extend the structuring approach (Reitmeier 2015) to capture the data and information flows by continuously evaluating the contextual factors with respect to the decision to be made (Fig. 2.13).

In (Reitmeier 2015) is distinguished for the classification of contextual factors in such of first degree, which are directly connected with the activity of the developer and thus also directly can affect the necessary activities, or even can be influenced by the developer. In addition to the aspects of the product and the process, the company's resources also play a role here. Not least, strategic considerations such as the importance of the development project for market, industry, customer or integration in the product management influence a role. These factors influence the adaptation of the development process to the development task. Boundary conditions of second order, which at a higher level can be derived from the macro environment or the company's strategy, however affect the basic design of the processes in the development.

2.3.3 An Approach for Systematic Analysis of Determining Factors for the Development Process

In addition to detection and description of the contextual factors it requires an analysis, which indications can be derived for the design of development processes. For this, a systemic approach after (Negele 1998), which in turn follows an idea of (Patzak 1982), is used. This approach of Negele will be introduced in the following chapter.

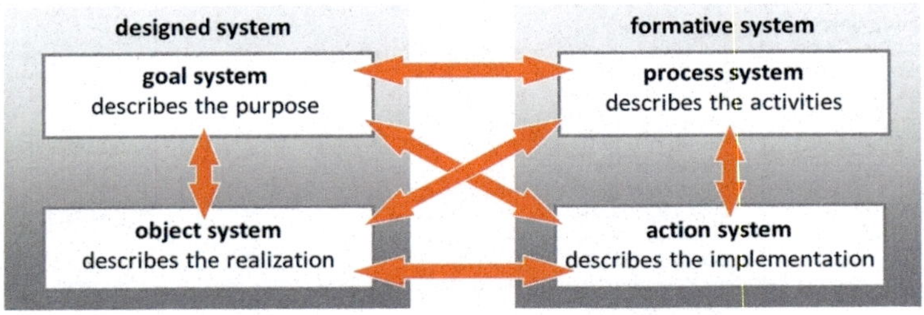

Fig. 2.14 The development of CPS using the ZOPH model after (Negele 1998)

In principle, it is distinguished between the system to be designed, here the CPS as product, and the formative system, that is the process of targeted activities that lead to this CPS. Both systems, the CPS and the development process cannot be considered independently. Considering further that the CPS initially is present only as an aggregation of targets, which are substantiated by development, it follows that the CPS is again divided into a target system and an object system. This opens up the possibility to address the uncertainties in development. On the other hand, in connection with the process is illustrated the extent to which the resulting CPS is correlated with the objectives. A similar approach is for the process to derive even. The formative system of the process is mainly determined only from the logical approach and has a generic character for the product group. The concrete development task can be derived only if demands are accurately specified, resources and responsibilities are defined and a termination occurs. For this reason, a specific project will be derived on the basis of the processes virtually, resulting in the distinction between a process and a system of action.

Finally, of course, it requires a definition of the system boundaries, which are characterized not least by contextual factors on the micro and macro level. The difficulty here is that this system boundaries often cannot be drawn strong and clear, because it is precisely the characteristic of the CPS that they interact closely with the environment or other systems in the environment. CPS must therefore be interpreted frequently in the context of a system of systems (SoS). Such blurred dividing line is, however, difficult to tackle in development. Therefore it is necessary to give special care on the delimitation and description associated in- and outgoing data, energy and material flows. Figure 2.14 illustrates this model approach after (Negele 1998). The name *ZOPH* of the approach results from the first letter of the german words of the partial models (**Z**iel, **O**bjekt, **P**rozess, **H**andlung).

This raises the question of how this mind set supports the process development and control of the development and the associated organization of data and information flows. For this purpose, the individual subsystems shall be further concretised in reference to (Negele 1998).

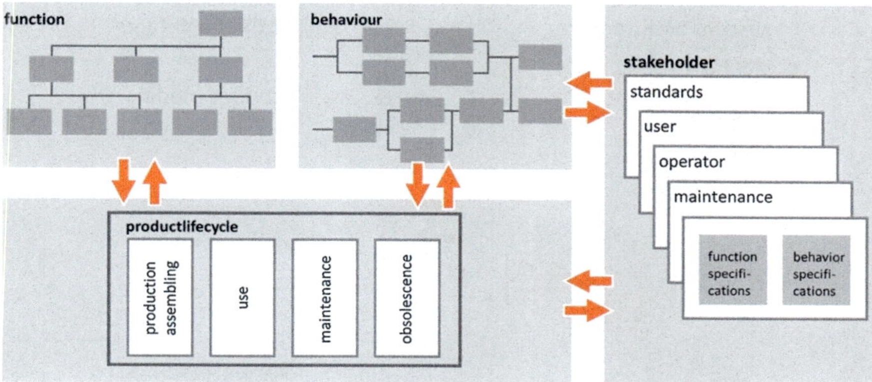

Fig. 2.15 Detailing of the target system

2.3.3.1 Goal System

The goal system classically summarizes all the requirements for the evolving CPS. As a product from the capital goods sector, here one needs to distinguish between load and performance specifications in principle. The challenge lies in the completeness of goal acquisition. Here are not only partial functions and their expression of importance, but also the description of the characteristics of the desired operations. For structuring the target system approaches for describing the system architecture can be used, which reflect both functional and structural aspects. Since generally in development always similar systems are in focus, may be used for the CPS also established product structures. Relationships between functioning synthesis and principle synthesis (Andreasen and Hein 1987) need to be considered especially for completely new developments. For adjustments, variant developments or incremental developments, these aspects are already included in the system architecture.

The complexity of the CPS results not only from the functional diversity but also from the diversity of the expected behaviour, which is not least determined by different stakeholders on the CPS to be developed. Therefore, the target system shall additional be analysed with respect to different types and forms of operations, what also integrates aspects of the product life cycle at the same time. This requires a detailed analysis of the stakeholders, as summarized in Fig. 2.15.

The goal system includes aspects of system definition, therefore the answer to the question of which functions can be realized only through the CPS and which arise from the interaction with other systems. One important result of the analysis to the target system must be to detect cross connections between functions, product structure, operations and stakeholders to present and show mainly conflicts between these. Ultimately, the target system is a reference for the object system, but also for the process system.

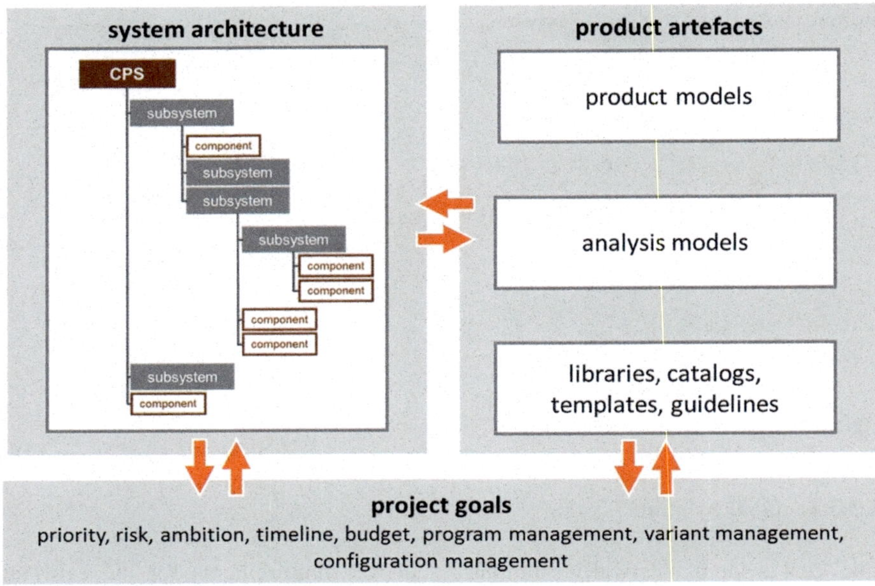

Fig. 2.16 Specification of the object system

The target system is manifested in the requirement description, which in turn is the basis for requirements engineering. In conjunction with the stakeholder description and hedging measures, risk assessments or reliability statements can be deduced.

2.3.3.2 Object System

The object system ultimately summarizes all results in the development process and therefore also all product artefacts in the various phases of the development process. This includes any form of additional information such as libraries, catalogues or templates that underlie the development in the different producing systems (Fig. 2.16). Ideally, the structuring is based on the product architecture, which generally underlies also the target system description. A consistent structuring of both, the object and the target system is the basis of the balance between demands and results of the development process. However, the orientation on product architecture takes generally into account the participating disciplines for component development.

The product artefacts that are collected in the object system are available, both as documents in data storage systems as well as hardware (prototypes, hardware-in-the-loop). The analysis of the object system therefore not only serves to flesh out the IT structures but also the identification of data-technical necessary interfaces between the product artefacts. The relationship between various product artefacts provide, at best, product data management systems and engineering workbenches. The relationship between data and information from other business sectors are ideally represented by ERP systems. The enrichment of product artefacts with information like date of creation, creator, etc. Versioning provides the one hand the connection to the system of action, on the other hand, these also can be assigned to phases within the development process and thus draw conclusions for the degree of maturity.

2.3.3.3 Process System

The process system defines in detail the activities and actions which are relevant for the specific development. Logical dependencies result from the physical relationships in the specialized domains as well as from the to-implement system functions and the expected operations. The structure of the process system should be such that the analysis of data and information flows can occur between the individual acts. The individual process steps are considered quasi as transformations of the initial state into a desired end state. For this transformation are usually available the experience and knowledge to the system, which provides guidelines and methods that take into account not only the physical relationships but also best practices. Both, methods and best practices include ultimately guidelines for action and thus provide the detailed procedures a framework for action. For the analysis of process steps therefore the proposed model Fig. 2.17 with reference to (Birkhofer et al. 2005) can be used.

A generic procedure description also results from the problem-solving cycle (see Sect. 2.2.1), because there with a classification of methods and the data characteristic is connected. As described above, the problem solving cycle is ultimately to go through at every stage of development. This results is not only a classification approach for methods (see Fig. 2.3) but also a statement of the maturity of the resulting system. Conversely, this means that each method must be adapted to the maturity level of the product. With the application of methods in practise usually IT-tools are connected, which data requirements and their evaluation results are directly attached to the methods or the best practices. For the description of the development processes, the development influencing tasks like procurement, purchasing or production and assembly now also have to be considered, as they either provide input or need input at a defined time. Via the process system one can recognize and specify procedural interfaces. The process system is manifested in the swim lane models mentioned above (Fig. 2.11).

2.3.3.4 Action System

The action system describes the organizational framework or the organizational impact parameters for the process design. The organizational structure results in

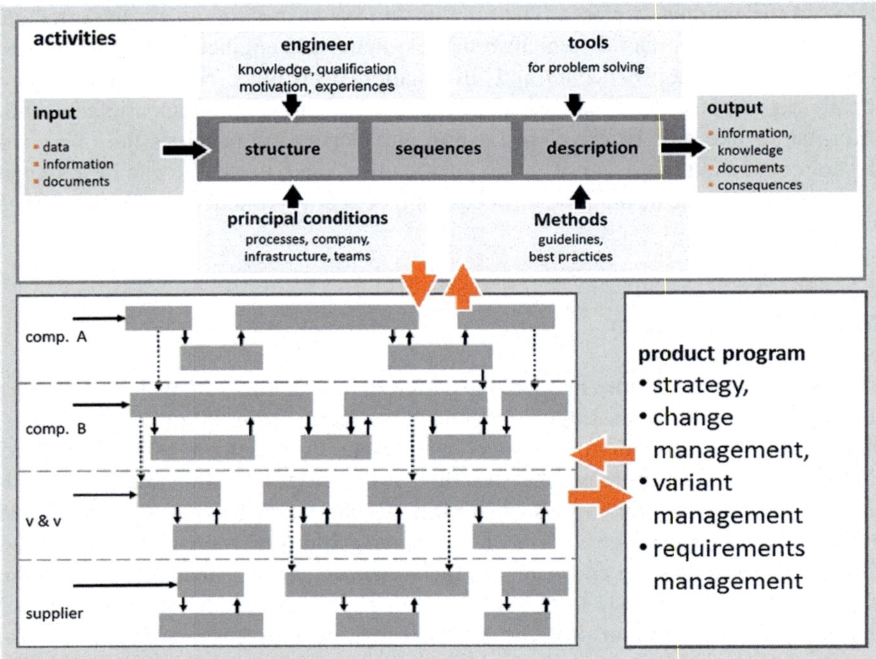

Fig. 2.17 Specification of the process system

responsibilities, role assignments and responsibilities for partial activities for both, the actual development work as well as to administrative supporting processes such as release mechanisms etc. In action system also material and personnel resources are defined, whereby the material resources available especially IT tools, licenses and their characteristics are includes, but also test stands, laboratory equipment, test equipment, etc. are covered. Since today a variety of development services is adopted by service providers or suppliers, these have to be interpreted as parts of the action system, which supports the development with resources and capacity.

The action system serves in addition to content issues also to scheduling specifications of development processes, depending on the development task and possible priorities of the company. Depending on the development task, deadlines and responsibilities must be assigned to the individual process steps. For the purpose of a multi-project management, his assignment must be accompanied by a project prioritization, which results from the program or product management. Thus, directly from within the action system, the processes are complemented by typical information to substantiate the workflow in terms of project management. The components of the action system are summarized in Fig. 2.18.

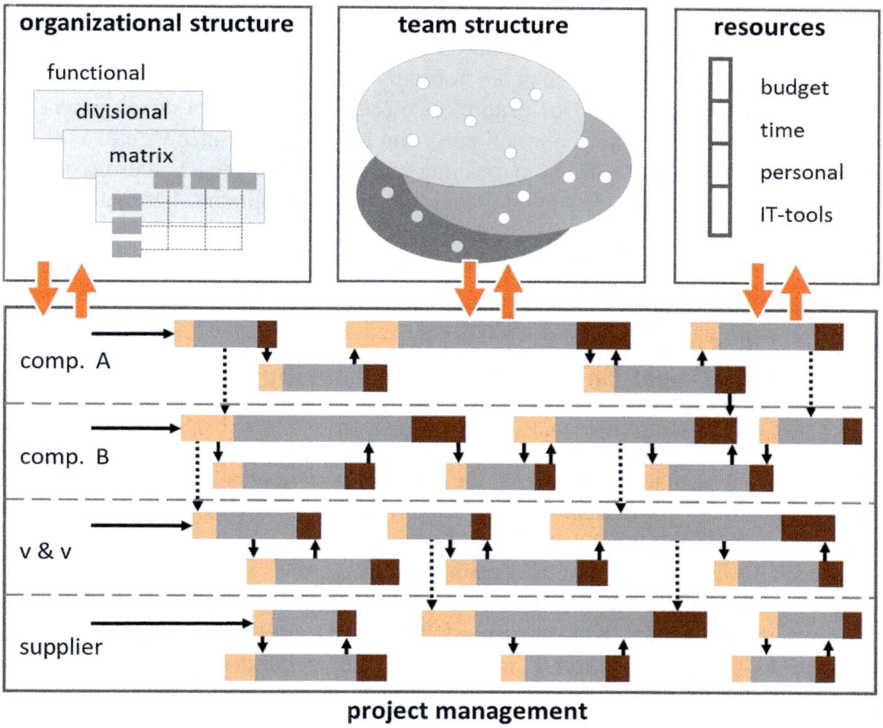

Fig. 2.18 Specification of the action system

2.3.3.5 Control of Development Tasks via the ZOPH Approach

The individual subsystems of the ZOPH model are not independent, as is already explained in Fig. 2.14. In terms of a holistic process management it is rather necessary to consider and develop all the subsystems equally. However, various views on the development process can be mapped over the subdivision into subsystems. The planning and integration of IT structures for integrated data and information flows, for example, requires not only the consideration of the typical system architectures but also of typical processes to tightly coordinate the individual IT tools. For supporting decision-making processes in development in turn the provision of all available data is required, but it also requires the knowledge of organizational, process and product-related interfaces that are clearly shown by the ZOPH approach. This is the basis on which to assess the impact of decisions and to inform those who are affected by the decision.

The main objective of the development is undoubtedly the CPS. This is also connected to a number of side objectives, such as quality, reliability etc. For this purpose usually consuming associated processes such as risk management or requirements management are implemented in product development, which are often to be considered as parallel to the actual product development process. Configuration management and change management on the other hand consider aspects of creating variants, which are normally anchored in program management. These associated processes, for example, determine the product development via modularization approaches, common parts and repeat parts concepts etc. With the analysis of the company within the meaning of the ZOPH approach, these processes can be mapped in a holistic way and coupled with development. Ultimately, these associated processes grab, even if they have different origins, back to specific elements in the subsystems and bring them goal-oriented together.

2.4 Model-Based Engineering for Mastering Complexity

The major challenge in mastering a CPS, nowadays lies in the system integration. For individual activities or aspects in development now a variety of powerful methods and tools are available, the overall system approach is, on the other hand, less pronounced supported.

Especially in the data and information flows are always media discontinuities to tackle, which easily lead to fractures in the reality in development. There are a variety of documents such as request lists, CAD models, feasibility analyses, calculations as well as Power Point or Word files generated which describe the resulting product both on functional as well as on structural plane. These are for individual development steps or specific decisions of importance, but are available in very different formats. In addition, they represent some very different perspectives on the CPS, which makes their interpretation and exploitation difficult. The cause of the resulting lack of development lies accordingly in the document-based management of the data and information flows, which neglect the data characteristic. The preparation of more or less unstructured data requires special effort, for which the creation of data exchange formats is often insufficient. An illustrated document-centric view in general supports only specific views on the CPS. The challenge arises not only from the different data formats and data structures but also from the fact that with the various activities (synthesis/analysis) different data characteristics are connected (properties/features), which complicates comparability and stringent further usage.

To overcome these development disruptions, and thus to ensure a certain completeness and efficiency of the development processes, increasingly model-based approaches are tried. With such a model-centric organization of data and information flows, the model becomes the source of relevant information by structuring the information according to a predefined scheme (Weilkins 2008).

Model-Based Engineering (MBE) should here be understood as an engineering approach, which uses models within the meaning of pre-structured data as a basis for the data and information flow. These models include all this data and information relating to the product life cycle, based on the requirements of the design, verification and validation of subsystems as well as on the overall system (NDIA 2010).

Model-based engineering may be understood as an overriding principle that originally sprung up in the development of software-intensive applications. The application of CPS is not only beneficial to master the complexity in development but also due to the system characteristic. The basic idea here is to continuously hedge the system behaviour at different levels throughout the entire development process by the use of these model approaches. For this purpose, firstly parameter-based models (for example, in Matlab Simulink) are constructed from the overall system, by which the system response can be displayed and analysed. These are successively refined, both, in terms of modelling approaches as well as by hardware components that are integrated (Albers et al. 2013). This ensures the system integration from the early stages of the development.

Meanwhile, the understanding of MBE is somewhat broader and also includes in many cases Model Based Systems Engineering (MBSE) and Business Process Modelling (BPM) (Lee 2008).

Difficulties or challenges in the dealing with models arise from their properties (Stachowiak 1973). Thus, models are only depictions of a system. This will entail contractions: by means of the attributes from the original, only those things are recorded, which are relevant for the defined viewing purpose. By assuming a certain substitution function, therefore models give only a specific limited view to the original again (Stachowiak 1973).

A further distinction needs to be made for models:

- Especially in engineering, the feasibility of models is presumed. Models are thus constructed under the use of method tools such that they can be used as basis for simulations. This opens the possibility to derive statements about behaviour, performance and function as early as the early stages of development.
- Increasingly descriptive models that allow a symbolic representation of the CPS with a defined syntax and semantic, gain in significance. These models are information models, in which feasibility generally does not exist (exceptions can be found in software development, where such a source code can be derived, see also (Rumbaugh et al. 1999)). In this context, UML or SysML models are exemplarily mentioned. In the end, even CAD-models, which initially support structure visualization and are based on a pure geometry-based view, are included. Simulations cannot be initialized directly but always the preparation of data for an analytical model is required.

The two different interpretations of the MBE are summarized once again in Fig. 2.19. This depiction illustrates that the model-based approaches on the basis of executable models correspond to a vertical integration. Accordingly, the challenge is to refine the models depending on the progress of development, or to link the different modelling approaches to make them accessible for simulations (Paetzold and Reitmeier 2010). MBSE other hand focuses on horizontal integration. The goal is here to compile all data and information for the CPS over the product lifecycle

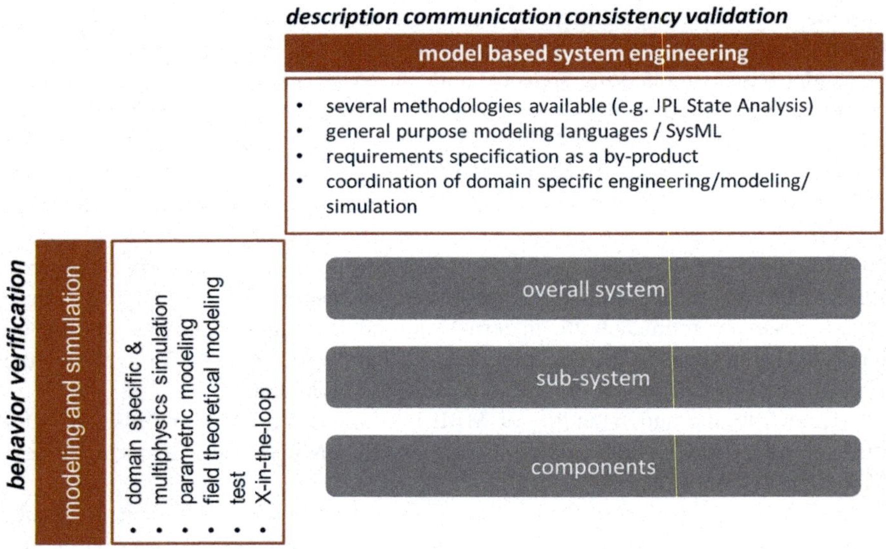

Fig. 2.19 Different views of the model-based engineering (Omiciuolo et al. 2015)

in order to ensure a holistic view on the resulting technical system or to assemble different views.

With vertical integration, especially the kind of modelling or the choice of the modelling paradigm is combined. From a systemic point of view one can distinguish both, on the structure level as well as on the function level or the behavioural level. A modelling based on the structure definition for the whole system is extremely difficult, what is caused by the domain-specific physical and related mathematical descriptions (geometry, source code, integrated circuits).

For the simulation of the system behaviour it is more likely to resort to parameter-based approaches to modelling, as these are based on the more generic functional description and are linked via input-output relations. With this, although one accepts a higher level of the abstraction model, however, succeeds a cross-domain modelling, which can be understood and interpreted in the domains. The challenges that are associated to this observation will not be further discussed here, but reference is made to additional literature (the problem of universal description language, for example (Panreck 2002); multi-domain vs. domain specific, for example (Sangiovanni-Vincentelli et al. 2009).

2.4.1 Model Based Systems Engineering

The second aspect mentioned in the context of model-based engineering is also treated under the concept of model-based systems engineering (MBSE).

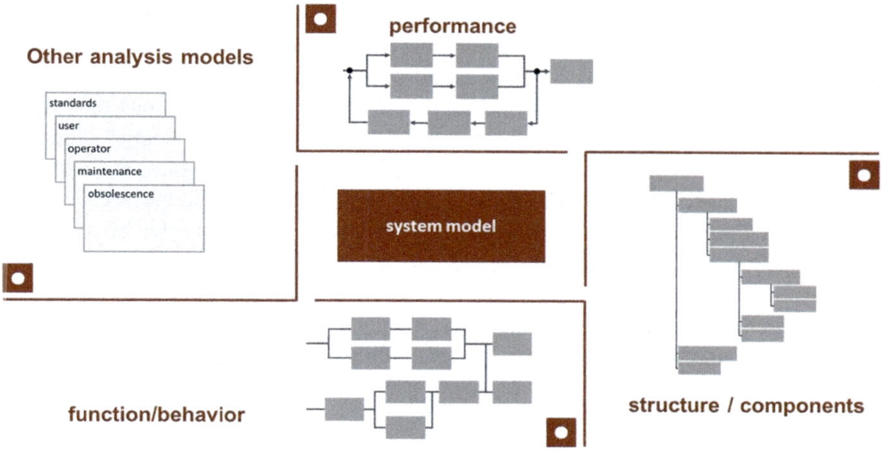

Fig. 2.20 Perspectives which shall support the system model

> *Model-based systems engineering (MBSE) is the formalized application of modelling to support system requirements, design analysis, verification and validation activities beginning in the conceptual design phase and continuing throughout development and later life cycle phases.* (INCOSE 2007)

The basic idea is to support the development process and especially the data and information flows through system models that depict not only aspects of the entire product life cycle but also the perspectives of different stakeholders. This is intended not only to support development processes but also to provoke balanced system solutions (Eigner et al. 2012). For this purpose, as a basis a unified modelling approach is used, which allows to map both, system models as well as development activities. MBSE thus should be understood as methodology; that means a summary of processes, methods and tools by which a defined goal, here the integral development of complex technical systems, is announced (Estefan 2007). The management of complex technical systems is supported by the fact that not only different perspectives on a CPS can be distinguished but also between function, structure, behaviour and performance. The possibility of hierarchy formation facilitates impact analyses or traceability of design changes (Fig. 2.20).

In order to deploy and support a methodology efficiently, three aspects must ultimately be integrated (Delligatti 2013):

- Methods which describe the development of the technical system,
- Languages that define grammar as a set of rules for system description and that are understood by all parties,
- Tools that translate the languages and that support in construction and interpretation of the models by providing a development environment and routines.

These aspects will be described briefly below.

Methods within the meaning of MBSE describe action rules for consistent configuration of system models. These methods are not only influenced by the purpose of the model but also by the mind sets and ways of abstraction of those who wish to use these models. Of course, the methods follow very much the logical and systemic approaches for development and the structuring of complex technical systems, as they are described in this chapter. Certainly, here play for example domain- or industry-specific priorities in the concretization of the methods such as for example domain-specific development and lifecycle models a role. Also specific aspects of the system Engineering management process, such as risk management issues or security issues, may affect the method in detail here. These methods reflect quasi a type of a modelling philosophy.

For MBSE, in the literature a number of methods can be identified. In the context of Systems Engineering as an integrated approach to development by CPS, one should particularly mention:

- OOSEM (Object-oriented Systems Engineering Method), which was designed as a top-down approach by the INCOSE, based on a functional analysis (Friedenthal 2014).
- SYSMOD method, a standard top-down approach developed by Tim Weilkiens, which is based on UML methods and expand them in the sense of systemic product development (Weilkins 2008).

Besides that, a number of other methods, such as IBM Telelogic Harmony SE, IBM Rational Unified Process for Systems Engineering (RUP-SE), JPL State Analysis (SA), Dori Object Process Methodology (OPM) exist. A detailed summary and explanation of these methods can be found in (Estefan 2007).

Modelling languages define the grammar, thus the nature of the elements and their connections by means of which a model must be built to represent defined contents intelligible. The language that is spoken, directly influences the way how the technical system is seen. The dilemma of the description language for interdisciplinary technical systems has long been known (for example, Panreck 2002). Domain-specific languages are generally not suitable to transmit content beyond the domain boundaries or to integrate domain-external matters into the model. Again and again, attempts have been made to develop general description languages, which on the one hand reflect the system holistically and are intelligible to another. This dilemma can also be seen for languages for MBSE. Exemplarily, reference is made to work, which indeed succeed in depicting geometry elements by using SysML (Eigner et al. 2014), but these are not readily to interpret as intuitive as a CAD model. They therefore require to "oneself empathize" into the language, to understand these.

For MBSE primary graphic-oriented languages are used, which not only define the element types which may be used in the model but also the relationships that are allowed between the element types. This is complemented by a notification which allows the display of elements and relations in diagrams. In the past, each of the developed method has often developed their own language, what leads to a corresponding diversity. In this context, *System Modeling Language (SysML),*

Unified Modeling Language (UML) and *Integration Definition for Functional Modeling (IDEFx)* are particularly mentioned.

All graphic languages have in common that charts are available for both, the structural view and the behavioural representation by which in the systemic sense both, structural and behavioural models are constructed and interrelated. The specific designations or aspects of hierarchy and points of view can, however, vary.

The tools for MBSE implement ultimately the language and modelling methods. They support the engineer by the availability of a graphical interface, through routines and libraries for creating and linking models and the automation to update them. Commercial importance have *Cameo System Modeling/MagicDraw (vendor: NoMagic), Enterprise Architect (vendor: Sparx Systems), Raphsody (vendor: IBM), Artisan Studio (vendor: Atego) or UModel (vendor: Altova).*

The benefits of using MBSE-approaches for interdisciplinary development are obvious. By provision of clear and accurate models can also interdisciplinary aspects be summarized in a model approach. The emerging information models are undoubtedly suitable to reflect and to summarize different perspectives on the CPS in different hierarchical levels. This also results in the ability to detect dependencies from both, a procedural, product-oriented as well as an organizational point of view and to derive interfaces more precise.

Because of these advantages is seen in both, academic as well as industrial environment, a considerable potential for development support by means of MBSE. Nevertheless, MBSE approaches are not yet established very wide. Challenges arise not only for the utilisation of the resulting information models and their integration into the data and information flows respectively the IT infra-structure but also in the creation of these models.

The models which are created on the basis of MBSE are not directly available for simulations. Although today parameter-based models which then e.g. are carried out by Matlab/Simulink (Sop Njindam 2015) can be derived from certain types of charts (block diagrams) commercial data interfaces are currently not yet provided for this. Now, there are some interesting approaches to integrate the MBSE-approaches into the IT structures of companies, for example, the coupling of requirements management with MBSE-approaches (e.g. Eigner et al. 2015a, b) with the objective of repeat and further use of the data and information.

For a successful integration of MBSE models into the data and information flows, and thus also in the IT structures of a company it is necessary that these reflect the product and process structures. This in turn presupposes that the development methods must be refined to such extent that typical development challenges and circumstances are addressed. The methods mentioned are, similarly as shown for the process models in this chapter, still relatively generic. For an effective use it is necessary to substantiate them and to adapt based on the specific development conditions. This not only requires an intensive analysis of the structures of the CPS but also of the development processes, which are reflected in the MBSE modelling methods and the models themselves.

In this context, also the question of the modelling depth for the technical systems which should be imaged is posed. In the literature currently the trend, to use MBSE approaches down to very low levels of detail, can be observed (e.g. Eigner et al. 2015a). However, the modelling depth depends not least on the model's purpose, especially if one is considering that the derivation of executable models out of MBSE models is associated with considerable effort. In addition, in each domain very powerful modelling and simulation tools are available, which on the one hand can depict the subject-specific contexts very well and in detail, and on the other, can be easily understood and interpreted by experts.

The strengths of MBSE lie in the possibility of an overall system view and the associated possibility to recognize interrelationships and interdependencies in the CPS and provide them for development processes. This aspect should therefore define the modelling depth for MBSE models.

References

Albers, A., Behrendt, M., Schroeter, J., Ott, S., Klingler, S.: X-in-the-loop: a framework for supporting central engineering activities and contracting complexity in product engineering processes. In: ICED 2013: 13th international conference on engineering design, Seoul, South-Korea (2013)

Andreasen, M.M.: Machine design methods based on a systematic appraoch. Dissertation, Lund University, Sweden (1980)

Andreasen, M.M., Hein, L.: Integrated Product Development. IFS, Bedford (1987)

Birkhofer, H., Jänsch, J., Kloberdanz, H.: An extensive and detailed view of the application of design methods and methodology in industry. In: Samuel, A., Lew-is, W. (Hrsg.) Proceedings of the ICED 2005, Melbourne (2005)

Blanchard, S.B., Fabrycky, J.W.: Systems Engineering and Analysis. Verlag Pearson New International Edition (2012)

Boehm, B.: Guidelines für Verifying and Validation Software Requirements and Design Specifications. In: Samet, P.A. (ed.) Euro IFIP 79. North-Holland Publishing Company, Amsterdam (1979)

Boehm, B.W.: A spiral model of software development and enhancement. Computer **21**, 61–72 (May 1988)

Chestnut, H.: Systems Engineering Methods. Wiley, New York (1967)

Delligatti, L.: SysML Distilled. Addison-Wesley (2013)

Ehrlenspiel, K., Meerkamm, H.: Integrierte Produktentwicklung – Denkabläufe, Methodeneinsatz, Zusammenarbeit. 5. Auflage. Carl Hanser, München (2013)

Eigner, M., Dickkopf, T., Schulte, T., Schneider, M.: mecPro^2 – Entwurf einer Beschreibungssystematik zur Entwicklung cybertronischer Systeme mit SysML. In: Schulze, S., Muggeo, C. (eds.) Tag des Systems Engineering, S. 163–172. Hanser Verlag, München (2015a)

Eigner, M., Eickhoff, T., Ernst, J., Eiden, A.: Systemübergreifendes Änderungsmanagement zwischen PLM und ERP. In: PLM Jahrbuch 2016 – Ein Leitfaden für den PLM Markt, pp. 76–79. Weka GmbH, Darmstadt (2015b)

Eigner, M., Gilz, T., Zafirov, R.: Neue Methoden, Prozesse und IT Lösungen für die virtuelle disziplinübergreifende Produktentwicklung. In: Berns, K., Schindler, C., Dreßler, K., Jörg, B., Kalmar, R., Zolynski, G. (Hrsg.) Proceedings of the 2nd commercial vehicle technology symposium (CVT 2012) (2012)

Eigner, M., Roubanov, D., Zafirov, R.: Modellbasierte Virtuelle Produktentwicklung. Springer, Berlin (2014)

Eisner, H.: Essentials of Project and Systems Engineering Management, 3rd edn. Wiley, New York (2008)

Estefan, J.A.: Survey of Model-Based Systems Engineering (MBSE) methodologies. Technical Report. California Institute of Technology (25 May 2007)

Forsber, K., Mooz, H.: The relationship of system engineering to the project cycle. Center for systems managements. Proceedings of the first annual meeting of the National Council for Systems Engineering and the 12th annual meeting of the American Society for Engineering Management, Chattanooga (20–23 October 1991)

Freisleben, D., Schabacker, M.: Wissensbasierte Projektnavigation in der Produktentwicklung. In: Proceedings of the 20th CAD-FEM Users' Meeting 2002, International Conference on FEM-Technology, BD. 2, S. 1–10 (2002)

Friedenthal, S.: Object oriented systems engineering. Process integration for 2000 and beyond: systems engineering and software symposium, 3rd edn. Lockheed Martin Corporation, New Orleans, LA (2014)

Gajski, D.D.: Construction of a large scale multiprocessor. Cedar Project, Laboratory for advanced Supercomputers, Dept. of Computer Sciences, University of Illinois at Urbana-Chambaign (Report/Department of Computer Sciences, No. UIUCDCS-R-83-1123) (1983)

Gausemeier, J., Michels, J.S., Orlik, L., Redenius, A.: Modellierung und Planung von Produktentstehungsprozessen, In: VDI-Berichte Nr. 1819, S. 245–256. VDI-Verlag, Düsseldorf (2004)

Gericke, K., Blessing, L.: Comparisons of design methodologies and process models across disciplines: a literature review. In: Culley, S.J., et al. (eds.) Design Processes, pp. 393–404. Design Society, Glasgow (2011)

Gericke, K., Meißner, M., Paetzold, K.: Understanding the context of product development. Proceedings of the 19th international conference on engineering design 2013, 19–22. August 2013, Seoul, Korea (2013)

Gericke, K., Moser, H.: Adapting a design approach: a case study in a small space company. In: Heisig, P., Clarkson, P.J., (eds.) Proceedings of 2nd international workshop on Modeling and Management of Engineering Processes MMEP, Cambridge UK (2012)

Haberfellner, R., de Weck, O.L., Fricke, E., Vössner, S.: Systems Engineering: Grundlagen und Anwendung. Orell Füssli Verlage; 13. Auflage (2015)

Hales, C., Gooch, S.: Managing engineering design, 2nd edn. Springer, London (2004)

Hammer, M.: Seven insights about processes. In: Proceedings of the conference on strategic power process ensuring survival creating competitive advantage. Boston, USA (2001)

Hitchins, D.K.: Systems Engineering: A 21st Century Systems Methodology. Wiley (2007)

Horvath, I., Gerritsen, B.: Cyper-physical systems: concepts, technologies and implementation principles. In: Proceedings of TMCE 2012, Karlsruhe (2012)

INCOSE: Technical operations. Systems engineering vision 2020, version 2.03. Seattle, WA: International Council on Systems Engineering, Seattle, WA, INCOSE-TP-2004-004-02 (2007)

INCOSE: Systems engineering handbook, version 3.2; INCOSE-TP-2003-002-03.2 (2010)

Kline, S.J.: Innovation is not a linear process. In: Research management, Vol. 26, Nr. 2, S.36–45 1995

Kossiakoff, A., Sweet, W.N., Seymour, S.J., Biemer, S.M.: Systems Engineering Principles and Practice, 2. Auflage. Wiley (2010)

Lee, E.: Cyper physical systems: Design challenges. In: 11th IEEE Symposium on Object Oriented Real Time Distributed Computing (ISORC) (2008)

Lindemann, U.: Methodische Entwicklung technischer Produkte – Methoden flexibel und situationsgerecht anwenden, 2. Auflage. Springer, Berlin (2007)

Meißner, M., Blessing, L.: Adapting a design process to a new set of standards – a case study from the railway industry. In: Marjanovic, D. (ed.) 8th international design conference – design 2004. Design Society, Glasgow (2004)

Miller, G.A., Galanter, E., Pribram, K.H.: Strategien des Handelns – Pläne und Strukturen des Verhaltens. Ernst Klett, Stuttgart (1973)

NASA Systems Engineering Handbook. NASA, SP-610S, Washington (1995)

NDIA, Wagner, Brockwell, Daniels, Loesh, Gosnell: NDIA Paper: use of a model-based approach to minimize system development risk and time-to-field for new systems. Final Report of the Model Based Engineering (MBE) Subcommittee NDIA, 10 Feb 2011 (2011)

Negele, H.: Systemtechnische Methodik zur ganzheitlichen Modellierung am Bei-spiel der integrierten Produktentwicklung. Dissertation, TU München (1998)
Oliver, D.W., Kelliher, T.P., Keegan, J.G.: Engineering Complex Systems with Models and Objects. McGraw-Hill (1997). ISBN:0-07-048188-1
Omiciuolo, M., Thiel, M., Förster, K.P., Paetzold, K., Foerstner, R.: General purpose modeling and domain specific simulation: a framework for space mechanisms design. IEEE SysCon 2015, Proceedings, Vancouver (2015)
Paetzold, K, Reitmeier, J.: Approaches for process attendant property validation of products. 1st International Conference on Modeling and Management of Engineering Processes MMEP 2010, Cambridge (2010)
Pahl, G., Beitz, W.: Konstruktionslehre – Grundlagen erfolgreicher Produktent-wicklung. Methoden und Anwendung. 7. Auflage. Springer, Berlin (2007)
Panreck, K.: Rechnergestützte Modellbildung physikalisch-technischer Systeme, Fortschritt-Berichte, VDI-Verlag, Düsseldorf (2002)
Parraguez, P.: A networked perspective on the engineering design process. PhD-Thesis, Techinical University of Denmark (2015)
Patzak, G.: Systemtechnik – Planung komplexer innovativer Systeme, Grundlagen, Methoden, Techniken. Springer, Berlin (1982)
Ponn, J., Lindemann, U. (eds.): Konzeptentwicklung und Gestaltung technischer Produkte. Springer, Heidelberg (2011)
Pugh, S.: Total design: Integrated Methods for Successful Product Engineering. Addison-Wesley, Wokingham, England (1991)
Rechtlin, E., Maier, M.W.: The Art of Systems Architecting. 2nd edn, CRC Press (2000)
Reitmeier, J.: Eigenschaftsorientierte Simulationsplanung - Ein Beitrag zur effizienten virtuellen Absicherung der Produktfunktionalität. Dissertation, UniBW München, Verlag Dr. Hut (2015). ISBN:978-3843921626
Rumbaugh, J., Jacobson, I., Booch, G.: The Unified Modeling Language Reference Manual. The Addison-Wesley Object Technology Series. Addison-Wesley (1999)
Sangiovanni-Vincentelli, A., Yang, G., Shukla, S.K., Mathaikutty, S.A., Sztipanovits, J.: Metamodeling: an emerging representation paradigm for system-level design. University of California Berkeley, IEEE Design & Test of Computers (2009)
Skalak, S.C., Kemser, H.-P., Ter-Minassian, N.: Defining a product development methodology with concurrent engineering for small manufacturing companies. J. Eng. Des. **8**(4), 305–328 (1997)
Sop Njindam, T.: A systemic approach to analyze failures of complex multidisciplinary systems on the basis of their weak emergent behaviour. Dissertation, UniBW München, Verlag Dr. Hut (2015)
Stachowiak, H.: Allgemeine Modelltheorie. Springer, Wien (1973)
Sztipanovits, J.: Composition of cyber-physical systems. In: 14th Annual IEEE Int'l. Conference and Workshop on the engineering of Computer-Based Systems (ECBS '07), pp. 3–6. IEEE Computer Society, Washington (2007)
VDI 2206: Entwicklungsmethodik für mechatronische Systeme, Hrsg. Verein Deutscher Ingenieure, Ausg. Juni (2004)
VDI 2221: Methodik zum Entwickeln und Konstruieren technischer Systeme und Produkte, Hrsg. Verein Deutscher Ingenieure, Düsseldorf, Ausg. November (1986)
VDI 2422: Entwicklungsmethodik für Geräte mit Steuerung durch Mikroelektronik, Hrsg. Verein Deutscher Ingenieure, Düsseldorf (1994)
Walker, R., Thomas, D.: A model of design representation and synthesis. 22nd design automation conference, Las Vegas (1985)
Weber, C.: CPM/PDD – An extended theoretical approach to modeling products and product development processes. In: Bley, H., Jansen, H., Krause, F.-L., Shpitalni, M. (eds.) Proceedings of the German-Israeli symposium on advances in methods and systems for development of products and processes. Fraunhofer, Stuttgart (2005)
Weilkins, T.: Systems Engineering mit SysML/UML: Modellierung, Analyse, Design, 2nd edn. dpunkt Verlag (2008)

Chapter 3
Cyber-Physical Product-Service Systems

Stefan Wiesner and Klaus-Dieter Thoben

Abstract Cyber-Physical Production Systems (CPPS) foster new processes and production methods for reducing "time to market", waste and failures, as well as improving quality and cost effectiveness. However, changes cannot be restricted to the technological side. An increasing share of services is offered with these systems in order to deliver new customized functions and other benefits. This trend has led to the introduction of Product Service Systems (PSS) as a promising framework describing the integrated development, realization and offering of specific product-service bundles as a solution. The integration of both CPPS and PSS concepts is becoming relevant for industry, because data monitoring, storage and processing allow creating a higher service layer able to deliver production systems with new "intelligent" behaviors and communicating capabilities. In this chapter, we use the term Cyber-physical Product-Service Systems (CPSS) for such an integrated approach. It gives a definition of CPS-based PSS and unveils the state-of-the-art for both concepts with major research issues for their integration. The evolution from products to solutions through servitization is shown, as well as the hardware, software, and service elements of CPSS, requiring an alignment of CPPS and service lifecycle models. Based on industrial use cases, this chapter also deals with challenges for engineering CPS-based PSS in terms of complexity, end user involvement with information exchange among stakeholders and linking views of multiple disciplines (mechanical engineering, information systems, service science etc.). This leads to implications for engineering processes, particularly cross-domain Requirements Engineering and design but also servitized Business Models enabled by CPS.

S. Wiesner (✉)
BIBA—Bremer Institut für Produktion und Logistik GmbH at the University of Bremen, Hochschulring 20, 28359, Bremen, Germany
e-mail: wie@biba.uni-bremen.de

K.-D. Thoben
BIBA—Bremer Institut für Produktion und Logistik GmbH at the University of Bremen, Hochschulring 20, 28359, Bremen, Germany

Faculty of Production Engineering, University of Bremen, Badgasteiner Straße 1, 28359, Bremen, Germany
e-mail: tho@biba.uni-bremen.de

© Springer International Publishing AG 2017

S. Biffl et al. (eds.), *Multi-Disciplinary Engineering for Cyber-Physical Production Systems*, DOI 10.1007/978-3-319-56345-9_3

Keywords Cyber-physical system • Product-service system • Extended product • Industrie 4.0 • Smart manufacturing

3.1 Introduction

Industrial companies are more and more facing complex customer needs, forcing them to analyze the underlying problem and create individually tailored solutions. In order to remain competitive on the market, such solutions still have to be based on economy of scale principles (Reichwald et al. 2009), often involving globally distributed partners (D'Aveni et al. 2010). In this way, the value chain is becoming an ecosystem of several partners with different competencies and able to share different knowledge, acting and interacting in a dynamic environment (Hintsa and Uronen 2012).

Furthermore, rapid technological developments have created possibilities for innovative production systems, enabling new processes and methods reducing "time to market" (Chang et al. 2013; Lee et al. 2010), waste and failures. Quality and cost effectiveness is improved with solutions meeting the customers' expectations (Kossiakoff 2011). For example, Follett (2014) enlists as significant emerging technologies a networked, intelligent world connected by Internet of Things (IoT), involving robotics, the usage of additive fabrication and 3D printing, and promoting the just-in-time production. The relevance of these technological advances has been recognized by being implemented in several governmental programs like the German "Industrie 4.0" paradigm (Kagermann et al. 2013) and the United States Smart Manufacturing Leadership Coalition (SMLC 2011). Among other application scenarios, these initiatives strongly advocate the implementation of cyber-physical technology in manufacturing, creating Cyber-Physical Production Systems (CPPS) (VDI/VDE 2014).

However, changes in production cannot be restricted to the technological side alone. As manufacturers increasingly demand support for all phases of the production system lifecycle, from development over assembly and distribution to operation and decommission, a wide range of services is provided in addition in order to deliver new customized functions and other benefits. Accordingly, production system provider's Business Models are also shifting from selling a system to the provision of integrated manufacturing solutions. This trend has originally been named as the "Servitization" of business (Vandermerwe and Rada 1988) and led to the introduction of Product Service Systems (PSS) as a promising framework describing the integrated development, realization and offering of specific product-service bundles as a solution for the customer (Goedkoop 1999; Baines et al. 2007). Offering of a PSS requires on the one hand additional competencies for the provision of the associated services and on the other hand a better understanding of the customer requirements. As the requirements now have implications on the whole production system life-cycle, an integrated development of the system, services, and the manufacturing processes is needed.

The integration of production systems and PSS concepts is becoming relevant for industry due to the diffusion of pervasive Information and Communication Technologies (ICT), which have strongly reduced the cost for additional sensors and cloud technologies, enabling data monitoring, storage and processing (Klocke et al. 2011). This allows creating a higher service layer able to deliver production systems with new "intelligent" behaviors and communicating capabilities (i.e. monitoring the manufacturing environment, interacting with the operator and other connected devices, being adaptable to the user and customer needs) (Yang et al. 2009). These developments represent new challenges for industrial companies in the design process, creating individual CPPS solutions for the customer, but also offering these solutions as a PSS with the appropriate services along the lifecycle. In this chapter, we use the term Cyber-physical Product-Service Systems (CPSS) (Gorldt et al. 2016) for such an integrated approach.

In an exemplary CPSS scenario, a customer could order a pre-defined machining capacity for his factory. He would work together with the CPSS provider to develop such a service specific according to his needs, including the suppliers of hard- and software components for the CPSS. External service providers could develop new machining configurations specifically for the CPSS and deliver spare parts and refurbished components according to customer requirements. The CPSS machining solution would automatically integrate into the manufacturing environment and process, while being monitored and maintained by the provider. In such a business model, the customer could either pay a fixed price for the machining capacity, a time-based rate for its availability or a fee for the actual usage of the machining and additional services.

The main prerequisite for successful and cost-effective development of CPSS offers is understanding their elements, life-cycle and the specific characteristics of the engineering process. A major challenge in this process is the integration of the product, service and ICT perspective, each having own models, methods and tools. Process complexity is further increased by the higher number of stakeholders, the coverage of the whole lifecycle (Blanchard 2004), and the dynamic of functionalities expected by the customers, leading to evolutionary changes during engineering. In this context, it is essential to identify and investigate the integrated phases of the CPSS engineering process. This begins with the ideation of product-service solutions with cyber-physical elements, continues with cross-domain Requirements Engineering (RE) and solution design, leading to the servitized Business Models (BM) through which they are provided to the customer.

In the further course of this chapter, first the research methodology used to investigate the above issues and the research objectives are described. The elements and lifecycle of Cyber-physical Product-Service Systems are derived from the established definitions of CPS and PSS. Based on this, the challenges for the engineering process are derived. Implications for ideation, Requirements Engineering and Business Model development are discussed, illustrated by an industrial use case. Finally, the results are summarized and the chapter is concluded.

3.2 Research Methodology and Objectives

For the combination of CPPS with PSS Business Models, it is important to establish the main characteristics of the resulting CPSS and their specific engineering challenges. This relates to a number of the fundamental research questions raised in Chap. 1:

RQ C2*: How shall several disciplines in product and production system engineering be linked to support the engineering of flexible and self-adaptable CPPS?* This leads to our first research question for this chapter, concerning the necessary link between production system engineering and service engineering to create self-adaptable CPPS that can be flexibly offered under a PSS concept, to be addressed in Sect. 3.3:

1. What is the potential of linking CPPS with the service perspective; i.e. what does it mean to apply a CPPS as the basis for a PSS offer?

RQ M2 (b)*: How can model-based methodologies for the application of CPPS by service providers or agents support information creation and processing in the different lifecycle phases of a CPPS?* This serves as starting point for our second research question for this chapter, dealing with the interaction points of service lifecycle and CPPS lifecycle phases, to be addressed in Sect. 3.4:

2. What are the elements and phases of CPPS and PSS lifecycles and how can their interaction be supported by a methodology towards a CPSS model?

RQ I2*: Which methods and technologies support assuring the required information quality for exchange of engineering project data and multi-disciplinary knowledge integration?* This is the basis for the third research question for this chapter, which deals with the necessary support for exchanging requirements and business knowledge among the involved disciplines, to be addressed in Sect. 3.5:

3. Which are the specific challenges and support needs for the information exchange in the engineering process resulting from integrated CPSS, particularly for Requirements Engineering and Business Model development?

In order to address these points, our methodological approach is based on an exploratory approach, combining a literature review and analysis of industrial use cases. The literature review has been conducted by analyzing scientific papers from multiple disciplines (systems engineering, information technology, business research), as CPS and PSS are a cross-domain research topic. For practical reasons, the search was limited to books, journal and conference papers in English and German language. As several expressions are used in literature to describe the concepts, we searched for literature on CPS, PSS, servitization, RE and BM. The search results were checked for relevance and redundancy by assessing the abstracts. Based on this, papers were selected for in-depth analysis of the content. Regarding the first research question, the analysis concentrated on the existing fundamental concepts for CPS, PSS, and first attempts for servitization of production systems.

For the second research question, the focus was shifted towards product and service lifecycle management approaches and integration attempts. Concerning the third research question, existing RE and BM approaches were analysed for their applicability on CPSS and open issues were extracted.

The work with the industrial use cases for CPSS had a different methodical approach. In order to compare the results of the literature review with real attempts to engineer CPSS, the authors were able to work together with companies developing such a solution in the frame of different research projects (PSYMBIOSYS 2014; Wiesner et al. 2014a). The researchers have been involved in the specification and development of the CPSS scenarios applying action research (Sein et al. 2011). With regards to the research question it could be studied, (1) how the responsible parties for production system and service engineering collaborated and where the weak points were, (2) how product and service lifecycles were managed and which methodologies were used, and (3) how information exchange in RE and BM development was supported and what were the drawbacks.

Evaluation of research results has been conducted in an iterative way. Initially, it was verified if the challenges for CPSS engineering faced by the industrial use cases are aligned with the challenges described in literature. Next, approaches found in literature to overcome the identified challenges were applied in the use cases and assessed for their usefulness. The outcomes were again validated during the literature review, leading to a consolidated result.

3.3 Elements and Definition of Cyber-Physical Product-Service Systems

This section identifies the specific characteristics of Cyber-Physical Systems and Product Service Systems and integrates them to the concept of Cyber-physical Product-Service Systems.

3.3.1 Cyber-Physical Systems

CPS integrate physical capabilities with Information and Communication Technology (ICT) and offer new ways of human–machine interaction using advanced sensors and actuators (see Fig. 3.1). They require the collaboration of different disciplines such as mechanical engineering, electrical engineering, and computer science for their realization. Additionally, for reaching the full potential of CPS, the system will also comprise the logistics and management processes, as well as internet services receiving, processing and analyzing data from the sensors and controlling the actuators, connected by digital networks and multi-modal human–machine interfaces. As such, CPS are open socio-technical systems with

Fig. 3.1 Cyber-Physical System

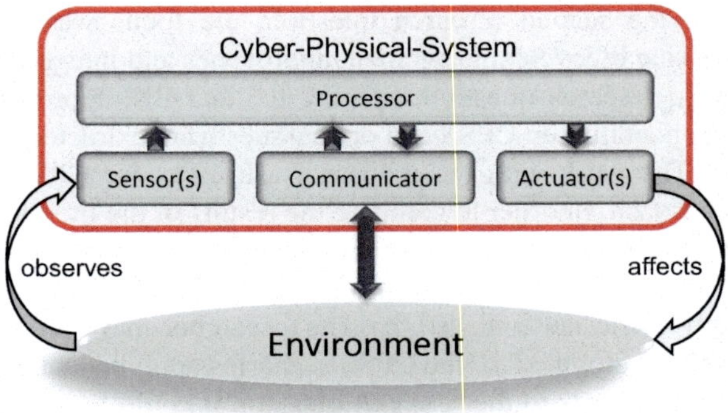

a functionality far exceeding controlled embedded systems (Geisberger and Broy 2012).

Several characteristics can be identified that describe CPS and distinguish them from other complex systems. The eponymous aspect of CPS is merging the physical and virtual world. CPS involve a multitude of parallel and interlinked sensors, computers, and machines, which collect and interpret data to decide on this basis and control real world physical processes. Thus, systems engineering needs to integrate processes and physical elements with information technology (Rajkumar et al. 2010). Secondly, CPS have dynamic system borders. Depending on application and task, different CPS are arranged into a system of systems for a limited time. Consequently, CPS have to be able to actively configure services and networks with other systems or part of systems, which may be unknown in the beginning, and provide new and composite components and services in a controlled way (Colombo et al. 2013). Furthermore, an important characteristic of CPS is their ability to adapt to environmental changes and application requirements. This requires continuous monitoring and assessment of environmental and application data (Wan and Alagar 2014). Additionally, in most cases, there will be no central control of a CPS. Decisions are made locally based on an assessment of the situation and lead to a cooperative learning process (Zhou et al. 2013). Finally, CPS have to interact with humans also on a physical level, which requires multi-modal control interfaces, recognition, and interpretation of human behavior and interactive decision making between the system and single persons or groups (Schirner et al. 2013).

3.3.2 Product-Service Systems

For a long time, manufacturers have considered the monetary profit from selling products as their main revenue stream. Services, if any, have been provided free (e.g. usage advice) or for a one-time fee (e.g. maintenance) and were strictly product focused. However, within the last decades product users increasingly demand holistic solutions for their individual problems, and their buying decision

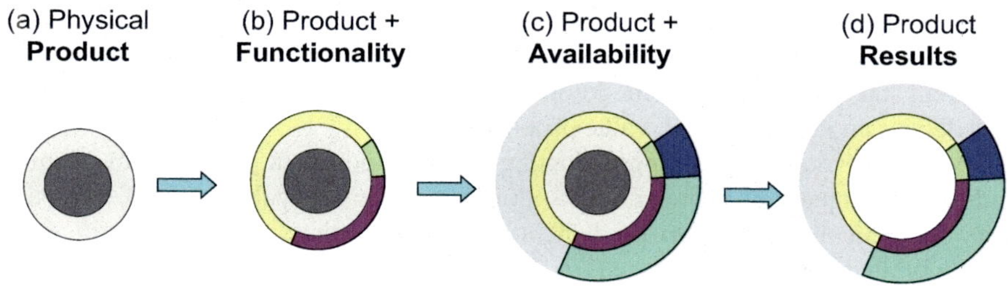

Fig. 3.2 Servitization of products—the Extended Product concept

is driven the expected benefits from using the solution rather than sales price only (Vandermerwe and Rada 1988). As an immediate consequence, especially the industry for complex products has started to deliver product-service bundles (see Fig. 3.2).

The Servitization process is a fundamental mean for manufacturing companies to find new business opportunities and involve new customer segments, increasing their market share (Wiesner et al. 2014a; Spohrer and Maglio 2010). Figure 3.2 illustrates this phenomenon from a customer perspective on the solution offered. In case **(a)**, the physical product comprises just a tangible offer and its shell (e.g. a production system plus accessories) to be bought by the customer. Case **(b)** describes a scenario, where the functionality of the tangible product is supported by a service layer (maintenance, spare part delivery etc.). In this case, the role of the tangible product is still dominating. Services ensuring functionality can be ordered *in addition* to the original product. In case **(c)**, the availability of the product for its purpose is guaranteed by the provider. While the tangible product might still be sold separately, there has to be some kind of service level agreement (e.g. defining the availability ratio of the production system). Of course, this requires pre-active maintenance or other measures to minimize product failure, which could be supported by the monitoring and communication capabilities of a CPS. Finally, case **(d)** is sharply decoupling the tangible product from the results of its application. The customer purchases services that are bundled for the solution of his specific problem (e.g. a specific manufacturing capacity). The tangible product, if still required, will not be sold but just used to provide the services. Payment models may include pay-per-use, pay-for-performance etc. This as well relies on the ability of the solution provider to monitor usage and dynamically reconfigure the product, both of which could be supported by CPS.

The PSS concept focuses on the bundling of products and services as a mix of tangible and intangible elements designed and combined to increase the value for customers (Goedkoop 1999; Meier et al. 2010). From the economic viewpoint, PSSs are able to create new market potentials and higher profit margins, and can contribute to higher productivity by means of reducing investment costs along the lifetime as well as reducing operating costs for the final users (Baines et al. 2007). Furthermore, PSS can increase resource productivity and closed-loop manufacturing

(McAloone et al. 2010), thanks to the service functionalities delivered. Value creations is realized through the extension of the current business network, involving different stakeholders having the knowledge and skills required to design, develop and deliver an integrated PSS value proposition.

Despite several methodologies having been proposed in literature to support industrial companies to design a PSS along its entire lifecycle, some of them are very theoretical and hard to implement in practice, others are very specific and have a limited applicability range (Garetti et al. 2012). Currently, integrating several existing methodologies to design a new PSS solution is being studied (Marilungo et al. 2015).

3.3.3 Cyber-Physical Product-Service Systems

In order to provide a holistic solution over the full system lifecycle to a customer, both the technological as well as the economic perspective have to be considered in a very early phase. The combination of CPPS functionalities with PSS business models has the potential to enable new and innovative production system offers. Picking up the example from Sect. 3.1, a manufacturer might require machining a small number of individual parts on an expensive special machine tool. Due to the high investment required, the manufacturer will not buy such a tool for his production system. At the same time, a production order at a third party might not make sense due to the small lot size or delivery time. CPS technology could now enable to share such a special machine tool among different manufacturers, dynamically integrating itself into the different production systems. Payment for such a shared resource would be based on a PSS model [see Fig. 3.2, case (d)], where the manufacturer would pay for the machining results, based on usage monitoring by the CPPS. Such an integration of product and service through CPS is our definition of a Cyber-physical Product-Service System (CPSS) (Gorldt et al. 2016).

In principle, the relation between the CPS and the PSS concept in a CPSS can be seen as interdependent, or symbiotic. When looking from a CPS perspective, the physical and ICT domains are complemented with service engineering for the development of the solution. This increases the number of stakeholders and adds additional domain-specific models, methods and tools to the development process. From the PSS perspective, the new cyber-physical functionalities are enablers of additional innovative services and Business Models (Hehenberger et al. 2016). As such, the extended technical opportunities have to be taken into account when ideating new services that enhance customer benefit. Furthermore, the possibility to measure usage and performance of the CPPS supports Business Models beyond a one-time sale, guaranteeing availability of the system or pay-per-use models.

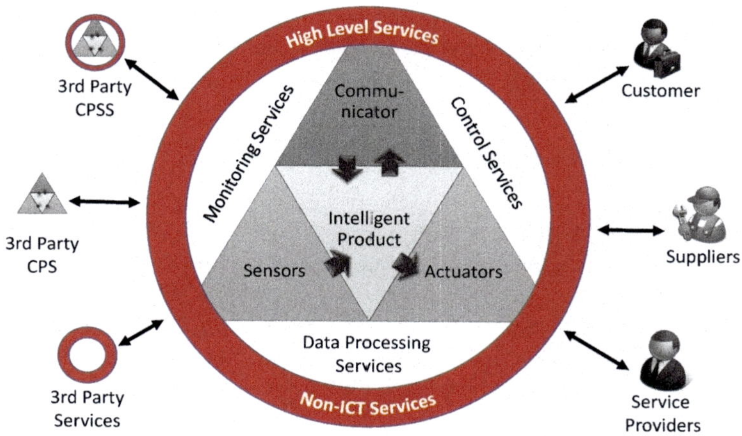

Fig. 3.3 The CPSS concept

The above considerations can be summarized to a first illustration of the CPSS concept. The core is an intelligent product, similar to the processor hardware of the CPS. Here, it is however not reduced to its computational capabilities, but comprises all the tangible aspects of the offer, such as design and haptics. Similarly, it is connected to sensors, actuators and communicators, which enable its interaction with its environment (see Fig. 3.3).

In the CPSS concept, the core components are used to offer "basic services" such as monitoring, control and data processing. This internal structure is obscured to the outside world by a halo of high-level and non-ICT services. In the halo, the basic services are aggregated with the non-ICT services to fulfill certain functionalities and tasks. These high level services are not hard-wired into the CPSS and can change dynamically to new demands. Through this service halo, the CPSS is able to collaborate with different entities to achieve its intended aim. Those entities can be other CPSS, CPS just providing basic services, or even third party services not directly developed for the system.

In the case of the special machine tool CPSS, the high level service for the customer would be the offered possibility to machine a varying amount of parts in a specific way. This can include non-ICT services, like transport of the parts, which is provided by a logistic service provider. Another element could be a third party service that optimizes part design for machining. Through communication with other CPS, e.g. in the production systems of the customers, the CPSS could automatically schedule orders according to various criteria. Finally, wear of components could be early detected through monitoring and trigger on-time spare part replacement by component suppliers. Customers can pay for access to these services and use them for their benefit. They can also work together with the CPSS provider to re-develop services specific to their needs. The same is true for suppliers of hard- and software. Finally, external service providers develop new services specifically for the CPSS, or use standardized interfaces to connect existing services.

3.4 Challenges for Integrating CPPS and PSS LifeCycles

Servitized Business Models extend the responsibility of the PSS provider to the whole life-cycle of the solution (Aurich et al. 2010). Moreover, Product Lifecycle Management (PLM) and Service Life-cycle Management (SLM) must be aligned to be able to create an integrated product-service offer for the customer. For CPPS offered as PSS, this means to understand the interaction between the elements of the production system lifecycle and the service lifecycle.

3.4.1 Product LifeCycle Management

PLM covers the whole lifecycle of a product from the first idea and concept to recycling and disposal. There are many different lifecycle models found in literature. However, the majority is based on three main lifecycle phases, Beginning of Life (BoL), Middle of Life (MoL), and End of Life (EoL) (Stark 2011), as shown in Fig. 3.4.

A similar three-phased structure is used in Chap. 4 of this book for a life-cycle oriented information integration approach for CPPS. However, for such complex products BoL, MoL and Eol are not connected by a linear information flow, but the three phases are arranged orthogonally. Pieces of information from every phase can be linked to information in any other phase, creating a network of microservices between the involved software tools. Thus, it addresses mainly ICT services and would have to be extended to support PSS offers. PLM concepts need to support multi-disciplinary development of products hardware and software, as well as methods and system functionalities for cross-domain engineering collaboration.

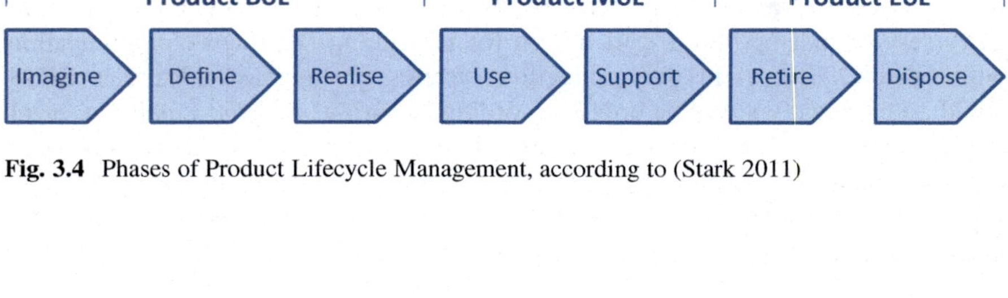

Fig. 3.4 Phases of Product Lifecycle Management, according to (Stark 2011)

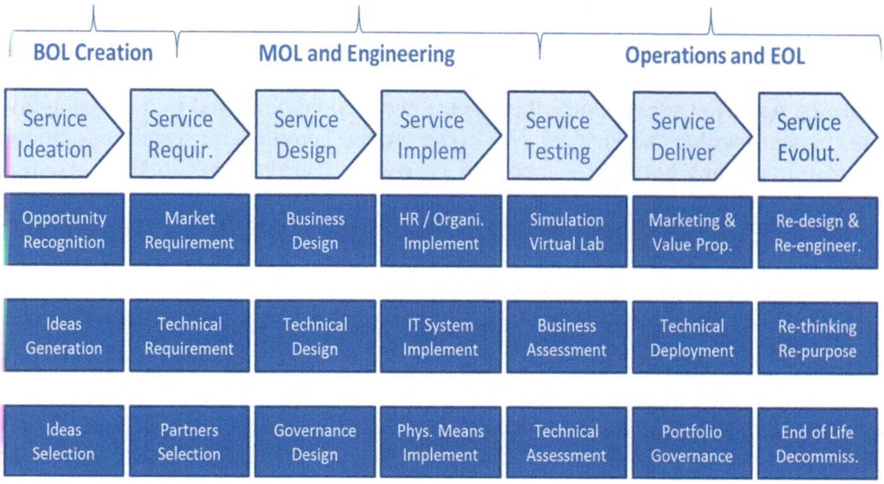

Fig. 3.5 Process model of Service Lifecycle Management (Freitag et al. 2013)

3.4.2 Service LifeCycle Management

SLM is a part of Service Science, Management and Engineering (SSME), which is a young field of research that addresses the open questions and challenges coming from the servitization process (Spohrer and Maglio 2010). Nevertheless there are already some SLM approaches, for instances from Freitag et al. (2013). In this case, the three main phases of the Service Lifecycle are service creation, service engineering and service operations management, which are further divided into several elements (see Fig. 3.5).

Service creation is the phase at the beginning of the Service Lifecycle Management. It mainly consists of two pillars: provision of conditions and ideation. The influences providing opportunities may be changing customer needs, new emerging technologies, transformations of the company environment, and other causes or drivers of change. For service ideation, they serve as triggers or stimuli. When a selection of service ideas is handed over to the first phase of service engineering, it comes to a structured evaluation of the service ideas based on market and technical requirements.

The service development process is a waterfall model for the development of new services (Spohrer and Maglio 2010). In this framework, the engineering phase consists of four elements: service requirements, service design, service implementation and service testing. In the requirements analysis the internal and external requirements are collected. The second element of the service development process is called service design, in which the new service is defined and described. In the third step, the implementation of the service also includes the operative realization of the described services concepts. Furthermore, the involved employees need to be trained as planned. The service should be tested by customers, by using a simulation tool, or at least by a checklist.

The first task in service operations is to acquire customers, respectively service projects. After the acquisitions, the service needs to be delivered to the customers. This happens within "service delivery". The support activities for service operations are also important, here for instances to evolve the service portfolio and to control the service operations.

3.4.3 Integration of PLM and SLM

Based on the targeted integrated design of PSS, PLM and SLM must also be integrated to provide the required interactions between product and service on an operational level. This is currently a focus of research (Garetti et al. 2012; Peruzzini et al. 2014). Meier and Uhlmann (2012) derive a PSS lifecycle directly from a product and service lifecycle. Relevant phases are planning, development, implementation, operation and dissolution, as shown in Fig. 3.6.

During the **planning phase**, customer needs and goals are determined for initiating the overall ideation process. Ideas for PSS solutions are identified, selected and specified to meet the customer needs. The resulting drafts are used for PSS development. In addition, first aspects of the Business Model are discussed.

The second phase is the **development** of the PSS. Stakeholder requirements are elicited and conceptual solutions are generated with adequate functionalities. The

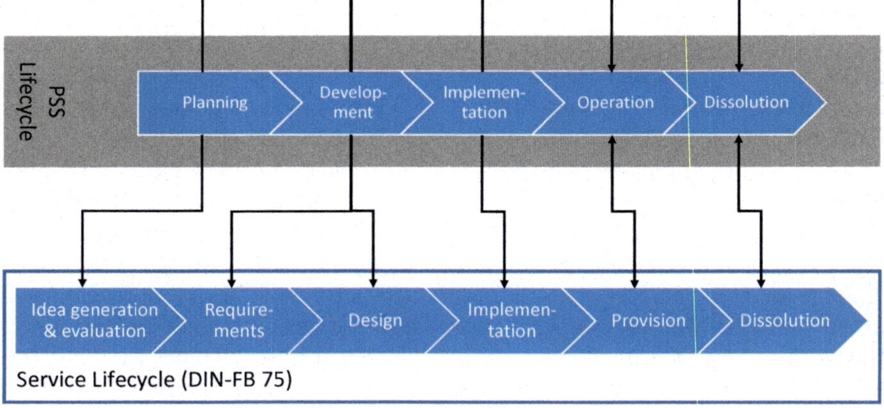

Fig. 3.6 PSS lifecycle according to (Meier and Uhlmann 2012)

product, service and ICT components are then configured and the responsibilities and resources between customers and suppliers are distributed in accordance with the concept. The result of the development phase represents a system model.

The **implementation phase** represents the third phase of the lifecycle. The tangible components are produced and integrated with ICT. In addition, the logistical processes for the delivery of products and services are designed. For future service provision, resource planning is carried out. Finally, the phase is completed with the commissioning of the PSS.

In the **operating phase**, the use of the PSS is in the foreground. Here service shares of the PSS are provided and dynamic adjustments to the PSS can occur. Therefore, knowledge from this phase is also relevant for PSS design. The aim is that a continuous improvement process can be ensured. A key component of the phase is maintenance. The last lifecycle phase represents the **dissolution**, in which the contractual relationship between client and provider is terminated. Tangible components of the PSS can be recycled or remanufactured accordingly.

However, there are some aspects of the model that can be criticized, especially in relation to CPSS. On the one hand, it is still depicting the lifecycles in a linear fashion, not showing the complex information exchange between the lifecycle phases. On the other hand, it aligns one product lifecycle with one service lifecycle, instead of recognizing the possibility of several service lifecycles being attached to different product lifecycle phases. The resulting engineering challenges are discussed in the next section.

3.4.4 Engineering Challenges

CPSS require the integration of a large number of different cyber, physical and service components. Stakeholders are typically separated spatially and organizationally and stem from multiple disciplines with own formalisms and tools. Various Authors (Baheti and Gill 2011; Baines et al. 2007) emphasize the need for theories and tools to design, analyze and verify the components at various levels, understand the interactions between systems and ensure safety and performance with minimal cost. Several challenges for engineering CPSS are identified below, which have to be addressed by suitable approaches for specific problem areas.

Products and services require different competencies, methods and tools to efficiently manage and perform the activities during their lifecycle. Usually production system providers have well-defined product development processes, but they lack sufficiently in structured service development processes. However, both manufacturing as well as service provision must be brought together and delivered in an appropriate way to offer an attractive product-service bundle to the customer (Spath and Demuß 2006). The basic assumption of many PLM approaches is that services and their lifecycles are aligned to the product. However, in many cases there is a strong need to have bi-directional coordination and interaction between PLM and SLM in a systematic way. Despite several methodologies have been proposed in literature to support industrial companies to integratively design a PSS (Garetti et al.

2012), some of them are very theoretical and hard to implement in practice, others are very specific and have a limited applicability range. Appropriate approaches and tools for supporting the development of PSS in an efficient way are missing (Marilungo et al. 2015). Production system providers could use the ICT capabilities of CPS to combine Product Lifecycle Management (PLM) with Service Lifecycle Management (SLM).

Many times, design weaknesses are only identified in manufacturing or operation of production systems. Information exchange between these phases has to be established to improve design iteratively. For CPSS, an approach is necessary that also includes feedback from the services during the lifecycle of the product, starting from engineering and ending at disposal. The current practice of a product-service development is still linear and hierarchical, meaning that the product is developed and manufactured before the services, hence being incapable of incorporating service requirements and constraints from all the stakeholders involved. CPSS providers must be able to identify ubiquitously the requirements and constraints arising from the services related to the product during the engineering phase, today still an ongoing challenge (Annamalai Vasantha et al. 2012). A collaborative environment for CPSS design, enabling communication with manufacturing, but also integrating knowledge form cross-disciplinary feedback loops, including customers is required. For CPPS, this means joint exploitation of (cyber-physical) data along the entire product-service lifecycle (not only from production system operation) to optimize the CPSS offering.

Current production systems are complex distributed systems that connect and coordinate intelligent machineries, sensors, actuators, control systems and manufacturing and business applications (Vyatkin 2013). Because of the still common application of sequential development approaches, many different views on the same system are generated along the lifecycle. How can these views, especially for the "real" production system and its "digital" representation, be reliably consolidated along the whole lifecycle? Multi-directional interoperability of the digital images of the same CPPS along every single phase of the lifecycle would benefit the pinpoint development of integrated services.

The design of CPSS requires knowledge that is usually scattered among different persons, departments or even organizations. Manufacturers are working closely with service providers, suppliers and customers to optimize designs of new product-service bundles before they are realized (Romero et al. 2012). To this end, both knowledge from the product side as well as the service side must be shared in an appropriate way, combined and utilized, in order to create an attractive CPSS for the customer. However, only about 4% of organizational knowledge is formalized (Bell 2006). Informal and unstructured knowledge, consisting of individual posts and discussions, ideas, comments and other interactions is difficult to codify and share, as it requires individual interaction to transfer. For production systems, knowledge sharing is mainly focused on re-using service knowledge to improve the product or services. The synthesis of knowledge across different domains of application, including methods of requirements analysis and modelling is needed (Broy et al. 2012).

Table 3.1 Problem areas and engineering challenges for CPSS

Problem area	Challenges
Integration of tangible and intangible components	• Alignment of product and service development • Structuring the service engineering process • Integration of PLM and SLM
Horizontal and vertical information sharing	• Exchange of key information between lifecycle phases • Cross-disciplinary feedback loops
Virtual representation of CPSS	• Creating different views on the same CPSS • Alignment of digital representations along the CPSS life-cycle
Knowledge management	• Management of explicit product-service knowledge • Capturing tacit product-service knowledge and stakeholder sentiment
Business Model innovation	• Development and classification of servitized Business Models • Assessment and mediation of effects on current Business Model

From the economic viewpoint, PSS are able to create new market potentials and higher profit margins, and can contribute to higher productivity by means of reducing investment costs along the lifetime as well as reducing operating costs for the final users (Baines et al. 2007). However, the transition to PSS might have an impact on existing Business Models. Generally, the PSS could be competitive, complementary or neutral to the existing business. These effects can occur externally on the market, e.g. if the PSS and the existing products compete for the same customer budgets ("cannibalization" effects), or internally if they apply the same resources, a situation that could produce conflicts or synergies. For production systems, services are fundamentally aligned to the physical product that is usually part of the existing business, so they should not be conflicting but supportive. However, if CPPS are used to enable use-oriented or even result-oriented solutions where the production system is no longer sold, this could lead to a decrease of sales in the existing business.

Table 3.1 above summarizes the engineering challenges for CPSS and classifies them into five problem areas:

3.5 Implications for the Engineering Process

This section describes the implications for the engineering process coming from the specific characteristics of CPSS. Two approaches are presented, which are addressing some of the challenges identified in Sect. 3.4. Cross-domain Requirements Engineering and design aims to capture the needs on tangible and intangible components and integrate them into a holistic CPSS design. Servitized Business Models enabled by CPS tries to classify CPSS innovation and its impact on the current Business Model.

3.5.1 Cross-Domain Requirements Engineering and Design

According to the challenges identified in the previous section, it is necessary to align the development of the tangible and intangible components of a CPSS. One of the earliest activities in both the product and the service lifecycle is establishing the requirements of the stakeholders towards the targeted system. A weak definition of requirements can slow down CPSS development and induce unnecessary costs for design changes (Boehm and Basili 2001). If incorrect requirements are identified, an unsuitable system architecture and implementation can result and the system may have missing or wrong functionalities.

Current approaches for Requirements Engineering do not provide full support for integrated CPSS development. The RE methodologies of the product, service and software disciplines focus on their respective domain. E.g., elicitation procedures in the product domain focus on technical requirements and methods such as checklist are not well suited for service requirements. Service engineering methodologies are not detailed enough to be used as the basis for CPSS development. Within software engineering methodologies, the representation of service requirements is not possible with the provided procedures and modeling techniques. First integrated approaches for PSS state the necessity of cross-domain knowledge, interfaces and interdisciplinary requirements (Berkovich et al. 2011). However, they are too vague and do not provide the procedures necessary in order to realize a PSS. The procedures are not explained in detail or similarly to service engineering, procedures of other domains are referenced (Annamalai Vasantha et al. 2012). In the following, suitable approaches to support the various RE activities for CPSS are presented.

In contrast to the design of complex, distributed systems such as CPPS, the additional inclusion of the service and business perspective lead to further clarification needs. The customer, production system manufacturer and service provider have different perceptions of requirements terminology and wording. In addition, for such complex systems the customer might not have a clear idea on how the product-service solution will look like (Gausepohl 2008). An approach to reduce the risk of capturing false requirements for the CPSS, the storytelling method can be applied for requirements elicitation (Vink 2015). It allows the participants to develop a use case story commonly (group storytelling), including the required services, and thus discuss the different perspectives as it is carried out. A narrative story telling helps to put the requirement in a specific context clarifying to the reader how they are to be understood.

In order to analyze the stories and extract an unambiguous, consistent and complete set of requirements for a CPSS, different modeling techniques can be applied. An older method that helps not only in identifying the human stakeholders, but also system restrictions and other mechanisms (including machinery, software etc.) is IDEF0. The storytelling can be combined with the IDEF0 generation in a workshop setting, in which the requirements engineer could establish AS-IS and TO-BE scenarios. In parallel, the process modelling can be done and all stakeholders and restrictions are captured.

To create a requirements specification that provides the developers with a complete description of the functionality of the CPSS, information on the connections between the components has to be recorded to serve as reference for requirements dependencies later. A change of a requirement on a specific service can lead to a number of changes in hard- and software of the system. Documentation has to establish the link from the single requirement towards the whole CPSS project, including information on the degree and moment of fulfilment. Methods for documentation of business requirements can be Business Process Model and Notation (BPMN) or data flow diagrams, showing the difference between AS-IS and TO-BE business processes. Describing the targeted CPSS behavior in terms of conditions or capabilities of the envisaged solution can be done through developing system models describing functionality and then documenting system requirements that capture the vision of the customer in technical terms. These models can be documented in SysML, among other modelling languages. First approaches of automatic transformation between business and functional notation are documented in literature (Wiesner et al. 2014a). Transformation between different domain specific models can be supported using methods such as semantic mediation, which enable conversion between model using ontologies (Hribernik et al. 2010). In this way, domain barriers can be greatly reduced or fully removed.

The completeness and correctness of the determined requirements is checked during requirements validation to ensure that the documented requirements accurately express the stakeholder's needs (Hull et al. 2005). Serious gaming can be used as a method to validate the requirements (Ribeiro et al. 2014). Using a simulation based game, mirroring the AS-IS and TO-BE scenarios of the use case, the player can create different CPSS configurations and see the outcome. This provides a basis for assessing the relevance and the importance of specific requirements and can be used to validate the requirements in a playful way.

Requirements traceability enables both assessing the effect of changes of stakeholder requirements to CPSS development as well as to check if every CPSS component is linked to a specific stakeholder requirement. In order to understand how product, service and business requirements and the CPSS design are connected and transformed into each other, lower-level requirements have to be linked with the higher-level requirements they originate from, so that each requirement can be traced to its information source (Wynn et al. 2011). The progress of PSS development can be monitored and the impact of changed requirement tested in this way. In addition, for dynamic systems like CPSS, requirements may change constantly. Changing environment or stakeholders induce changes all along the life-cycle and impact the development process (Lim and Finkelstein 2011). To ensure that such modified requirement are fed back into PSS development, a change management process has to be established (Huang et al. 2011).

Finally, it has to be evaluated if the CPSS complies with the requirements specification or not. This confirmation that the PSS fully satisfies the documented requirements is conducted in requirements qualification. Deviations from requirements can be detected e.g. by requirement reviews, design inspections, component tests and trials, which has to start early in order to avoid late design changes and rebuilds (Hull et al. 2005). This is done by first testing the individual components functions, then the integrated CPSS and finally the fulfilment of stakeholder requirements.

3.5.2 Servitized Business Models Enabled by CPS

A Business Model (BM) describes the rationale of how an organization creates, delivers, and captures value (Osterwalder and Pigneur 2013). According to this definition, BMs in the manufacturing industry have focused on the fabrication or assembly of more or less customized products and have generated revenue from their sales. With the introduction of CPPS, the provider has to support more and more all phases of the production system lifecycle. In particular, this includes more value added service propositions like training, system integration and consulting. With CPSS, manufacturing becomes even no longer the differentiating process.

CPS technology can be utilized for service provision and to develop closer relationships to the customer. A manufacturing enterprise that changes from the fabrication of products to offering CPSS and transforms its supplier base into an ecosystem of network partners will have to analyze and adapt the elements in all building blocks to create a new and competitive BM. Another main challenge is to align the new and unknown BM with the existing BM to avoid cannibalization. Wiesner et al. (2014b) have developed an eight step approach to develop servitized BMs (see Fig. 3.7).

During the first phase of the methodology, the current strategy and BM of the company are analyzed in detail. The analysis of the strategy based on Porter (2008) gives an indication if the company's fundamental strategy is targeting cost leadership, differentiation or selling niche products. A competitor's analysis makes potentials and market boundaries visible (Bergen and Peteraf 2002). If the company is aware of its current strategy, it can be mapped out using the Strategy Canvas as an analytical tool (Mauborgne and Kim 2005). It describes strategic factors that are relevant for the competition within an industry. The current BM is analyzed in the next step and is mapped out on the Business Model Canvas (Osterwalder and Pigneur 2013) (see Fig. 3.8). If the manufacturer is not able to identify its own strategy, it is necessary to analyze the current BM of the company first and then extracting the strategic factors out of the BM and create a Strategy Canvas.

Phase two of the methodology starts with recognizing macro-economic factors that can pose opportunities or threats for the BM of the manufacturer. Therefore, a simplified STEEP-Analysis (Fleisher and Bensoussan 2015) is used to capture future trends that are likely to affect the business. In order to elaborate how

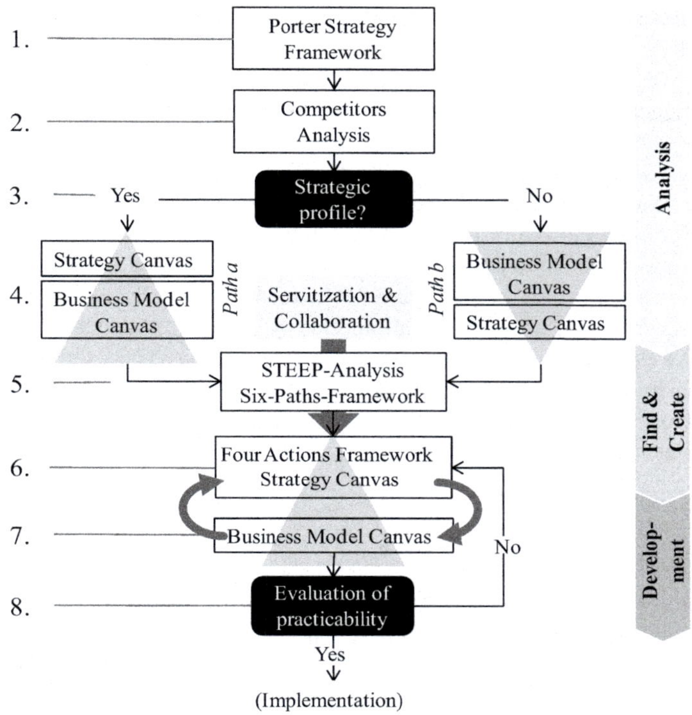

Fig. 3.7 Steps of the approach for Business Model development

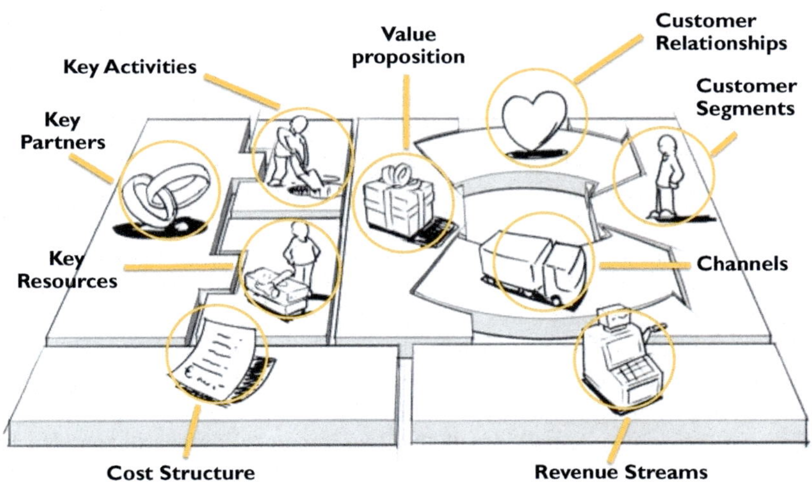

Fig. 3.8 The Business Model Canvas (Osterwalder and Pigneur 2013)

to maximize opportunities and minimize threats for a new BM, the Six Paths Framework is applied. Using poster representations of the six paths, company representatives discuss in a creative process how to change the business according to the STEEP factors. Is a shift into another *strategic group* or even into an *alternative industry* possible? Can a new *group of buyers* be created? What are *complementary product and service offerings*? Are buyers targeted on a *functional or emotional level*? Can external trends be *shaped over time*? To create a new Strategy Canvas, the Four Actions Framework is used as tool to reconstruct strategic elements and to develop a new value curve in the Strategy Canvas (Mauborgne and Kim 2005).

In phase three, a new strategy and BM are developed based on the superior vision of servitization and collaboration. Out of the new Strategy Canvas, a new Business Model Canvas is created. The new BM is now visualized and becomes comprehensible. The creation of the new strategy and the BM is interrelated and is understood as an iterative process. Finally, the impact of the new BM on the company business is evaluated.

3.6 Industrial Use Case

The case company is a Spanish SME that manufactures and assembles high added-value machine tools. It evolved from a small mechanic business that produced pieces for other machine-tools manufacturers to offering production systems worldwide. The high maintenance costs during guarantee period of the systems sold abroad and the growing competition coming from low-cost manufacturing countries is driving them to improve the average machine availability and reliability, reducing the manufacturing costs of the machines, and offering new added-value services to the clients.

The machines are very specific and sometimes they require complex maintenance tasks, which must be performed by specialized workers trained and certified by the company. This implies often the physical presence of the technicians on-site, even if customers are thousands of kilometers away. Furthermore, apart from the planned or usual maintenance operations, when machines have unexpected breakdowns they request urgent attendance with the associated high maintenance costs. In this context, the company realized the opportunity of rethinking the basic maintenance service provided to its customers to a high added-value service. The aim of this new business oriented service is not just to reduce maintenance costs, but also to earn revenues from providing the service, improving at the same time machine availability and reliability, which is a key point from the customer's perspective.

The case company has applied both the Requirements Engineering and Business Model development approach presented in Sect. 3.5 in order to implement the targeted CPSS offer. In a first step, a story was created describing the intended scenario. The company needs their machines to be smarter and more autonomous, once deployed and set up at customer's facilities they shall have no unexpected breakdowns leading to production stops. It would need to be able to predict anomalies in the machine in order to react earlier than the problems really occur. In case of a breakdown, they will be able to take the proper decision avoiding unfortunate damages and undesired stops. Furthermore, the problem solution should be provided fast and reliable without stopping the machine. It should use fewer resources, both human and economic, by avoiding physical intervention of a technician on-site.

Comparing the AS-IS with the desired TO-BE scenario, requirements for product as well as service developments have be derived. On the one hand, an embedded device is needed that will be physically connected via Ethernet to all the machines of the company in the future. It shall be able to query data using different kind of communication protocols used in the industry, such as different standards under OPC umbrella, LSV-2 etc. The device also executes predefined rules to generate and send corresponding alarms warning about possible breakdowns and failures, including associated info helping in the traceability and identification of potential causes. On the other hand, requirements for intelligent maintenance services have been derived. The service functionality takes the information from the mentioned device and links it with the whole maintenance operations lifecycle system. Once the client confirms a malfunction, the company receives historical information with the alarm message and forwards it to the maintenance service staff. After analyzing it, they check the needed assets to fix the detected problem. Once the customer confirms by email, the selected maintenance operation is performed. The customer is able to provide feedback on the quality of the maintenance service provided, which will be taken into account for future operations.

In order to offer the intended CPSS in a profitable Business Model, the BM development approach was applied in a workshop, facilitating the different steps of the approach. Every method was introduced to the participants and the results were documented continuously. The current BM of the company is based on selling customized machine tools to metal works in different industries. Table 3.2 shows a summary of the results of the methodology application:

Communication problems with foreign customers and the development of smart intelligent products have been identified as issues and trends during the workshop.

Table 3.2 Case study Business Model development

BM building block	AS-IS BM	PSS BM
Value proposition	Custom machine tools	Carefree production
Customer segments	Metalwork	Expansion to BRIC
Channels	Direct and distributors	Maintenance platform
Customer relationship	Personal, co-creation	Trust and confidentiality
Key resources	Design skills	Decision supp. system
Key activities	Manufacturing and customization	Coordination of maintenance partners
Key partners	Suppliers and training	Ecosystem of maintenance providers
Cost structure	Manufacturing	Maintenance platform
Revenue streams	Sales	Availability fees

Thus, for the new Strategy Canvas, reaction time for maintenance requests was created as new strategic factor. The new BM features a carefree production with guaranteed machine availability, realized with an ecosystem of local maintenance providers. Additional revenue is targeted by regular fees for the carefree production service.

3.7 Summary and Conclusions

Production system engineering is evolving from a centralized development process for individual systems and components towards the orchestration of distributed software, hardware and business processes for a common purpose. The scale and complexity of the objects targeted by systems engineering is constantly growing, reflected by the emergence of CPPS and PSS offers. The integration of both perspectives creates huge potentials in terms of functionality and revenue, but also new challenges for engineering and Business Model innovation. In the following, the answers to the initial research questions, but also strengths and limitations of the work are presented.

3.7.1 Research Questions Answered

Regarding the first research question, the paper has presented the potential of applying a CPPS as the basis for a PSS offer in Sect. 3.3. The new cyber-physical functionalities are enablers of additional innovative services. Furthermore, the possibility to measure usage and performance of the production system supports Business Models beyond a one-time sale, guaranteeing availability of the system or pay-per-use models.

Through answering the second research question in Sect. 3.4, the elements and phases of both product and service lifecycles could be identified, as well as the engineering challenges resulting from their interaction. The main problem areas are the integration of tangible and intangible components, horizontal and vertical information sharing, virtual representation of CPSS, Knowledge Management and Business Model innovation.

The third research question has been answered in Sect. 3.5, showing implications for the engineering of CPSS and existing approaches. For Requirements Engineering and solution design, suitable methods and tools for various RE activities are presented. Integration of the product, ICT and service perspective could be achieved using the storytelling approach. The comparison of AS-IS and TO-BE scenarios delivers meaningful requirements, which can be validated using gamified approaches. For servitized Business Models enabled by CPS, an eight-step approach was presented to create CPSS BMs. The approach features the Business Model

Canvas as well as methods from the Blue Ocean Strategy. Both, the RE and the BM approach have been evaluated in an industrial use case described in Sect. 3.6.

3.7.2 Strengths and Limitations

The strength of the presented work is clearly the conceptual integration of the CPPS and PSS concepts, aligning the technological and the service perspective on a production system. A first conceptual model of such a CPSS could be established, as well as engineering challenges resulting from the integration of product and service lifecycles. These challenges could further be classified into five problem areas. In order to support CPSS engineering, approaches for two of these problem areas (integration of tangible and intangible components, Business Model innovation) could already be presented. Furthermore, these approaches have already been applied in several industrial use cases.

Limitations of the work lie as well in its conceptual approach. Being exploratory, the research cannot claim to be exhaustive, neither for the challenges, nor for the potential approaches. The CPSS model must be detailed and open issues clearly addressed. Methodologies to solve the challenges in all problem areas are needed, and the existing approaches must be extended and completed for practical use. Finally, the conducted case studies could only give a qualitative assessment of the approaches for RE and BM innovation. A quantitative evaluation is missing and should be conducted to show comparable benefits.

A full CPSS Requirements Engineering framework would help to make the development of CPSS more cost effective and faster, while retaining a high system quality. Future work in this area should include the specification of a requirements structure, which helps to manage changing requirements and predicts the emerging properties of a CPSS. As it might not be possible to replace domain specific models in all cases, future work should deal with the implementation of interfaces that are able to translate between different models without information loss or delay. Concerning the development of servitized Business Models, a method to predict and quantify indicators such as cost and expected revenues could provide decision support to choose between several alternative Business Models.

References

Annamalai Vasantha, G.V., Roy, R., Lelah, A., Brissaud, D.: A review of product–service systems design methodologies. J. Eng. Des. **23**(9), 635–659 (2012). doi:10.1080/09544828.2011.639712

Aurich, J.C., Mannweiler, C., Schweitzer, E.: How to design and offer services successfully. CIRP J. Manuf. Sci. Technol. **2**(3), 136–143 (2010). doi:10.1016/j.cirpj.2010.03.002

Baheti, R., Gill, H.: Cyber-physical systems. Impact Control Technol. **12**, 161–166 (2011)

Baines, T.S., Lightfoot, H.W., Evans, S., Neely, A., Greenough, R., Peppard, J., Roy, R., Shehab, E., Braganza, A., Tiwari, A., Alcock, J.R., Angus, J.P., Bastl, M., Cousens, A., Irving, P., Johnson, M., Kingston, J., Lockett, H., Martinez, V., Michele, P., Tranfield, D., Walton, I.M., Wilson, H.: State-of-the-art in product-service systems. Proc. Inst. Mech. Eng. Part B J. Eng. Manuf. **221**(10), 1543–1552 (2007). doi:10.1243/09544054JEM858

Bell, S.: Lean Enterprise Systems: Using IT for Continuous Improvement, Wiley series in systems engineering and management. Wiley-Interscience, Hoboken, NJ (2006)

Bergen, M., Peteraf, M.A.: Competitor identification and competitor analysis: A broad-based managerial approach. Manag. Decis. Econ. **23**(4–5), 157–169 (2002). doi:10.1002/mde.1059

Berkovich, M., Leimeister, J.M., Krcmar, H.: Requirements engineering für product service systems. Wirtschaftsinf. **53**(6), 357–370 (2011). doi:10.1007/s11576-011-0301-3

Blanchard, B.S.: System Engineering Management, 3rd edn. Wiley, Hoboken, NJ (2004)

Boehm, B., Basili, V.R.: Top 10 list [software development]. Computer. **34**(1), 135–137 (2001). doi:10.1109/2.962984

Broy, M., Cengarle, M.V., Geisberger, E.: Cyber-physical systems: imminent challenges. In: Hutchison, D., Kanade, T., Kittler, J., Kleinberg, J.M., Mattern, F., Mitchell, J.C., Naor, M., Nierstrasz, O., Pandu Rangan, C., Steffen, B., Sudan, M., Terzopoulos, D., Tygar, D., Vardi, M.Y., Weikum, G., Calinescu, R., Garlan, D. (eds.) Large-Scale Complex IT Systems. Development, Operation and Management, vol. 7539, pp. 1–28. Springer, Berlin (2012)

Chang, W., Yan, W., Chen, C.-H.: Customer requirements elicitation and management for product conceptualization. In: Stjepandić, J., Rock, G., Bil, C. (eds.) Concurrent Engineering Approaches for Sustainable Product Development in a Multi-Disciplinary Environment, pp. 957–968. Springer, London (2013)

Colombo, A.W., Karnouskos, S., Bangemann, T.: A system of systems view on collaborative industrial automation. In: 2013 IEEE International Conference on Industrial Technology (ICIT 2013), pp. 1968–1975, 2013

D'Aveni, R.A., Dagnino, G.B., Smith, K.G.: The age of temporary advantage. Strateg. Manag. J. **31**(13), 1371–1385 (2010). doi:10.1002/smj.897

Hintsa, J. and Uronen, K.: Cassandra D1.1 – FINAL – Compendium (2012)

Fleisher, C.S., Bensoussan, B.E.: Business and Competitive Analysis: Effective Application of New and Classic Methods, 2nd edn. Pearson Education, Upper Saddle River, NJ (2015)

Follett, J.: Designing for Emerging Technologies. O'Reilly Media, Sebastopol, CA (2014)

Freitag, M., Kremer, D., Hirsch, M., Zelm, M.: An approach to standardise a service lifecycle management. In: Zelm, M., van Sinderen, M., Pires, L.F., Doumeingts, G. (eds.) Enterprise Interoperability, pp. 115–126. Wiley, Chichester (2013)

Garetti, M., Rosa, P., Terzi, S.: Life cycle simulation for the design of product–service systems. Comput. Ind. **63**(4), 361–369 (2012). doi:10.1016/j.compind.2012.02.007

Gausepohl, K.A.: Investigation of storytelling as a requirements elicitation method for medical devices. Masters Thesis in Industrial and Systems Engineering, Virginia Polytechnic Institute (2008)

Geisberger, E., Broy, M.: agendaCPS: Integrierte Forschungsagenda Cyber-Physical Systems. SpringerLink: Bücher. Springer, Berlin (2012)

Goedkoop, M.J.: Product service systems, ecological and economic basics. Publikatiereeks produktenbeleid, nr. 1999/36 [Ministry of Housing, Spatial Planning and the Environment, Communications Directorate]. Distributiecentrum VROM [distr.], The Hague, Zoetermeer (1999)

Gorldt, C., Wiesner, S., Westphal, I.: Product-Service Systems im Kontext von Industrie 4.0: Auf dem Weg zu CPSS. In: Gronau, N. (ed.) Industrie 4.0 Management 1/2016: Product-Service Design, Erstauflage, neue Ausgabe (2016)

Hehenberger, P., Vogel-Heuser, B., Bradley, D., Eynard, B., Tomiyama, T., Achiche, S.: Design, modelling, simulation and integration of cyber physical systems: methods and applications. Comput. Ind. **82**, 273–289 (2016). doi:10.1016/j.compind.2016.05.006

Hribernik, K.A., Kramer, C., Hans, C., Thoben, K.-D.: A semantic mediator for data integration in autonomous logistics processes. In: Popplewell, K., Harding, J., Poler, R., Chalmeta, R. (eds.) Enterprise Interoperability IV, pp. 157–167. Springer, London (2010)
Huang, H.-Z., Li, Y., Liu, W., Liu, Y., Wang, Z.: Evaluation and decision of products conceptual design schemes based on customer requirements. J. Mech. Sci. Technol. **25**(9), 2413–2425 (2011). doi:10.1007/s12206-011-0525-6
Hull, E., Jackson, K., Dick, J.: Requirements Engineering, 2nd edn. Springer, London (2005)
Kagermann, H., Helbig, J., Hellinger, A., Wahlster, W.: Umsetzungsempfehlungen für das Zukunftsprojekt Industrie 4.0: Deutschlands Zukunft als Produktionsstandort sichern ; Abschlussbericht des Arbeitskreises Industrie 4.0. Forschungsunion; Geschäftsstelle der Plattform Industrie 4.0, Berlin (2013)
Klocke, F., Kratz, S., Auerbach, T., Gierlings, S., Wirtz, G., Veselovac, D.: Process monitoring and control of machining operations. IJAT. **5**(3), 403–411 (2011). doi:10.20965/ijat.2011.p0403
Kossiakoff, A.: Systems engineering principles and practice, Wiley series in systems engineering and management, 2nd edn. Wiley, Oxford (2011)
Lee, S., Park, G., Yoon, B., Park, J.: Open innovation in SMEs – an intermediated network model. Res. Policy. **39**(2), 290–300 (2010). doi:10.1016/j.respol.2009.12.009
Lim, S.L., Finkelstein, A.: Anticipating change in requirements engineering. In: Avgeriou, P., Grundy, J., Hall, J.G., Lago, P., Mistrík, I. (eds.) Relating Software Requirements and Architectures, pp. 17–34. Springer, Berlin (2011)
Marilungo, E., Peruzzini, M., Germani, M.: An integrated method to support PSS design within the virtual enterprise. Proc. CIRP. **30**, 54–59 (2015). doi:10.1016/j.procir.2015.02.021
Mauborgne, R., Kim, W.C.: Blue Ocean Strategy: How to Create Uncontested Market Space and Make the Competition Irrelevant. Harvard Business School Press, Boston, MA (2005)
McAloone, T.C., Mougaard, K., Restrepo, J., Knudsen, S., others: Eco-innovation in the value chain. In: Marjanović, D. (ed.) Design 2010: DS 60: Proceedings of DESIGN 2010, the 11th International Design Conference, pp. 855–864, Dubrovnik, Croatia, Zagreb, 2010
Meier, H., Roy, R., Seliger, G.: Industrial product-service systems—IPS2. CIRP Ann. Manuf. Technol. **59**(2), 607–627 (2010). doi:10.1016/j.cirp.2010.05.004
Meier, H., Uhlmann, E. (eds.): Integrierte Industrielle Sach- und Dienstleistungen. Springer, Berlin (2012)
Osterwalder, A., Pigneur, Y.: Business Model Generation: A Handbook for Visionaries, Game Changers, and Challengers. Wiley, Hoboken, NJ (2013)
Peruzzini, M., Marilungo, E., Germani, M., others: Sustainable product-service design in manufacturing industry. In: DS 77: Proceedings of the DESIGN 2014 13th International Design Conference, pp. 955–964, 2014
Porter ME (2008) On Competition, Updated and expanded ed. The Harvard business review book series. Harvard Business School Pub, Boston, MA
PSYMBIOSYS: Product-Service sYMBIOtic SYStems. http://www.psymbiosys.eu/ (2014). Accessed 19 Apr 2016
Rajkumar, R., Lee, I., Sha, L., Stankovic, J.: Cyber-physical systems. In: Sapatnekar, S. (ed.) The 47th Design Automation Conference, p. 731, 2010. doi:10.1145/1837274.1837461
Reichwald, R., Piller, F., Ihl, C.: Interaktive Wertschöpfung: Open Innovation, Individualisierung und neue Formen der Arbeitsteilung, 2., vollständig überarbeitete und erweiterte Auflage. Gabler Verlag/GWV Fachverlage GmbH, Wiesbaden (2009)
Ribeiro, C., Farinha, C., Pereira, J., Mira da Silva, M.: Gamifying requirement elicitation: practical implications and outcomes in improving stakeholders collaboration. Entertain. Comput. **5**(4), 335–345 (2014). doi:10.1016/j.entcom.2014.04.002
Romero, D., Rabelo, R.J., Molina, A.: On the management of virtual enterprise's inheritance between virtual manufacturing & service enterprises: supporting "dynamic" product-service business ecosystems. In: 2012 18th International ICE Conference on Engineering, Technology and Innovation (ICE), pp. 1–11, 2012
Schirner, G., Erdogmus, D., Chowdhury, K., Padir, T.: The future of human-in-the-loop cyber-physical systems. Computer. **46**(1), 36–45 (2013). doi:10.1109/MC.2013.31

Sein, M.K., Henfridsson, O., Purao, S., Rossi, M., Lindgren, R.: Action design research. MIS Q. **35**(1), 37–56 (2011)

SMLC, Smart Manufacturing Leadership Coalition: Implementing 21st Century Smart Manufacturing: Workshop Summary Report, Washington, DC (2011)

Spath, D., Demuß, L.: Entwicklung hybrider Produkte – Gestaltung materieller und immaterieller Leistungsbündel. In: Bullinger, H.-J., Scheer, A.-W. (eds.) Service Engineering, pp. 463–502. Springer, Berlin (2006)

Spohrer, J.C., Maglio, P.P.: Toward a science of service systems. In: Maglio, P.P., Kieliszewski, C.A., Spohrer, J.C. (eds.) Handbook of Service Science, pp. 157–194. Springer, Boston, MA (2010)

Stark, J.: Product Lifecycle Management: 21st Century Paradigm for Product Realisation, Decision engineering, 2nd edn. Springer, London (2011)

Vandermerwe, S., Rada, J.: Servitization of business: adding value by adding services. Eur. Manag. J. **6**(4), 314–324 (1988). doi:10.1016/0263-2373(88)90033-3

VDI/VDE: Industrie 4.0 – Gegenstände, Entitäten, Komponenten. Status report (2014)

Vink, J.: Storytelling: conceptualize, define, design, discover, implement. http://designresearchtechniques.com/casestudies/storytelling/ (2015). Accessed 11 Dec 2015

Vyatkin, V.: Software engineering in industrial automation: state-of-the-art review. IEEE Trans. Ind. Inf. **9**(3), 1234–1249 (2013). doi:10.1109/TII.2013.2258165

Wan, K., Alagar, V.: Context-aware security solutions for cyber-physical systems. Mobile Netw. Appl. **19**(2), 212–226 (2014). doi:10.1007/s11036-014-0495-x

Wiesner, S., Guglielmina, C., Gusmeroli, S., Doumeingts, G. (eds.): Manufacturing Service Ecosystem: Achievements of the European 7th Framework Programme FoF-ICT Project, neue Ausg. Bremer Schriften zur integrierten Produkt- und Prozessentwicklung, vol. 78. Mainz, G, Aachen (2014a)

Wiesner, S., Padrock, P., Thoben, K.-D.: Extended product business model development in four manufacturing case studies. Proc. CIRP. **16**, 110–115 (2014b). doi:10.1016/j.procir.2014.01.014

Wynn, M.T., Ouyang, C., ter Hofstede, A., Fidge, C.J.: Data and process requirements for product recall coordination. Comput. Ind. **62**(7), 776–786 (2011). doi:10.1016/j.compind.2011.05.003

Yang, X., Moore, P., J-S, P., Wong, C.-B.: A practical methodology for realizing product service systems for consumer products. Comput. Ind. Eng. **56**(1), 224–235 (2009). doi:10.1016/j.cie.2008.05.008

Zhou, K., Ye, C., Wan, J., Liu, B., Liang, L.: Advanced control technologies in cyber-physical system. In: 2013 5th International Conference on Intelligent Human-Machine Systems and Cybernetics (IHMSC), pp. 569–573, 2013

Chapter 4
Product Lifecycle Management Challenges of CPPS

Detlef Gerhard

Abstract In the chapter *Product Lifecycle Management (PLM) Challenges of CPPS,* data and information management issues arising from the advanced use of modern product development and engineering methods are addressed. These advanced methods are required for engineering processes of smart systems and individualized products with high complexity and variability. Emphasis is put on challenges of the life-cycle oriented information integration of products and the respective *Cyber-Physical Production Systems (CPPS).* Furthermore, the chapter addresses data and information management problems coming from integration of the use and operation phase of products and systems in terms of forward and backward information flows.

Keywords Product lifecycle management (PLM) • Cyber-physical production systems (CPPS) • Information management • Model based systems engineering (MBSE) • Digital Twin

4.1 Introduction

Gill (2010) coined the term Cyber-Physical Systems (CPS) around 2006 and describes them as *"physical, biological, and engineered systems whose operations are integrated, monitored, and/or controlled by a computational core. Components are networked at every scale. Computing is deeply embedded into every physical component, possibly even into materials. The computational core is an embedded system, usually demands real-time response, and is most often distributed"*. Cyber-Physical Production Systems (CPPS) is a special term that depicts the introduction of the concept of CPS in the production domain in order to make production processes in general or production systems in particular "smarter"; this can be seen similarly to concepts in other domains, e.g. smart mobility, smart home, smart grid.

D. Gerhard (✉)
Mechanical Engineering Informatics and Virtual Product Development (MIVP) Research Group, Technische Universität Wien, Getreidemarkt 9/307, 1060, Wien, Austria
e-mail: detlef.gerhard@tuwien.ac.at

© Springer International Publishing AG 2017

S. Biffl et al. (eds.), *Multi-Disciplinary Engineering for Cyber-Physical Production Systems*, DOI 10.1007/978-3-319-56345-9_4

CPS as the entity of "smartness" combined with physical processes and objects that uses the *Internet of Things* (IoT) as a communication platform form CPPS in the sense of production value-added chains.

Within the domain of production of goods and products including associated services, CPS based technical systems have to be taken into account twofold: On the one hand side, products themselves are incorporating CPS concepts and on the other hand production facilities for product manufacturing and assembly as well. Hence, with the introduction of CPS, processes of product development, production system development, production (including production system commissioning) and product use (including maintenance, repair and overhaul processes) move closer together, are even strongly interlinked. This truly indicates the given complexity.

Product[1] Lifecycle Management (PLM) is the general concept to consistently create and manage all information related to products (systems/components). In particular, engineering information linked to corresponding engineering and production processes, as well as operation and usage phase is addressed. The major aim is to generate a sophisticated information basis for business and value generation based on products or systems to be produced. This concept comprises aspects of management, organization, and IT solutions and is typically realized with different types of business information software applications, e.g., Product Data Management (PDM), Enterprise Resource Planning (ERP), Manufacturing Execution Systems (MES), and Maintenance/Service/Asset Management.

PLM focuses on three major phases: *engineering, production, and operation* of products. The engineering phase reaches from conceptual design of a product including all components up to detailed engineering. Complex technical systems, such as CPPS, are developed with an extensive utilization of different engineering tools and methods. In particular, the model based approach to product development called Model Based Systems Engineering (MBSE) is on the rise. MBSE depicts *"the formalized application of modeling to support system requirements, design, analysis, verification and validation activities beginning in the conceptual design phase and continuing throughout development and later life cycle phases"* (INCOSE 2007). This leads to many interlinked models from different so called authoring tools (e.g. Computer Aided Design CAD, simulation, software engineering) representing various required engineering domains, which have to be managed and maintained. With an increasing complexity of engineering projects and associated models, significant emphasis has to be put on interoperability and the ability to capture the semantics of data in order to be able to efficiently interface different systems and build tool chains. In industrial applications, predominantly PDM systems cover information management tasks of the engineering phase. This system category also supports engineering processes in the sense of workflow management. For this task, many procedural models for dividing this phase into several subsequent steps are commonly used in industry in order to realize product development processes in

[1] The term "product" is used synonymously for any kind of consumer product, machine or technical system in general, which requires a development and engineering process.

a systematic way, e.g. (VDI2221 1993) or (VDI2206 2004). Particularly, release and change management processes together with configuration management and versioning of information captured in documents and models is supported.

During production, the transition from the conceptual and virtual world to the materialization of a product with its parts and components takes place. This phase typically also starts with conceptual tasks, such as production system engineering and operations planning, e.g. Numerical Control (NC) and Programmable Logic Controller (PLC) programming. Commissioning of a production system is the transition to operation. Further, it involves all resource and production planning tasks resulting from order processing and additionally the respective control functions. This is a fundamental difference to the engineering phase. Whereas engineering focuses on generic product definition without consideration of capacities and resources, production focuses on the specificity of a single item instance or lot released for production with given constraints in terms of dates and resources (personnel, machines, material, etc.). Therefore, production can be seen as a lifecycle phase that is "orthogonal" to engineering (VDI2219 2002). The engineering phase does not stop with the beginning of the production phase. It rather continues creating further releases of a product reflecting improvements, variants, derivatives, etc. According to the corresponding hierarchical structure for industrial automation (IEC62264-3 2014), ERP systems are the category of IT systems that cover the customer orders processes by orchestrating all company's activities like commercial, financial, purchasing, logistics, production, etc. (Ben Khedher et al. 2011). Hence, they are predominantly used in industrial applications for planning, controlling, and information management purposes of the lifecycle phase production on level 4 of IEC62264 functional hierarchy. Furthermore, on level 3 specialized MES systems or ERP manufacturing operations management modules cover the required IT functions, e.g. for detailed planning or production data collection. Manufacturing Execution Systems (MES) are used to deal with detailed production planning and control tasks. MES is much closer to the shop floor activities and therefore it requires more specific production related information because of its shorter planning intervals.

Again orthogonal to the previous phases engineering and production of the lifecycle, there is the operation phase of a product. Each produced single item instance is used or operated differently after production and shipment. This holds true for consumer products, e.g. household appliances, which are produced in a considerable lot size of identical items as well as for customized one-of-a-kind special purpose machines. Hence, in terms of product information management requirements, each item instance has to be treated separately, sometimes even components of the item instance. Processes that have to be supported during operation phase are maintenance (service), repair, and overhaul (MRO) processes, depending on the type of product. On the one hand side, these processes imply new service orders and generate business processes, which have to be managed and supported. This is typically done with software modules of ERP systems or special service management software tools. On the other hand, there is a link to the upstream product lifecycle phase. In particular, tracking of production steps or

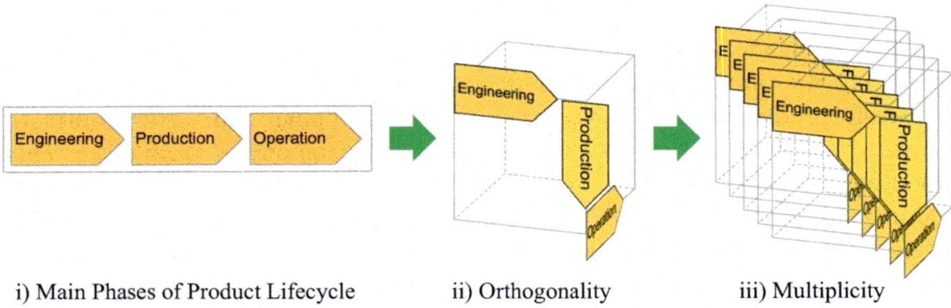

Fig. 4.1 Different views (schematic) on product lifecycle phases

engineering tasks, relevant for a situation that occurs during operation is required. At this point, the orthogonality of the different phases becomes obvious since very heterogeneous types of information have to be linked and dealt with: For instance, information about a particular product in operation with information of a production lot of a particular component and corresponding generic engineering information.

Figure 4.1 schematically shows the different views towards the product lifecycle, as indicated in the previous paragraph. Part (1) depicts a rather simplified view on the three main phases, (2) emphasizes on the aspect that the different phases have to be seen as orthogonal to each other, and (3) indicates, that the orthogonal phases of the product lifecycle are characterized by massive multiplicity and different maturity status. The latter imposes even increased complexity in terms of chronological interdependencies of products and components in development, production or operation as well as their associated processes.

In order to derive challenges and requirements for PLM in the context of CPPS, it is necessary to analyze product engineering and production processes one level deeper. In particular, it is required to distinguish different product types, production concepts, and production types, since this distinction is the basis for the required product information management approach. As already mentioned in Chap. 1, four main different production concepts, reflecting the procedure during order processing, can be differentiated (Schuh 2006): Make-to-Stock (MTS), Assemble-to-Order (ATO); Make-to-Order (MTO); Engineer-to-Order (ETO). This differentiation of product types and their production concepts is necessary to express the degree of dependencies among the product components and the production system. Below, aspects that typically apply to the different concepts are outlined and described in terms of main characteristics for PLM.

MTS determines a production concept that is typically applied for consumer products, which are produced in larger volumes (series or mass production) without major variants that have to be managed. Such products as for instance hand machining tools are in general not subject to extensive after sales services. Therefore, the application of CPS within those products, particularly to collect usage and operation data, is still an exception. Nonetheless, usage phase data could be collected indirectly based on customer feedback, but collected usage data in the operation phase is not directly fed back to the previous phases or used for maintenance business creation. MTS products are developed without customer orders based on market research. Due to the high production volume, a specialized production system has to be engineered in parallel. This production system respectively the engineering process, can be seen as an ETO product or system itself. Therefore, many companies producing mass products on the one hand side, are also producers of production facilities with separate business units. In order to effectively manage both processes, the two different information flows have to be tightly linked together (Gerhard and Lutz 2011).

The ability to directly optimize the production process based on process data that is collected and analyzed in real-time, is the major goal of CPPS engineering in ETO production processes, not so much CPS based adaptability and re-configurability of production systems due to the large production volumes. Furthermore, a major objective of improvement is to shorten commissioning and ramp-up phase because this can save significantly time and money. Virtual Product Engineering (VPE) methods are widely adopted in industry, i.e., the complete description of complex technical systems and their characteristics as computer model together with integration and optimization of IT tools for domain-spanning, multi-disciplinary information management.

Virtualization of production system engineering, often referred to as *"Digital Factory"*, is not as widely adopted in industrial applications but truly on the rise since there is a lot of potential for speeding up time-to-production. From product geometry and material properties defined in engineering design, there is a direct link to required manufacturing operations and programming of NC machining tools in case of material shaping operations. In addition, clamping devices, fixtures, jigs, chucks as well as quality assurance and measurement devices have a direct geometrical link to products or parts of products themselves. Therefore, release and change processes of products and production system items are closely interlinked. The Digital Factory approach goes even beyond this and includes additionally the virtual representation of assembly (ergonomics), intra-logistics, and material handling processes. With respect to the chronological order, two main stages have to be considered, firstly engineering of a production system (typically an ETO process, concurrently started at a certain maturity status of product engineering) and secondly production system operation (including ramp up). The aim is to build a so called *"Digital Twin"* or *"Cyber Twin"* during production system engineering, i.e., computerized companions of physical artefacts that can be used for various engineering purposes and use data from sensors to represent their near real-time status, working condition, position, etc. (Tuegel et al. 2011). A production system—as generally most complex technical systems—faces constant changes and adaptions during operation due to maintenance and improvement procedures. In addition to the Digital Twin representing the results of the development engineering processes, an up-to-date Digital Twin representation is the goal during operation phase. This way, all possible changes and modifications can be verified in advance and also tracked. Furthermore, a Digital Twin of each component can be used for capturing condition monitoring records and synthesizing future steps to provide

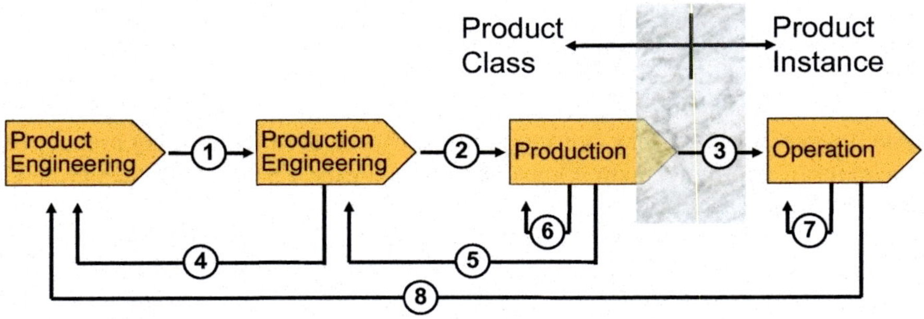

Fig. 4.2 Mayor information flows—example ETO system

self-awareness and self-prediction (Lee et al. 2015). To realize this, a fundamental distinction has to be realized within the respective PLM solution. Whereas all engineering processes take only product classes into account, for each Digital Twin representation, a product instance representation is required representing runtime and operation. This is depicted in Fig. 4.2 below. The different information flows are detailed in a later paragraph. For simplification reasons, phases are shown in a strictly subsequent manner though they are partially overlapping and concurrent.

Whereas MTS and ETO both define the end points, ATO and MTO are located in the center of the production concepts spectrum. ATO is oriented closer to MTS since this concept describes preproduction of standard products with manufacturer-specific variants. MTO is oriented closer to ETO since this concept describes production of standard products with customer-specific variants that are partly composed of pre-defined components and partly made up newly created components. There are two main drivers, that have major influence on production: The combination of IT and "ordinary" products leads to *"Smart Products"* with embedded systems providing an added value to both, customers and producers. Producers, in particular, can extend ordinary products to *Product Service Systems (PSS)*, as described in Chap. 3. Smart products require extensively multi-disciplinary engineering of the product itself as well as the production system and furthermore intelligent backend information technology for supporting the use phase. The trend towards individualized products with customer specific requirements is moving production towards mass customization and lot-size-one concepts.

These concepts for production processes are mainly addressed with CPPS approaches, though CPPS approaches for ATO and MTO production have mainly different goals compared to MTS production. Within MTS production, the product is rather invariant. The production process can be adjusted and optimized with respect to a single product. Short commissioning and ramp-up plays a vital role as well as optimization of the whole process utilizing sensor data and machine feedback with respect to completion, quality or errors in a direct control loop. This goal is mirrored in the *"Overall Equipment Effectiveness" (OEE)* Key Performance Indicator (KPI) coined by Nakajima (1982). This KPI particularly reflects on the measures for optimization of mass production. OEE is beneficial in high-volume and highly automated process-based manufacture where capacity utilization is a high priority. Deployment of OEE in low-volume job shops is not very beneficial (Charaf and Ding 2015).

ATO and MTO production concepts demand extensive flexibility in manufacturing and assembly processes and a by far greater collaboration of product engineering and production system engineering. Variant rich and customized products require particularly configuration of product structures and linked manufacturing and assembly operations. Production according to these concepts typically is realized by manufacturing shops providing different NC machining tools for special operations and flexible manufacturing cells for small and medium batch production of parts. The major goal for CPPS approaches in this case is to realize an intelligent, resilient and self-adaptable system of interconnected machines and manufacturing cells capable of producing customized products with a high degree of automation and thereby at competitive costs. In this scenario, in addition to manufacturing operations, automated material handling plays a vital role. This insight is not new (Sethi and Sethi 1990) but nonetheless still an issue that is currently not sufficiently tackled using computer aided engineering methods (Seibold and Furmanns 2015).

With a PLM view on CPPS in ATO and MTO production environments, again engineering, production, and operation/use have to be taken into account and adequately supported in different ways. In the conceptual stage, based on the product definition, the different manufacturing and assembly steps have to be planned using a variety of CAD methods. This is done in general on a product class level though partially single items have to be tracked on instance level. Production engineering in this case is reduced to operations planning and NC programming, i.e., an existing set of machining tools and material handling systems has to be customized and set up in the sense of a flexible manufacturing shop but a special production system does not have to be engineered and build for this kind of products. Afterwards, during execution time of the respective production orders of parts and assemblies (typically in smaller batches), the required production information has to be provided and actual production data has to be captured. ERP and MES system build the runtime environment for production execution and respective information management, top-down in the sense of planning and target values as well as bottom-up in the sense of shop floor data collection. Production execution is the transition from the virtual to the real product and also the transition from the product class view to the product instance view. Each of the produced product instances during its use phases is operated in a different way and under certain environmental conditions. Nowadays, many of the more sophisticated products are equipped with embedded systems and are capable to a certain extent to capture usage information. One prominent example for such products are cars. They are equipped extensively with controllers and embedded systems. In addition to the functions vital for operating the car, they provide capabilities for capturing data that can be used for classical maintenance and service processes, even for new services supporting drivers or car holders in everyday tasks, navigation, etc.

4.2 State of the Art and Challenges of PLM in the CPPS Context

As indicated above, the vision of CPPS depicts production facilities consisting of smart machines that are connected and able to connect ad-hoc with smart products and objects in order to autonomously exchange information, trigger actions and control each other. CPPS concepts are based on the "Internet of Things (IoT)" concept coined 1999 by Ashton (2009), i.e., every physical object, machine, product or object has a virtual representation. Gunes et al. (2014) elaborates on the following challenges to CPS: Interoperability, Predictability, Reliability, Sustainability, Dependability, and Security. These have also to be taken into account in engineering and PLM for CPPS. Interoperability has several aspects: combining and incorporating heterogenic components of technical systems scaling in size, throughput or other dimensions. Predictability reflects on accuracy of intended outcomes in terms of behavior that autonomous systems show based on inferencing and reasoning in particular contexts. From an engineering point of view, this leads to challenges with respect to robust and stable performance of a technical system, i.e., predictability of a guaranteed behavior and reliant operation. Correct functioning, availability, safety, and maintainability during operation has to be assured, especially if CPPS are self-adaptable or reconfigurable according to changing contexts and dynamic tuning. Particularly, issue tracking based on the right product lifecycle information management becomes difficult since the origin of many issues might be software based. This aspect also leads to security challenges and questions of integrity or reliability, i.e., if privacy and confidentiality of information can be guaranteed and if information is correct and trustful.

In addition to conventional automation and control technology for production facilities, the aim of CPPS is to design "smart" systems that embody so called self-x capabilities (e.g., self-configuration, self-organization, self-optimization) in order to be able adapt autonomously to unforeseen states on machine level as well as non-intended situations on production system level due to failures, lack of material, etc. Even though autonomous interaction on micro-level is intended and required for the implementation of advanced production concepts in modern environments, predictability and controllability of the whole production system on macro-level has to be assured. Machine operators as well as production planners in charge need to have control over the production system. A supplying company of a machine tool, a flexible production cell or a material handling system has to guarantee certain levels of function and behavior. Therefore, behavior and logic needs to be represented using model based descriptions.

Consequently, the challenges of PLM approaches for CPPS are threefold:

- Processes and methods to support systematic multi-domain engineering
- System and information modelling (model representation)
- Information management, particularly data linkage and data analytics.

4.2.1 Processes and Methods

CPPS are complex technical systems characterized by networked structures, non-linear behavior and means multi-causation, multi-variability, multi-dimensionality, interdependence with the environment, and openness. Therefore, product design as well as production engineering tasks have to be addressed in a systematic way. Complexity in this particular case has the following facets (in causal order):

- Product and production system complexity
- Process and organization complexity
- IT landscape and tool complexity.

Complexity of products is caused by multiple instances and variants of a base product to meet requirements with respect to customization demands and differentiation of the target markets. Besides the mechanical components, nearly all products consist of electronics, embedded systems with sensors and actuators and have firmware/software driven controllers. Because there are many different domains of expertise involved in engineering tasks, process complexity also increases through dissemination over locations (countries, cultures) and distribution within the supply chain (organizations). Collaborative engineering processes require even more extensive use and support of advanced IT systems. The diversity and dynamics of the relationships between project partners, manufacturers, vendors, and suppliers leads to highly sophisticated IT landscapes with docents of data formats, representing different semantics.

CPPS, have to be engineered and designed within a multi-domain environment of virtual product development tools. Outcomes of such a development process determines corridors of operation and limitations to autonomous behavior of a CPPS. The product and system development process today is well supported with different tools for e.g. software engineering, CAD, electronic design automation, and simulation. Integration of the domain specific processes for mechanical design, electric/electronics engineering and software development is the main challenge in creating complex technical systems as CPPS in a robust and reliant way. Especially, there is a gap within the information flow between early phases, i.e., engineering design, and later phases, i.e., production (Gerhard and Weilguny 2008). Both phases are in general well supported by different IT tools but the integration and information flow between is still missing the required level of maturity. This problem particular increases with the trend to customization and individualization of products and product service systems. CPPS require a systematic approach towards the different engineering design tasks. Requirements engineering methods help to identify the core of given challenges and further guide through the development process (Cheng and Atlee 2007). Modern CAx systems offer extensive functions to solve geometric modelling and design tasks, but for multi-domain modelling, there is always an information loss when exchanging data between different models.

By expanding the range of functions and system limits, smart products and production systems have more complexity than conventional products. Existing development methods and state of the art tools are not fully suitable to support the specific characteristics and requirements of communication-capable, autonomous decisive CPPS. Methods and a tool chains for the coherent virtual representation of a production system being in operation (product manufacturing) and linked to the product development process are required. In the context of CPPS, this is necessary in order to be able to rapidly respond to required changes and simulate the new behavior of a customized system in advance. Typically, engineering design processes end with the completed definition of a product ready for production and ready to be utilized after production. The requirements of future technical systems such as CPPS go one step further. System functionality from production and operation phase has to be captured by corresponding system models as well, though the semantics of information models in engineering design and production still is quite different. Only in this way, it is possible to collect data from operation and use phase and utilize this models for simulation and optimization of the production process or use phase of a product. Different simulation models are required to capture e.g. cycle times, output or quality data of production systems. Having those models in place, different scenarios of operation can be simulated (e.g. maximum speed vs. average speed or eco mode) in order to find an optimal operation behavior in terms of time, material or energy consumption, friction, loss, wear, etc.

Since PLM in particular focuses on engineering processes, one major challenge is to find a comprehensible way to support engineering design and development process of CPPS on a methodical level. The question is if it is a feasible approach to enhance or adapt the procedural VDI 2206 V-model in terms of cascaded systems of systems modelling. Nattermann and Anderl (2010) for instance proposed a W-model approach for systematic engineering of adaptronic systems. This model suggests the use of a special data management layer, which provides not only a central control of the records of all disciplines but is also able to analyze the discipline-specific records and to synchronize across disciplines. This data management system should be capable to capture state and behavior of the system under development at any time and to ensure compatibility of the discipline-specific components and subsystems.

There is a direct linked to the research questions formulated in Chap. 1: How can model-based methodologies support information creation and processing in the different life cycle phases of a CPPS and how shall several disciplines in product and production system engineering be linked to support the engineering of flexible and self-adaptable CPPS? The question is how virtual engineering support for an integrative CPPS hardware/software co-design, verification, validation, and testing can be realized, given the multitude of different tools and methods. Particularly not tackled adequately so far are methods and technologies that support the links between product, production technology, and production systems engineering, i.e., horizontal, vertical and life cycle integration within production systems and digital links between engineering and operation phases.

4.2.2 Model Representation

The central concept embodied in multi-disciplinary engineering and model-based design is that the 3D product model is the most appropriate vehicle for delivering all of the detailed product information necessary for downstream processes and operations to perform their portion of the product creation (Quintana et al. 2010). CAD models are enriched with explicit and implicit knowledge which needs to be extracted, formalized and managed for re-use in different contexts. However, extracting knowledge encapsulated in CAD models remains a challenge and does not cover at all the complex systems engineering requirements. (PROSTEP 2015) gives a comprehensive over relevant standards for the different procedural steps of the VDI 2206 V-model, but the V-model more or less stops at the start of production and does not cover production and operation.

Nonetheless, for information modelling in engineering processes, there have been considerable developments within *STEP—STandard for Exchange of Product Model Data* (ISO 10303 1994). ISO 10303 is an international standard for the description of physical und functional features of product data. STEP interface definitions and data formats allow data exchange between different CAx systems and aims to represent all data of the whole product lifecycle of a product. In addition to geometric data, this data can e.g. comprise production planning information, bills of materials, simulations, design studies, and much more. To deal with all kinds of lifecycle data, ISO 10303 consists of an extensive collection of so called *Application Protocols* (AP). Each AP is adapted for a special purpose, for instance ISO 10303-242 *"Managed Model Based 3D Engineering"* provides relevant data models merging AP203 and AP214, which are currently the most implemented for CAD data exchange between existing commercial CAD systems. AP 239 *"Application Protocol: Product Life Cycle Support"* furthermore includes the representation of a product through life including product requirements and their fulfilment, the identification of the configuration of a product for a given role, and the specification of effectivity constraints applied to configuration of a product. AP 233 specifies the representation of systems engineering data and defines the context, scope and information requirements for various development stages during the design of a system.

For modelling lightweight representations of geometry together with *product manufacturing information (PMI)*, the newer JT standard (ISO 14306 2012) is interesting to consider. To capture model information beyond geometry, e.g., requirements, logic, function, and physics, System Modeling Language (SysML) (see also Chap. 2 and SYSML 2007) is adopted increasingly. This virtual representation of artefacts is a means to integrate and organize the multitude of models necessary to describe all the aspects of the system with the aim to support interoperability between the domains and their data. Since many different disciplines are involved, complexity handling in engineering and manufacturing (Tolio et al. 2010) is the main issue that is tackled with these standardization approaches.

Self-x functionalities in the sense of smart systems or applicable for a smart factory are functionalities take into account the context of (a group of) CPPS and react accordingly to this context, facilitating the adaptive autonomic behavior of CPPS. Going more into detail, research questions of Chap. 1 contain the challenges of how to define or model self-x functionalities of CPPS as well as means for capturing context, behavior and state of artefacts? Furthermore, strategies and algorithms for modelling and simulation of anticipatory system behavior and layer-crossing integration of self-x actions have to be developed. Monostori (2014) states in a comprehensive survey the following R&D challenges for CPPS (among others): For context-adaptive and (at least partially) autonomous systems, methods for comprehensive, continuous context awareness as well as for recognition, analysis and interpretation of plans and intentions of objects are necessary. Furthermore, the development of new methods is required, which support the fusion of the real systems with the virtual representation in order to reach the goal of an intelligent production system which is robust in a changing and uncertain environment.

4.2.3 Information Management and Integration

In industrial applications today, product development processes, production planning processes, and order based production planning and control are still to a large extent disconnected. This holds in particular true for data generated during the use phase of products. For CPPS and smart production, a closed loop information management is crucial, spanning the whole lifecycle from product concept and design to production system planning, to order management and production, and finally to product operation or usage. The international standard IEC 62264 (IEC 62264-3 2007) defines models and transactions for the integration of ERP and MES. Its main objective is the integration of business planning and logistics systems to manufacturing operations management systems. While ERP systems operate in time frames of months, weeks and days, the detailed production planning is done using much shorter periods like shifts, hours, minutes, seconds and even sub-seconds. VDI 5600 guideline (VDI 5600 2007) offers a problem-oriented description of MES and its application potentials. The main tasks of MES like detailed scheduling and process control, equipment and material management, etc. are defined and the role of MES for enterprise processes is highlighted. Similar to IEC 62264, VDI 5600 contains recommendations for the interface management between MES and machines/terminals/sensors on the manufacturing level. Both standards mainly address the so called *Automation Pyramid* from shop floor to top floor in a vertical integration direction, respectively along the production value chain in a horizontal integration direction. The integration along the product lifecycle from very early engineering design stages to production and operation stage or vice versa is not covered.

The main challenge is that generated data and relevant information at the various stages of the product lifecycle is quite different in terms of the three orthogonal aspects of data storage, processing, and analysis (3Vs of Big Data), i.e., volume, variety, and velocity (Laney 2001). Particularly, there is a variety of formats capturing different semantics, not just relationally structured data representing different models but unstructured content. Content that is tagged with metadata and hierarchical file system data. In engineering design, information about the product in general is created, in early phases there are abstract models, later more tangible and concrete content. In later stages of the product lifecycle, data or information flows can be highly inconsistent with peaks and the meaning can also change over time. The generated information is to a large extent unstructured and stochastic but not model driven. Data or information flows can be highly inconsistent with peaks and the meaning can also change over time. The more an integration of the different information aspects over the product lifecycle is possible the more efficient processes become. The use of cognitive computing approaches, ontologies and semantic technologies in the integration of data, information, as well as knowledge throughout the complete design cycle of a product has the potential to substantially improve both the product and associated development processes (Welp et al. 2007).

Forward and backward information flows have to be distinguished. Design engineers usually have a good understanding of the product they are developing, but any approach to integrate information required in later process stages (e.g. environmentally relevant information or production relevant information) into early product development and design stages often fails since it adds to workload or complexity of the process. Ostad-Ahmad-Ghorabi et al. (2013) tackle this issue introducing ontological approach to set up primary parameters systematically for particular product categories driving e.g. the environmental performance of products. The aim of ontological approaches in general is to enable the management and (re-)use of heterogeneous data along the product development process. Different information and different views concerning a product structure facilitates the work of each phase. Yet, the lack of contextual relationships within the structures avoids linking the correct data between them. Therefore, similar information must often be entered multiple times and a cross comparison between the information from other domains or an automated comparison of the various activities in concurrent engineering processes become virtually impossible. With the use of semantic technologies and ontologies the continuity of information can be achieved through a coherent semantic structure associated with views on different areas of the development process (Gerhard 2012). Ontologies serve as a neutral or intermediate layer, in which semantic web technologies can be used to build queries or filters that provide specific data of heterogeneous sources. Ontologies are thus of great importance when encoding design knowledge as well as integrating software systems for facilitating semantic interoperability (Chandrasegaran et al. 2013).

Semantic technologies also play a vital role in data analytics. Smart factories collect vast amounts of data from different sources: Product design data such as bill-of-materials (BOM) and CAD-files; production process data such as CAM-files, machine scheduling and QC measurements; logistics data including demand forecasts; and data from a multitude of sensor constantly monitoring machine parameters. Currently, ERP and MES systems used in manufacturing operations do not adequately mine this data to identify useful patterns and draw conclusions for operations or engineering processes. This is because current data mining techniques are typically not suitable for time series data and therefore, are of limited use in making predictions (Gröger et al. 2012). Additional data objects or information model enhancements of existing software tools or standards are required to capture and interface engineering as well as run time data of complex CPPS so they can be indexed and retrieved for reuse across different products or projects. This comprises versioning and means to align different (multi-domain) development paths. Still the research question remains, e.g. if it is possible to find algorithms to analyze data patterns from manufacturing data for re-use at different stages of the product creation process, transforming information from data to knowledge level.

4.3 PLM Forward and Backward Information Flows in CPPS

With the above introduced contents and challenges, it is clear that an enhanced use of CPS in products and production systems imposes new approaches to look at the way product related models and documented information have to be managed along the product lifecycle. As stated before, the lifecycle phases cannot be seen just as subsequent stages but the orthogonality has to be acknowledged.

IT systems and software solutions for engineering information modeling and management represent “virtual” product and production engineering information linked to a product class as well as order based production planning and real time data of the production process as well as individual usage data. In other words, they have to cover a complex patchwork of different views in terms of functionality and models semantics, i.e.,

- Development/engineering, planning, production, operation/use
- Requirements, features/working principles, logic/behavior, geometry/shape
- Structure of products (systems), modules/components and parts/elements
- Mechanics, hydraulics, pneumatics, E/E, control/software
- Manufacturing, assembly, testing, packaging, transportation
- Customer, supplier, service owners/operators
- Building/infrastructure, energy.

A unified and coherent model description of all necessary information (in a knowledge domain as CPPS) is unrealistic. Requiring all applications to share a common standardized data model to be truly integrated is not a feasible solution. Hence, data linkage in a federated manner, semantic technologies, and cognitive computing approaches can be seen as enablers to introduce new agility and expanded scope to enterprise applications, such as for instance:

- Automated extraction of metadata to transform unstructured data into a fully classified resource and synthesize it with existing structured data

- Enrichment of structured data with qualitative data from vast "unstructured" sources like sensor-captured production data or usage data from emails, blogs, chats, and social Web pages
- Identification of embedded meanings and relationships within and across resources through data analytics
- Natural language processing to interpret imprecise requests and offer spelling corrections, close matches, and related content
- Creation of innovative and tailored apps that seamlessly merge content and functionality from diverse sources such as databases, mapping services, and WWW resources.

This is necessary in order to perform data specialization tasks in forward direction (early lifecycle phases to late phases) and data generalization tasks in backward direction (vice versa).

Forward integration of engineering information in the sense of "Design For X" (DfX) is the concept comprising all endeavors towards making the right decisions in the product development process on basis of sufficient and universally applicable knowledge basis. The aim is to take into account impacts that decisions in early phases of the product development process have on later phases. Particularly, concepts of Design for Manufacturing, Assembly, and Service are relevant for CPPS and in many companies in place in order to ensure high quality at optimized cost and time efforts in the production or operation phase. Forward integration nowadays means predominantly manual processing of data, e.g. using CAD/CAM data in order to extract required information for operations planning. PDM systems support these tasks to a certain coverage since they provide easy access to required information but they do not provide assistance in terms of e.g. supporting operations planning. Especially the inherent semantics of product defining data to be used at later stages is still a weak point. Therefore, to a large extent, operations planning relies on the experience of planning engineers. Knowledge that can be derived from past projects and tasks is not taken into account systematically in many cases and the potential of intelligent knowledge re-use is not addressed. In the forward direction new questions arise since products become more sophisticated integrating embedded IT systems and/or IoT technologies.

In the backward direction, feedback information from usage and operation phase collected on a single item or instance basis, which is in general less structured needs to be aggregated and generalized to be used earlier phases, i.e., engineering design or production, in the sense of knowledge management. Backward information integration in terms of PLM also requires new approaches in order to leverage opportunities and adequately support processes related e.g. to PSS. The semantics of information models in engineering design and production still is quite different. Instance based information from the use phase of a product has to be captured, generalized and mapped to product class information of product engineering or production engineering phase in order take benefit in terms of knowledge management. In the operation phase of a product or machine (maintenance and support phase), the inherent task of evaluating if design requirements are met is to take a close look at the performance of the product and actual use. A closed feedback loop is the idea that the output is looked at with respect to a desired goal and then the inputs are modified in order to change the output to close the gap between what the output is producing and what is desired. Feedback loops can be direct and internal or indirect and external to a system. If a product does not meet its design requirements in actual usage or if actual usage surfaces additional requirements, respective information should be fed back into a base of knowledge so that engineering can understand the gap between the requirements they thought could be fulfilled and what actually occurred. Conclusions can be drawn and requirements for future versions of the product can be adapted.

Particularly for the backward information flow, it is important to distinguish the type and the instances of an information object over its lifecycle, i.e., to have unique identifiers both in the digital (virtual) and in in the physical world.

- Instantiation of product data: For each product in the field, there has to be a separate instance of product data created. This dedicated instance will be maintained over the lifetime of the product, to stay up to date, even when parts are changed during maintenance (e.g. for long life products like machining centers).
- Instantiation of usage data: Data collected during the usage of products have to be associated to the specific instance of the product instead of its generic model.

Figure 4.2 of the previous section gives an example (ETO production concept) for the different information flows. Briefly the main information flows can be depicted as follows:

1. Engineering design data are used to generate production system and operations planning data
2. Planning data serve as the basis for production control (target values)
3. Actual data of production are the basis for product operation/use including MRO processes
4. Feedback data to improve the product
5. Feedback data to improve production system and operations planning
6. Actual data for direct optimization of the production process (target-performance comparison)
7. Actual data for direct optimization of the operation and MRO support
8. Actual data for improving and further developing the product.

Benefits resulting from the forward and backward information integration are to a large extent company and use case specific. For each use case, the first step is to figure out who benefits from the delivered information, and therefore, in what form and where the information has to be presented, e.g. is the information already necessary/useful in the production planning phase, or is it important later in the physical production process. For example, simulated milling operation times stored in the PDM system can be used to calculate target times for operations planning, or assembly instructions can be displayed on terminal screens in manual assembly lines. After identification of the required data, suitable approaches for data structures and data processing have to be defined. Concerning backward integration, it is important to identify, which data from the MES or other data sources is accessible and useful for feedback in the PDM data backbone. That can be raw data (e.g. from sensors) or already processed data (e.g. key performance indicators of machines). It has to be clarified, which data has an impact to the generic data in the PDM system, and how information can be viable created out of all the collected data. For example, if production introduces new cutting inserts for milling operations resulting in higher duration of the tool and less tooling time in a production process, this information can be fed back to operations planning. This means that raw production data has to be mined and analyzed with respect to the deviation of planned and actual values taking into account possible outliers.

A software architecture comprising so called authoring tools for different engineering tasks on the one hand side and comprehensive engineering and business information management software systems on the other hand side has to take application diversity into account. Ther e are ongoing research activities to investigate and develop an approach for multidisciplinary life-cycle oriented information integration in systems engineering within the *Open Services for Lifecycle Collaboration* (OSLC) working group of OMG (OSLC4MBSE 2013). The focus of major activities still is in the engineering domain, but engineering, production and usage have to be treated holistically in the context of CPPS, which goes beyond this viewpoint. Especially, many different concepts have to be mixed and supported with software tools, like e.g. model based engineering, document oriented information flows, time related data and time series processing, location based data and geospatial processing, database oriented transaction handling and posting entries to ledger accounts. Hence, a principle IT system architecture needs to support a strongly federated network approach of nodes performing particular tasks, in which the nodes themselves follow an approach that can be compared to an onion with a shell like structure incorporating the concept of microservices. Microservices is an emerging trend in the cloud era: briefly *"microservices are small, autonomous services that work together"* (Newman 2015) to achieve a common or requested functionality. Similar to the Service Oriented Architecture (SOA) approach, microservices are independently deployable, small, and modular services that communicate loosely coupled over HTTP protocol typically through REST APIs with simple semantic standards that can map to any data model using JSON as a data exchange format. This concept also supports, that contextual information from third party systems can be provided via a persistent linked data layer that overcome system and organizational boundaries. Figure 4.3 below depicts the rationale behind a principle IT system architecture suitable for PLM in the context of CPPS. As stated before, the different phases of the product lifecycle have to be seen orthogonal to each other.

In all phases, many different software tools (depicted as dots of the network in the respective colors in Sect. 2 of the figure) generate information that is linked to another portion of information, e.g. structured or unstructured information, simulation models of the engineering phase as well as sensor data of the production phase that cannot be captured in models. At the outer area of the network (colored in red), there are even dots representing system boundaries or transition to other

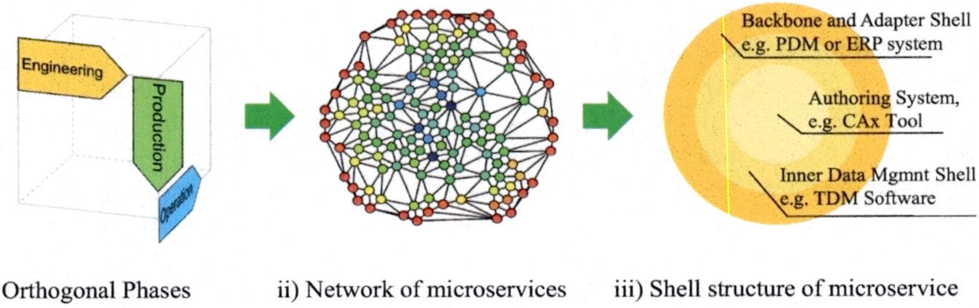

Fig. 4.3 Principle IT system architecture for PLM in the context of CPPS

domains, e.g. smart grid energy management systems. The example given in Sect. 3 of Fig. 4.3 reflects the viewpoint of mechanical engineering: A CAD system as core of required software functionality for a particular process task is wrapped in a Team Data Management (TDM) environment with high integration depth supporting collaborative engineering work. The TDM system again is enveloped in a PDM or ERP backbone. On this level the microservice approach comes into play, i.e., communication to other services through defined API leads to a federated approach of exchanging required information within the given plethora of systems. Beyond the expressed example, the same holds true for different engineering domains, e.g. a CAE system wrapped in a specialized simulation data management tool or a software engineering tool which organizes the software development work within source code management environment. Section 3 of Fig. 4.3 could also be colored blue or green in order to represent a microservice of the production or operation phase or even in mixed colors if there is no clear assignment possible. The major differences are the type of generated or captured information and the point in time, which the information represents. Nonetheless, information of the three different life cycle phases is truly interlinked.

Incorporating a microservices concept likely leads to a situation, that instead of less large and established enterprise applications suddenly a landscapes with a variety of small, fast-changing services emerges, which all have to be configured, managed, and monitored. This issue can be tackled using a so called container technology like *"Docker"*. Docker uses *"containers"*, which capture everything that is needed to run a chosen software (e.g. code, runtime, system tools, system libraries, binaries, dependencies, etc.). Docker containers represent one encapsulated unit of functionality to the "external world". In this way, it is assured that the code will run in any selected environment the same way (Mouat 2015). Docker containers are rather lightweight in comparison to hypervisor techniques, since virtualization is done on operating system level, encapsulated from the rest of the host system. The feasibility of this concept is underpinned by the fact that most of large public clouds have made their systems compatible to Docker (e.g. AWS Elastic Beanstalk, Google AppEngine, IBM Cloud, Microsoft Azure, Rackspace Cloud). With this support and adoption, Docker will probably become the most prevalent system used to create cloud applications (Matthias and Kane 2015).

4.4 Summary and Outlook

Digital and real worlds merge. The development of products that are networked within their operational environment and the support of service-oriented business models requires the linkage of traditional product data with the digital shadow of the delivered product configuration (Cyber Twin) as well as the use and association with data of production and operation. A shift from divided designs of physical systems, control subsystems and software architecture to integrated and optimized design can be observed with respect to the process of product and production systems engineering. Concerning operation of production systems, human and information-centric operation moves to highly-automated, autonomous, and coordinated frameworks. Engineers have to be better supported in their development work through system-spanning information links, and assistant functionality that utilizes advanced information analytics and cognitive computing approaches. Thus, IT system strategies supporting operation and product lifecycle information management are changing from centralized to federated, decentralized, and configurable approaches.

Previously, the focus was on the modeling of all necessary artifacts and preparation of all necessary documents for the design and manufacture of a system. In the field of mechanical design, the result virtually consists of a complete digital mock-up on product class level. In the software domain, the result of development activities is a static program code, which e.g. evaluates captured sensor data and system status and possibly performs actuator actions or executes user interaction. With CPPS, modeling of systems in operation as well as the continuous documentation of all MRO operations and changes in the operation is necessary. From the mechanical engineering viewpoint, a functional mock-up is required on product instance level. In the software field, adaptive program code (for example, PLC, CNC) to map self-x functionalities.

In particular, the role of PLM in the context of *Smart Systems, CPPS and Industrie 4.0* approaches will change dramatically towards product information management on a single item instance basis. Today's PDM systems are complex technical IT systems that require considerable effort for customizing and implementation in specific contexts of manufacturing enterprises. System deployment and operation of a PDM system are complex and costly issues and not only few projects heavily struggle. The networking of products and services on the Internet of Things (IoT) raises the question whether ordinary PLM approaches are not completely overburdened and will become redundant with *Linked Open Data, Big Data or self-learning systems*. PLM approaches have to be adopted in order to support the companies optimally in their *digital transformation* processes.

The benefit of a particular PLM solution heavily depends on the processes to be supported within a company and therefore on the production concept, e.g. ETO or MTO. After sales and customer service play an increasingly important role in the context of PSS. Customized and individualized products have to be produced in smart factories incorporating CPPS approaches with the goal to keep required efforts low. Data from the operation phase of the products has to be linked with the engineering data. Necessary are on the one hand side PLM concepts that support multi-disciplinary development of products with high degree of integrated control technology and software and on the other hand side methods and system functions for cross-domain engineering collaboration. Nonetheless, PDM systems have to be able to manage complex product configurations, including electronics and software and also depict changes in the configuration during operation. Therefore, PLM solutions need to support data structuring approaches that go beyond centrality of the traditional *Bill of Material (BOM)* concept coming predominantly from mechanical engineering focus.

Monolithic approaches with one single leading system for PLM in general do not meet the given demands, particularly because engineering is done to a large extent collaboratively in joint ventures together with development partners embedded in a supply network. A modular IT architecture with best of breed solutions ensures a flexible and user-friendly working environment. It is essential that PLM solutions are dynamically adaptable reflecting the ongoing changes of data models together with the process changes in the organizations. Similar to the way the world-wide web works, federated approaches that link the distributed digital models are required for the product related information management.

References

Ashton, K.: That 'internet of things' thing. RFiD J. **22**, 97–114. http://www.rfidjournal.com/articles/view?4986 (2009). Retrieved 25 Jul 2016

Ben Khedher, A., Henry, S., Bouras, A.: Integration between MES and product lifecycle management. In IEEE International Conference on Emerging Technologies and Factory Automation (ETFA'11), pp. 8–13. Toulouse, 2011

Chandrasegaran, S.K., Ramani, K., Sriram, R.D., Horvath, L., Bernard, A., Harik, R.F., Gao, W.: The evolution, challenges, and future of knowledge representation in product design systems. Comput. Aided Des. **2013**, 204–228 (2013)

Charaf, K., Ding, H.: Is overall equipment effectiveness (OEE) universally applicable? The case of Saint-Gobain. Int. J. Econ. Fin. **7**(2), 241–252 (2015)

Cheng, B.H.C., Atlee, J.M.: Research directions in requirements engineering. In: Proceedings of Future of Software Engineering, pp. 285–303. IEEE Computer Society, Washington (2007)

Gerhard, D.: The role of semantic technologies in future PLM. In: Fathi, M. (ed.) Integration of Practice-Oriented Knowledge Technology: Trends and Prospectives, pp. 157–170. Springer, Berlin (2012)

Gerhard, D., Lutz, C.: Rechnerunterstütztes Konfigurieren und Auslegen kundenindividueller Produkte. ZWF Z. wirtschaftlichen Fabrikbetrieb. **3**, 103–104 (2011)

Gerhard, D., Weilguny, L.: Applied feature technology – review of developing a generic solution facilitating data-consistency and enabling knowledge-based engineering. In: CIRP Design Conference 2008, Twente, NL, 2008

Gill, H.: Cyber-physical systems – beyond ES, SNs, and SCADA. In: Presentation in the Trusted Computing in Embedded Systems (TCES) Workshop. http://repository.cmu.edu/cgi/viewcontent.cgi?article=1724&context=sei (2010). Retrieved 25 Jul 2016

Gröger, C., Mitschang, B., Niedermann, F.: Data mining-driven manufacturing process optimisation. In: Proceedings of the World Congress on Engineering, vol. III, pp. 1–7, 2012

Gunes, V., et al.: A survey on concepts, applications, and challenges in cyber-physical systems. KSII Trans. Internet Inf. Syst. **8**(12), (2014)

IEC 62264-3: Enterprise-control system integration – Part 3: activity models of manufacturing operations management. Beuth, Berlin (2014)

INCOSE: International council on systems engineering: systems engineering vision 2020 INCOSE-TP-2004-004-02 ver. 2.03. http://oldsite.incose.org/ProductsPubs/pdf/SEVision2020_20071003_v2_03.pdf (2007). Retrieved 25 Jul 2016

ISO 10303: Industrial Automation Systems and Integration – Product Data Representation and Exchange. International Organization for Standardization (ISO), Geneva (1994)

ISO 14306: Industrial Automation Systems and Integration – JT File Format Specification for 3D Visualization. International Organization for Standardization (ISO), Geneva (2012)

Laney, D.: 3-D data management: controlling data volume, velocity and variety. http://blogs.gartner.com/doug-laney/deja-vvvue-others-claiming-gartners-volume-velocity-variety-construct-for-big-data/ (2001). Retrieved 25 Jul 2016

Lee, J., Bagheri, B., Kao, H.A.: A cyber-physical systems architecture for Industry 4.0-based manufacturing systems. Elsevier Manuf. Lett. **3**, 18–23 (2015)

Matthias, K., Kane, S.P.: Docker Up & Running, Shipping Reliable Containers in Production. O'Reilly Media, Sebastopol, CA (2015)

Monostori, L.: Cyber -physical production systems: roots, expectations and R&D challenges. Variety management in manufacturing. Proceedings of the 47th CIRP Conference on Manufacturing Systems. Proc. CIRP. **17**(2014), 9–13 (2014)

Mouat, A.: Using Docker, Developing and Deploying Software with Containers. O'Reilly Media, Sebastopol, CA (2015)

Nakajima, S.: TPM tenkai, JIPM Tokyo. https://en.wikipedia.org/wiki/Seiichi_Nakajima (1982). Retrieved 25 Jul 2016

Nattermann, R., Anderl, R.: Approach for a data-management-system and a proceeding-model for the development of adaptronic systems. In: Proceedings for the ASME International Mechanical Engineering Congress & Exposition (IMECE), Vancouver, 2010

Newman, S.: Building Microcervices, Designing Fine-Grained Systems, p. 18. O'Reilly Media, Sebastopol, CA (2015)

OSLC4MBSE: Working Group. http://www.omgwiki.org/OMGSysML/doku.php?id=sysml-oslc:oslc4mbse_working_group (2013). Retrieved 25 Jul 2016

Ostad-Ahmad-Ghorabi, H., Rahmani, T., Gerhard, D.: Forecasting environmental profiles in the early stages of product development by using an ontological approach. In: Abramovici, M., Stark, R. (eds.) Smart Product Engineering, pp. 715–724. Springer, Berlin/Heidelberg (2013)

PROSTEP: White Paper Datenmanagement für Smart Systems Engineering (Smart SE). ProSTEP iViP (2015)

Quintana, V., Rivest, L., Pellerin, R., Venne, F., Kheddouci, F.: Will model-based definition replace engineering drawings throughout the product lifecycle? A global perspective from aerospace industry. Comput. Ind. **61**(5), 497–508 (2010)

Schuh, G.: Produktionsplanung und -steuerung Grundlagen, Gestaltung und Konzepte, 3rd edn. Springer, Berlin (2006)

Seibold, Z., Furmanns, K.: Dezentrale Koordinationsmechanismen für Multifunktionalität und Wiederverwendbarkeit. In: Bauernhansl, et al. (eds.) Industrie 4.0 in Produktion, Automatisierung und Logistik, pp. 1–17. Springer, Berlin (2015)

Sethi, A.K., Sethi, S.P.: Flexibility in manufacturing: a survey. Int. J. Flex. Manuf. Syst. **2**, 289–328 (1990)

SYSML: The SysML specification, v 1.0. http://www.sysml.org (2007). Retrieved 25 Jul 2016

Tolio, T., Ceglarek, D., El Maraghy, H.A., Fischer, A., Hu, S.J., Laperrière, L., Newman, S.T., Váncza, J.: SPECIES – co-evolution of products, processes and production systems. CIRP Ann. Manuf. Technol. **59**(2), 672–693 (2010)

Tuegel, E.J., Ingraffea, A.R., Eason, T.G., Spottswood, S.M.: Reengineering aircraft structural life prediction using a digital twin. Int. J. Aerospace Eng. **2011**, Article ID 154798 (2011). http://dx.doi.org/10.1155/2011/154798. https://www.hindawi.com/journals/ijae/2011/154798/

VDI 2206: VDI-Guideline 2206 – design methodology for mechatronic systems. Beuth, Berlin (2004)

VDI 2219: VDI-Guideline 2219 – information technology in product development – introduction and usage of PDM systems. Beuth, Berlin (2002)

VDI 2221: VDI-Guideline 2221 – systematic approach to the development and design of technical systems and products. Beuth, Berlin (1993)

VDI 5600: VDI-Guideline 5600 – manufacturing execution systems (MES). Beuth, Berlin (2007)

Welp, E.G., Labenda, P., Bludau, C.: Usage of ontologies and software agents for knowledge-bases design of mechatronic systems. In: Proceedings of the 16th International Conference on Engineering Design (ICED 07), Paris, 2007

Part II
Production System Engineering

Chapter 5
Fundamentals of Artifact Reuse in CPPS

Arndt Lüder, Nicole Schmidt, Kristofer Hell, Hannes Röpke, and Jacek Zawisza

Abstract Recent research and development activities within the field of production system engineering and use focus on the increase of production system flexibility and adaptability. One common issue of those approaches is the consideration of hierarchical and modular production system architectures where the individual components of the system are equipped with certain functionalities and information. Up to now there is no common understanding about what a component can constitute, i.e. which parts of a production system can be regarded as components within the hierarchy and which functionalities and information are assigned to it. This gap will be closed within this and the subsequent two chapters.

They will at first discuss the relevant layers of components in a production system, then the types of information required to be assigned to a component on the different layers to establish a digital representation of the component, and at last the description means exploitable to represent the identified information in the different life cycle phases of a production system.

This chapter in particular will consider hierarchies of production system components and their life cycle. Based on a literature survey and practical experiences candidates for hierarchy layers and their identification criteria are named. In addition, main life cycle phases of production systems are discussed.

Keywords Production system hierarchy • Industrie 4.0 component • Administration shell • Life cycle phase

---

A. Lüder (✉) • N. Schmidt
Faculty Mechanical Engineering, Otto-von-Guericke University, Magdeburg, Germany
e-mail: arndt.lueder@ovgu.de; nicole.schmidt@ovgu.de

K. Hell • H. Röpke • J. Zawisza
Volkswagen Aktiengesellschaft, Wolfsburg, Germany
e-mail: kristofer.hell@volkswagen.de; hannes.roepke@volkswagen.de; jacek.zawisza@volkswagen.de

© Springer International Publishing AG 2017
S. Biffl et al. (eds.), *Multi-Disciplinary Engineering for Cyber-Physical Production Systems*, DOI 10.1007/978-3-319-56345-9_5

5.1 Introduction

Production systems all over the world are facing similar challenges. At first, the number of competitors, aiming to gain visibility by the same customers, is rising. This can be seen by the increasing growth of global trade by 6% every year—for the last 25 years (Schmale 2015). Thus, producing companies are trying to offer more products, product types, and variants of the same product type to optimally fulfil customer requirements. For example, car manufacturer Volkswagen offered 10 different car models in 1990 and 19 ones in 2016 (not including car model derivates like limousines and convertibles). In parallel, the producing companies are reducing the life cycle of products being able to respond to changing customer requirements much faster. Whereas the first generation of the VW Golf has been available for 9 years, the sixth generation was only available for five.

The second challenge is based on production system technology development. The pursuit of product variety is applied within the customer market and also in the market of production technology and production system equipment including technology breakthroughs like additive production technologies. Hence, producing companies have to monitor trends in production technology and production system equipment development to identify possibilities, which can improve their production systems regarding quality and economical aspects.

Facing both challenges forces production systems to increase adaptability and flexibility related to producible product portfolio, order volumes, and production system resources while trying to improve the economic impact of the production system over its complete life cycle.

To cope with this problem different development directions are considered. An interesting one from the automation point of view is the *Industrie 4.0* initiative. It aims (among others) at developing methods, tools, and systems enabling production system owners and developers to handle system complexity and to increase resource efficiency within the process of improving production system adaptability and flexibility (Kagermann et al. 2013). Therefore, a new production system architecture and a new production system component structure are under development. A first result of this development is the Reference Architecture Model *Industrie 4.0* (RAMI 4.0), (Kagermann et al. 2013). This model combines the production system life cycle with the control hierarchy and the value streams relevant for production.

One key element of RAMI 4.0 is the *Industrie 4.0* component. As indicated in (Vogel-Heuser et al. 2015) the implementation of *Industrie 4.0* components can utilize the CPPS paradigm resulting in a hierarchy of CPPS nesting one CPPS within another. Thereby, each CPPS is an Industrie 4.0 component on its own.

In (VDI 2015) structural, functional, and information related requirements on an *Industrie 4.0* component are collected especially including the need to combine objects of the physical world and the virtual world, which are related to the same component of a production system. Thus, an *Industrie 4.0* component shall have

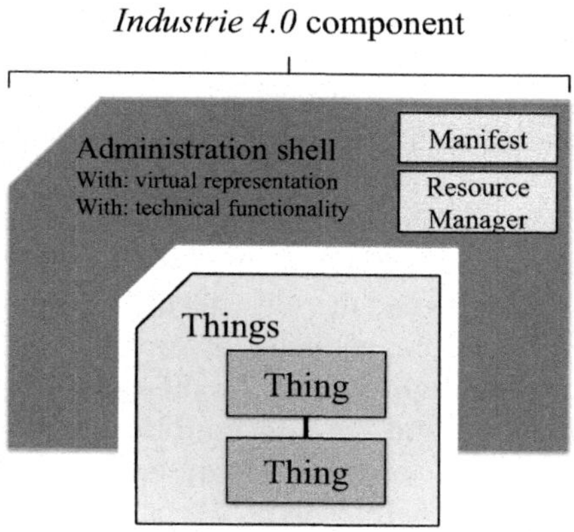

Fig. 5.1 *Industrie 4.0* component following (Kagermann et al. 2013)

a virtual representation of itself containing all relevant information related to the physical, functional, and behavioral properties of the represented physical object (see Fig. 5.1).

In order to properly design *Industrie 4.0* components the virtual representation has to be filled with information. But depending on the granularity of the component within the hierarchical system architecture as well as depending on the life cycle phase this information might be completely different.

Following research question RQ M1 described in Chap. 1, it is of interest to identify requirements and architectures for *Industrie 4.0* component modelling addressing the multi-disciplinary nature. Considering research question RQ M2, it is necessary to have a look at the information creation and use of any *Industrie 4.0* component throughout its complete life cycle. Finally, following research question RQ C2, it is relevant to link such multi-disciplinary information enabling a digital shadow of the *Industrie 4.0* component.

A representative example is an industrial robot applied within a welding shop of a car manufacturer which consists of several welding cells. Both, the robot and the welding shop, can be considered as an *Industrie 4.0* component. During engineering phase, relevant aspects for a robot are its mechanical and electrical construction. Their corresponding digital shadow can be obtained from 'mechanical engineering information' (MCAD) and 'electrical engineering information' (ECAD). The individual instances of these information types are usually stored as drawings like TIFF or JPEG or as special engineering files like STP or JT. For the complete welding shop no detailed mechanical and electrical engineering is made, instead information types like a list of used resources, resource throughput Key Performance Indicators (KPIs), and costs KPIs are of interest, which can be coded by description means like CSV files. During the operation phase the robot motions are controlled exploiting detailed robot control program instances, while at welding shop level the order assignment to the welding shop is relevant based on B2MML file instances. And finally, during the removal of the production system, within its End-of-Life phase, KPI of interests is the robot's degree of abrasion, to decide about its reusability, and meterials, to consider recycling possibilities. In contrast, for the welding shop, its transferability to another location or country can be discussed.

Up to now there is no available approach, enabling engineers to decide about the right information set to be covered by an *Industrie 4.0* component and to decide about applicable implementation technologies. These open issues will be tackled within this and the subsequent two chapters.

Therefore, the main research question of these chapters is the following: **What are the requirements on the capabilities of *Industrie 4.0* components to create, manage, and use information along its complete life cycle.**

To answer this main research question three research questions need to be addressed beforehand:

Research question 1 What are relevant layers of *Industrie 4.0* components in a production system? Which layers can be considered at general and in which of them can *Industrie 4.0* components be found?

Research question 2 What is the life cycle of an *Industrie 4.0* component? Which main phases are necessary in this life cycle to distinguish an Industrie 4.0 component from their information content? What are general capabilities to reuse information between life cycle phases of the same or of different Industrie 4.0 components?

Research question 3 What types of information must be assigned to an *Industrie 4.0* component on the different layers to be covered by the virtual representation throughout all life cycle phases of a production system? Are these information types characteristic for the different layers?

Research question 4 Which description means are exploited to represent the information types in these life cycle phases? Are there special description means related to the different layers?

The named research questions are interrelated to the general research questions identified in Chap. 1. In addition, they will be addressed not only in this but also in the subsequent two chapters.

This chapter focuses on the discussion of Research questions 1 and 2. It will develop a hierarchy of potential *Industrie 4.0* components, describe their life cycle and will give insights into the dependencies of life cycles of different *Industrie 4.0* components. Thereby, the research questions RQ M1 and M2 of Chap. 1 are addressed. It should be noted, that the *Industrie 4.0* component hierarchy subsumes elements like components. In this context the same words have a different meaning.

The following Chap. 6 discusses Research question 3. Here, the different identified life cycle phases of *Industrie 4.0* components are considered in detail. For each phase, information types relevant within them are collected and discussed. Thereby, the research questions RQ M2 and C2 of Chap. 1 are targeted.

Finally, Chap. 7 is considering Research question 4. Here, description means for the different information types are named. Out of them, a set of description means is selected potentially being able to represent all relevant information for an *Industrie 4.0* component and, thereby, being a starting point for the implementation of the digital shadow of *Industrie 4.0* components. Thus, it will be related to the research questions RQ M1 and C2 of Chap. 1.

Since an exhaustive discussion of all possible kinds of production systems goes beyond the scope of this chapter, the research questions will be only validated against the background of discret manufacturing processes and the mixed-model-manufacturing-lines applied within them. This decision is based on the one hand on the scope of the *Industrie 4.0* approach and on the other hand on the practical experiences of the authors.

To answer the research questions the chapter is structured as follows. In Sect. 5.2 the applied research approach is described. The first research question is answered in Sect. 5.3. Therefore, the identified layers hosting *Industrie 4.0* components are described. Afterwards, in Sect. 5.4, the life cycle of a production system is sketched. Here, the main phases are given. With a summary the chapter ends.

5.2 Approach

In order to identify requirements on capabilities of *Industrie 4.0* components to create, manage, and use information along their complete life cycle the following three research steps have been conducted.

At first, the different layers of manufacturing systems needed to be identified. Therefore, various kinds of hierarchical structures of production systems have been discussed on the basis of a literature survey. In addition, the structure of different production systems used in automotive industry has been evaluated on a practical level. As a result, a generic hierarchical production system architecture has been developed that characterizes the functionality of each production system hierarchy layer. This generic architecture has been verified by mapping several other production systems to it, based on expert discussions.

In the next step, a general model of a production system life cycle, consisting of three general life cycle phases and its sub-phases, has been applied to identify relevant information considered within them.

As a result, a classification space for *Industrie 4.0* components has been developed covering the three dimensions production system hierarchy, life cycle phases, and information description means. Within this space the different *Industrie 4.0* components can be placed (see Fig. 5.2).

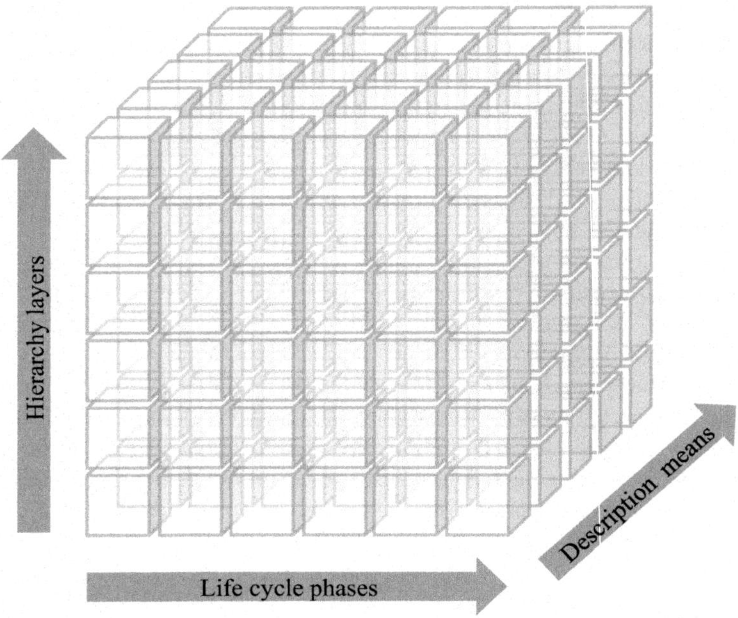

Fig. 5.2 Classification space for Industrie 4.0 components

5.3 Generic Production System Architecture

5.3.1 Literature Review

This section provides an overview of the different production system architectures described in recent literature. The architecture types are analyzed with respect to their layer structure. The main findings of this literature review are described below.

First of all, technical systems can be classified by various views, also named aspects (DIN 2010), which are divided into the following four categories:

- Local aspect (focused on spatial relations between objects),
- Function aspect (focused on functional context of objects),
- Product aspect (focused on constructional relation between objects), and
- Further aspects (with focus on different aspects of objects).

The literature review showed that there is no consensus on which of these views shall be applied for the hierarchy definition of production systems, instead depending on the field of interest researchers have defined diverging layer classifications that can be mapped to four named aspects:

- Local aspect: The first set of generic production system architectures was derived from real production systems that are currently used in various industries. These architectures are based on a factory planning point of view and consist of layers such as networks, sites, segments, systems, cells, and stations. Generally, these layers represent a specific manufacturing resource of a production system. Two examples of this category can be found in Wiendahl et al. (2007) and Scholten (2007).

- Function aspect: The second set takes a function-oriented approach. The layers of these generic production system architectures represent technological manufacturing functions. Exemplary layers are cells, main function groups, function groups, and sub-function groups. Three representatives of this category can be found in Kiefer (2007) and Meling (2012).
- Product aspect: The third set of architectures focuses on the individual components of a production system that define its physical behavior. This mechatronic device-centered perspective defines layers such as function units, devices, and device functions. Three representatives of this view point can be found in Lindemann et al. (2009) and DIN (2013).
- Further aspects: The last set of architectures combines the above mentioned aspects to an integrated structure. Two examples of this category can be found in VDI (2015) and NA35 (2003).

In summary, a variety of different layer types is defined in literature which represents both functions and physical objects. The physical objects range from large physical system such as production networks to small and tangible objects such as mechatronic devices and mechanical parts.

The architecture of production systems for manufacturing industry must meet certain requirements for reuse:

- vertical integration,
- consistency in engineering, and
- functional connection.

The comparison of requirements for reuse with existing production system architectures demonstrates that current production system architectures do not focus on component reuse (see Table 5.1). In addition, it is hard to give an exhaustive representation of the practically applied layers inside a production system. Furthermore also RAMI 4.0 does not provide any element classification enabling reuse of elements throughout all different parts of a production system.

Finally, the architecture of interest has to represent different sets of information in the different layers. Therefore, the major challenge is to define the least number of layers with the differentiable information volumes. Finally, there is the need to complement existing production system architectures.

5.3.2 Hierarchy Layers

Based on the literature review and practical experiences gained within expert interviews, an all-embracing and object-oriented set of production system hierarchy

Table 5.1 Requirements for reuse of production system elements

		Scholten	Wiendahl	Kiefer	Meling	Lindemann	ISA 95	RAMI 4.0	Namur
General	Applicable for automotive flow production	2	4	4	4	2	2	2	0
	Applicable for greenfield- & brownfield-projects	4	4	4	4	4	4	4	4
	Consideration of existing design artifacts and processes	4	0	4	4	4	2	4	0
Vertical integration	Clear definition of all elements in automotive flow production	2	0	4	2	2	2	2	0
	Clear definition of layers in automotive flow production	2	2	2	0	0	2	2	2
	Presentation of all relevant layers of production systems	0	0	2	0	0	0	0	0
	Identical structure for all automotive production departments	0	2	0	0	2	2	0	0
	Unambiguous attribution of elements to the layers of an production system with a one-to-one relationship	2	0	0	0	2	2	2	0
	Allocation of comparable elements of different production departments to the same layer	0	0	0	0	0	0	0	0
Consistency in engineering	Applicable throughout the life cycle of a production system and its components	2	0	2	2	2	4	2	0
	Universally usable libraries	2	0	4	2	4	4	2	0
	Usage of mechatronic devices	4	0	4	4	4	4	4	0
	Transferability from OEM to plant engineering and many more	4	0	2	2	2	2	4	2
	Production department reference of data and design decisions to layers	0	0	2	2	2	2	0	0
Functional connection	Map the functionalities of an automotive flow production	2	0	4	4	4	2	2	4
	Unambiguous attribution of processes/functions to the layers	2	0	4	4	4	2	2	4
	Libraries are arranged according to the fulfilled functions	0	0	2	2	2	0	0	2

0 non-compliant, 2 partially compliant, 4 fully compliant

layers has been identified. This set of hierarchy layers serves to locate *Industrie 4.0* components in a production system. For the identification of the objects, which make up each layer, different criteria were defined. The primary criterion is the technical functionality of each object. In the context of *Industrie 4.0*, technical functionality refers to the function an object contributes to the overall function of a production system. This can be a value-adding function, a function that supports value-adding functions, or a function that is required to supervise, control, diagnose, and maintain a production system. Additionally, the following subordinate characteristics of a production system have been considered: the hardware modularity, the control architecture, the control information, the relations to human labor, the relevance of a production system within the different engineering phases and activities, and the relations to the complexity of the product that is manufactured within a production system. The set of hierarchy layers that has been identified based on these criteria is shown in Table 5.2 and Fig. 5.3.

To assign an *Industrie 4.0* component to one of the defined hierarchy layers, two alternative matching procedures can be applied. The first alternative is to apply a bottom-up approach. Starting with the lowest hierarchy layer (no. 1) the considered component is compared to the functional characteristics of each layer, until a layer is found whose characteristics exceed the required functionality. Alternatively, it is also possible to match an *Industrie 4.0* component to a specific hierarchy layer using

Table 5.2 Hierarchical structure model

	Layer	Characterization
9	Production network	Includes the strategic and long term sales planning of the products of a company
8	Factory	Considers the entire range of activities and facilities required to produce end products or an intermediate good
7	Production line	Enables a distinction among different sections of a production system which are producing and processing components of the end product using different manufacturing technologies
6	Production line segment	Connection of different Work Units via buffers. Thereby, disturbances within the product/component/material flow can be controlled
5	Work unit	Executes the smallest non-divisible process in producing the end product
4	Work station	Represents the manufacturing related realization of a set of value adding and auxiliary functions required for the smallest non-divisible process
3	Function group	Represents the technical realization of one value adding or auxiliary function required for the smallest non-divisible process
2	Component	Enables the smallest non-divisible process
1	Construction element	Are not enabling the fulfillment of the smallest non-divisible process directly but enable the Component to provide their functionality

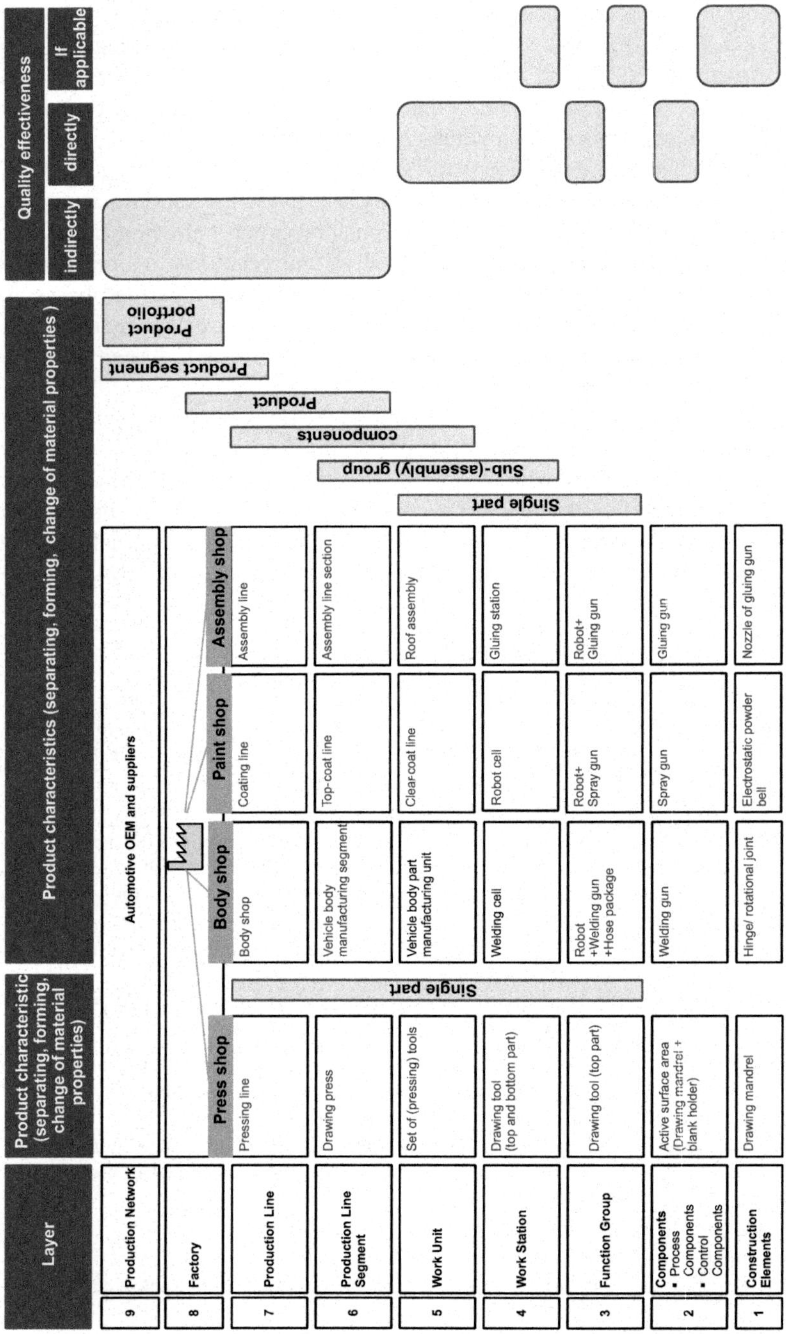

Fig. 5.3 Product system hierarchy level structure including exemplary layer objects and quality effectiveness evaluation

a top-down approach. Every hierarchy layer can be understood as an aggregate of all lower layers. Thus, a component can be assigned to the highest layer (no. 9) that matches its functional characteristics.

To identify the information that is required to perform the matching process described above, a detailed process analysis of production system life cycle phases was conducted. The analysis is based on the procedure described in Schäffler et al. (2013) and focuses on the functional characteristics that determine product related value-adding processes.

The value adding functionality of each *Industrie 4.0* component serves as the main criterion to determine the associated hierarchy layer. However, certain manufacturing resources of production system (e.g. press shop, body shop, assembly shop) exhibit functional characteristics that can be assigned to multiple layers. Therefore the functionality criterion is companied by resource life cycle information that ensures a conclusive layer distinction. Thereby, a clear separation of layers and a clear assignment of Industrie 4.0 components to layers get possible.

The architecture depends on the complexity of the production system. Complex production systems partially require supplementing layers. Furthermore the architecture can contain cross directional structures like work groups or protective circles. Cross directional structures are not useful for defining layers mainly due the lack of unambiguous attribution.

This criteria enrichment process results in a detailed characterization of each hierarchy layer that allows a practical distinction of *Industrie 4.0* components used in manufacturing (particularly automotive industry) environments. The function-based identification criteria are presented below.

Construction Elements

The smallest and non-dismountable functional parts in a production system are defined as construction elements. Construction elements empower the functionality of components. They are able to influence product characteristics, but cannot individually initiate value-adding processes. One example demonstrating this element-component-relationship is a rotary spray painting bell and a paint atomizer. The bell embodies an individual construction element of the paint atomizer component.

Components

Components can be subdivided into process and control components. Control components process electric signals. They are often used for diagnostics and are not directly involved in value-adding activities. Process components support manufacturing processes and influence product quality characteristics. These components can be subdivided into value-adding components and non-value-adding components. Value-adding components empower core value adding processes. A list of these processes such as gluing or welding can be found in DIN (2003) and VDI (1990). Non-value-adding components exclusively support to perform the process realization. One example for such components are robots that are used as mounting devices for value-adding components such as welding tools.

Generally, all components that serve material-handling purposes are non-value-adding components. Process components feature a predetermined range of functions that can be parameterized and adapted to individual needs.

Function Group
Function groups are able to individually perform manufacturing processes or supportive tasks (DIN 2003; VDI 1990). Generally, solely value-adding function groups determine product quality characteristic. However, in certain circumstances non-value-adding function groups also influence product quality. A handling device, for instance, that positions parts to perform a gluing process is also responsible for the quality of the gluing application. Faulty positioning might result in rework. Additionally, specialized multi-functional devices are able to perform both value-adding and non-value-adding activities. Most steel presses reshape and cut steel and also have material handling capabilities.

Value-adding function groups are usually used to manufacture single parts and typically incorporate some kind of robotic application that performs manufacturing processes such as gluing or welding. Such processes need to be coordinated and supervised by a control unit.

Work Station
Work stations aggregate multiple function groups that are required to manufacture single parts or (sub)-assembly groups. They usually combine value-adding and non-value-adding processes. For instance, multifunctional gluing work stations that are used in automotive assembly for bonding window frames to the windshield both position parts and perform the actual gluing process. Work stations such as automated gantry manipulators or picking stations do not perform any value-adding activities.

Work Unit
Work units perform multiple interrelated value-adding activities that are required to manufacture a component of a final product. To complete a work order, all functional activities that are performed within one work station are repeated "m"-times (REFA 1993). Work units define the product characteristics of a specific part or component. All manufacturing activities are completed in a closed and spatially connected system (e.g. door welding, sunroof assembly). Thus, a work unit embodies the highest hierarchy layer that is able to directly influence product quality and that determines the product characteristics of single parts or components. The range of work station functions contains value-adding and non-value-adding activities.

Production Line Segment
Production line segments offer a defined and homogeneous range of functions and provide a connected manufacturing structure. This structure is used to manufacture part modules and components. The scope of a production line segment depends on the vertical range of in-house manufacturing. Product quality characteristics are not directly determined on this hierarchy layer. The control parameters of production line segments are, for instance, assembly or body shop welding sequences.

Production Line

Production lines serve as autonomous and closed systems that are able to produce physical output (Laucht 1995). The range of functions provided by a production line exhibits homogenous characteristics and enables to manufacture components, final products, or product segments. The characteristics and scope of a production line depend on its spatial organization patterns. Usually, production lines require personnel planning. They are often used in a just in time or just in sequence production setting that is synchronized with internal and external entities that supply intermediate goods.

Factory

Factories provide the entire range of functions that is required to manufacture an intermediate good, a final product, a product segment, or an entire product portfolio (REFA 1993). However, the range of specific functions of a factory depends on individual company related characteristics. Factories require both tactical and operational planning. This includes tasks such as freight cost optimization, requirements analysis, forecasting, material supply planning, and production scheduling.

Production Network

Production networks are responsible for strategic long-term distribution/sales planning (<10 years) and intermediate-term tactical planning (<3 years). In addition, also operational interdisciplinary tasks such as order management, product data management, and technical feasibility testing are performed on this hierarchy layer.

5.4 Production System Life Cycle

To efficiently identify the relevant information needed to be assigned to the components of the different layers it is essential to understand completely the life cycle of a production system with its different phases and the dependencies among them.

The life cycle of production systems has been considered by several authors. Attri and Grover (2012) have provided a detailed survey on different models for production system life cycles summarizing them to the main phases initiation of system, design and development of system, operation of system, revision of system, and termination of system. Wiktorsson (2013) has a more limited view on the production system life cycle with only production system engineering and production system operation as main phases. Beyond this view also the End-of-Life of the production system is essential following the reduction of a production system life cycle and the need of sustainable use of natural resources.

Thus, the authors decided to focus on the three general but essential life cycle phases of a production system: engineering, operation and maintenance, and End-of-Life (VDMA 2013; VDI 2005). They are aware that there are much more distinguishable engineering phases. As Lindemann has identified in (Lindemann 2009), the life cycle of products (and thereby also production systems) is a hierarchy of different phases, activities, and actions. This hierarchy can be considered in different level of detail. In addition, a cyclic life cycle structure is reached if redesign is also taken into consideration. The selected very low granularity follows the aim of enabling a broad application of the identification criteria for Industrie 4.0 components.

The engineering phase covers all activities related to the production system before its complete physical existence. It ends with a completely built up/installed, commissioned (see Chap. 15), and ramped-up production system. Within this phase all engineering activities related to a production system, as named e.g., in Lüder et al. (2011), are included. Examples of the engineering activities are the product design including the definition of the bill of material and bill of operation of the product to be produced with the production system, the planning of the overall production process and the technologies used within, the functional engineering covering activities like detailed mechanical and electrical engineering of the production system components, and the control system design and programming as well as the commissioning and test of the production system.

The subsequent operation and maintenance phase of the production system covers the complete period of the use of the production system to manufacture products. During this phase the production system is controlled, monitored, repaired if necessary, and partially rebuilt if modernization is necessary in order to adapt the production system to different sets of products, production technologies, etc..

The last phase of the production system life cycle is the End-of-Life phase of the production system. It covers the period of the production system between the end of the production of the last product and the restoration of the so-called greenfield; It is the complete removal of the production system. During this phase the production system is disassembled, possibly completely or partially reused, and/or recycled.

The resulting structure of the life cycle of the production system is depicted in the upper part of Fig. 5.4.

But the use of information is not only relevant within the life cycle of the actual production system—The information can also be used within the life cycle of other production systems [see VDI/VDE (2014)]. With respect to this chapter the impact of the life cycle of one production system on the life cycle of another's production system life cycle is considered.

As shown in Fig. 5.4 different cases of utilization of information are considered ranging from the use of engineering information in the same engineering phase (① in Fig. 5.4) over use of information of the operation phase within the engineering and operation phases of another production system (⑤ and ⑥ in Fig. 5.4) to the use of engineering information in the End-of-Life phase of the same production system (⑧ in Fig. 5.4) or End-of-Life information in the engineering of another production system (⑩ in Fig. 5.4). The End-of-Life phase—as the last phase—has more cases than the other phases since, at first, it can use the information created in the previous phases and, secondly, it directly connects the life cycle of the actual production system with the life cycle of other production systems.

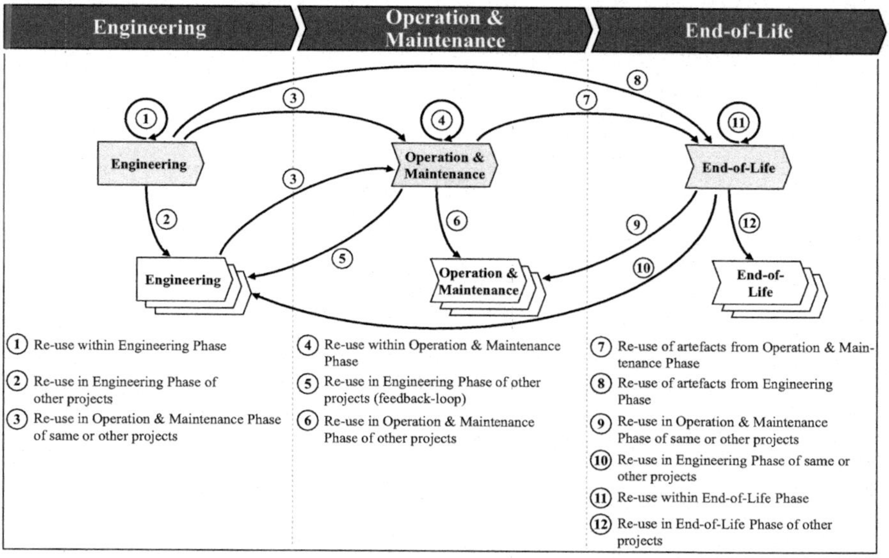

Fig. 5.4 Reuse of physical and information artifacts in production system life cycle

For each of the nine identified layers and each phase of the three life cycle phases, the relevant information shall be named completing the matrix depicted in Fig. 5.5 and answering the second research question.

5.4.1 Characteristics of Engineering Phase

To identify the information relevant for *Industrie 4.0* components in engineering and to assign this information to the presented granularity layers of the components, a detailed look at different engineering approaches and tasks is essential. The focus of the research is on the analysis of the dependencies between the granularity layers and the life cycle phases. In order to analyze the dependency in more detail, the point of time in the life cycle at which the artifacts are generated or are used should be considered. Therefore, the engineering phase has to be divided in smaller parts to be able to assign the identified artifacts more precise to the point of time at which they are generated or are used. Thus, the characteristics of the engineering are described in the following paragraphs to divide the phase into meaningful subphases.

The guideline VDI 2221 (VDI 1993) is an important general approach for engineering. This guideline separates the whole process into different tasks. The tasks end with defined results. The first results have an abstract character, e.g. function structures, and become more and more specific in later phases, e.g. fixed layouts. Thus, the process can be divided in four phases which are different depending on industry or use case.

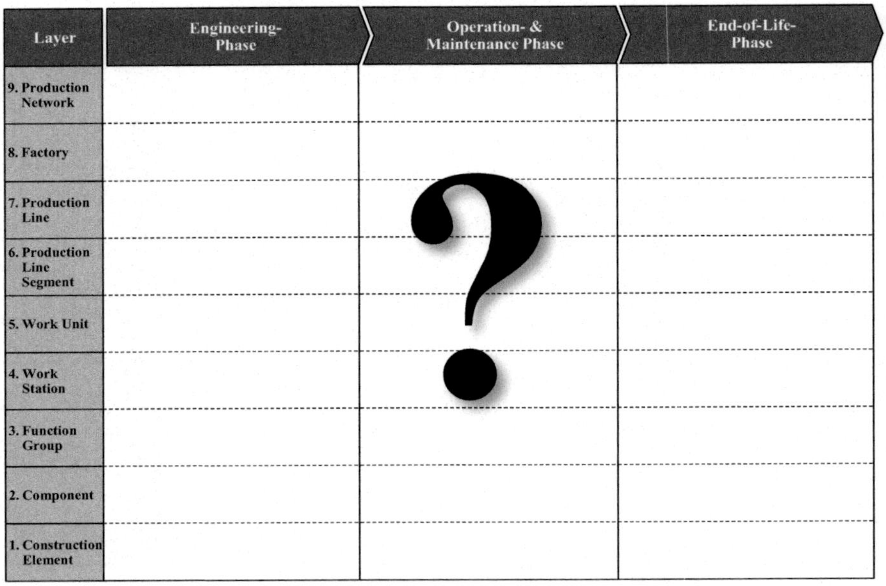

Fig. 5.5 Artifacts during the different life cycle phases need to be identified

Exploiting the general guideline VDI 2221 (VDI 1993), other engineering approaches like NA35 (2003), Höffken and Schweitzer (1991), VDIa (2010), Gepp (2014), and Urbas and Krause (2012), project management guidelines like Reschke and Svoboda (1984) and several practical approaches in plant realization, the engineering process can be characterized in a more specific way (see Fig. 5.6).

The engineering phases contain defined tasks. The phase of concept is focused by invest decisions, general studies and proposition. If this concept is feasible, rough planning will be started. The phase of rough planning is focused by dimensioning of production system and definition of processes. The next phase is detailed planning during which the defined processes are assigned to related single functions. Also, components relevant to perform these functions are preselected. As a result, specifications are ready to call for tenders. Specifications contain among others a fixed budget, layouts, a value chain process, a preselection for components, requested suppliers, relevant components to perform needed support functions. The following phase of design is often operated by a general contractor. In this phase, the necessary components are defined. There are certain parts and tools which have to be newly designed for a special purpose at this stage. The phase of design can be divided in separate disciplines, i.e. mechanical, electrical, information technology and fluidics design, see more in VDI (2006). As a result, plans and documentation needed to fabricate the designed plant are finished. In the following two phases of engineering, fabrication and commissioning, the components and equipment are ordered, assembled to a functioning plant and commissioned. Commissioning ends

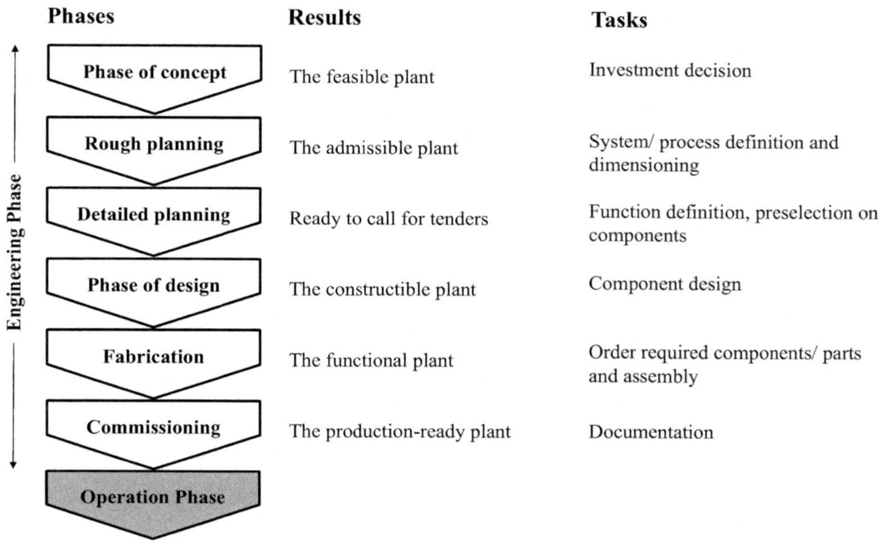

Fig. 5.6 Phases in plant engineering based on Gepp (2014), Höffken and Schweitzer (1991), Urbas and Krause (2012), Reschke and Svoboda (1984), and NA35 (2003)

with the rework of documentation due to actual realization and performance level. The project is finished upon completion of final report and customer acceptance.

A challenge in present engineering projects is related to market pressure. Time-to-market has to be as short as possible to be competitive. Thus, there are special demands on information quality, handling, availability and transfer due to parallelization of tasks. Much information created in one task has to be transferred to other disciplines at a defined date in order to conclude the other tasks. Parallelization of tasks and transfer of information and data is supported by tools of the *digital factory*, see more in VDI (2011). Vice versa, the needed information transfer makes high demands on applied tools of *digital factory*. To deal with the challenge, identification of relevant information and its relations is necessary, especially in engineering.

Exploiting the design methodology for mechatronic systems VDI 2206 (VDI 2006) given in Fig. 5.7, a first characteristic of suitable information can be deviated.

Engineering starts with system design therefore needs information relevant for the whole system, e.g. requirements, economical data, and proposition of production system. In the domain-specific tasks, information related to these special purposes is necessary. The information is highly dependent due to parallel design of mechatronic components in each discipline. The information important at this stage contains data for single components and function groups rather than system information. Towards the end of engineering, during the system integration, the relevant information emerges to system data, as the components are assembled and interact in a system.

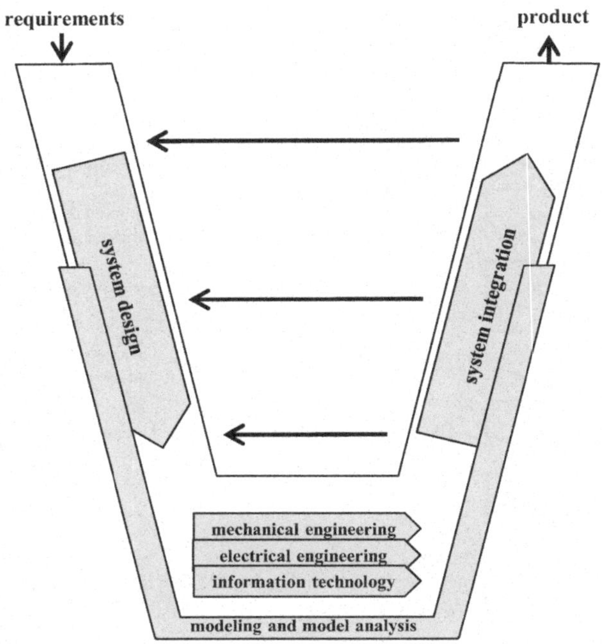

Fig. 5.7 V model for engineering of mechatronic products (VDI 2006)

To summarize, in this section, an overview of general engineering approaches and of the importance of the information identification and the handling in current digital-supported engineering projects was given. The next Sect 5.4.2 deals with the characteristics of the Operation and Maintenance Phase and the related demand on information artifacts in this phase.

5.4.2 *Characteristics of Operation and Maintenance Phase*

The engineering phase of a production system concludes with the construction, commissioning and finally the ramp-up of the physical production system (Schenk et al. 2014). When the production system has eventually transitioned into a steady state, it can be fully exploited (Attri and Grover 2012). In this phase, usually called use phase or operation phase [see Dencovski et al. (2010), Schenk et al. (2014), and Attri and Grover (2012)], the participating elements of the system need to communicate with each other in order to be productive (VDI 2015).

The operation phase is characterized by a various set of control activities. First, on one side of the spectrum, sensors and actuators need to be controlled in order to perform physical processes on field level. This is usually realized by implementing programmable logic control (PLC). Second, there are control activities designated to the allocation and supervision of manufacturing resources that are used for the processes on field level. Those activities are performed by manufacturing execution systems (MES). Finally, at the other end of the spectrum, there are activities for planning and control of business and management tasks. The goal of these control activities is the optimization of the underlying processes, while at the same time considering the available resources. These tasks are executed by Enterprise Resource Planning Systems (ERP). For further information, see Lüder (2006), ZVEI (2010), and Spath et al. (2013).

Production planning and control activities in operation phase have been described by a broad set of authors. One of the most common and most frequently cited models is the "Aachen Model of Production Planning and Control" originally published by Luczak et al. (1998). It splits the activities into the three main categories network, core and cross-sectional activities (see Fig. 5.8).

Since its first publication, the Aachen Model has been frequently updated and is now continued by Schuh and Stich (2012), ensuring that its core principles are still valid today. However, the model lays its focus mainly on the planning part. Other authors have addressed this issue and developed models with a stronger focus on production control. Two important contributors in the context of manufacturing industry are Lödding (2008) and Dörmer (2013). Lödding (2008) concentrates on the configuration aspect of manufacturing control and thereby puts the emphasis on control tasks, such as order authorization and capacity control. This allows tracking of key performance indicators such as delivery reliability, material inventory, lead time and equipment utilization. Dörmer (2013) on the other hand adds crucial elements for highly individualized products in the manufacturing industry. Among them are material delivery as part of the supply chain and resequencing of production orders to optimize production efficiency and effectiveness.

In addition to the above-mentioned activities, the different aspects of the automation pyramid need to be considered in order to break down production

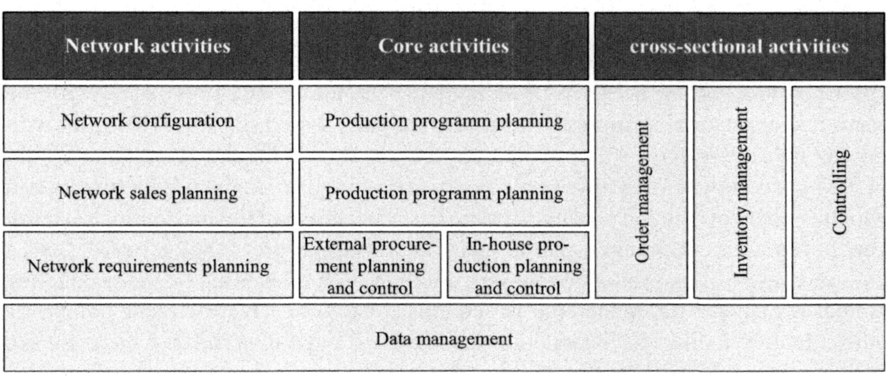

Fig. 5.8 Production planning and control activities following (Schuh and Stich 2012)

activities to field level and guarantee functionality, e.g. by ensuring compatibility between the different systems and real-time capability of critical components. For further information, see Vogel-Heuser et al. (2013) and Diedrich et al. (2011).

The foregoing shows that the control activities and therefore the information relevant in operation and manufacturing phase are numerous and can be very different from each other. Nevertheless, in order to implement the required functionalities into *Industrie 4.0* components they have to be integrated by the available implementation technologies for control systems [see e.g. Lunze (2008), Kiehl (2007), Kis (2014), Yang et al. (2014), and Leitao and Karnouskos (2015)]. A step towards this goal is described in the following section.

5.4.3 Characteristics of End-of-Life Phase

Devaluations bring the production system's operation phase to an end (Seliger et al. 2001). When all inherent flexibility and reconfiguration potentials are exhausted as well as all upgrade and update possibilities (Wiendahl et al. 2015), the production system has to be decommissioned, disassembled, removed, reused, recycled, and disposed (Weber 2008; VDI 2005) to make room for another production system, then, it is engineered to meet the changed requirements which come along with the demand of producing new products or a new product mix for the customers. This last phase of the production system life cycle is called End-of-Life phase (EoL phase).

The essential purpose of the EoL phase is to make room for a new production system. But how it is removed depends on the further purpose and how the production system (incl. its components and material) shall be treated in the end of its life. Hence, the EoL phase is use case specific but general treatments, further called *EoL Scenarios,* can be identified.

This chapter only focuses on those scenarios (representing the recovery of physical artifacts—including disassembly) and excludes the (decommissioning) activities which bring the production system into a safe state prerequisite to recover the physical artifacts.

To depict the scenarios in its entirety the production system life cycle (as described in Chap. 3) needs to be extended to get an overall loop—here, with the earth and its natural resources as source and sink (see Fig. 5.9). It begins with mining these resources, which are processed to raw material for manufacturing out of it single components. Those components are aggregated to the production system which is, then, producing products for the customers (use of the production system). Due to requirement changes, like trends or technical progress (Seliger et al. 2001), the use phase of this production system configuration ends. Finally, the production system is disposed or incinerated which closes the loop. This life cycle comprises other life cycles, like the life cycle of a component or of material, but these are out of the scope of this article.

Even though, just disposing or incinerating the production system is also a loop, all inherent value of the production system and its components is lost, because all

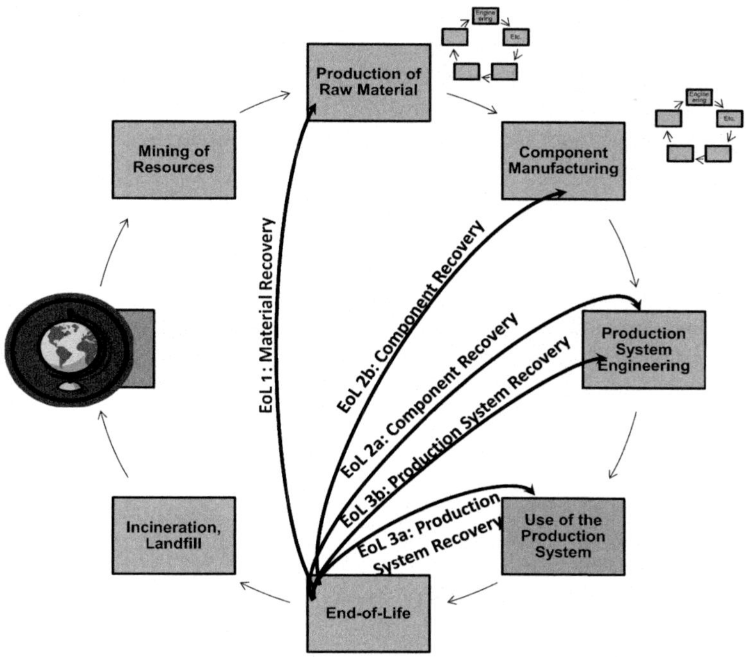

Fig. 5.9 Extended life cycle of a production system: End-of-Life scenarios of production systems [based on VDI (2002), Duflou et al. (2008), Pahl et al. (2007), and Seliger et al. (2001)]

phases (production of raw material, component manufacturing etc.) has to be done a second time. For that reason, this is not considered in this article. Instead the following EoL scenarios are considered:

EoL 1—Material Recovery: This comprises the recovery of material, the components are made of, by dissolving the component and its structure (Duflou et al. 2008). It is applied when devaluations had brought the operation phase of a production system to an end, and the technology of the components of the production system is obsolete or the component is too worn to consider it for a second life.

EoL 2—Component Recovery: This comprises the reuse of used components of the production system, because the life span of those is longer than the life span of the production system (Huber 2001). It is applied when devaluations had brought production system's operation phase to an end, but the technology of the components is still up-to-date. This scenario can be divided into direct Component Recovery (EoL 2a) and Component Recovery after remanufacturing (EoL 2b); Remanufacturing comprises the activities disassembly, inspection, cleaning, reprocess, testing, reassembly, and storage.

EoL 3—Production System Recovery: This comprises the reuse of the used production system. It is applied when the production system needs to be relocated due to devaluations, i.e. changed requirements caused by globalization, but the technology of the production system is still appropriate for its purpose. This scenario can be divided into direct Production System Recovery (EoL 3a) and Production System Recovery after remanufacturing (EoL 3b).

Decommissioning and disassembling activities lead up to those recovery scenarios.

Since natural resources are used extensively, pollutant emissions are generated in high amount, and the ecological awareness of the people is increasing, nowadays, (VDMA 2015) a consideration of manufacturing from a sustainable point of view is inevitable (Schultmann 2004). So, it is of economic and environmental interest to choose an appropriate loop to close the life cycle of a production system. This loop should be kept small to recover the inherent value of the production system and its components and material (Duflou et al. 2008; VDMA 2015). The more inherent value can be remained by recovery, in general, the better, because less energy is consumed, less resources are depleted, and less waste is generated (Duflou et al. 2008).

5.5 Summary and Outlook

This chapter has analyzed what are relevant layers of *Industrie 4.0* components in a production system. In this context nine layers for *Industrie 4.0* components were identified and characterized by generalizing an exemplarily automotive manufacturing environment and validating this structure in various industrial systems. Figure 5.10 shows a comparison of hierarchy layers. The defined layers allow an unambiguous attribution of elements to the layers of a production system with a one-to-one relationship. Consequently, the developed hierarchy is applicable for production systems with comparable complexity and research question 1 is answered.

The developed hierarchy serves as a foundation for the reusability and modularization of *Industrie 4.0* components. First of all, the modularization aspect refers to the construction phase. Components that have been previously used can be easily integrated in new construction concepts. This process leads to lower equipment development cost. Secondly, modularization also offers saving potentials with regard to maintenance activities. Outdated or broken modules can be exchanged with minimal effort. In general, the reusability of engineering artifacts creates a tremendous equipment cost reduction potential. The experience gained from tested and successfully implemented components gradually improves the engineering quality of components implemented in the future (e.g. equipment availability/down time).

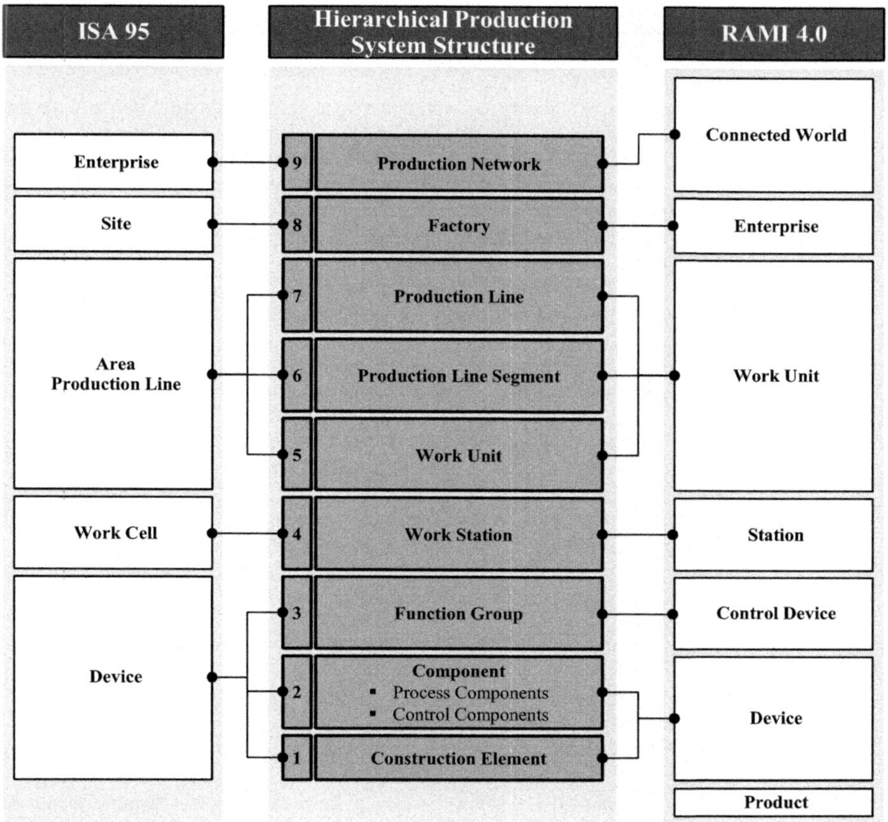

Fig. 5.10 Comparison of hierarchy layers

Furthermore, this chapter provides the life cycle of an *Industrie 4.0* component. This creates the necessary conditions to identify the relevant information needed to be assigned to the components of the different layers in Chap. 6.

References

Attri, R., Grover, S.: A comparison of production system life cycle models. Front. Mech. Eng. **7**(3), 305–311 (2012)

Dencovski, K., Löwen, U., Holm, T., Amberg, M., Maurmaier, M., Göhner, P.: Production system's life cycle-oriented innovation of industrial information systems. InTech, http://cdn.intechopen.com/pdfs-wm/10837.pdf (2010). Accessed 01.04.2016

Diedrich, C., Lüder, A., Hundt, L.: Bedeutung der Interoperabilität bei Entwurf und Nutzung von automatisierten Produktionssystemen. at – Automatisierungstechnik. **59**(7), 426–438 (2011)

DIN 8580: Fertigungsverfahren – Begriffe, Einteilung. Beuth Verlag, Berlin (2003) (in German)

DIN 81346-1: Industrielle Systeme, Anlagen und Ausrüstung und Industrieprodukte – Strukturierungsprinzipien und Referenzkennzeichen. Beuth Verlag, Berlin (2010) (in German)
DIN 62264-1: Integration von Unternehmensführungs- und Leitsystemen. Beuth Verlag, Berlin (2013) (in German)
Dörmer, J.: Produktionsprogrammplanung bei variantenreicher Fließproduktion. Untersucht am Beispiel der Automobilendmontage. Springer Gabler, Wiesbaden (2013) (in German)
Duflou, J.R., Seliger, G., Kara, S., Umeda, Y., Ometto, A., Willems, B.: Efficiency and feasibility of product disassembly – a case-based study. CIRP Ann. Manuf. Technol. **57**(2), 583–600 (2008)
Gepp, M.: Standardisierungsprogramme als Ansatz zur Steigerung der Wirtschaftlichkeit im industriellen Anlagen-Engineering. Schriftenreihe Innovative Betriebswirtschaftliche Forschung und Praxis, vol. 418. Dr. Kovac Verlag, Hamburg (2014)
Höffken, E., Schweitzer, M. (Hrsg.): Beiträge zur Betriebswirtschaft des Anlagenbaus. Schmalenbachs Zeitschrift für betriebswirtschaftliche Forschung Sonderheft. Düsseldorf, Verl.-Gruppe Handelsblatt, vol. 28 (1991)
Huber, A.: Demontageplanung und -steuerung: Planung und Steuerung industrieller Demontageprozesse mit PPS-Systemen. Dissertation, Otto-von-Guericke University Magdeburg (2001) (in German)
Kagermann, H., Wahlster, W., Helbig, J. (eds.): Recommendations for implementing the strategic initiative INDUSTRIE 4.0 – Securing the future of German manufacturing industry. Final report of the Industrie 4.0 Working Group, April 2013. http://www.acatech.de/fileadmin/user_upload/Baumstruktur_nach_Website/Acatech/root/de/Material_fuer_Sonderseiten/Industrie_4.0/Final_report__Industrie_4.0_accessible.pdf. Accessed 2016-02-01
Kiefer, J.: Mechatronikorientierte Planung automatisierter Fertigungszellen im Bereich Karosseriebau. PhD Thesis, Schriftenreihe Produktionstechnik, vol. 43, Universität des Saarlandes (2007) (in German)
Kiehl, E. (ed.): Antriebslösungen – Mechatronik für Produktion und Logistik. Springer, Berlin (2007)
Kis, T.: Planning and Scheduling in the Digital Factory, KOMSO Challenge Workshop – Math for the Digital Factory. In: Proceedings, Berlin, Germany, May 2014
Laucht, O.: Flexibilisierung der manuellen Großmontage. PhD Thesis, TU Braunschweig (1995) (in German)
Leitão, P., Karnouskos, S.: Industrial Agents: Emerging Applications of Software Agents in Industry, pp. 153–170. Elsevier, Waltham, MA (2015)
Lindemann, U.: Methodische Entwicklung technischer Produkte. Springer, Berlin (2009) (in German)
Lindemann, U., Maurer, M., Braun, T.: Structural Complexity Management – An Approach for the Field of Product Design. Springer, Berlin (2009)
Lödding, H.: Verfahren der Fertigungssteuerung. Grundlagen, Beschreibung, Konfiguration, 2. erw. Aufl. Springer (VDI), Berlin (2008) (in German)
Luczak, H., Eversheim, W., Schotten, M. (eds.): Produktionsplanung und -steuerung. Grundlagen, Gestaltung und Konzepte, 1. Aufl. Springer (VDI), Berlin (1998)
Lüder, A.: Strukturen zur verteilten Steuerung von Produktionssystemen, Habilitationsschrift. Fakultät Maschinenbau, Otto-von-Guericke Universität Magdeburg (2006)
Lüder, A., Foehr, M., Hundt, L., Hoffmann, M., Langer, Y., Frank, St.: Aggregation of engineering processes regarding the mechatronic approach. In: 16th IEEE International Conference on Emerging Technologies and Factory Automation (ETFA 2011), Toulouse, France, Proceedings-CD, September 2011
Lunze, J.: Automatisierungstechnik – Methoden für die Überwachung und Steuerung kontinuierlicher und ereignisdiskreter Systeme. Oldenbourg Verlag, München (2008)
Meling, F.: Methodik für die Rekombination von Anlagentechnik. PhD Thesis, TU Munich (2012) (in German)
NAMUR: Automatisierungstechnik – Methoden für die Überwachung und Steuerung kontinuierlicher und ereignisdiskreter Systeme. In: Abwicklung von PLT-Projekten, NA 35, Namur-Arbeitsblatt, 24.03.2003. NAMUR Arbeitskreis 1.1 “Planungsmethodik” (2003)

Pahl, G., Beitz, W., Feldhusen, J., Grote, K.-H.: Engineering Design – A Systematic Approach. Springer, London (2007)
REFA: Ausgewählte Methoden der Planung und Steuerung. Carl Hanser Verlag, Munich (1993) (in German)
Reschke, H., Svoboda, M.: Projektmanagement. Konzeptionelle Grundlagen. Ges. für Projektmanagement, Munich (1984)
Schäffler, T., Foehr, M., Lüder, A., Supke, K.: Engineering process evaluation – evaluation of the impact of internationalisation decisions on the efficiency and quality of engineering processes. In: 22nd IEEE International Symposium on Industrial Electronics (ISIE 2013), Taipei, Taiwan, Proceedings, May 2013
Schenk, M., Wirth, S., Müller, E.: Fabrikplanung und Fabrikbetrieb – Methoden für die wandlungsfähige, vernetzte und ressourceneffiziente Fabrik, 2nd edn. Springer Vieweg, Berlin (2014)
Schmale, C.: Welthandel: Beunruhigende Signale! http://www.godmode-trader.de/artikel/welthandel-beunruhigende-signale,4412211 (2015). Accessed Mar 2016
Scholten, B.: The Road to Integration. ISA (2007)
Schuh, G., Stich, V.: Produktionsplanung und -steuerung 1 – Grundlagen der PPS, 4th edn. Springer, Berlin (2012)
Schultmann, F.: Industrielles Produktions- und Logistikmanagement. In: Haasis, H.-D., Spengler, T.S. (eds.) Produktion und Umwelt – Festschrift für Otto Rentz. Springer, Berlin (2004) (in German)
Seliger, G., Basdere, B., Keil, T.: e-Cycling platform for profitable re-use. In: IEEE International Symposium on Assembly and Task Planning, Fukuoka, Japan, Proceedings, 28–29 May 2001
Spath, D. (Hrsg.), Ganschar, O., Gerlach, S., Hämmerle, M., Krause, T., Schlund, S.: Produktionsarbeit der Zukunft – Industrie 4.0. Studie, Fraunhofer-Institut für Arbeitswirtschaft und Organisation. Fraunhofer-Verlag, Stuttgart (2013)
Urbas, L., Krause, A.: Process Control Systems Engineering. Oldenbourg Industrieverlag GmbH, Munich (2012)
VDI 2860: Montage- und Handhabungstechnik – Handhabungsfunktionen, Handhabungseinrichtungen; Begriffe, Definitionen, Symbole. Beuth Verlag, Berlin (1990) (in German)
VDI 2221: Methodik zum Entwickeln und Konstruieren technischer Systeme und Produkte. Beuth Verlag, Berlin (1993)
VDI 2243: Recycling-oriented Product Development. Beuth Verlag, Berlin (2002)
VDI 2884: Purchase, Operating and Maintenance of Production Equipment using Life Cycle Costing (LCC). Beuth Verlag, Berlin (2005)
VDI 2206: Design methodology for mechatronic systems. Beuth Verlag, Berlin (2006)
VDI 4499: Digital Factory. Beuth Verlag, Berlin (2011)
VDI/VDE-Gesellschaft Mess- und Automatisierungstechnik – Fachausschuss 7.21 "Industrie 4.0", "Status Report: Reference Architecture Model Industrie 4.0 (RAMI4.0), July 2015. http://www.zvei.org/Publikationen/GMA-Status-Report-RAMI-40-July-2015.pdf
VDI/VDE: Industrie 4.0 –Wertschöpfungsketten. VDI/VDE Gesellschaft Mess- und Automatisierungstechnik, Statusreport, April 2014
VDIa 3695: Engineering of industrial Plants, Evaluation and Optimization, Part 1. Beuth Verlag, Berlin (2010)
VDMA 34160: Prognosemodell für die Lebenszykluskosten von Maschinen und Anlagen. Beuth Verlag, Berlin (2013)
VDMA Arbeitsgemeinschaft Großanlagenbau: Lagebericht 2014/2015: Weltweite Krisen meistern – lokale Chancen nutzen, Beiträge zum Industrieanlagenbau. Technical Report, VDMA Arbeitsgemeinschaft Großanlagenbau (2015) (in German)
Vogel-Heuser, B., Bauernhansl, T., ten Hompel M. (eds.): Handbuch Industrie 4.0 – Produktion, Automatisierung und Logistik. Springer, Berlin (2015) (in German)
Vogel-Heuser, B., Diedrich, C., Broy, M.: Anforderungen an CPS aus Sicht der Automatisierungstechnik. Automatisierungstechnik **61**(10), 669–676 (2013)
Weber, K.H.: Dokumentation verfahrenstechnischer Anlagen. Springer, Berlin (2008) (in German)

Wiendahl, H.-P., ElMaraghy, H.A., Nyhuis, P., Zäh, M.F., Wiendahl, H.-H., Duffie, N.A., Brieke, M.: Changeable manufacturing – classification, design and operation. Ann. CIRP. **56**(2), 783–809 (2007)

Wiendahl, H.-P., Reichardt, J., Nyhuis, P.: Handbook Factory Planning and Design. Springer, Berlin (2015)

Wiktorsson, M.: Consideration of legacy structures enabling a double helix development of production systems and products. In: Henriques, E., Pecas, P., Silva, A. (eds.) Technology and Manufacturing Process Selection, Springer Series in Advanced Manufacturing, pp. 21–32. Springer, London (2013)

Yang, C., Vyatkin, V., Pang, C.: Model-driven development of control software for distributed automation. IEEE Trans. Syst. Man Cybern. Syst. **44**(3), 292–305 (2014)

ZVEI - Fachverband Automation (Hrsg.): Manufacturing Execution Systems (MES). Branchenspezifische Anforderungen und herstellerneutrale Beschreibung von Lösungen. ZVEI, Frankfurt (2010)

Chapter 6
Identification of Artifacts in Life Cycle Phases of CPPS

Arndt Lüder, Nicole Schmidt, Kristofer Hell, Hannes Röpke, and Jacek Zawisza

Abstract Recent research and development activities within the field of production system engineering and operation focus on the increase of production system flexibility and adaptability. One common issue of those approaches is the consideration of hierarchical and modular production system architectures where the individual components of the system are equipped with certain functionalities and information. Up to now there is no common understanding about what a component constitutes, i.e. which parts of a production system can be regarded as components within the hierarchy and which functionalities and information are assigned to it. This gap will be closed within this, the prior, and the subsequent chapter.

They will at first discuss the relevant layers of components in a production system, then the types of information required to be assigned to a component on the different layers to establish a virtual representation of the component, and at last the description means exploitable to represent the identified information in the different life cycle phases of a production system.

This chapter in particular will consider in detail the information sets relevant for a production system component along the life cycle of a production system. Relevant artifacts are identified for each of the three main life cycle phases described in Chap. 5, assigned to the different layers of the production system hierarchy, and discussed against main cases of information reuse within the life cycle of production systems. Through this, it is intended to enable an identification of hierarchy layers based on relevant information sets.

Keywords Industrie 4.0 component • Administration shell • Life cycle information • Virtual representation

---

A. Lüder (✉) • N. Schmidt
Faculty Mechanical Engineering, Otto-von-Guericke University, Magdeburg, Germany
e-mail: arndt.lueder@ovgu.de; nicole.schmidt@ovgu.de

K. Hell • H. Röpke • J. Zawisza
Volkswagen Aktiengesellschaft, Wolfsburg, Germany
e-mail: kristofer.hell@volkswagen.de; hannes.roepke@volkswagen.de; jacek.zawisza@volkswagen.de

© Springer International Publishing AG 2017
S. Biffl et al. (eds.), *Multi-Disciplinary Engineering for Cyber-Physical Production Systems*, DOI 10.1007/978-3-319-56345-9_6

6.1 Introduction

The main research questions of the previous chapter, of this chapter and the subsequent chapter is the following: **What are the requirements on the capabilities of *Industrie 4.0* components to create, manage, and use information along its complete life cycle?**

In this chapter, the aim is to identify relevant artifacts in the life cycle phases of CPPS. As the *Industrie 4.0* component (I4.0 component) concept is a possible form of realization for future CPPS in production systems (see Chap. 5), the relevant information for I4.0 components will be analyzed.

The demands on the technical realization of the administration shell, which should contain the relevant information, have not been examined so far. One important step to define these demands is to identify the types of information dependent on the granularity of an I4.0 component and on the life cycle phase in which the information in generated. This chapter focusses on information necessary for current production systems which are also important for future production system. These future systems will require additional information for their comprehensive functionalities. But currently, it is not possible to estimate all the required additional information. Due to this fact, the types of information are identified which must be considered even in future I4.0 component based production systems.

Derived from this gap and to answer the main research question, four subordinate research questions need to be addressed (see Chap. 5 for the overall view). This Chap. 5 deals solely with the Research question 3, which is:

Research question 3: What types of information must be assigned to an *Industrie 4.0* component on the different layers to be covered by the virtual representation throughout all life cycle phases of a production system? Are these information types characteristic for the different layers?

In this chapter, description means for the different information types are identified. Out of them, a set of description means is selected potentially being able to represent all relevant information for an *Industrie 4.0* component and, thereby, being a starting point for the implementation of the digital shadow of *Industrie 4.0* components. Thus, it will be related to the research questions RQ M1 and C2 of Chap. 1.

In addition, as the identified information sets are based on the consideration of the different life cycle phases of a production system it will also contribute to research questions RQ M2.

As a limitation of this chapter, the management process to handle the information is not considered here. An analysis of the required management process can be found in Chap. 2. Another limitation of this chapter is the not considered management information, which is necessary to handle the management process that supports the use of I4.0 components. This topic is specially addressed in Chap. 6.

To systemically cover the life cycle of an I4.0 component, three main sections are distinguished. Section 6.2 addresses the engineering phase, Sect. 6.3 addresses the operation and maintenance phase and Sect. 6.4 addresses the end-of-life phase. Each of these sections includes life cycle specific cases (presented in Chap. 5) which envision certain possibilities by the use of I4.0 component information. This chapter ends with a summary and an outlook in Sect. 6.5. Based on the results of this chapter, the following Chap. 7 will focus further on the demands of the virtual representation of I4.0 components.

6.2 Engineering Phase

In this section, the identified relevant information of *Industrie 4.0* components (not to be confused with the Component Layer in Chap. 5) during the engineering phase is presented. The phase of engineering is the beginning of the production system life cycle, described in Chap. 5.

After having a deeper understanding of the demands and the use of *Industrie 4.0* components in engineering (see Chap. 5), in Sect. 6.2.1, the approach for the identification of relevant artifacts, generated in the engineering phase, is presented. In Sect. 6.2.2, the identified information needed for major engineering tasks is described. In Sect. 6.2.3 three cases are presented (see Chap. 5). These cases are examples of effective usage and new opportunities of cyber-physical *Industrie 4.0* components in engineering exploiting the available relevant information artifacts.

6.2.1 Approach for the Identification of Artifacts in the Engineering Phase

In order to identify the relevant types of information for an *Industrie 4.0* component, a detailed analysis of the engineering processes has been executed. The process analysis followed the method which is described in (Schäffler et al. 2013). This method addresses the special needs for the artifact identification in the engineering phase. As a result, the engineering processes have been modeled as a network of engineering activities executed by humans, creating and exchanging engineering artifacts, and exploiting engineering tools.

In a second step, the relevant information types have been mapped to the layers of the previously defined generalized hierarchical production system architecture within the engineering phase (see Chap. 5). Thus, the engineering information relevant for an *Industrie 4.0* component on a certain hierarchy layer, to be covered by a virtual representation, could be identified. The information types have been analyzed which resulted in the identification of main criteria that characterize each production system hierarchy layer for the engineering phase.

6.2.2 Identification Criteria for Artifacts in Engineering Phase

To answer the research question for engineering phase, the relevant engineering information has to be assigned to the identified different layers of the production system hierarchy. The results are based on the process analysis, mentioned in Sect. 6.2.1.

Here, the identified artifacts and tools as well as their assignment are presented. As there is a large amount of different engineering artifacts, only major artifact types will be named here. A more detailed description of the assignment can be found in (Hell et al. 2016).

6.2.2.1 Requirements

Initial requirements coming from the product design, like number of parts or specified joining processes, coming from economical departments like choice of location, and coming from legal authorities.

6.2.2.2 Layouts and Visualizations

Block layouts define the set of manufacturing resources at Work Unit Layer within a production system and put them in a logical interrelation.

2D layouts represent the placement of resources within a factory building. There are conceptual layouts, rough layouts and other 2D layouts for more detailed information, like a transport system related 2D conveyor system layout.

3D layouts provide a more detailed representation of the resources of a single Work Unit. They remain in a conceptual state covering Work Stations, Function Groups and their geographical locations. There are for example 3D rough layouts, 3D layouts including electronics.

6.2.2.3 Basic Specifications

The basic specifications contain general definitions of production system components. They cover for example the component quantity structures, interrelation structures between manufacturing processes and resource structures like cycle time.

6.2.2.4 Behavior Models

Behavior Models describe the production system behavior ranging from abstract models like Gantt charts down to simulation based decision models.

6.2.2.5 CAD Construction

The CAD construction contains detailed mechanical and electrical construction of the production system often named MCAD and ECAD.

The part list is a register of all components or elements of the production system which have to be purchased.

Simulation models are usually developed to validate availability and functionality of interacting components of production system, e.g. virtual commissioning or accessibility analysis. Control programs subsume the complete set of software developed to control the production system, e.g. human machine interfaces (HMI), programmable logic controllers (PLC), and robot programs.

The power supply concept represents the detailed engineering of the energy supply for elements of the production system.

The fluidic plans cover the engineering of the hydraulic and pneumatic systems. The safety concept contains the detailed engineering of relevant safety related features.

The identified information can be assigned to the different layers of the production system hierarchy by considering the engineering tasks within the engineering life cycles and the hierarchy levels they address. As a result the assignment presented in Fig. 6.1 can be concluded.

At the Construction Element Layer, the most detailed engineering information is relevant, i.e. part lists, mechanical and electrical specification, CAD construction, and electrical construction.

Information assigned to the Component Layer contains basic specifications and behavior models like joint locations, 3D layouts, part lists, mechanical and electrical specification, CAD construction, control programs, electrical construction, detailed behavior models, and simulation models.

At Function Group Layer, the engineering information is more abstract like basic behavior models, 3D layouts, specifications, control programs, safety concept, and simulation models.

At Work Station Layer, also rough and detailed engineering information can be mapped. Here, basic behavior models, 3D plans, 3D layouts, mechanical and electrical specifications, control programs, safety concept, detailed behavior models, and simulation models are relevant.

At Work Unit Layer, the detail level of engineering information decreases; mapped artifacts are basic behavior models, specifications, 3D layouts, and safety concepts.

At Production Line Segment Layer, only 2D layouts are still relevant. Finally, at Production Line Layer, requirements and 2D layouts are considered.

At Factory and Production Network Layer, the analysis has not provided engineering information of interest. Considering usual engineering processes, only requirements and economical and technical constraints might be relevant on these layers.

To summarize the research for the engineering phase, we executed an identification of relevant information and its assignment to *Industrie 4.0* components on

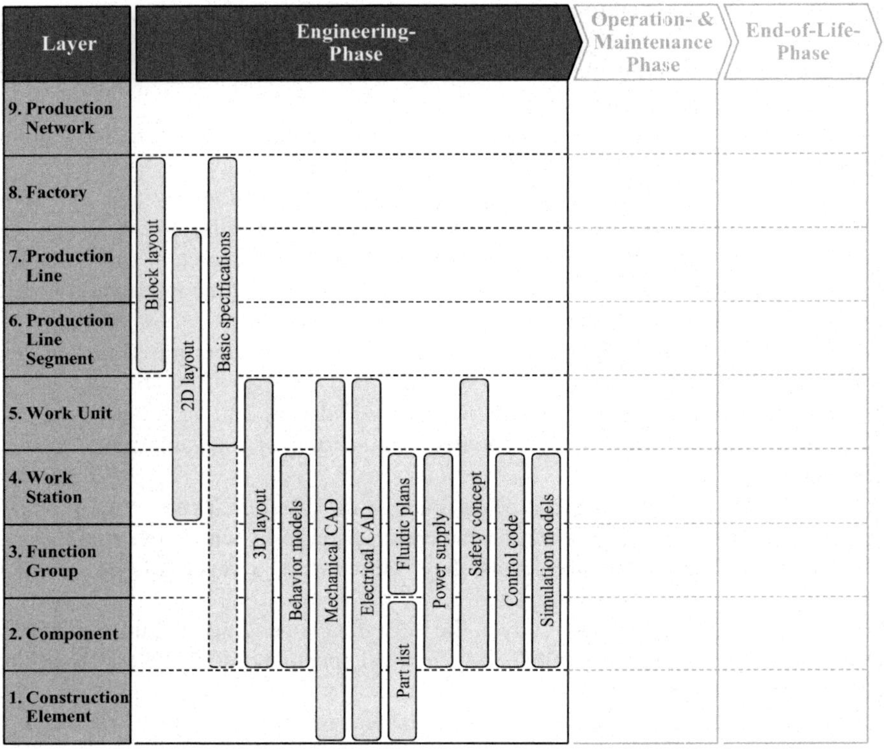

Fig. 6.1 Layer mapping of engineering phase artifacts

different hierarchy layers. Some information cannot be mapped to a single layer, but spans several ones. Nonetheless, there are characteristic sets of information types on the different layers relevant for use of *Industrie 4.0* components in the engineering phase.

6.2.3 Usage of Engineering Phase Artifacts

The engineering of production systems can be very complex and may cover several different engineering steps as indicated in (Lüder et al. 2011). It is out of the scope of this chapter to discuss the different versions of engineering processes in detail. A possible insight is given for example in Chaps. 2, 4, and 9 or in a more general view in (Lüder et al. 2011).

In this subsection, three cases are presented. All of them deal with the reuse of engineering artifacts (see cases 1–3 in Chap. 5). The difference is the direction of their information flows as well as sender and addressee.

Case no. 1: Reuse of Engineering Artifacts Within Engineering Phase
Case no. 2: Reuse of Engineering Artifacts in Engineering Phase of Other Projects
Case no. 3: Reuse of Engineering Artifacts in Operation and Maintenance Phase Within One Project or in Other Projects

The analyzed cases come from a sunroof assembly plant. The cases, covering the engineering phase, focus on the plant's robots with their related wear-data and other information artifacts.

Case no. 1: Reuse of Engineering Artifacts Within Engineering Phase
Considering the availability and correct assignment of the entire identified relevant engineering artifacts of each *Industrie 4.0* component on different production system layers and its information storage in a systematic repository, reuse of these information engineering artifacts becomes one promising case. Based on reuse of artifacts in software engineering (see Sametinger 1997), the method of reuse in mechatronic engineering as considered in (VDI 2010) becomes feasible by effective usage of related artifacts to cyber-physical *Industrie 4.0* components.

To exemplify the reuse of robot wear-data within the phase of engineering, which is available in the *Industrie 4.0* administration shell on Component Layer, the task of device approval is the first step to emphasize. The approval contains a test of devices and components under defined condition. Thus, an offering of devices, integrable in plant design, is made. Depending on operational purposes, the engineer chooses devices out of the offering. An operational purpose can be the amount of planned movements of a robot. Then, a risk analysis can be performed by comparing the planned movement during operational phase to the amount of movements successfully tested in the approval of devices or to the specified durability given by the robot supplier. In this case, the artifact of robot wear-data, especially maximal durability, is reused within engineering phase to approve the robot, to choose an adequate robot depending on its operation purpose and to perform a risk analysis based on this artifact.

Case no. 2: Reuse of Engineering Artifacts in Engineering Phase of Other Projects
Another case is the reuse of robot wear-data within the engineering phase, but another project. It is the same phase, but in another project with time offset, i.e. a subsequent project. If the artifact is available in the administration shell that contains the robot on Component Layer, this artifact can be reused to improve engineering time and quality. Availability includes the aspects data existence, systematical data storage, data validity, data traceability, data findability, and easy accessibility to relevant data in subsequent projects.

Exploiting the continuous availability of robot wear-data of a previous project's engineering phase, the decision on the robot type during Component selection in a subsequent project with comparable operation purposes becomes easier and more reliable. In addition, former results of a performed risk analysis can also be reused in subsequent projects to improve the process of device approval.

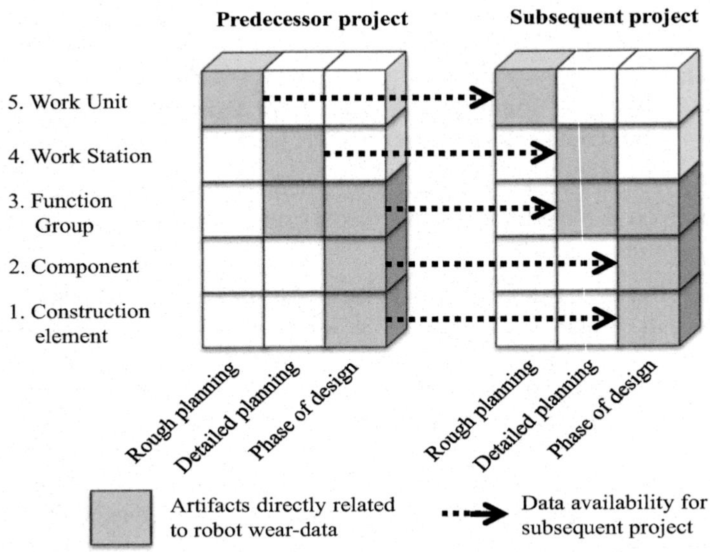

Fig. 6.2 Necessary data availability for *Industrie 4.0* components exemplified through robot wear-data

Figure 6.2 shows the layer-based reuse of wear-data related to the *Industrie 4.0* concept in a subsequent project in more detail. On Layers 4 and 5 the production process data is available to compare the original operation purpose to the one of a subsequent project. On Layer 3, information about Function Group Components is available. Thus, this information can be used for Component selection in a subsequent project. On Layers 1 and 2, approvals of Components and Construction Elements are available, which contain the tested and permitted maximum amount of movements. This information can be used in engineering of a subsequent project to support the phase of design. The date of necessary availability of relevant information in subsequent projects depends on the engineering task. Thus, information located on different production system layers is needed at different time slots, as shown in Fig. 6.2.

Case no. 3: Reuse of Engineering Artifacts in Operation and Maintenance Phase of Same and Other Projects

Case No. 3 deals with the information flow of engineering artifacts from engineering phase to operation phase (see case 3 in Chap. 5).

An example for this artifact reuse would be the transfer of a robot maintenance plan or of an instruction about change of robot gear related to its use in the operation phase. These artifacts are created in the engineering phase, but are reused in the operation phase in order to improve maintenance. The transfer of these artifacts within one project, from engineering to operation phase, is a common example of comprehensive information artifact reuse. The presented maintenance information could be stored in the administration shell of *Industrie 4.0* components, which are mapped to Component Layer (Layer 2).

A related case, and therefore included in No. 3, is the reuse of engineering artifacts of a predecessor project in the operation phases of subsequent projects, i.e. of other projects, recognizable by a dotted arrow (see case 3 in Chap. 5). An example is the transfer of revised instructions about change of a robot gear related to its use in operation phase. This revised instruction could be the result of a new robot strain analysis in the engineering phase of a subsequent project, which caused a modified robot approval. This information, which is created in a subsequent project, would improve the operation phase of a predecessor project.

6.3 Operation and Maintenance Phase

The virtual representation of *Industrie 4.0* components is designed to interpret and process specific control information appropriately, in order to perform a certain task during the operation and maintenance phase (VDI/VDE 2015). However, the type of control information may vary broadly, depending on the task being performed and the layer(s) it takes place on in the hierarchical production system structure (see Chap. 5). In this section, typical control activities and information will be identified and classified, that are required to enable *Industrie 4.0* components' full potential. As a result, a characterization of each layer of the hierarchical production system structure will be presented from a production control point of view.

Therefore, this section is structured as follows: First, Sect. 6.3.1 will give a brief overview of the approach chosen to identify the tasks performed within the operation and maintenance phase of a production system. Then, in Sect. 6.3.2, typical control information and the characteristics of each layer of the hierarchical production system structure will be described. Finally, in Sect. 6.3.3, a case from the automobile industry will be presented. It demonstrates that a single control decision may require information on numerous levels and that information artifacts can be used within and between different life cycle phases.

6.3.1 Approach for the Identification of Artifacts in the Operation and Maintenance Phase

The literature review in Sect. 6.4.2 has shown that control activities and the associated information artifacts in the operation and manufacturing phase are numerous and can vary greatly in their goals as well as in their behavior (e.g. real time requirements).

Against the background of these chapter's research questions, it is a requirement to assign the different control activities and the associated information to the layers of the hierarchical production system model proposed in Chap. 5.

To consider the different aspects of the automation pyramid (see e.g. Vogel-Heuser et al. 2013; Diedrich et al. 2011) a bandwidth of control information was taken into account. This includes control of sensors/actuators and processes on field level, the allocation of manufacturing resources in production planning systems (MES-Systems), as well as the coordination and optimization of enterprise processes in ERP-Systems (Lüder 2006). Furthermore, there is a vast number of implementation technologies available for control systems that allow the implementation of required functionalities into *Industrie 4.0* components (see e.g. Lunze 2008; Kiehl 2007; Kis 2014; Yang et al. 2014; Leitão and Karnouskos 2015), which were also considered in the analysis. With this in mind, the control activities applied in a representative set of companies in the manufacturing industry were analyzed and assigned to the different layers of the model. These companies include press shop, body shop, paint shop, assembly shop and in-house logistics in the automobile industry, a hot strip mill and a cross cutter in the steel industry, a stone crusher in the mining sector, a roof truss manufacturing system in the wood industry, a micro cuvette manufacturing system in medical technology, a logistics center, a solar park and a gas turbine facility, both in the power industry. In this context, it is important to notice that not all of the discussed control tasks and information are mandatory in all application scenarios and industries. Some of them can be left out, depending on the complexity and quality requirements of the product and production system.

The performed research results in an amount of control tasks and information that can be assigned to the different layers of the proposed hierarchy model, and are presented in the following subsection.

6.3.2 Identification Criteria for Artifacts in Operation and Maintenance Phase

An important identification criterion for the layers of the hierarchical production system structure is the corresponding control information of each layer. Since there are different tasks to be performed on each layer, the input, throughput and output information varies greatly. However, sets of control information can be identified that share the same sort of characteristics and therefore can be assigned to the various layers. These characteristics and typical types of control information of each layer are described in detail in this subsection.

Based on the literature review in Sect. 6.4.2 and a detailed analysis of existing manufacturing systems from the automobile sector, a broad spectrum of control tasks and information was identified and assigned to the layers of the hierarchical production system structure. The results are summarized in Tables 6.1, 6.2, 6.3, 6.4, 6.5, 6.6, 6.7 and 6.8.

Table 6.1 Control tasks and relevant control information on component layer

Layer	Control tasks	Control information
2. Component	– Switching operations	– Switch-on and -off commands
	– Condition monitoring and asset-management	– KPIs of process components
	– Communication of a representation of an activity status to surrounding systems (condition monitoring, asset management) – No communication within each element	– No control information

Table 6.2 Control tasks and relevant control information on function group layer

Layer	Control tasks	Control information
3. Function group	– Parametrization of process components	– Process parameters
	– Product identification	– Product ID data: vehicle identification number, order number
	– Condition monitoring and asset management	– Interpreted sensor and actuator data
	– Manual manipulation of specific functionalities (HMI)	– Robot control: trajectories, start and hold points, operating and breaking speed
	– Sensor and actuator control	– Sensor and actuator data

Table 6.3 Control tasks and relevant control information on work station layer

Layer	Control tasks	Control information
4. Work station	– Process coordination and supervision, consisting of...	
	• Simultaneous status detection of multiple function groups	– Status information
	• Simultaneous control of multiple function groups	– Control commands
	– Quality control and quality data acquisition, production status documentation	– Quality data
	– (Semi-)manual manipulation (HMI)	– Program fetch, manual control commands

The control activities on the various layers of the production system structure range from controlling individual equipment components, such as electric motors, welding guns and pumps, to planning of factory allocation and strategic production programs. They can be implemented using different approaches like multi-agent systems (MAS) or as a service in a service oriented architecture (SOA). For further details, see Chaps. 8 and 14. The characteristics of each layer and the associated control information are described in the following paragraphs, starting at the bottom layer (see Röpke et al. 2016; Zawisza et al. 2016).

Table 6.4 Control tasks and relevant control information on work unit layer

Layer	Control tasks	Control information
5. Work unit	– System operation and management (HMI)	– Cycle times, variant configuration, maintenance cycles, status information
	– Production and machine data logging	– Output, standalone availability KPIs, malfunction evaluation, energy consumption, NC data, tool data
	– Visualization at the level of the executed process	– Status and error signals, filling levels, cycle and process times

Table 6.5 Control tasks and relevant control information on production line segment layer

Layer	Control tasks	Control information
6. Production line segment	– Equipment control and supervision (HMI)	– Cycle time controlling: movement, process, and auxiliary times, availability and degree of utilization of chained up equipment, TPM-alarms, benchmarking
	– Material flow between work units, sequencing/order picking, route planning, transport control, physical resequencing	– Product variant specific production sequences, deviations of actual and target position, quality of pearl chain/order sequence
	– Material inventory management and stock call-offs	– Messages for incoming and outgoing goods, work in progress

Table 6.6 Control tasks and relevant control information on production line layer

Layer	Control tasks	Control information
7. Production line	– Utilization forecast and takt simulation	– Personnel requirement, shift scheduling, takt time deviations
	– Delivery instructions: internal and external short term call-offs, JIT and JIS call-offs	– Planned demand of next 5–15 workdays, material stock, operating times for suppliers, vehicle order status, order data
	– Monitoring of production lines (possibly HMI)	– Target and actual output, system status, (available/disturbed), buffer state for product variants

6.3.2.1 Construction Element

There are two forms of Construction Elements: active and passive ones. Active Construction Elements, in contrast to their passive counterparts, are able to represent a certain activity status that can be communicated to surrounding elements. However, communication within each element is not possible and therefore Construction Elements cannot be characterized by control parameters.

Table 6.7 Control tasks and relevant control information on factory layer

Layer	Control tasks	Control information
8. Factory	– Net dependent requirements determination, call-off forecasting	– Vehicle program, technical and sales restrictions, manufacturability, parts stock/availability
	– Weekly and daily production scheduling, detailed material supply planning, freight cost optimization, authorization of production, definition of manufacturing sequence	– Production line restrictions, product mix, line allocation, capital commitment costs, factory readiness
	– Factory monitoring and analysis, virtual resequencing	– Production relevant order data, factory KPIs

Table 6.8 Control tasks and relevant control information on production network layer

Layer	Control tasks	Control information
9. Production network	– Long term sales planning, volume planning of vehicles (primary requirements planning), strategic production program planning, factory allocation	– Volume of vehicles, investment and capacity plan, factory costs
	– Volume planning of vehicle properties, gross dependent requirements planning, demand and capacity management, bottleneck management	– Installation rates (especially of heavy items), internal and external capacities, costs and prices, network KPIs, profitability calculations
	– Product data management, order management, manufacturability testing	– Part-numbers, part status, primary properties numbers, technical rules and product hierarchies, sales rules

6.3.2.2 Component

Components can be distinguished between Process and Control Components. The latter provide driver functionalities for active Construction Elements. Process Components on the other hand feature a static or predetermined range of functions that can be parametrized and adapted to individual needs. The common element shared by process and Control Components is an I/O information processing system that is required to perform basic functionalities, like sensor and actuator control, switching operations, and diagnostics. Consequently, control information on this layer is characterized by sensor and actuator data, such as oscillation data as a function of time and voltage, switch-on and -off commands (e.g. of an electric engine in a conveyor belt), as well as Key Performance Indicators (KPIs) of Process Components like the stroke rate of a welding gun. An overview is shown in Table 6.1.

6.3.2.3 Function Group

The above mentioned basic processes usually require several Components to work together. The parametrization of these Components is realized by Function Groups. A typical example is the control of a welding robot in a car body welding shop, where the robot, the welding gun, and the cable assembly need to be parametrized in order to approach specific coordinates and place a welding spot. The welding current of the welding gun in Ampere and the holding period in seconds are controlled on this layer. Other examples for process parameters are the volume of a glue bead in volume per distance (dm^3/m) of a gluing application and the contact pressure in force per surface (N/mm^2) of a suction gripper.

Due to the diverse product mix manufactured in modern production systems, an important function is also product identification, such as vehicle identification numbers and separate order numbers. They can e.g. trigger a status message at a capture point or inform the equipment about the next to build product variant.

Furthermore, condition monitoring and asset management are performed on Function Group Layer as well, in order to maintain equipment in the desired state. Relevant control information in this context are interpreted sensor and actuator data, like e.g. the oscillation information of rolling bearings and oil data of a top or bottom part of a drawing tool in press shop. Finally, HMIs allow the execution of functionally specific tasks on this layer. In the example of robot control, this may include changing robot trajectories, start and hold points, as well as operating and breaking speed. An overview is shown in Table 6.2.

6.3.2.4 Work Station

Work stations integrate multiple Function Groups in order to successfully perform a manufacturing process. This circumstance makes coordination and (automated) supervision indispensable and requires work stations to simultaneously control the process parameters of the participating Function Groups. For this, both, simultaneous state determination and control of the subordinate Function Groups are needed. As a result, status information and control commands are required at the same time. Using the example of a car body welding shop, the necessary status information includes the location of the welding gun ("defined coordinates reached") and the status of the clamping system holding down the parts ("clamping elements closed"). Corresponding control commands are e.g. "keep holding down clamps" and "initiate welding current". This kind of control information allows further functionalities like quality assurance and data logging of control parameters that are required for quality and warranty claim purposes. Typical examples are parameters like torque, angle, and cycle time of bolting sequences for security-related connections, volume and speed of adhesive application, as well as coordinates, deviations, and tolerances of car body geometry. An overview is shown in Table 6.3.

Similar to the Function Group Layer, it is also possible to manipulate work stations (semi-)manually through HMIs. The control information is associated to manual control commands and program fetches, e.g. to reset a work station to home position after a malfunction.

6.3.2.5 Work Unit

The process steps, which are determined through the product itself, are being realized by sequencing the manufacturing steps of multiple Work Stations on the higher level of Work Units. Consequently, system operation and management are major activities on this layer. They deal with process information such as cycle times, variant configurations, maintenance cycles, and status information of the Work Unit as a whole. Furthermore, supervision functionalities including production and machine data logging are performed on this layer, resulting in control information like output, availability, malfunction evaluation, and energy consumption, as well as NC and tool data. Availability, however, is measured for stand-alone Work Units only and includes KPIs like Overall Equipment Effectiveness (OEE), Mean Time To Repair (MTTR) in seconds, and Mean Time Between Failures (MTBF) in hours. Other characteristic control parameters are status and error signals (e.g. welding gun: "operational/disturbed", photoelectric safety switch: "open/interrupted"), filling levels of process and auxiliary materials, as well as cycle and process times. Usually this data is gathered and visualized for the system operator, which is the third functionality on Work Unit Layer. An overview is shown in Table 6.4.

6.3.2.6 Production Line Segment

Main activities on Production Line Segment Layer are divided into three categories. The first category is concerned with operational equipment control and supervision. This includes controlling activities like tracking of cycle time, process times, auxiliary process times, and movement times. Furthermore, it contains monitoring the availability and utilization of chained equipment, which allows setting of Total Productive Maintenance (TPM)-alarms in case of deviations and creating benchmarks, e.g. of energy consumption of Work Units in the Production Line Segment. The second category of control activities is built around material flow. An overview is shown in Table 6.5.

This means material flow between work stations, as well as supply of subassemblies connected to them. Supporting activities like order picking, route planning (e.g. of tugger trains), transport control, and physical resequencing of orders are also included. Those activities are characterized by control information of the production sequence and the included product variants (e.g. body shop or assembly line sequence), deviations between target and current position, as well as quality of pearl chain or other sequence measuring KPIs. The last category of activities includes material inventory management and stock call-offs. Typical parameters are signals of incoming and outgoing goods as well as work in progress.

6.3.2.7 Production Line

The segments on Layer 6 are grouped on Production Line level. Accordingly, control activities on this level are of general character. The main planning activities include forecasting utilization and simulating cycle time on the basis of the scheduled production program. Typical control parameters are personnel requirements, shift planning schedules, and cycle time deviations between different orders in the product mix that have to be balanced. The operational part of the activities on this layer is concerned with delivery instructions for internal and external suppliers (short term call-offs) as well as just-in-time (JIT) and just-in-sequence (JIS) call-offs. Required control information features, among others, short term planned demand (next 5–15 workdays), material inventory, and supplier timing schedules. Since JIS call-offs require knowledge about the exact status and timing of the products they are made for, vehicle order status and order data like technical features of the order are required as well.

Finally, monitoring of the entire production line is part of the control activities on this layer, too. Consequently, predominant control parameters are target and actual output, the overall system status (available/disturbed), as well as buffer states for the product variants in the current production program. An overview is shown in Table 6.6.

6.3.2.8 Factory

The Factory Layer combines the processes of multiple production lines in order to produce a complete product. To achieve this, three types of control activities are required on this layer. First, there are tactical planning activities ensuring adequate parts supply in the short term (weeks to months). Important representatives of those activities are net dependent requirements determination and call-off forecasting. The related control parameters are, among others, the production program of the vehicles planned, the technical and sales restrictions, manufacturability, as well as material stock and parts availability. Second, the level is characterized by operational preparatory and short term production planning activities (hours to days). They include breaking down the production program to weeks and days, detailed material supply planning, as well as authorization of production and definition of the manufacturing sequence (e.g. start of body shop sequence or color batches for paint shop). Typical control information includes production line restrictions, product mix and production line assignment, if there are multiple options to produce a product. Third, operational supervision activities aiming at ensuring production effectiveness and efficiency can be found on factory level as well. Monitoring and analysis of factory KPIs are summarized under this category, as well as virtual resequencing of orders throughout the factory. Relevant control parameters for the former contain cost and manufacturing KPIs such as lead time, delivery reliability, production costs, and pearl chain/order sequence quality, as well as quality KPIs like first pass yield and deviations. The later include information about product variants and the combination of primary properties numbers. An overview is shown in Table 6.7.

6.3.2.9 Production Network

At the top of the production system structure lays the Production Network Layer. In contrast to the mainly operative control activities at factory level, the production network is mostly designated to strategic and tactical planning activities with a horizon of up to 10 years ahead. On the one hand, this includes activities such as long term sales planning, strategic production program planning, and factory allocation.

On the other hand, more detailed activities like planning of vehicle properties considering heavy items, demand and capacity management, as well as bottleneck management of internal and external resources are performed on this layer. These activities are associated with control information such as volume of vehicles per country and year, investment and capacity plans, factory costs, as well as installation rates of technical features, internal and external capacities, network KPIs (e.g. delivery reliability, indirect procurement costs, etc.), and profitability calculations (e.g. sales costs, direct costs, etc.).

Most operational tasks on Production Network Layer, however, are connected to data management. This is a task that has to be performed network-wide, since the products or subassemblies of the network may be produced or used in multiple factories around the world. This is especially relevant if multiple products are using carry over parts from other products inside the network. Consequently, typical activities are product data management, client order management, and manufacturability testing. The control information needed for these activities include part numbers or primary properties numbers for complicated and individualized parts like dashboards, part status (valid/invalid) of each part, technical rules and product hierarchies, as well as sales rules that allow or prohibit certain configurations.

In summary, the analysis shows that the spectrum of control activities and parameters required during Operation and Maintenance Phase of a production system is very broad. An overview of the information artifacts in this phase can be found in Fig. 6.3. Although the control activities and the related parameters are linked through a complex network of decisions, it is possible to analyze them closer by narrowing the field of observation to (quasi-)separate decisions. The following subsection demonstrates this by analyzing a case from the automobile industry. An overview is shown in Table 6.8.

6.3.3 Usage of Operation and Maintenance Phase Artifacts

The various control activities and related parameters in a production system are linked in a complex network of interactions and can spread throughout all levels of the hierarchical production system structure. This subsection illustrates how

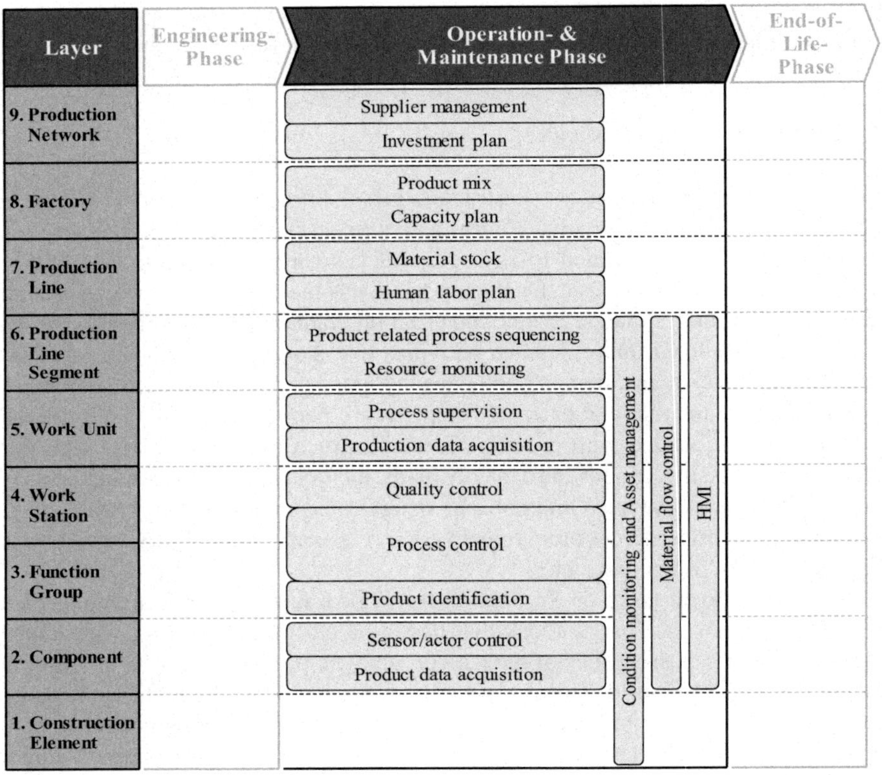

Fig. 6.3 Information artifacts during operation and maintenance phase

certain information artifacts can be used, both within and between different life cycle phases. To pick up the example of Sect. 6.2.3, the focus is put on robot wear data as an information artifact. For this purpose, first of all the control process of the case, a sunroof assembly in an automobile assembly shop, will be described. In a second step, the three cases of artifact reuse in Operations and Maintenance Phase will be analyzed (see Chap. 5, cases 4–6). These cases are:

Case no. 4: Reuse of Artifacts Within Operation and Maintenance Phase

Case no. 5: Reuse of Artifacts in Engineering Phase of Other Projects (Feedback-Loop)

Case no. 6: Reuse of Artifacts in Operation and Maintenance Phase of Other Projects

The regarded case consists of a sunroof assembly process with an adjacent route planning and transport control process for tugger trains. A similar configuration can be found in state of the art automobile factories around the world (see Fig. 6.4).

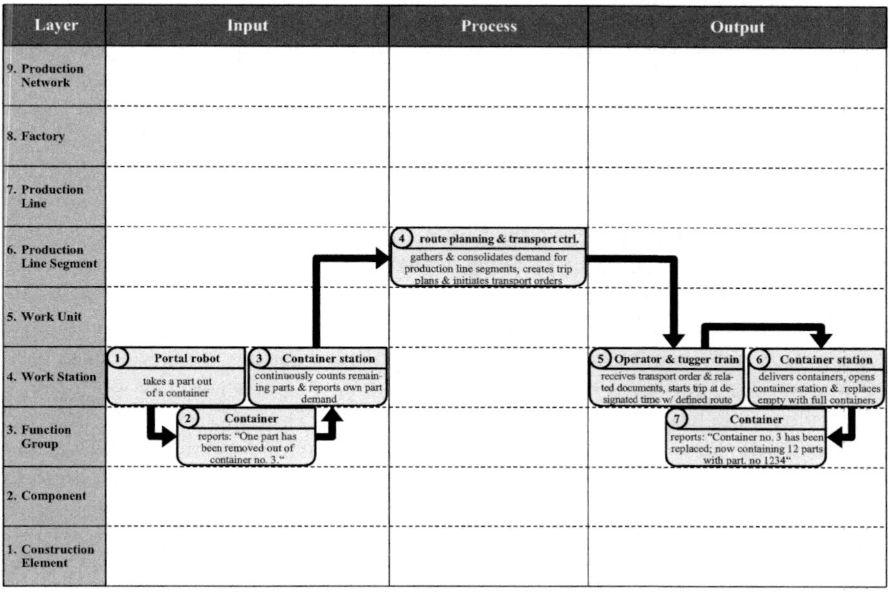

Fig. 6.4 Control process for route planning and transportation control for restocking parts for sunroof assembly

For a more complex case see (Zawisza et al. 2016), where a Just-in-time call-off between an automobile OEM and a supplier is described.

The control process of the sunroof assembly begins with a robot taking a sunroof out of a container and installing it into a car on the assembly line with the help of other robots and workers. This process takes place on Work Station Layer of the production system structure (Layer 4). Typically, there is no direct information exchange between the robot and the container. However, the robot can be equipped with a sensor, enabling it to determine how many sunroofs are left inside the container. Usually this is achieved using a photoelectric sensor. The robot can, thus, report to the container station that a sunroof has been removed (Layer 3, Function Group). The container station on the Work Station Layer holds various containers for different product variants. It continuously counts the total remaining sunroofs for each product variant and, when necessary, orders new sunroofs. The call-off for required parts usually contains information about the type of container to be replaced, the part-no. of the demanded parts and the time they have to be delivered (e.g. "Container no. 3 is empty; deliver additional parts with part-no. 1234; remaining parts last for 24 min. of production").

Since a tugger train usually doesn't supply only one installation at a time, the part orders are collected on a higher layer, i.e. the Production Line Segment Layer. Here the information considering demand is gathered and consolidated for the whole production line segment and this is also where the route planning and transport control takes place. Using the consolidated information on this layer, a software algorithm calculates an ideal trip plan and creates a schedule for each available tugger train, and then initiates the transport order. In the next step, this order is being forwarded to the individual tugger trains and their operators on Work Station Layer. It contains all the information the operator needs, including a bill of materials, a route plan, a schedule etc. The operator then starts the trip at the scheduled time and uses a defined route to fulfill the assignment. He or she then delivers the containers to the defined location at the defined time, so that a bottleneck is avoided. Then the operator has to open the container station and replace an empty with a full container. This occurs on Work Station Layer and implicates interaction between the operator and the container station using the HMI of the latter. Finally, the container reports back to the container station that the demanded part has been restocked (e.g. "Container no. 3 has been replaced; now containing 12 parts with part no. 1234"), which closes the loop.

To describe how information artifacts of Operation and Maintenance Phase can be reused, the following paragraphs distinguish between reuse within Operation and Maintenance Phase (case no. 4) and reuse in other phases of different projects (cases no. 5 and 6).

Case no. 4: Reuse of Artifacts Within Operation and Maintenance Phase
As shown above, a large quantity of information is generated and collected during the described processes in Operation and Maintenance Phase. However, some necessary information in this phase can neither be measured directly via the systems sensors nor can it be retrieved from its actuators and the corresponding information flow. This applies for robot wear and tear data, too. As a workaround, the existing information artifacts of Operation and Maintenance Phase can be reused in a way that they serve as a direct indicator for a robots wear and tear.

As an example the number of assembled sunroofs in the above case is equal to the number of load cycles performed by the robots in the work station. Therefore, robot wear and tear can be measured indirectly by counting the load cycles of a certain task a robot performs using the integrated counter. In this case, maintenance personnel can compare the current number of load cycles with the estimated lifetime from the engineering Phase and, based on this information, decide whether an inspection or a part replacement is needed. In an *Industrie 4.0* scenario this information is stored in the administration shell and therefore the *Industrie 4.0* component is able to trigger an inspection or part replacement by itself. Furthermore, based on the data from the engineering Phase, it is able to predict for how long it will be able to perform its tasks given the specific requirements like speed and precision, allowing predictive maintenance and planned stops.

On the other hand, the load cycle data, usually provided by a counter, can also be reused in a different context. In the sunroof assembly case for example it allows triggering a call-off for the timely delivery of required parts to a work station and thereby enables the *Industrie 4.0* component to control its part supply and related activities.

Another example for reuse of artifacts in this life cycle phase is the ability of an *Industrie 4.0* component to perform process or quality control by itself. This is possible through comparing data from Engineering and Operation and Maintenance Phase and correcting occurring deviations, e.g. checking the correct positioning of a welding spot or the precise assembly of a car sunroof.

Case no. 5–6: Reuse of Artifacts in Engineering and Operation and Maintenance Phase of Other Projects
The artifacts generated in Operation and Maintenance Phase are the basis for an effective and efficient production process within the life cycle phase itself. However, those artifacts can have a deep impact on other life cycle phases and other projects as well.

On the one hand, the information generated during production can be reused in the engineering Phase of other projects to ensure the responsible engineers get a reliable feedback on their work and thereby improve future engineering decisions in new projects. For instance, if an *Industrie 4.0* component in the above scenario detects a deviation in the precision of the sunroof assembly over time (e.g. due to wear and tear), this information should be reused in the engineering Phase to improve future production systems and their design. This methodology can be applied to continuously validate various engineering specifications over the lifetime of a production system and implement necessary adjustments.

On the other hand, the artifacts used during production can also be a valuable input during the Operation and Maintenance Phase of other projects in a company. Since different factories or dependencies of a company may experience very different conditions (e.g. humidity, temperature, production schedule, product types, etc.) an information exchange across factory limits could be of great interest. Therefore, the information artifacts generated in Operation and Maintenance Phase should be available throughout a company and even between different cooperating companies in order to continuously improve performance. This can include information like a malfunction log, best practices for maintenance and quality assurance purposes, and version management to handle different combinations of hardware with a different firmware status. In the case of robot wear and tear, factories can benefit greatly from each other's experience by exchanging and analyzing the development of observed behavior in different locations over time. As a consequence, parameters like cycle times, velocities, and maintenance cycles of equipment could be adjusted on the base of a much larger database than it is today, so that problems can be detected earlier and have less negative effect on productivity.

6.4 End-of-Life Phase

This section intends to identify relevant requirements and information emerging from End-of-Life (EoL) phase which need to be covered by an *Industrie 4.0* component to be applicable in this life cycle phase. The EoL phase is the last phase of the production system life cycle, described in Chap. 5.

After having a closer look at the demands and use of *Industrie 4.0* components in EoL (see Chap. 5), Sect. 6.4.1 will describe the approach used to gather relevant requirements and information necessary within the EoL phase. The requirements, identified as relevant for the EoL phase, will then be mapped to each hierarchy layer (see Sect. 6.4.2), which will result in a characterization of each layer of the hierarchical production system structure from the EoL point of view. In Sect. 6.4.3, cases are presented which shall illustrate the benefits and opportunities when *Industrie 4.0* components are used in the EoL phase.

6.4.1 Approach for the Identification of Artifacts in the Engineering Phase

To identify requirements emerging from EoL phase, a literature survey on different recovery (including recycling and reuse), disassembly, EoL, and EoL scenarios related publications was chosen. The survey resulted in over one hundred requirements for the EoL phase, which have been categorized into the following categories, representing a 'production system level of detail' according to VDI 2243 (VDI 2002): General requirements (G), production system specific requirements (P), component specific requirements (C), and material specific requirements (M).

In a next step, the categories were mapped to each hierarchy layer for a characterization of each layer of the hierarchical production system structure (see Chap. 5) from the EoL point of view. Using these categories—representing a 'production system level of detail'—as identification criterion for an *Industrie 4.0* component on the different layers of the hierarchical production system structure is an abstract but viable approach.

6.4.2 Identification Criteria for Artifacts in End-of-Life Phase

A possible identification criterion for the layers of the hierarchical production system structure is the corresponding 'production system level of detail' named in Sect. 6.4.1: General requirements (G), production system specific requirements (P), component specific requirements (C), and material specific requirements (M).

Examples for general requirements are:

- Compliance with statutory provisions (Ruhland 2006; Steinhilper and Rieg 2012; VDI 2002; Huber 2001; Industrie 2016; Schiffleitner et al. 2012), or
- Avoidance of environmental impact (VDI 2002; Hartel and Spath 1994).

Examples for product specific requirements are:

- Description of production system and connection structure—(Steinhilper and Rieg 2012; PAS 2004; VDI 2002; Schultmann et al. 2002; Pahl et al. 2007; Hubig

2001; Ruhland 2006; Feldmann et al. 1999; Rosemann et al. 1999; Duflou et al. 2008; Hartel and Spath 1994; Simolowo and Onovughe 2013; Huber 2001; Willems et al. 2003), or
- Application of a functional and modular structure—(Steinhilper and Rieg 2012; VDI 2002; Pahl et al. 2007; Hubig 2001; Ruhland 2006; Duflou et al. 2008; Willems et al. 2003).

Examples for component specific requirements are:

- Enabling of an easy and non-destructive disassembly structure—(Steinhilper and Rieg 2012; Pahl et al. 2007; VDI 2002), or
- Enabling of remanufacturing processes—(Steinhilper and Rieg 2012; Pahl et al. 2007; Hubig 2001; Duflou et al. 2008; Obst et al. 2013; Lindahl et al. 2006).

Examples for material specific requirements are:

- Identification of material (also regarding hazard potential)—(Steinhilper and Rieg 2012; PAS 2004; VDI 2002; Pahl et al. 2007; Hubig 2001; Ruhland 2006), or
- Documentation of deterioration occurred during use time of the production system—(Schultmann et al. 2002; Hubig 2001; Ruhland 2006; Duflou et al. 2008; Feldmann et al. 1999; Simolowo and Onovughe 2013; Obst et al. 2013; Huber 2001; Seliger et al. 2001).

Using the categorized requirements, the assignment of those categories to each hierarchy layer within the hierarchical production system structure was done. The categories reflect the level on which the decision has to be made whether a requirement is fulfilled or not. For this, information is necessary—see Fig. 6.5.

General information (G) has to be provided on Layer 8 and 7. On Layer 8 this could be complying with specific laws and restrictions, like determination of mandatory disassembly amounts, incineration with energy recovery, or CO_2 emissions. On Layer 7 this could be complying with current component/material/substance prohibitions related to the applied manufacturing technology of the Production Line.

To finally provide information, this might need to be broken down to lower layers. But the aggregation of information of the lower layers has to be done on the superordinated layer.

Production system specific information (P) can be found on Layer 8, 7, 6, 5, down to Layer 4, since the structure of the overall production system is influenced by the design of each *Industrie 4.0* component on each of those layers, mostly following a functional and modular design approach. This information is relevant mainly for EoL 3—Production System Recovery (see Chap. 5, Fig. 5.9).

Component specific information (C) is placed on Layer 5 down to Layer 2—Work Units and their structure are usually engineered in their entirety. Here, an easy disassembly as well as remanufacturing process is important, so that the EoL scenarios can be enabled. This information is relevant mainly for EoL 2—Component Recovery (see Chap. 5, Fig. 5.9).

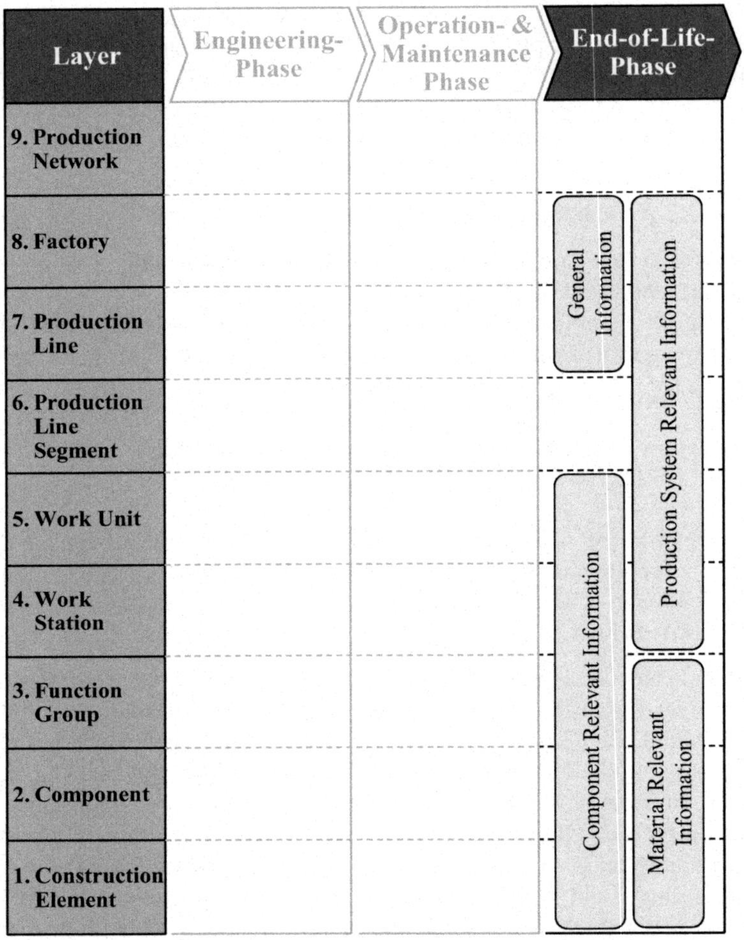

Fig. 6.5 Categories of 'production system level of detail' as identification criterion assigned to hierarchical production system structure

Material specific information (M) can be found on Layer 3, 2, and 1, because information about material with hazard potential is relevant here—in general the material identification. This information is relevant mainly for EoL 1—Material Recovery (see Chap. 5, Fig. 5.9).

Besides the loops which represent the recovery of physical artifacts coming from the EoL phase, there are also loops which represent the reuse of information artifacts emerging from that phase. To also describe these the production system life cycle is reduced back to the life cycle which is described in Chap. 5—only considering engineering phase, operation and maintenance phase, and EoL phase.

In doing so, the scenarios EoL 1 and EoL 2b (see Chap. 5, Fig. 5.9) are excluded on behalf of clarity.

Both types of loops are explained with cases in the following subsection.

6.4.3 Usage of End-of-Life Phase Artifacts

In this subsection two main cases are presented (see cases 7–12 in Chap. 5) which illustrate the benefits and opportunities when *Industrie 4.0* components are used in the EoL phase. They deal with recovery of physical artifacts and reuse of information artifacts.

Case no. 7: Reuse of Artifacts from Operation and Maintenance Phase
Case no. 8: Reuse of Artifacts from Engineering Phase
Case no. 9: Reuse in Operation and Maintenance Phase of Same or Other Projects
Case no. 10: Reuse in Engineering Phase of Same or Other Projects
Case no. 11: Reuse within End-of-Life Phase
Case no. 12: Reuse in End-of-Life Phase of Other Projects

The cases focus on industrial robots with their related wear-data and other information artifacts.

Case no. 7–10: Reuse Artifacts from the Same Project and Reuse It in Another Project

For this case the EoL scenario Component Recovery (EoL 2a) is chosen to illustrate how a component could be recovered, when the production system is decommissioned.

Component Recovery can be found in the automotive industry, e.g. when a series expires and the corresponding production system becomes obsolete. This production system is then decommissioned, disassembled, and used components can be utilized, on the one hand to equip other production systems in other factories at other locations (e.g. with a different layout producing a different product); or on the other hand to store the used components as spare parts in case that those types of components are still in use within the company, so that a provision of these spare parts is reasonable.

According to Fig. 6.5 in Sect. 6.4.2, Component Recovery can be found on Layer 5 to 2 of the generic production system architecture. For this case, Layer 3 "Function Group" is chosen, in particular an industrial robot. Given that the EoL relevant information for the robot is stored in its administration shell, the robot knows how it can get decommissioned, disassembled, and how it can get recovered based on its current state.

In case the robot has evaluated by itself that it can be reused or further used, the robot could provide the following information and guidance/recommendations to the personnel involved in the EoL phase:

- The robot needs to be decommissioned this way.
- The robot can get disassembled in xx minutes by undoing xx connections.
- Due to the robot's time in use it should be cleaned/maintained.
- This part of the robot can get replaced: This would be the optimal trade-off between costs to invest and life span expansion.
- Due to robot's actual wear status it cannot fulfil the 'welding skill' anymore. It is too worn out. Instead the robot could 'insert workpieces'—for this the robot can be less precise.

This self-description of the robot and its evaluation, that it can get used a second time, is again the input for engineering tools of the engineering phase of other projects (see Chap. 5, Fig. 5.4) on the one hand. Here, engineers could select this self-description from a component library in which the virtual representation of these real components is stored. Even construction site management tools could use this self-description—not for erection (Dreher et al. 2013) but for removal. On the other hand, this self-description could be used to automatically evaluate if the robot meets the requirements of being a spare part for a specific, already operating production system.

This case deals with recovery of physical artifacts and reuse of information artifacts. In addition to the presented cases, in (Schmidt et al. 2016), a generic model for EoL scenarios of production systems is developed as a basis to enable the EoL phase being supported by future software tools.

Case no. 11–12: Reuse End-of-Life Phase Experience
This case deals with reuse of information artifacts. These artifacts contains the documentation of experiences gained by the personnel during the EoL phase of a production system, which are either used within the same EoL phase (to optimize workflows, disassembly activities etc.) or in EoL phases of other production systems.

6.5 Summary and Outlook

In this Chap. 6, the life cycle of a production system has been analyzed in detail to identify artifacts covering all information sets relevant for the different layers of production system hierarchy. Based on this, a mapping of information sets to hierarchy layers and vice versa has been presented, being able to represent all relevant information for an *Industrie 4.0* component and, thereby, being a starting point for the implementation of the digital shadow of *Industrie 4.0* components. Thus, it is providing answers to research questions RQ M1 and C2 of Chap. 1.

To be able to provide these answers, the production system life cycle was analyzed in three main phases: the engineering phase, the operation and maintenance phase and the end-of-life phase. For each phase, a suitable approach was selected to identify the relevant artifacts. The types of information could be identified for the analyzed automotive industry. Also, the relevant information for each layer could be detected.

As a result, the identified artifacts as well as the information contained differ dependent of the granularity of an I4.0 component and the point of time at which the artifacts are generated or are used. There is no holistic concept that is valid over all life cycle phases. The sum of the types of information and its granularity are characteristic for the different layers, but one kind of information type can appear on different layers. The artifacts types of information, in order to virtual represent an I4.0 component at a certain layer, are various. Due to the additional information needed for future CPPS, the demands on the administration shell to handle the various types, will become even higher.

To expand the validation of the conducted research, an identification of relevant types of information, dependent on the different layers and the point of time, could be undertaken in additional industries. Here, the process industry, the energy industry or batch production can be mentioned.

The dependencies between the information artifacts were not considered in this chapter, but it is undisputed, that there are many links in between. In Chap. 12, dependencies between the artifacts will be addressed.

Another important aspect in order to effectively use the information artifacts of an I4.0 component is the quality assurance of the data. It is clear that the data must be correct and complete. Chap. 16 deals particularly with the topic of data quality assurance, missing in the Chap. 6.

Based on the results of this chapter, the following Chap. 7 analyzes the description means necessary to hold the information throughout the life cycle of CPPS.

References

Diedrich, C., Lüder, A., Hundt, L.: Bedeutung der Interoperabilität bei Entwurf und Nutzung von automatisierten Produktionssystemen. Automatisierungstechnik. **59**(7), 426–438 (2011)

Dreher, S., Nürnberger, A., Kägebein, S., Schoch, A.: Digitiales Baustellenmanagement für Produktionsanlagen. ZWF Jahrg. **108**, 1–2 (2013, in German)

Duflou, J.R., Seliger, G., Kara, S., Umeda, Y., Ometto, A., Willems, B.: Efficiency and feasibility of product disassembly – a case-based study. CIRP Ann. Manuf. Technol. **57**(2), 583–600 (2008)

Feldmann, K., Trautner, S., Meedt, O.: Innovative disassembly strategies based on flexible partial destructive tools. Annu. Rev. Control. **23**, 159–164 (1999)

Hartel, M., Spath, D.: Öko-Portfolio: Methode zur Beurteilung der Recyclingeignung technischer Serienprodukte, pp. 371–392. VDI-Berichte, 1171, VDI Düsseldorf (1994, in German)

Hell, K., Hillmann, R., Lüder, A., Röpke, H., Zawisza, J., Schmidt, N., Calà, A.: Demands on the virtual representation of physical Industrie 4.0 components. In: Conferenza INCOSE Italia su Systems Engineering (CIISE 2016), Turin, Italy, 14–16 November 2016

Huber, A.: Demontageplanung und -steuerung: Planung und Steuerung industrieller Demontageprozesse mit PPS-Systemen. Dissertation, Otto-von-Guericke University Magdeburg (2001, in German)

Hubig, M.-A.: Erarbeitung einer Methode zur deduktiven Ableitung Strategischer Recyclinganforderungen mit Hilfe der Szenariotechnik. Seminar Paper, University Kaiserslautern (2001, in German)

AG Forschung and Innovation (Plattform Industrie 4.0): Aspekte der Forschungsroadmap in den Anwendungsszenarien. Report, BMWi (2016, in German)

Kiehl, E. (ed.): Antriebslösungen – Mechatronik für Produktion und Logistik. Springer, Berlin (2007)

Kis, T.: Planning and scheduling in the digital factory, KOMSO challenge workshop – math for the digital factory. In: Proceedings, Berlin, Germany, May 2014

Leitão, P., Karnouskos, S.: Industrial Agents: Emerging Applications of Software Agents in Industry, pp. 153–170. Elsevier, Waltham, MA (2015)

Lindahl, M., Sundin, E., Östlin, J., Björkman, M.: Concepts and definitions for product recovery, analysis and clarification of the terminology used in academia and industry. In: Brissaud, D., et al. (eds.) Innovation in Life Cycle Engineering and Sustainable Development, pp. 123–138. Springer (2006)

Lüder, A., Foehr, M., Hundt, L., Hoffmann, M., Langer, Y., Frank, St.: Aggregation of engineering processes regarding the mechatronic approach. In: 16th IEEE International Conference on Emerging Technologies and Factory Automation (ETFA 2011), Toulouse, France, Proceedings-CD, September 2011

Lüder, A.: Strukturen zur verteilten Steuerung von Produktionssystemen, Habilitationsschrift. Fakultät Maschinenbau, Otto-von-Guericke Universität Magdeburg (2006)

Lunze, J.: Automatisierungstechnik – Methoden für die Überwachung und Steuerung kontinuierlicher und ereignisdiskreter Systeme. Oldenbourg Verlag, München (2008)

Obst, M., Holm, T., Bleuel, S., Claussnitzer, U., Evertz, L., Jäger, T., Nekolla, T.: Automatisierung im Life Cycle modularer Anlagen: Welche Veränderungen und Chancen sich ergeben. atp edition. **55**(01–02), 24–31 (2013, in German)

Pahl, G., Beitz, W., Feldhusen, J., Grote, K.-H.: Engineering Design – A Systematic Approach. Springer, London (2007)

PAS 1049: Transmission of Recycling Relevant Product Information Between Producers and Recyclers – The Recycling Passport. Beuth Verlag, Berlin (2004, in German)

Röpke, H., Lüder, A., Hell, K., Zawisza, J., Schmidt, N.: Identification of "Industrie 4.0" component hierarchy layers. In: Submitted to IEEE International Conference on Emerging Technologies and Factory Automation (ETFA 2016), Berlin, Germany, September 2016

Rosemann, B., Meerkamm, H., Trautner, S., Feldmann, K.: Design for recycling, recycling data management and optimal end-of-life planning based on recycling-graphs. In: International Conference on Engineering Design (ICED 1999), pp. 1–6, Munich, Germany, 24–26 August 1999

Ruhland, K.: Methoden und Werkzeuge zur recyclinggerechten Automobilentwicklung. Dissertation, University Kaiserslautern (2006, in German)

Sametinger, J.: Software Engineering with Reusable Components. Springer, New York (1997)

Schäffler, T., Foehr, M., Lüder, A., Supke, K.: Engineering process evaluation – evaluation of the impact of internationalisation decisions on the efficiency and quality of engineering processes. In: 22nd IEEE International Symposium on Industrial Electronics (ISIE 2013), Taipei, Taiwan, Proceedings, May 2013

Schiffleitner, A., Bley, T., Schneider, R., Wimpff, D.-P.: Stakeholder perspectives on business model requirements for a sustainability data exchange platform across supply chains. In: Joint International Conference and Exhibition "Electronics Goes Green", pp. 9–12, Berlin, Germany, Proceedings, September 2012

Schmidt, N., Lüder, A., Hell, K., Röpke, H., Zawisza, J.: A generic model for the end-of-life phase of production systems. In: IECON 2016: The 42nd Annual Conference of IEEE Industrial Electronics Society, Florence, Italy, 24–27 October 2016

Schultmann, F., Fröhling, M., Rentz, O.: Demontageplanung und -steuerung mit Enterprise-Resource- and Advanced-Planning-Systemen. Wirtschaftsinformatik (WI-Aufsatz). **44**(6), 557–565 (2002, in German)

Seliger, G., Basdere, B., Keil, T.: e-Cycling platform for profitable reuse. In: IEEE International Symposium on Assembly and Task Planning, Fukuoka, Japan, Proceedings, 28–29 May 2001

Simolowo, E., Onovughe, E.: Automation of generation of models for disassembly process planning for recycling. In: World Congress on Engineering 2013, vol. III (WCE 2013), London, UK, Proceedings, 3–5 July 2013

Steinhilper, R., Rieg, F. (eds.): Handbuch Konstruktion. Carl Hanser Verlag, Munich (2012, in German)

VDI 2243: Recycling-Oriented Product Development. Beuth Verlag, Berlin (2002)

VDI/VDE—GMA—Fachausschuss 7.21: Industrie 4.0. Referenzarchitekturmodell Industrie 4.0 (RAMI4.0). https://www.vdi.de/fileadmin/user_upload/ (2015). Last access 25 Oct 2016

VDI 3695: Engineering of Industrial Plants. Evaluation and Optimization. Subject Methods, Part 3. Beuth Verlag, Berlin (2010)

Vogel-Heuser, B., Diedrich, C., Broy, M.: Anforderungen an CPS aus Sicht der Automatisierungstechnik. Automatisierungstechnik. **61**(10), 669–676 (2013)

Willems, B., Seliger, G., Duflou, J., Basdere, B.: Contribution to design for adaptation: method to assess the adaptability of products (MAAP). In: EcoDesign2003: Third International Symposium on Environmentally Conscious Design and Inverse Manufacturing, pp. 589–596, IEEE, Tokyo, Japan, Proceedings, 8–11 December 2003

Yang, C., Vyatkin, V., Pang, C.: Model-driven development of control software for distributed automation. IEEE Trans. Syst. Man Cybern. Syst. **44**(3), 292–305 (2014)

Zawisza, J., Hell, K., Röpke, H., Lüder, A., Schmidt, N.: Generische Strukturierung von Produktionssystemen der Fertigungsindustrie. 17. Branchentreff der Mess- und Automatisierungstechnik (Automation 2016), Baden-Baden, Germany, Proceedings, VDI-Verlag, June 2016 (in German)

Chapter 7
Description Means for Information Artifacts Throughout the Life Cycle of CPPS

Arndt Lüder, Nicole Schmidt, Kristofer Hell, Hannes Röpke, and Jacek Zawisza

Abstract Recent research and development activities within the field of production system engineering and use focus on the increase of production system flexibility and adaptability. One common issue of those approaches is the consideration of hierarchical and modular production system architectures where the individual components of the system are equipped with certain functionalities and information. Up to now, there is no common understanding about what a component can constitute, i.e. which parts of a production system can be regarded as components within the hierarchy and which functionalities and information are assigned to it. This gap will be closed within this and the two the prior chapters.

They will at first discuss the relevant layers of components in a production system, then the types of information required to be assigned to a component on the different layers to establish a digital representation of the component, and at last the description means exploitable to represent the identified information in the different life cycle phases of a production system.

This chapter, in particular, will consider the artifacts and description means related to them in each of the three life cycle phases on each layer of the hierarchical production system structure as proposed in Chap. 5. Furthermore, the artifacts are clustered and generic artifact classes are derived from the fragmented information artifact landscape. Finally, description means are assigned to the artifact classes, paving the way for future research on this topic.

Keywords Model-based engineering • Description means • Information artifact classes • Life cycle phases • Production system hierarchy

A. Lüder (✉) • N. Schmidt
Faculty Mechanical Engineering, Otto-von-Guericke University, Magdeburg, Germany
e-mail: arndt.lueder@ovgu.de; nicole.schmidt@ovgu.de

K. Hell • H. Röpke • J. Zawisza
Volkswagen Aktiengesellschaft, Wolfsburg, Germany
e-mail: kristofer.hell@volkswagen.de; hannes.roepke@volkswagen.de; jacek.zawisza@volkswagen.de

© Springer International Publishing AG 2017

S. Biffl et al. (eds.), *Multi-Disciplinary Engineering for Cyber-Physical Production Systems*, DOI 10.1007/978-3-319-56345-9_7

7.1 Introduction

Due to the ever growing and accelerating trend towards industrialization and digitalization, the number of information artifacts, as well as the associated data formats and description means, are growing at an even higher speed (Vogel-Heuser 2015). However, as of today, there is no multi-disciplinary data modeling concept that is also consistent throughout the life cycle of a CPPS. The *Industrie 4.0* component addresses this topic, among others, by providing a structure for information elements to enable the implementation of functionalities of physical and virtual assets (Plattform Industrie 4.0 2016). Nevertheless, it is still unclear which overall requirements *Industrie 4.0* components will have to fulfill, considering the depiction of information artifacts and description means.

Therefore, the main research goal of Chaps. 5, 6, and 7 is to define the requirements on the capabilities of *Industrie 4.0* components in order to be able to create, manage, and use information along its complete life cycle. The two previous chapters already laid the groundwork to answer this question by providing a hierarchy and life cycle model to structure the emerging information (see Chap. 5) and by assigning information artifacts to the hierarchy layers and life cycle phases (see Chap. 6).

Against this background, in this chapter, the work of the previous two chapters is continued and the remaining research questions are answered, which are:

Which description means are exploited to represent the information types in the life cycle phases of a CPPS? Are there special description means related to the different layers?

These questions are closely linked to the research questions RQ M1 and M2 of Chap. 1, which state that it is of interest to identify requirements and architectures for *Industrie 4.0* component modeling considering their multi-disciplinary character and to have a look at the information creation and use related to it.

Therefore, in the course of this chapter, it will be shown which description means can be assigned to which generic types of engineering artifacts that are used throughout the life cycle of CPPS. In order to achieve this, the authors gathered and evaluated a characteristic set of information and the artifacts they are coded within. The first step was a literature review with focus on the work of (Foehr et al. 2012), where a Delphi-based expert survey was carried out. In a second step, the authors of this chapter complemented the data with further expert interviews and assigned a student work (Hell et al. 2016) to the topic in order to extend the representativeness of the work. The results are discussed here and, although the artifacts and description means in this work are not completely exhaustive, it is important to state that they are representative for a large part of the data types and formats used in the regarded setting.

In the foregoing, the focus is laid on interdependent information sets throughout the life cycle phases. For a detailed description on dealing with heterogenous information on project level using semantic web technologies, see Chap. 12.

Another important aspect in this context is data quality assurance, which deals with securing the faultlessness of (engineering) data throughout the life cycle. For further information on this topic, see Chap. 16.

To answer the described research questions, this chapter is structured as follows: first, a disambiguation is provided in order to delimit the topic of this chapter. In a second step, description means for the identified artifacts are identified, which are exploited to represent information coded within these artifacts. Thereby, it is possible to analyze whether there are special description means related to the different layers or not. In the process, description means are assigned to each layer of the hierarchy model and life cycle phases provided by Chaps. 5 and 6. This makes it possible to define classes of artifacts and corresponding generic description means in the next step. Finally, the contributions of this particular chapter as well as the overall contributions of Chaps. 5, 6, and 7 are discussed and an outlook to future research is given.

7.2 Disambiguation: Description Means, Information Handling Methods, and Tools

An important point, here, is that the distinction between description means, information handling methods, and information handling applications and tools shall be considered (Schnieder 1999). The focus of this chapter is on the description means, neither on information handling methods nor tools. For further reading about methods and tools the reader can refer to Chap. 9.

Description means enable the expression of problems and their solutions in more or less formal ways. Usually, it can be distinguished between the information covered by description means (semantics) and their representation (syntax) (Diedrich et al. 2011). For example, the behavior of a production system component can be expressed on an information level by a Petri net where the places cover local states of the component while transitions express the state evolution. On the representation level, places are represented by circles with annotations and transitions by bars with annotations. Nevertheless, description means shall be independent from any methodology and any technical solution for their creation, management, and use.

Methodologies and any technical solutions for creation, management, and use of artifacts are represented by information handling methods (usually covering the Engineering, Operation and Maintenance, and End-of-Life phase activities as named in the previous chapter) and information handling applications and tools (used within the same phases). Usually, they are strongly correlated with the description means applied.

Within the following, the focus will be on the description means and not on the methodologies and applications/tools. The aim of this Chapter is to name and assign types of relevant description means to the artifacts identified in the previous

Chapter. As the number of artifacts and the number of applicable description means is apparently high, only representative examples will be considered.

7.3 Description Means for Artifacts

As it can be seen within the following figure, the artifacts named within Chap. 6 can be assigned to the three identified life cycle phases as well as to the different layers of the identified hierarchy of a production system in Chap. 5 (see Fig. 7.1). The description means within the three phases are described in the following.

7.3.1 Description Means During Engineering Phase

Initially, during engineering phase, the production system is designed in a kind of top-down approach starting with the requirement specifications leading to the details specification about how to install and commission the production system. Usually, engineering information is created here, which are stored and exchanged as files, database content, or paper documents in an appropriate way.

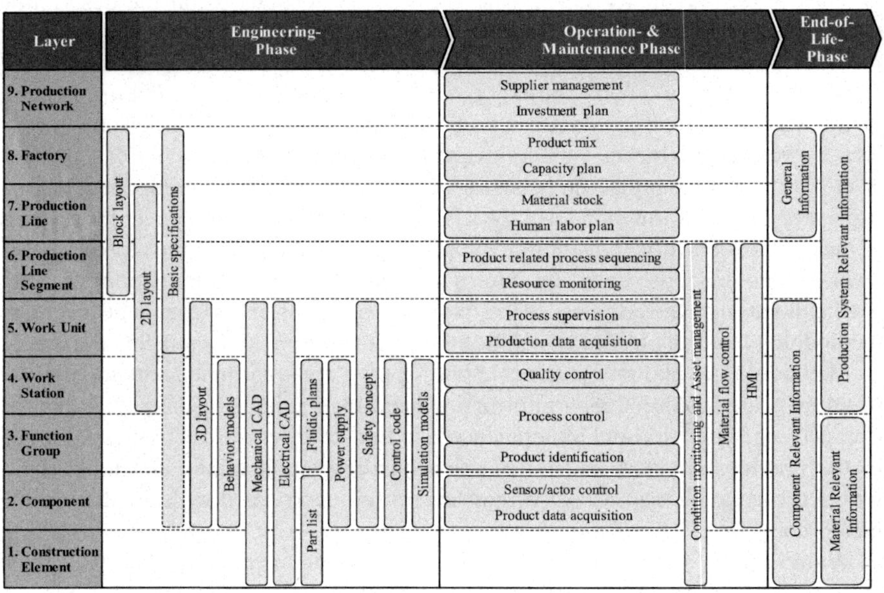

Fig. 7.1 Production system life cycle phases and phase-specific information

On the Factory and Production Network layers, requirements and legal, economic, and technical constraints (so-called propositions) are usually assigned. In most cases, these are text-based documents exchanged as PDF files. In some cases, companies have started to apply more formal ways within the requirement specification exploiting modeling tools like IBM DOORS or modeling languages like SysML. In this case, the exchanged documents are XML based (following XMI in the SysML case) or the information is stored in databases.

On Production Line and Production Line Segment layers, the production system structure is designed in general. Thus, 2D layouts are relevant which are exchanged as CAD drawings using files like TIFF, JPG, and PDF which are accompanied by text-based documents like PDF.

On Work Unit layer, there are basic behavior models, 2D layouts, mechanical and electrical specifications, 3D layouts, and safety concepts. For modeling basic behaviors usually Gantt Charts or similar high-level models are applied given in spreadsheet based documents like Microsoft Excel. Mechanical and electrical specifications are stored in a textual way or based on dedicated MCAD and ECAD tools with its appropriate data formats like JT or STEP. 3D layouts are given, in general, as CAD files. Safety concepts are described by text and/or in structures defined by legal regulation organizations exploiting XML.

On Work Station layer, behavior models, 3D geometry models, mechanical and electrical specifications, control programs, fluidic plans, powers supply concepts, safety concept, electrical construction, and simulation models are applied. Again related to 3D modeling as well as mechanical, electrical and fluidic constructions CAD data are created using special CAD files like JT and STP or dedicated XML structures. Comments and additions to these documents are given in text-based documents like PDF. Behavior models and simulation models usually cover more detailed behavior descriptions given by timing diagrams, SysML diagrams, or automatons as well as by dedicated models for simulation tools like automatons and Petri nets stored as legacy files or XML. Finally, control programs are given in appropriate control code following e.g. IEC 61131 stored as PLCopen XML.

On Function Group layer, the engineering information is more detailed covering basic behavior models, 3D layouts, mechanical and electrical specifications, control programs, fluidic plans, power supply concepts, safety concepts, electrical construction, detailed behavior models, and simulation models. The 3D data and construction specifications are again given as CAD files. Behavior and simulation models are usually given as detailed timing diagrams (in the specification case), automatons, or dedicated simulation models like Simulink or Modelica. For the other named information types XML-based or text based documents are often applied to represent the information.

Again, on Component Layer, the level of detail increases, especially related to the structure and behavior specifications like 3D layouts and constructions, control programs, power supply concepts, safety concepts, electrical construction, detailed behavior models, and simulation models. In addition, detailed part lists will be relevant defining the purchase parts of the production system. CAD based construction models are usually given as CAD files also used for mechanical and electrical constructions. Control programs are based on the relevant control programming languages, which are component type and vendor dependent, or they are stored as more or less user organization based XML formats like PLCopen XML. Behavior and simulation models are either given as dedicated simulation models like Simulink or Modelica or specified by automation based models like state charts. Especially for part lists Excel based CSV documents are relevant.

On the lowest layer, the Construction Element layer, the most detailed engineering information is relevant. This include part lists and mechanical and electrical specifications which are given as text-based documents, CSV files or XML-based files, and CAD-based construction given as CAD specific files like JT and STP.

An overview of layer specific artifacts and description means during Engineering Phase is given in Table 7.1.

7.3.2 Description Means During Operation and Maintenance Phase

Within the use phase, the identification of description means is a bit more complicated as here the time of validity and the way of artifact collection/transmission are relevant for the description means definition. Artifacts on the higher hierarchy layers have a validity period of hours and days while the artifacts on the lowest hierarchy layers are real-time control data with a validity period of minutes, seconds, and below. Thus, in contrast to engineering data, there is a drastic shift within the description means.

On the Production Network Layer, there is volume and cost planning information, economical Key Performance Indicators (KPIs) as well as technical and sales rules which usually are coded within text documents like PDF and XML-based documents.

On the Factory Layer, the relevant artifacts are very similar containing long-term production program and manufacturing planning, as well as technical and sales restrictions which mostly are given as text-based documents like PDF, economic, manufacturing, and logistics related KPIs given as XML structures like KPI XML, and material stock/availability, and order data coded as database content.

On Production Line Layer, the use related data mostly cover KPIs for staff, material, and resource allocation possibly coded by XML structures like KPI XML as well as resource planning, order data, supplier orders modeled as text-based documents like PDF, spreadsheet-based documents like CSV and database content like SAP systems.

The only difference between Production Line Segment Layer and Production Line Layer is the more detailed focus on resources. Thus, there are resource related KPIs, resource alarming, order data, and material logistics data which can be represented by spreadsheet based documents like CSV, XML structures like KPI XML, database content, and, as new means especially for alarming, PLC data.

Table 7.1 Layer-specific artifacts and description means during engineering phase

	Layer	Artifacts	Usable description means
9	Production network	Requirement specifications, legal, economic, and technical constraints	Text-based documents like PDF, XML structures
8	Factory	Requirement specifications, legal, economic, and technical constraints	Text-based documents like PDF, XML structures, database content
7	Production line	2D-layouts	CAD drawings using files like TIFF, JPG, PDF, text-based documents like PDF
6	Production line segment	2D-layouts	CAD drawings using files like TIFF, JPG, PDF, text-based documents like PDF
5	Work unit	Basic behavior models, 2D-layouts, mechanical and electrical specifications, 3D-layouts, and safety concepts	Gantt Charts or similar high-level behavior models in Excel, text-based documents, CAD files like JT or STP, XML-based files
4	Work station	Behavior models, 3D-geome-try-models, mechanical and elec-trical specifications, control programs, fluidic plans, powers supply concepts, safety concepts, electrical construc-tion, simulation models	CAD files like JT and STP, XML structures, text-based documents like PDF. Impulse diagrams, SysML diagrams, automatons, and Petri nets as legacy or XML files, IEC 61131 code as PLCopen XML
3	Function group	Behavior models, 3D-layouts, mechanical and electrical specifications, control pro-grams, fluidic plans, powers supply concepts, safety con-cepts, electrical construction, detailed behavior models, and simulation models	CAD files like JT and STP, XML structures, text-based documents like PDF. Impulse diagrams, SysML diagrams, automatons, and Petri nets as legacy or XML files, simula-tion models like Simulink or Modelica, IEC 61131 code as PLCopen XML
2	Component	Behavior models, 3D-layouts, part lists, mechanical and elec-trical specification, CAD con-struction, control programs, powers supply concepts, safety concepts, electrical construc-tion, detailed behavior models, simulation models	CAD files like JT and STP, XML structures, text-based vendor dependent documents, simulation models like Simulink or Modelica State charts, IEC 61131 code as PLCopen XML, CSV files
1	Construction element	Part lists, mechanical and electrical specifications, CAD-based construction	Text-based documents, CSV files, XML-based files, CAD specific files like JT and STP

On Work Unit Layer, the real-time control impact drastically increases and the importance of KPIs decreases. The relevant data on this layer contains especially maintenance related resource KPIs, resource state information and resource alarming, order related data, and production process control data. Thus, here are with spreadsheet-based documents like CSV, XML structures like KPI XML, database content, and PLC data the same description means as on Production Line Segment Layer.

On Work Station Layer, there are the same artifacts and description means relevant as on Work Unit Layer.

The most relevant shift within the used description means can be found between Work Station Layer and Function Group Layer. On Function Group Layer as well as on Component Layer, the use data are related to field control. They subsume resource and component state and alarming information as well as production process control data which are modeled as PLC, RND, CNC, etc. data.

Finally, on the Construction Element Layer, there are no direct use phase related data. Nevertheless, it can be assumed that there might be construction element related data relevant for maintenance activities like durability of materials. But they are collected on Work Station Layer as resource KPIs. An overview of layer specific artifacts and description means during Operation and Maintenance Phase is given in Table 7.2.

7.3.3 Description Means During End-of-Life Phase

The identification of artifacts for the End-of-Life phase of production systems is still an open field of research. However, with general information (relevant on layers 8 and 7), production system related information (relevant on layers 8–4), component-specific information (relevant on layers 5–2), and material-specific information (relevant on layers 3–1), four general information- types are known.

General information is mostly related to guidelines and regulations defined/developed by legal advisors, standardization organizations, or vendor organizations. They are usually provided as text-based documents like PDF, sometimes including more formal representations like UML diagrams and XML structures.

Production system related information is mostly based on the re-use of engineering information as well as engineering-like information describing the current state of the structure and behavior of the production system. This may include 2D- and 3D-plans and -layouts, behavior models, and part lists, which are modeled by CAD-drawings using files like TIFF, JPG, PDF, text-based documents like PDF, XML-based structures, Gantt and timing diagrams, as well as spreadsheet-based documents like CSV.

The component-specific information is very similar to the production system related information, but usually contains a higher level of detail. They subsume mechanical and electrical specifications and constructions, behavior models, safety concepts, and part lists, which are modeled by CAD-drawings using files like

Table 7.2 Layer-specific artifacts and description means during operation and maintenance phase

	Layer	Artifacts	Usable description means
9	Production network	Volume and cost planning, economical KPIs, technical and sales rules	Text-based documents like PDF, XML structures like KPI XML, database content
8	Factory	Program and manufacturing planning, economic, manufacturing and logistics related KPIs, technical and sales restrictions, manufacturability, parts stock/ availability, order data	Text-based documents like PDF, XML structures like KPI XML, database content
7	Production line	Staff, material and resource allocation KPIs, resource planning, order data, supplier orders	Text-based documents like PDF, table-based documents like CSV, XML structures like KPI XML, database content
6	Production line segment	Resource KPIs, resource alarming, order data, logistics data	Table-based documents like CSV, XML structures like KPI XML, database content, PLC data
5	Work unit	Resource KPIs, resource state and alarming, order data, production process control data	Table-based documents like CSV, XML structures like KPI XML, database content, PLC data
4	Work station	Resource KPIs, resource state and alarming, order data (like quality), production process control data	Table-based documents like CSV, XML structures like KPI XML, database content, PLC data
3	Function group	Resource state and alarming, production process control data	PLC, RND, CNC, etc. data
2	Component	Component state and alarming, production process control data	PLC, RND, CNC, etc. data
1	Construction element	–	–

TIFF, JPG, PDF, text-based documents like PDF, XML-based structures, timing or automaton diagrams, as well as spreadsheet-based documents like CSV. The focus of this information is on the modular and hierarchical structure of the production system and the way it is assembled during installation. Thereby, the way of dis-assembling can be defined. In addition, the component-specific information covers use-phase related information, describing the way of the utilization of components. This is necessary to estimate the current economic value of a component and its capability to be reused in other production systems. The named information mostly covers resource related KPIs, which are modeled as text-based documents like PDF, XML structures like KPI XML, spreadsheet-based documents like CSV, database content, or even as PLC data.

Finally, the material specific information focuses on the material identification within the used production system elements as well as on the description of the deterioration of the used material during use time. This is based on the use of device, material, and other documentations which are given as text files for example as PDF and on resource/component related KPIs which are modeled as XML structures like KPI XML, spreadsheet-based documents like CSV, database content, or again as PLC data.

The named mappings of useable description means are summarized in Table 7.3. As can be seen, the number of currently applied description means is high. It changes with the considered layer being more formal and can be automatically evaluated better on the lower hierarchy layers. In the future, it is advisable to reduce the number of applied description means.

7.4 Artifact Classification

In order to identify description means potentially usable for the artifacts in a future scenario, the artifacts need to be classified and the description means need to be assigned to each artifact class. By this, a potential unification of artifacts can be reached. It might enable the representation of all identified information types within/of *Industrie 4.0* components by a uniform representation.

One possible classification of the identified artifacts is given in Fig. 7.2. It represents the three main life cycle phases of the production system, indicates artifact classes applicable, and maps the artifacts named above to the artifact classes.

Considering the available description means named in the sections above, generalized XML-based data formats like AutomationML (Drath 2010) and (AutomationML 2009) can be identified as possible candidates for a data format covering all artifacts. Nevertheless, following (Schmidt and Lüder 2015) AutomationML will not cover all relevant information sets. Therefore, it needs to be extended by text files like PDF, CAD drawings using files like TIFF or JPG, and control information represented in databases like OPC UA or in control devices like PLC, RND, or CNC. (Lüder et al. 2014) presents a possible integration approach.

Table 7.4 summarizes the combination of the five named information description means to represent all artifact classes.

7.5 Summary and Outlook

The goal of this Chapter was to define description means that can be exploited to represent the information types in the three life cycle phases and whether they are related to the different hierarchy layers as described in Chap. 5.

In order to achieve this, in a first step usable description means were assigned to the artifacts on each layer of the underlying hierarchy structure. Since the

Table 7.3 Layer-specific artifacts and description means during EOL-Phase

	Layer	Artifacts	Usable description means
9	Production network	–	–
8	Factory	General information, production system related information	Text-based documents like PDF, UML diagrams, CAD drawings using files like TIFF, JPG, PDF, XML-based structures, Gantt and Impulse diagrams, table-based documents like CSV
7	Production line	General information, production system related information	Text-based documents like PDF, UML diagrams, CAD drawings using files like TIFF, JPG, PDF, XML-based structures, Gantt and Impulse diagrams, table-based documents like CSV
6	Production line segment	Production system related information	Text-based documents like PDF, CAD drawings using files like TIFF, JPG, PDF, XML-based structures, Gantt and Impulse diagrams, table-based documents like CSV
5	Work unit	Production system related, component-specific information	Text-based documents like PDF, CAD files like TIFF, JPG, PDF, XML-based structures, Gantt, Impulse, or automaton diagrams, table-based documents like CSV, database content, PLC data
4	Work station	Production system related, component-specific information	Text-based documents like PDF, CAD files like TIFF, JPG, PDF, XML-based structures, Gantt, Impulse, or automaton diagrams, table-based documents like CSV, database content, PLC data
3	Function group	Component-specific information, material specific information	CAD drawings using files like TIFF, JPG, PDF, text-based documents like PDF, XML-based structures like KPI XML, Impulse or automaton diagrams, table-based documents like CSV. database content, PLC data
2	Component	Component-specific information, material specific information	CAD drawings using files like TIFF, JPG, PDF, text-based documents like PDF, XML-based structures like KPI XML, Impulse or automaton diagrams, table-based documents like CSV. database content, PLC data
1	Construction element	Material-specific information	Text files like PDF, XML structures like KPI XML, table-based documents like CSV, database content, PLC data

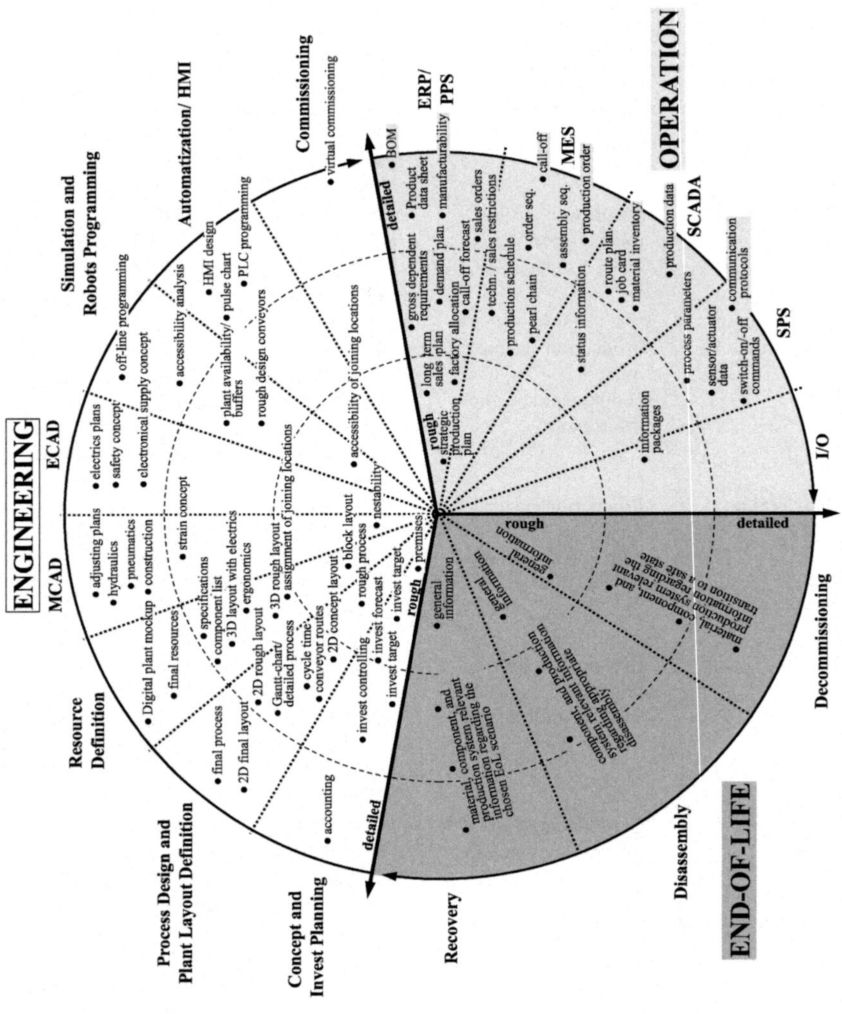

Fig. 7.2 Examples of relevant artifacts in the three life cycle phases

Table 7.4 Description means assignment to artifact classes

Artifact classes	Description means				
	XML-based structures like AutomationML	Text files like PDF	CAD drawings using files like TIFF, JPG	Database content like OPC UA	PLC, RND, CNC, etc. data
Concept and invest planning	X	X			
Process design and plant layout definition	X		X		
Resource definition	X		X		
MCAD	X		X		
ECAD	X		X		
Simulation and robots programming	X		X		
Automation/HMI	X			X	X
Commissioning	X	X	X	X	X
ERP/PPS	X	X			
MES	X			X	
SCADA	X			X	
PLC				X	X
Field-I/O					X
Decommissioning	X	X	X	X	X
Disassembly	X	X	X	X	X
Recovery	X	X	X	X	

information and description means can vary greatly between the three specified life cycle phases of CPPS, it was necessary to match artifacts and description means for engineering, operation, and EoL-Phase. Thereby, it was shown that both, the assignment of description means to hierarchy layers as well as to life cycle phases is possible.

However, since the information used in the various scenarios can be interdependent, incompatibility of description means used throughout the life cycle is a possible outcome. Therefore, in order to make description means compatible in the future, in this Chapter possible artifact classes were defined and assigned to description means, making them more consistent.

What are the requirements on the capabilities of *Industrie 4.0* components to create, manage, and use information along its complete life cycle?

Considering the main research question of the Chaps. 5, 6, and 7 the intention of this work was to provide assistance for engineers to decide about the right information set to be covered by an *Industrie 4.0* component and to decide about applicable implementation technologies. Therefore, the main research question of this work has been the identification of requirements on the capabilities of *Industrie 4.0* components to create, manage, and use information along their complete life cycle.

To answer this question three underlying research questions have been addressed beforehand. At first, nine different layers of *Industrie 4.0* components in a production system have been presented in Chap. 5. For each layer, the relevant functionalities of an *Industrie 4.0* component were used as criteria for identification.

At second, the different types of information to be assigned to an *Industrie 4.0* component on the different layers have been discovered establishing a kind of virtual representation of the component. Therefore, the main life cycle phases of a production system have been reviewed in Chap. 6.

Finally, description means applicable to represent the identified information types in these life cycle phases are given in Chap. 7.

As a result, the named three Chapters together are able to provide

- Candidates for *Industrie 4.0* components,
- Candidates for meaningful information sets relevant for these *Industrie 4.0* component candidates following the different layers of components in a production system, and finally
- Candidate technologies for representing these information sets in the administration shell of the *Industrie 4.0* component.

The three Chapters failed to provide information-based identification characteristics for *Industrie 4.0* components. Figure 7.2 depicts this problem. Here it gets visible, that for example the Layers Work Station, Function Group, and Component contain similar engineering information just with a different granularity. Also, the runtime information are very similar. Thus, these information set relevant for an *Industrie 4.0* component is not necessarily layer characterizing. Here, further research can provide a more detailed view on the problem.

In addition, this work still leaves another essential question open. Especially Tables 7.1, 7.2, and 7.3 indicate that there are various information sets relevant for an *Industrie 4.0* component. But these information sets are not independent from each other. In contrast, they are strongly dependent, requiring on the one hand consistency between the information sets as well as sometimes enabling the generation of some information exploiting other information sets.

The representation of these dependencies and generation processes are beyond the scope of this paper but need to be considered in detail in future work.

Furthermore, the requirements of model-based engineering make it necessary to integrate proper methods and tools throughout the engineering processes. Since this is also a workflow-management topic, Chap. 11 discusses the adoption of this model-driven systems engineering approach considering the requirements for CPPS.

References

AutomationML Association: AutomationML web page (2009). www.automationml.org

Diedrich, C., Lüder, A., Hundt, L.: Bedeutung der Interoperabilität bei Entwurf und Nutzung von automatisierten Produktionssystemen. Automatisierungstechnik. **59**(7), 426–438 (2011)

Drath, R. (ed.): Datenaustausch in der Anlagenplanung mit AutomationML—Integration von CAEX, PLCopen XML und COLLADA. Springer, Berlin (2010)

Foehr, M., Lüder, A., Steblau, A., Lüder, M.: Analyse der praktischen Relevanz verschiedener Beschreibungsmittel im Entwurfsprozess von Produktionssystemen Entwurf komplexer Automatisierungssysteme (EKA 2012). Magdeburg, Deutschland, Proceedings, pp. 61–72 (2012)

Hell, K., Hillmann, R., Lüder, A., Röpke, H., Zawisza, J., Schmidt, N., Calà, A.: Demands on the Virtual Representation of Physical Industrie 4.0 Components. Conferenza INCOSE Italia su Systems Engineering (CIISE 2016), November 14–16, Turin, Italy (2016)

Lüder, A., Schmidt, N., Rosendahl, R., John, M.: Integrating different information types within AutomationML. 19th IEEE International Conference on Emerging Technologies and Factory Automation (ETFA 2014), Barcelona, Spain, Proceedings (2014)

Plattform Industrie 4.0: Struktur der Verwaltungsschale. Fortentwicklung des Referenzmodells für die Industrie 4.0-Komponente. http://www.plattform-i40.de/I40/Redaktion/DE/Downloads/Publikation/struktur-derverwaltungsschale.pdf?__blob=publicationFile&v=8 (2016). Accessed October 2016

Schmidt, N., Lüder, A.: AutomationML in a Nutshell. AutomationML consortium. http://www.automationml.org (2015). Accessed September 2016

Schnieder, E.: Methoden der Automatisierungstechnik. Vieweg Verlag (1999)

Vogel-Heuser, B.: Herausforderungen und Anforderungen aus Sicht der IT und der Automatisierungstechnik. In: Vogel-Heuser, B., Bauernhansl, T., ten Hompel, M. (eds.) Handbuch Industrie 4.0—Produktion, Automatisierung und Logistik, pp. 37–48 (in German). Springer, Berlin (2015)

Chapter 8
Engineering of Next Generation Cyber-Physical Automation System Architectures

Matthias Foehr, Jan Vollmar, Ambra Calà, Paulo Leitão, Stamatis Karnouskos, and Armando Walter Colombo

Abstract Cyber-Physical-Systems (CPS) enable flexible and reconfigurable realization of automation system architectures, utilizing distributed control architectures with non-hierarchical modules linked together through different communication systems. Several control system architectures have been developed and validated in the past years by research groups. However, there is still a lack of implementation in industry. The intention of this work is to provide a summary of current alternative control system architectures that could be applied in industrial automation domain as well as a review of their commonalities. The aim is to point out the differences between the traditional centralized and hierarchical architectures to discussed ones, which rely on decentralized decision-making and control. Challenges and impacts that industries and engineers face in the process of adopting decentralized control architectures are discussed, analysing the obstacles for industrial acceptance and the new necessary interdisciplinary engineering skills. Finally, an outlook of possible mitigation and migration actions required to implement the decentralized control architectures is addressed.

Keywords CPS in production • Future automation systems • Industrial systems engineering • Migration strategy • System architectures

---

M. Foehr (✉) • J. Vollmar • A. Calà
Siemens AG Corporate Technology, Erlangen, Germany
e-mail: matthias.foehr@siemens.com; jan.vollmar@siemens.com; ambra.cala.ext@siemens.com

P. Leitão
Polytechnic Institute of Bragança, Bragança, Portugal
e-mail: pleitao@ipb.pt

S. Karnouskos
SAP, Walldorf, Germany
e-mail: stamatis.karnouskos@sap.com

A.W. Colombo
University of Applied Sciences Emden/Leer, Emden, Germany
e-mail: awcolombo@technik-emden.de

© Springer International Publishing AG 2017

S. Biffl et al. (eds.), *Multi-Disciplinary Engineering for Cyber-Physical Production Systems*, DOI 10.1007/978-3-319-56345-9_8

8.1 Introduction

Production systems are complex systems composed of various, often engineering discipline specific, subsystems. One important subsystem to be considered is the automation system. Due to the close interaction between the automation system and other system components like actuators and sensors the whole system and its environment needs to be considered when dealing with the automation system architecture. From this perspective the purpose (e.g., product to be produced) and system goal (e.g., output capacity) are main influencing factors. But also the overall system architecture (e.g., structure of production system, layout, IT-Systems) and the functional and non-functional requirements (e.g., degree of automation) towards the automation system have an impact on the automation infrastructure. Last but not least the available technology and hardware must be taken into account.

For today's systems the environment, system goals and system architecture are considered stable over the whole life-cycle of the production system. Changes occur when product changes (e.g., new model of car, new chemical substance) or requirements change (e.g., new safety regulation) but they normally have no impact on the architecture as such, except from the software and run-time aspects of the automation system. If changes occur, the production is stopped and the system is changed and production is re-started after modification. These downtimes, even if planned, are resulting in a loss of production capacity and finally a loss of money. This is also reflected in the classical automation system architecture as discussed in Sect. 8.2.

The question that arises is: are the above mentioned influencing factors going to remain stable also in the future? To answer it, the German National Academy of Science and Engineering (ACATECH, 2011) investigated four future scenarios of Cyber-Physical Systems (CPS) application with a time horizon until the year 2025.

One of these scenarios "Cyber-Physical System for the factory of the future" describes the characteristics and challenges for production systems. Production systems shall be able to react virtually in real time to changes in the market and the supply chain using CPS, which cooperate with ultra-flexibility even beyond company boundaries. Therefore a future industrial system architecture is needed that will focus on key aspects as identified in Kagermann et al. (2013), specifically:

- Allow flexibility and reconfiguration (with no downtime)
- Enable high production system resilience (deal with uncertainties)
- Enable continuous, automatic production optimization
- React faster and more automated to evolving customer and production demands
- Support for highly individualized production and small batches/lot sizes (lot size 1)

The posed requirements are reflected in several key research questions (RQs). However, most notably this chapter pertains aspects that tackle or enable approaches targeting the following RQs (see Chap. 1):

- Modelling the structure and behaviour of Cyber-Physical Production Systems (CPPS) (RQ M1)
- Information integration in and across value chains (RQ I1)
- Description of plug-and-play capabilities and interfaces for engineering and run time (RQ I3)
- Modelling of CPPS flexibility and self-adaptation capabilities (RQ C1)
- Linking discipline-specific engineering views for flexible and self- adaptable CPPS (RQ C2)

The contributions of this chapter, are strongly liked to the emerging domain of CPS, and especially in their utilization in production systems. The discussions pertaining this chapter focus on providing an overview of automation system needs and evolution, how these are migrated to an new information-driven interoperable and service-enabled infrastructure, and what key considerations as well as challenges lie ahead. The intention in this chapter is not to provide a new model-based approach but to understand why and how the already existing methods and tools that enable production system flexibility and self-adaptation of CPPS are not adequate or too poorly implemented in industrial practice.

Based on discussed key requirements new automation system architectures are emerging in different research approaches which will be described in more detail in Sect. 8.2. As the design of completely new production systems, also referred to as green-field, is of secondary importance since a high number of legacy systems already exists, adequate migration strategies are needed to transform and migrate from existing automation system architectures to future ones. This transformation is described in Sect. 8.3. Furthermore the way to engineer these future automation systems has to be re-thought. This must include new methods and tools for engineers to design, implement and support such systems. Also educational programs have to be up-dated to ensure availability of experts that are capable to deal with these new systems architecture and new engineering paradigms. In Sect. 8.4 a closer look is taken upon these aspects. Finally, Sect. 8.5 gives an outlook and presents the main conclusions.

8.2 The Evolution of Automation System Architectures

Today companies are facing new market challenges in the manufacturing industry. In response to new requirements, innovative forms of manufacturing are recently introduced accordingly to the German "Industry 4.0" paradigm (Kagermann et al., 2013). The need of new manufacturing approaches is influenced by several aspects, namely market competitiveness, technology innovation, and customer requirements.

The global competition requires shortened delivery time and time-to-market, smaller lot sizes and shorter product life-cycle. Meanwhile, rapid changes in process technology force the fast integration of new functions into existing systems that are subject to obsolescence. Furthermore, customer expectations include not only lower prices but also more variety, higher quality and faster delivery of the product. In order to dynamically react to continuous changes of the business environment, the view on production system control must evolve.

The traditional production control systems are not able to support industries in overcoming such issues (Delsing et al., 2012). Centralized and hierarchical control architectures are characterized by rigid and top-down communication flows that do not enable the easy integration of new modules and, therefore, cannot cope with sudden and rapid changes. Considering all these aspects, new challenges for industries arise (Karnouskos et al., 2014a), and the next efforts attempt to introduce in industry a new production approach characterized by flexibility to different processing tasks, adaptability to changing production environment, and reconfigurability to enable these changes, while maintaining the security, safety and stability provided by classical production systems.

8.2.1 Classical Automation System Architectures

Traditional automation control systems are generally structured hierarchically or centralized, due to the complexity of automation tasks and interactions between components. According to the ISA-95/IEC 62264 (ANSI/ISA, 2010) standard, the main automation tasks are split in different layers of a pyramidal structure as shown in Fig. 8.1. The ANSI/ISA (2010) standard defines a model for exchanges of information between systems in five abstraction levels: *Level 0*—Field, *Level 1*—Control (PLC), *Level 2*—Process Control (SCADA), *Level 3*—Manufacturing Execution (MES) and *Level 4*—Enterprise Management (ERP).

The applications located on the different levels typically consider different time frames that range from months, weeks and days for the higher levels to hours, minutes, seconds and milliseconds for the lower levels. The first three levels perform the control function to execute the technological production processes. The field level uses actuators and sensors to measure, determine and display the equipment

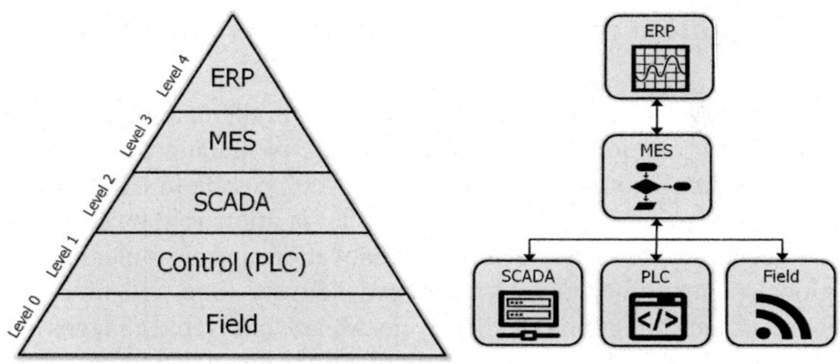

Fig. 8.1 ISA 95 hierarchical view of automation infrastructures

data, while the control and process control levels are related respectively to the control of the product/process technology and to monitor the overall production system. Level 3 comprehends the activities of coordination and management of the production execution and, especially, the integration of different applications with respect to the main data and work flows. Level 4 is the highest level and represents the overall business management of the enterprise, including economical and logistic activities.

In system architectures structured according to ISA-95 the decision control is distributed among these hierarchical levels. This kind of structure has the advantages of predictability and robustness, as well good global optimization. It can be effective for small systems due to the characteristics of easy development and maintenance, and also adequate for systems running in very stable and structured environments. However, it is not adequate for emerging self-x automated manufacturing systems because of the insufficient adaptability and flexibility to production changes and the reduced performance in case of a single point of failure.

Analysing the scenario of "Cyber-Physical System for the factory of the future" (ACATECH, 2011), a "real-time" reaction of the production system to market changes cannot be performed by a hierarchical automation control system. In order to react more quickly to customer demands and environment changes, a more seamless integration of the automation pyramid's levels is required to change the production equipment and functions accordingly. The production units need to cooperate and organize themselves to optimize the production systems, saving time and costs. Capabilities, such as flexibility, adaptability and reconfigurability, are limited in a rigid communication structure with no cross-layer interoperability (Delsing et al., 2012), therefore, the traditional hierarchical ISA-95 structure needs to be transformed into a modular and flexible automation system architecture with decentralized control systems. The envisioned future production systems that possess self-x features, are cost efficient and easy to integrate at mass scale, cooperate in a cross-layer manner, interact with multiple stakeholders etc., justify the trend towards a distributed approach that is hardly or too costly to be realized with traditional approaches.

8.2.2 Emerging Automation System Architectures

The Cyber-Physical Systems (CPS) concept represents one of the key enablers of innovation in production systems accordingly with the Industry 4.0 paradigm. CPS focuses on the integration of logical and physical processes to control distributed physical systems, using cyber technologies (mechatronics, communication and information) (Lee, 2008; Leitão et al., 2016a). Since decades multi-agent systems (MAS) and service-oriented architecture (SOA) have been considered as the main approaches for implementing CPS and developing decentralized control systems in industry (Leitão and Karnouskos, 2015b). Several projects (Leitão et al., 2016b) have demonstrated their benefits. MAS is one of the most common approaches to realize decentralized control architectures by means of intelligent, modular and distributed agents that can be interconnected with physical hardware devices (Leitão et al., 2016a); and SOA is an architectural model for organizing and utilizing distributed capabilities in order to enable all components to communicate and interact via services (MacKenzie et al., 2006).

Next to these paradigms other concepts, such as plug-and-produce technology, web services and cloud manufacturing, have been investigated to build flexible and reconfigurable manufacturing control systems. During the last years a significant amount of research has been conducted and, recently, several European funded projects have reported important developments in this field and presented results at high technology readiness levels.

The GRACE—Integration of process and quality control using multi-agent technology—project (Castellini et al., 2011) developed, implemented and validated a cooperative MAS to integrate process control with quality control at local and global level. The MAS architecture was designed to manage the planned changes of set-point in production processes and the large variety of unforeseen disturbances and changes in process parameters and variables. Self-adaptation procedures and optimization mechanisms for process and product parameters were implemented and integrated into control and diagnostic systems at local level, in terms of individual agents, and global level, considering the data gathered in all the production system.

In parallel, the IDEAS—Instantly deployable evolvable assembly systems—project (Onori et al., 2013) developed a fully distributed and pluggable environment capable to self-organize itself and control at the shop floor level using agent technology. The IDEAS assembly system ran with a multi-agent control setup and could be reconfigured on-the-fly assuring the integration of different self-configured modules at the shop floor in runtime. Moreover, the self-diagnosis capability of each module permits to have a distributed diagnosis and the entire system is capable of checking the propagation of problems and re-adapt whenever a component (module) is plugged without requiring programming effort in order to manage unpredicted behaviours.

Taking the experience from these projects, the PRIME—Plug and produce intelligent multi-agent environment based on standard technology—project (Antzoulatos et al., 2014) has gone one step forward to support assembly systems in distributed reconfiguration and monitoring. It developed a multi-agent architecture using plug-and-produce principles for module integration, including legacy equipment, and methods for rapidly configuring production systems through innovative human-machine interaction mechanisms. The PRIME approach is based on standard technologies (JADE multi-agent software framework, Vaadin and Cassandra database) and languages (JAVA and OPC-UA programs for interfacing and data exchange) for the integration and networking of heterogeneous control system from different equipment suppliers to support system evolution linked to process performance and product volume variability.

The I-RAMP[3]—Intelligent reconfigurable machines for smart plug-and-produce production—project (Goncalves et al., 2014) focused on the transformation of conventional production equipment into network-enabled devices (NETDEVs). The NETDEV interface enables the integration of plug-and-produce devices and sensors and actuators at MES level for work flow optimization and production data assessment, using standardized communication and collaboration mechanisms.

The SOCRADES—Service-Oriented Cross-layer infRAstructure for Distributed smart Embedded Systems—project (Colombo and Karnouskos, 2009; Karnouskos et al., 2010; Colombo et al., 2010) used the Service-Oriented Architecture paradigm at device and application levels to build a design, execution and management platform for innovative industrial automation systems. The project focused on designing and implementing a cross-layer infrastructure that would enable the integration of industrial automation systems and devices up to the MES/ERP level (Karnouskos et al., 2007, 2009). The approach was driven by open standards, service-based integration, and collaboration among the various stakeholders, setting the stage for the next generation of automation systems (Colombo and Karnouskos, 2009).

The IMC-AESOP—Industrial Monitoring and Control ArchitecturE for Service-Oriented Process—project (Colombo et al., 2014b,a) used as a starting point the SOCRADES approach and extended it to realize cloud-based industrial CPS. Driven by key emerging information and communication technologies in industrial automation, and with a strong focus on the cloud (Karnouskos and Somlev, 2013), the project envisioned and realized an architecture (Karnouskos et al., 2014b) for industrial CPS automation infrastructures. The results have been demonstrated in the next generation cloud & service based SCADA/DCS (Karnouskos and Colombo, 2011) for monitoring and control, including visions for their design, implementation, collaboration, and migration The architecture enables cross-layer service-oriented collaboration both at horizontal and vertical levels by utilizing service-oriented integration and the cloud.

The Self-Learning—Reliable Self-Learning production systems based on context-aware service—project (Stokic et al., 2011) proposed the service-oriented integration of different auxiliary processes into the main control. The processes are represented as services that fully interoperate in a Web Services platform. The Self-Learning system enables the reconfiguration of machines and processes based on user experiences acquired during the system runtime.

The FLEXA—Advanced flexible automation cell—project (Webb and Asif, 2011) developed a flexible manufacturing system based on web services architecture that connects the cell controller to ERP/MES.

The SelSus—Health Monitoring and Life-Long Capability Management for Self-Sustaining Manufacturing Systems—project (Sayed et al., 2015) proposed a new paradigm for highly effective, self-healing production systems to maximize their performance over longer lifetimes using web-based services for multi-modal data acquisition techniques to validate, update and document all information on failure modes or degradation states.

The CassaMobile—Flexible Mini-Factory for local and customized production in a container—project (Friedrich et al., 2014) developed a new kind of local, flexible and environmentally friendly production system for highly customized

Table 8.1 Overview of technologies in emerging automation system architectures

Technologies \ Research Projects	GRACE	IDEAS	PRIME	I-RAMP3	SOCRADES	IMC-AESOP	Self-Learning	FLEXA	SelSus	CassaMobile	ARUM
Multi-agent systems	●	●	●		◑	◑					●
SOA/Web services					●	●	●	●	◑	●	●
Cloud					◑	●			●		
Plug-and-Play		●	●	●	●	●				●	
Self-* features	●	●	●	◑	◑	◑	●		●	●	

Legend: ● covered; ◑ partially covered

parts based on a combination of different manufacturing processes. The production is based on a modular architecture that includes mechanical and control system adaptation by means of a SoA system.

One of the current trends in the future automation control research is to integrate these solution concepts in the same architecture. One example is the ARUM—Adaptive Production Management—project (Leitão et al., 2013), which combined holonic multi-agent systems with services architecture using Enterprise Service Bus (ESB) to improve planning and control systems.

These projects show a transformation of the centralized architecture into a distributed control system using different technologies, as shown in Table 8.1. High levels of autonomy and cooperation of individual entities have been reached via multi-agent systems in which agents have their own intelligence and interact with each other optimizing their behaviour iteratively (Wooldridge, 2002; Leitão et al., 2016b). Service-oriented architecture technologies enable the integration of components that provide services to other components they are linked to, creating an Internet of Services for the production system. Web Services contain components description and exchange data information enhancing the vertical collaboration between device level and enterprise level. Moreover, hosting these services in a Cloud it is possible to rapidly compose new industrial application just by selecting and combining the information stored inside (Colombo et al., 2014b). Plug-and-play technologies are investigated to build modular structures that improve components interoperability and reusability to satisfy the requirement of rapid reconfigurability of the system (Antzoulatos et al., 2014). In addition, self-* capabilities support equipment integration, control and monitoring, as well as cooperation and adaptation.

Each of these projects provided an individual solution for flexible and reconfigurable distributed control architectures involving multi-agent systems (Leitão et al., 2016b), standard communication protocols, web services and Cyber-Physical components. However, these solutions solve only narrowed specific problems neglecting other technological issues. In order to facilitate a wider industrial uptake, the future industrial system architecture should be a result of the integration of these

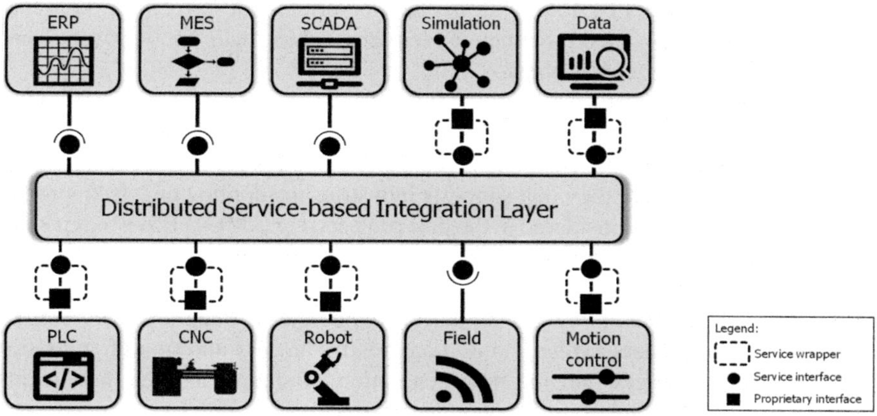

Fig. 8.2 Automation system integration vision over a common (service) infrastructure

technologies in a unique form (e.g., as shown in Fig. 8.2), covering the architecture, assets and process aspects of the overall production system. As an example, the SOCRADES project has demonstrated largely feasibility of this vision using web services for cross-layer integration and collaboration among devices, systems and other stakeholders (Colombo et al., 2010; Karnouskos et al., 2009; Taisch et al., 2009). A recent survey of acceptance factors of agent systems in industry (Leitão and Karnouskos, 2015a) sheds some additional light on key aspects that should be investigated at large when engineering of future industrial automation systems is to be considered.

In Fig. 8.2 a vision of automation system integration over a common service-based infrastructure is proposed. A key role in this new vision is performed by the distributed service-based integration layer that aims to ensure the transparent, secure and reliable interconnection of the diverse heterogeneous hardware devices e.g., robotic cells and Programmable Logic Controllers (PLC), and software applications e.g., MES and SCADA/DCS (Karnouskos and Colombo, 2011). Current Business systems and higher-level applications (i.e., ERP and MES etc.) are typically fully service-based in their interactions with other systems. As such, integration with such systems is possible via services, and commonly via Internet technologies such as web services. However, any proprietary system, not providing service based interfaces, needs to be integrated via a service wrapper that translates proprietary interfaces in standard service based interfaces in order to connect the system to the other software applications and industrial hardware devices. An important innovation of this integration layer, e.g. developed in the PERFoRM project (PERFoRM, 2016a), is its distributed and cloud approach, instead of the centralized ones that can be mostly found nowadays and can act as a single point of failure as well as a limitation for the system scalability. For this purpose, this distributed integration layer handles the interconnection of these heterogeneous production components by following the service-orientation principles, i.e., each one is exposing their functionalities as services, which will be discovered and requested by the other components.

Since the implementation of new control technologies will have a direct impact on the production, the implementation of a new decentralized control architecture is not sufficient to achieve the exploitation of Cyber-Physical Production Systems (CPPS). A migration strategy that supports industries in adopting new technologies has been only partially considered in the past projects, e.g., IMC-AESOP envisioned the next generation SCADA/DCS systems (Karnouskos and Colombo, 2011) and investigated an approach to migrate SCADA and DCS systems to SOA (Delsing et al., 2011). At present, it is required a set of guidelines for engineers, equipment developers and end users to plan, support and realize an easy and smooth migration of the existing factories into the new generation of smart factories, taking into consideration both technical and economical issues.

8.3 The Transformation of Automation System Architectures

8.3.1 Towards Information-Driven Automation Systems

Business continuity and agility form the core modus operandi of modern global enterprises (Karnouskos, 2009), and efforts that yield results of more efficient automation systems are well-justified. In order to achieve the pursued agility and continuity, business processes performed in highly distributed production systems need to be efficiently integrated with a sophisticated shop-floor infrastructure that is capable of responding to dynamic adaptations in a timely manner (Karnouskos, 2011).

The prevalence of CPS and the advanced capabilities they offer, mean a drastic reshaping of the future automation system architectures. The increased complexity and sophistication of involved systems, make it very hard to follow monolithic and one-size-fits-all approaches, and make the transition towards modular, dynamic, and open systems imperative (Colombo and Karnouskos, 2009; Karnouskos, 2011). Over the last years, significant efforts have been realized towards service oriented architectures and systems that interact with them (Colombo et al., 2014b). The CPS principle pushes such limits even further, as CPS themselves as well as constellations of them and larger systems of systems need to adhere to similar design patterns and principles.

In such sophisticated infrastructures, emphasis is put on interaction of the CPS with its surrounding environment, which may dynamically change, and which is based on open technologies and interaction patterns rather than closed systems and proprietary software. Hence the integration aspects gain importance, and its focus is significantly expanded for large infrastructures and highly heterogeneous landscapes composed of thousands of devices, systems, services that need to

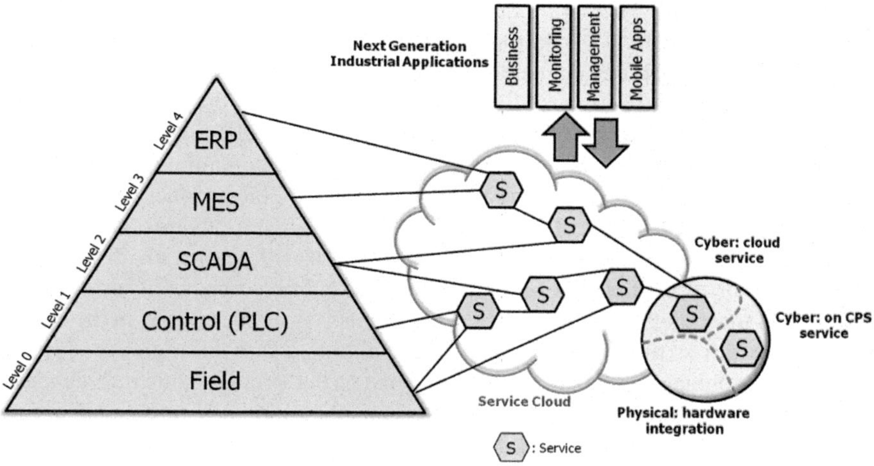

Fig. 8.3 Transitioning towards a SOA-based information-driven architecture by offering key functionalities as services (Karnouskos et al., 2014b; Colombo et al., 2014b)

interact, cooperate and realize their goals in an efficient manner (Colombo et al., 2013).

Considering envisioned architecture transitions such as the one shown in Fig. 8.3, the high-level changes imposed upon engineering of future automation systems are becoming easier to recognize. Figure 8.3 advocates that in parallel to traditional hierarchical architectures in industrial infrastructures, selected functionalities at different levels (e.g., as defined by the ISA-95 paradigm), can be exposed as a collection of CPS services. The latter, may exist in the CPS, traditional systems, as well as the cloud, giving rise to a highly heterogeneous, dynamic, and adequately performant ecosystem of services (Colombo et al., 2013; Karnouskos and Somlev, 2013). Upon such services, applications can cherry-pick the functionalities they need in order to rapidly and efficiently fulfil their goals.

It is important to notice that this transformation of automation systems is performed mainly at the virtual IT level and not in the physical counterpart of the system, which simplifies the migration from the existing automation production systems running currently in the factories to the future ones. Additionally, according to the McKinsey's report (McKinsey, 2015), the implementation of "industry 4.0" solutions will bring significant benefits with only about 40–50% of replacement of equipment.

Considering the proposed innovative automation systems architectures described in Sect. 8.2, one can identify some similarities among them. They build upon the distribution of control functions over intelligent, modular and cooperative entities providing modularity, flexibility, robustness, scalability and reconfigurability, which are at large weak aspects of traditional monolithic architectures. The distributed approach addresses the need to have adequate automation system architectures to tackle the scenario of "Factory of the Future", while being in-line with the guidelines defined by the "Industry 4.0" platform. These architectures also present intelligence and adaptation capabilities embedded in the distributed nodes and in the emergent system behaviour, and some exhibits evolution and self-* properties, such as self-organization, self-adaptation, self-optimization and self-healing.

The deployment of these new decentralized, smart automation architectures in industrial environments need to be performed in a smooth manner, transforming the solutions based on the traditional hierarchical ISA-95 automation structure into solutions based on a network of CPS (ACATECH, 2011; Leitão et al., 2016a). This transformation effort should consider the integration of heterogeneous robotics and automation machinery, as well as the existing legacy systems running in the current industrial solutions to avoid discontinuity and aiming a smooth migration. For this purpose, the plugability is simplified by considering proper industrial standards for protocols and technologies that enable easy integration and interaction among systems and services, while avoiding the creation of "technology islands".

8.3.2 Migration Strategies

The envisioned next generation of industrial automation architectures provide tangible benefits and are a good match for newly established infrastructures (greenfield projects) e.g., can be deployed in a new plant. However, the vast majority of existing infrastructures are brown field projects as they already have constraints in place (e.g., integration with legacy systems and processes), and need to go through migration stages, that will enable the smooth transition from existing systems to the sophisticated infrastructure envisioned.

Current lifetime of production facilities are long, and changes are infrequent and limited. However, this is increasingly changing and in conjunction with the prevalence of software and computational processing at the heart of the 4th industrial revolution, changes are going to be not sporadic but an integral part of the day-to-day business, transitioning towards a DevOps culture. As such, it can be considered that these changes will be applied through incremental migration steps, during the whole lifecycle of devices, services, systems and landscapes. This is especially important as plant operators typically invest multiple millions into their production systems. A change over to decentralized control by a complete revamp of the automation system in one big shot does not only yield a high risk of failure but also annihilates high amounts of investment before they repaid. A stepwise approach of system changeover can bring in small portions of the new distributed control at a time, reducing risks and also allowing to change over the system in accordance with investment ability of the plant operator. Hence, migration strategies are expected to play a pivotal role to the success of the envisioned infrastructures.

Considering the migration to an information-flat and service-based infrastructure as shown in Fig. 8.3, the steps that need to be undertaken are depicted at high level

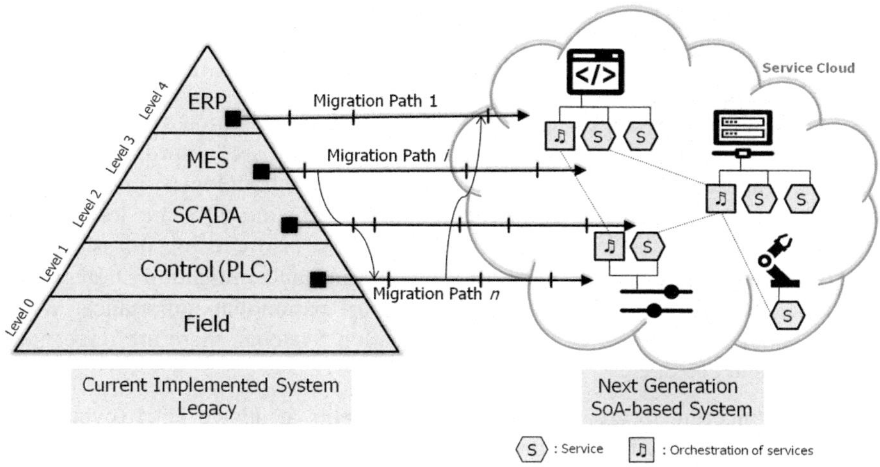

Fig. 8.4 Migration of complex functionalities and cross-layer dependencies to a full SOA-based Infrastructure (Colombo et al., 2014b; Delsing et al., 2011)

in Fig. 8.4. The different system characteristics prevalent in each ISA-95 layer, will need to be captured step-wise in services and be made available. However, as there are several inter-dependencies, the potential migration paths have to be assessed and a migration has to be done step-wise. In doing so, partially the new functionalities will become available to applications and services. Such migration will also unleash at system level emergent behaviours as a result of the dynamic interactions among the different devices and systems. Top-down and bottom-up approaches will need to be analysed in detail (Delsing et al., 2011), and the resulting migration strategies can be highly complex, depending on the preconditions, requirements and goals. More detailed examples with respect to migration and its challenges can be found in Colombo et al. (2014b). Figure 8.4 makes it clear that the migration is not an one-time operation, but rather a continuous one, that the automation industry will have to get accustomed to.

The software industry has long experience with step-wise development, release and upgrade of systems, and can manage such step-wise changes quite well. However, when it boils down to CPS infrastructures with strict operational and timing requirements, things are challenging. In addition, any migration strategies have a multitude of goals that go beyond technology and include, cost-effectiveness, resource-efficiency, agility, deterministic behaviour, operational easiness, business continuity etc. (Karnouskos, 2009). Due to their cross-disciplinary nature, applied at enterprise level, such migration strategies pose some risk which needs to be managed. However, once the envisioned architectures and modus operandi are in place, such incremental migratory actions are expected to be easier to realize.

8.4 Considerations on Future Automation System Architectures

The transition from the existing traditional industrial automation systems, mainly based on Product-Life-Cycle (PLM) into the new CPS based approaches, should be smooth and requires a rethinking of engineering methodologies, integration of methods and tools from the different domains where the CPS are located and best practices. Since such changes have to consider also existing infrastructures and business continuity, consideration of migration and mitigation strategies to overcome the identified challenges is seen as of paramount importance. When talking about Engineering CPS-based Automation Systems, there are three main task clusters to consider:

- The Engineering to create new CPS components at device level (cyber- and physical views) and the Engineering to build the System of CPS.
- The Engineering to reconfigure or adapt an existing CPS, to operate it and to manage its evolution, both at device and system levels
- The Engineering to design, implement, operate and manage autonomous/smart CPS components within an intelligent automation infrastructure

8.4.1 Rethinking of Automation Systems Engineering

It is important to recognize that all the parts involved in future automation system architectures, will not be under the control of a single authority, and technology, and therefore, the integration, interaction and operation will need to be done via open interfaces exposed by the services (Karnouskos, 2011; Colombo et al., 2014b). Taking into consideration the goals of a CPS, as discussed in Sect. 8.1, the engineering effort to adapt a CPS during run-time must be minimal. This means "Zero Engineering" during run-time must be prepared and implemented. The ability to reconfigure existing elements and to integrate new elements have to be a "built-in" capability of the CPS on system level. Engineering such systems has to cope with continuous updates of the infrastructure (both in hardware and software) and to provide high resilience for the CPS.

Aspects such as Systems of CPS integration and dynamic reconfiguration require a set of complementary engineering tasks, which are strongly related to the major characteristics to be covered by an adequate Systems-of-Systems engineering approach, i.e.,

- engineering evolvability at system level due to plug-and-play integration and live removal of CPS components;
- dynamic requirements engineering to support incremental live validation of structural and behavioural modifications of the system (understanding and managing "emergency";

- control re-configuration for several control systems that are strongly coupled;
- last but not least the integration of the human factor in each of the phases of the life cycle without loosing the System-of-Systems view perspective

As the different parts of such a system will evolve independently, good practices for engineering, upgrading, operating and maintaining them need to be followed.

The core idea behind the amalgamating the physical and virtual (digital) worlds, is to seamlessly gather useful data and information about objects of the physical world, transform it to knowledge, and empower various industrial applications (Karnouskos, 2011; Colombo et al., 2014b). The emerging engineering systems, operating in highly sophisticated infrastructures as discussed, are expected to enable the elimination of many existing pain points, but unavoidably it will create others. The new ones will require engineers to draw on knowledge from multiple disciplines (Broy and Schmidt, 2014; Karnouskos et al., 2014a) if they want to effectively capitalize on the new capabilities.

The automation engineers dealing with Industrial Systems of Cyber-Physical Systems have to possess a much wider set of skills to understand how the different constituent systems interact, both in structural and behavioural manner, as well as a solid background on Information, Communication, Control Technology and their fusion.

As such, engineering effective solutions implies e.g., technical excellence, understanding of hardware and software components in the infrastructure, knowledge of industrial operational context, understanding of interactions at device and system level, risk estimation, understanding of the impact of engineering decisions e.g., to safety, security, dependability, etc.

8.4.2 Directions and Challenges

The described transformation into the future industrial automation systems, and their industrial adoption, presents several challenges, which can be aggregated in 6 major clusters (Leitão et al., 2016a):

- *CPS Capabilities*, which comprises the modularization and servification of CPS systems, the development of CPS as System of Systems (SoS), their optimization and real-time monitoring and control, as well as the consideration of advanced (big) data analytics.
- *CPS Management*, which includes the security and trust in the management of large scale CPS, aiming to achieve industrially mature solutions.
- *CPS Engineering*, which comprises the safe programming and validation, the resilient risk mitigation, and methods and tools for the CPS and Systems-of-CPS life-cycle support, which are crucial challenges for the industry. A challenge is the need to apply new methods within the engineering of these systems (e.g., collaborative workflow generation and processing).

- *CPS Ecosystems*, which includes the design and deployment of collaborative, autonomic, self-* and emergent CPS, as well as the integration of Humans in the Loop, many of them being expected to be matured only in the long-run.
- *CPS Infrastructures*, which are related to interoperability services, and mitigation and migration strategies to support the transformation of current automation systems into the future CPS ones.
- *CPS Information Systems*, which considers artificial intelligence, data transformation and data analytics to capitalize the huge amount of collected data to reach actionable knowledge.

A brief analysis of reported research and innovation results demonstrated over the last 15–20 years allows to better understand how such actions can be realized by combining CPS, Internet-of-Things and Internet-of-Services technologies. Embedding at large industrial agents and Service-oriented based automation (SOCRADES, 2016; Taisch et al., 2009; Leitão et al., 2016b) is one innovation approach to be highlighted. In fact, agents may act as enablers for CPS-based industrial system architectures and contribute in terms of technology/solution maturity, methodologies and tools, human in the loop, smooth migration and self-* properties, and standardization (Leitão et al., 2016b; Leitão and Karnouskos, 2015a)

Another important dimension for the fully industrial adoption of CPS-based automation systems architectures is the standardization (Kagermann et al., 2013; IEC, 2015), since the standards compliance may affect the development, installation and commissioning of industrial applications. In fact, standardization can support the deployment of CPS, and particularly the smooth migration of these systems, by easily interfacing with existing legacy systems, plugging devices and systems, and adapting their behaviour and relationships on-the-fly. The integration of humans in the loop is seen as a key factor to achieve flexibility (Kagermann et al., 2013), and not more as an obstacle for the complete system automation, as sustained in the past, and particularly during the advent of Computer Integrated Manufacturing (CIM) paradigm.

The Reference Architecture Model Industrie 4.0 (RAMI4.0) standard (DIN, 2016) presents the major architectural specifications for Cyber-Physical components (labelled as I4.0-component) and the set of rules for engineering Industry 4.0 compliant architectures. Aspects related to the CPS-integration within an ISA95-compliant architecture, the different phases of the life-cycle of the CPS components and systems of CPS are considered as the base for supporting the engineering of CPS-based industrial systems. In this sense, something that has to be highlighted is the specification of the six digitalization-layers, which cover the full process of building a Cyber-Physical component, starting with the mechatronics (assets) and going through the integration, communication, information, function and business layers. A set of communication and information layers based on the use of Internet technologies, and the exposition of automation function as services in an Internet-of-Services fashion, enable the I4.0-components (CPS-component) to engage into business relationships with other components within a system of CPSs.

Additionally, the implementation of the new generation of automation systems will demand new challenges for vocational and academic training and continuing professional development, as sustained by the "industry 4.0" high-level working group in its recommended actions (Kagermann et al., 2013; Karnouskos et al., 2014a). In fact, nowadays, engineers need to integrate multidisciplinary and cross-domain knowledge, focusing more on the understanding of system of systems perspective than in a deeply topic domain. In parallel, the penetration of Information and Communication technologies into traditional mechatronics, hydraulics, pneumatic systems, are continuously re-shaping the world, and require an integrative learning process.

The engineering-students are no more dealing only with the physical but predominantly with the cyber part of complex engineering systems, which implies that their acquired knowledge quickly becomes obsolete (some times in less time than the student takes to get the undergraduate degree). Therefore, they need to learn different topics to be able to compete in the future (more systems/system of systems understanding instead of pure (deep) domain knowledge). As example, new engineers have to cope with new paradigms and concepts (e.g., modelling, semantics, (crowd) collaboration, interoperability, self-organization and self-diagnosis) and emergent technologies (e.g., Internet-of-Things, Big data, Machine-to-Machine, advanced data analytics, cloud computing and augmented reality).

Considering all the raised concerns, educating engineers, in the "Industry 4.0" context, means learning how to design, develop, test, deploy, and operate a traditional engineering environment that is being digitalized in both, its structural but also in its behavioural/functional aspects.

The implementation of strategies for the smooth migration from traditional automation systems into the new generation of distributed automation systems are crucial since legacy systems will continue running and will co-exist with the new systems (Leitão et al., 2016b; Karnouskos et al., 2014a). As an example, during the implementation of the GRACE MAS system in the Whirlpool's factory plant producing washing machines (Leitão et al., 2015), the lower control level using PLCs running IEC61313-3 programs was preserved to ensure the real-time control and the MAS solution was placed at the higher control level to introduce intelligence and adaptation to the system performance. However, this is an emergent topic that deserves a significant research in the near future to establish the proper strategies to ensure a smooth migration transforming the existing running systems into Industrie 4.0 compliant systems. These migration strategies should consider the technical perspective, as briefly described in Sect. 8.3.2, but also a deeply study of the impact of economical and social perspectives.

8.5 Conclusion and Outlook

There is a need for flexibility, resilience and optimization in industrial settings, that can not be adequately tackled with traditional approaches. Although significant steps have been realized by concepts and utilization of key technologies such as MAS, SOA, Cloud, CPS, significant efforts are still needed to tackle additional challenges related to their engineering and interaction in emerging cooperative production systems. The intention in this chapter is not to provide a new model-based approach but to understand why and how the already existing methods and tools that enable production system flexibility and self-adaptation of CPPS are not adequate or too poorly implemented in industrial practice. At the end, the successful applications of such concepts and technologies will not only be determined by the ability to deal with technology problems, but effectively cover also all other associated aspects that enable continuous business growth and effectiveness.

One of these aspects is about the availability and quality of information. As (sub-) systems are not considered as monolithic building blocks any more, but are seen in their environment of strongly interconnected systems of systems the view on information availability needs to be altered. This altered view needs to reflect not only the system itself, but also its role within its environment, lifecycle, functional hierarchy, etc. This aspect has already been described in Chaps. 5, 6 and 7 of this book.

The increased integration of the cyber and physical aspects of systems, also leads to new challenges for system applications (Lee, 2008; Leitão et al., 2016a; Broy and Schmidt, 2014; Karnouskos et al., 2014a). In the past optimization and improvements have been targeted mainly on isolated parts of the system. Hence, improved production processes and technologies has led to new or improved assets (see top left in Fig. 8.5) or improved control approaches and technologies in the system architecture (see top right in Fig. 8.5). To bring these improvements into

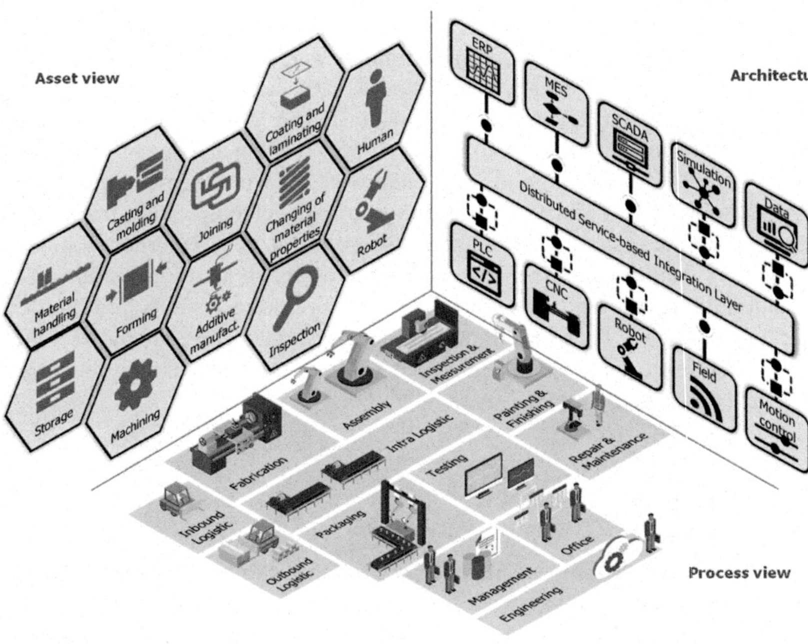

Fig. 8.5 PERFoRM (2016b) project multi-view on production systems

already existing production systems basically meant to interchange an existing building block (e.g., production asset, IT system) with a new one.

Nowadays these improvements are still possible, but they will not allow to sufficiently address all challenges which are arising from the new complexity of self-* systems and distributed intelligence. In fact the introduction of these concepts requires a change in the heart of each system as they lead to changes in multiple areas and are not isolated only to system building blocks. The integration of assets and IT will allow to improve the whole value adding process (see bottom in Fig. 8.5).

To do this in an efficient and cost-effective way which is suitable for plant operators, new migration methods have to be researched and mitigation strategies need to realized, as discussed in Sects. 8.3 and 8.4. As an example of such an effort, the PERFoRM (2016b) project does not focus on the development of new technologies for tackling flexibility, resilience and optimization needs, but to the re-use of existing developments and their harmonization as also already shown in Sect. 8.2. Additionally a strong focus is set to the development of suitable migration methods and mitigation of existing obstacles in order to create an environment and guidelines for industry to apply decentralized automation system architectures.

This approach, as also proposed within this chapter, allows to re-use already developed technologies and especially to capitalize on the money already spent for this research. Additionally, as a side effect, it stops the ongoing diversification in developed solutions and thus a further diversification of similar technologies to be harmonized or even standardized later on. A downside of this approach is that it can only utilize technologies that already passed at least a conceptional stage at which they are recognized as an already available technology.

Acknowledgements The authors would like to thank the European Commission for the support, and the partners of the EU Horizon 2020 project PERFoRM (2016b) for the fruitful discussions. The PERFoRM project has received funding from the European Union's Horizon 2020 research and innovation programme under grant agreement No 680435.

References

ACATECH: Cyber-Physical Systems: driving force for innovation in mobility, health, energy and production. Tech. rep., ACATECH – German National Academy of Science and Engineering. https://goo.gl/Q6WFQN (2011)

ANSI/ISA: ANSI/ISA-95.00.01-2010 (IEC 62264-1 Mod) Enterprise-Control System Integration – Part 1: Models and Terminology. http://www.isa.org (2010)

Antzoulatos, N., Castro, E., Scrimieri, D., Ratchev, S.: A multi-agent architecture for plug and produce on an industrial assembly platform. Prod. Eng. Res. Devel. **8**(6), 773–781 (2014); doi:10.1007/s11740-014-0571-x

Broy, M., Schmidt, A.: Challenges in engineering Cyber-Physical Systems. Computer **47**(2), 70–72 (2014); doi:10.1109/mc.2014.30

Castellini, P., Cristalli, C., Foehr, M., Leitão, P., Paone, N., Schjolberg, I., Tjonnas, J., Turrin, C., Wagner, T.: Towards the integration of process and quality control using multi-agent technology. In: IECON 2011 – 37th Annual Conference of the IEEE Industrial Electronics Society (2011); doi:10.1109/iecon.2011.6119347

Colombo, A.W., Karnouskos, S.: Towards the factory of the future: a service-oriented cross-layer infrastructure. In: ICT Shaping the World: A Scientific View, pp. 65–81. European Telecommunications Standards Institute/Wiley, New York (2009)

Colombo, A.W., Karnouskos, S., Mendes, J.M.: Factory of the future: a service-oriented system of modular, dynamic reconfigurable and collaborative systems. In: Springer Series in Advanced Manufacturing, pp. 459–481. Springer, London (2010); doi:10.1007/978-1-84996-119-6_15

Colombo, A.W., Karnouskos, S., Bangemann, T.: A system of systems view on collaborative industrial automation. In: 2013 IEEE International Conference on Industrial Technology (ICIT) (2013); doi:10.1109/icit.2013.6505980

Colombo, A.W., Bangemann, T., Karnouskos, S.: IMC-AESOP outcomes: paving the way to collaborative manufacturing systems. In: 2014 12th IEEE International Conference on Industrial Informatics (INDIN) (2014a); doi:10.1109/indin.2014.6945517

Colombo, A.W., Bangemann, T., Karnouskos, S., Delsing, J., Stluka, P., Harrison, R., Jammes, F., Martínez Lastra, J.L. (eds.): Industrial Cloud-based Cyber-Physical Systems: The IMC-AESOP Approach. Springer, New York (2014b)

Delsing, J., Eliasson, J., Kyusakov, R., Colombo, A.W., Jammes, F., Nessaether, J., Karnouskos, S., Diedrich, C.: A migration approach towards a SOA-based next generation process control and monitoring. In: IECON 2011 – 37th Annual Conference of the IEEE Industrial Electronics Society (2011); doi:10.1109/iecon.2011.6120045

Delsing, J., Rosenqvist, F., Carlsson, O., Colombo, A.W., Bangemann, T.: Migration of industrial process control systems into service oriented architecture. In: IECON 2012 – 38th Annual Conference on IEEE Industrial Electronics Society, Montreal, QC (2012); doi:10.1109/iecon.2012.6389039

DIN: Reference Architecture Model Industrie 4.0 (RAMI4.0). Tech. rep., Deutsches Institut für Normung (DIN). http://www.din.de/en/wdc-beuth:din21:250940128 (2016)

Friedrich, J., Scheifele, S., Verl, A., Lechler, A.: Flexible and modular control and manufacturing system. Procedia CIRP **33**, 115–120 (2014); doi:10.1016/j.procir.2015.06.022

Goncalves, G., Reis, J., Pinto, R., Alves, M., Correia, J.: A step forward on intelligent factories: a smart sensor-oriented approach. In: Proceedings of the 2014 IEEE Emerging Technology and Factory Automation (ETFA) (2014); doi:10.1109/etfa.2014.7005227

IEC: White paper factory of the future. Tech. rep., IEC – International Electrotechnical Commission (2015)

Kagermann, H., Wahlster, W., Helbig, J.: Securing the future of German manufacturing industry: recommendations for implementing the strategic initiative industrie 4.0. Tech. rep., ACATECH – German National Academy of Science and Engineering. http://goo.gl/oc3X4n (2013)

Karnouskos, S.: Efficient sensor data inclusion in enterprise services. Datenbank-Spektrum **9**(28), 5–10 (2009)

Karnouskos, S.: Realising next-generation web service-driven industrial systems. Int. J. Adv. Manuf. Technol. **60**(1–4), 409–419 (2011); doi:10.1007/s00170-011-3612-z

Karnouskos, S., Colombo, A.W.: Architecting the next generation of service-based SCADA/DCS system of systems. In: 37th Annual Conference of the IEEE Industrial Electronics Society (IECON 2011), Melbourne (2011); doi:10.1109/iecon.2011.6119279

Karnouskos, S., Somlev, V.: Performance assessment of integration in the cloud of things via web services. In: 2013 IEEE International Conference on Industrial Technology (ICIT) (2013); doi:10.1109/icit.2013.6505983

Karnouskos, S., Baecker, O., de Souza, L.M.S., Spiess, P.: Integration of SOA-ready networked embedded devices in enterprise systems via a cross-layered web service infrastructure. In: 2007 IEEE Conference on Emerging Technologies and Factory Automation (EFTA 2007), pp. 293–300 (2007); doi:10.1109/efta.2007.4416781

Karnouskos, S., Guinard, D., Savio, D., Spiess, P., Baecker, O., Trifa, V., de Souza, L.M.S.: Towards the real-time enterprise: service-based integration of heterogeneous SOA-ready industrial devices with enterprise applications. IFAC Proc. Vol. **42**(4), 2131–2136 (2009); doi:10.3182/20090603-3-ru-2001.0551

Karnouskos, S., Savio, D., Spiess, P., Guinard, D., Trifa, V., Baecker, O.: Real-World Service Interaction with Enterprise Systems in Dynamic Manufacturing Environments. Springer Series in Advanced Manufacturing, pp. 423–457. Springer, New York (2010); doi:10.1007/978-1-84996-119-6_14

Karnouskos, S., Colombo, A.W., Bangemann, T.: Trends and challenges for cloud-based industrial cyber-physical systems. In: Industrial Cloud-based Cyber-Physical Systems: The IMC-AESOP Approach, pp. 231–240, Springer, New York (2014a); doi:10.1007/978-3-319-05624-1_11

Karnouskos, S., Colombo, A.W., Bangemann, T., Manninen, K., Camp, R., Tilly, M., Sikora, M., Jammes, F., Delsing, J., Eliasson, J., Nappey, P., Hu, J., Graf, M.: The IMC-AESOP architecture for cloud-based industrial Cyber-Physical Systems. In: Industrial Cloud-Based Cyber-Physical Systems, pp. 49–88. Springer, New York (2014b); doi:10.1007/978-3-319-05624-1_3

Lee, E.A.: Cyber physical systems: design challenges. In: 11th IEEE International Symposium on Object and Component-Oriented Real-Time Distributed Computing (ISORC), pp. 363–369 (2008); doi:10.1109/ISORC.2008.25

Leitão, P., Karnouskos, S.: A survey on factors that impact industrial agent acceptance. In: Industrial Agents: Emerging Applications of Software Agents in Industry, pp. 401–429. Elsevier, Amsterdam (2015a); doi:10.1016/b978-0-12-800341-1.00022-x

Leitão, P., Karnouskos, S. (eds.): Industrial Agents: Emerging Applications of Software Agents in Industry. Elsevier, Amsterdam (2015b)

Leitão, P., Barbosa, J., Vrba, P., Skobelev, P., Tsarev, A., Kazanskaia, D.: Multi-agent system approach for the strategic planning in ramp-up production of small lots. In: 2013 IEEE International Conference on Systems, Man, and Cybernetics (2013); doi:10.1109/smc.2013.807

Leitão, P., Rodrigues, N., Turrin, C., Pagani, A.: Multi-agent system integrating process and quality control in a factory producing laundry washing machines. IEEE Trans. Ind. Inf. **11**(4), 879–886 (2015); doi:10.1109/tii.2015.2431232

Leitão, P., Colombo, A.W., Karnouskos, S.: Industrial automation based on cyber-physical systems technologies: Prototype implementations and challenges. Comput. Ind. **81**, 11–25 (2016a); doi:10.1016/j.compind.2015.08.004

Leitão, P., Karnouskos, S., Ribeiro, L., Lee, J., Strasser, T., Colombo, A.W.: Smart agents in industrial Cyber–Physical Systems. Proc. IEEE **104**(5), 1086–1101 (2016b); doi:10.1109/jproc.2016.2521931

MacKenzie, C.M., Laskey, K., McCabe, F., Brown, P.F., Metz, R.: Reference model for service oriented architecture 1.0. http://docs.oasis-open.org/soa-rm/v1.0/ (2006)

McKinsey: Industry 4.0: How to navigate digitization of the manufacturing sector. Tech. rep., McKinsey Digital. https://www.mckinsey.de/files/mck_industry_40_report.pdf (2015)

Onori, M., Maffei, A., Durand, F.: The IDEAS plug and produce system. In: International Conference on Advanced Manufacturing Engineering and Technologies, NewTech (2013)

PERFoRM: Definition of the system architecture. Tech. rep., Deliverable D2.2, PERFoRM project. http://www.horizon2020-perform.eu/files/documents/D2_2_public.pdf (2016a)

PERFoRM: Production harmonizEd Reconfiguration of Flexible Robots and Machinery (PERFoRM) project, European Commission, horizon 2020 programme. http://www.horizon2020-perform.eu (2016b)

Sayed, M.S., Lohse, N., Sondberg-Jeppesen, N., Madsen, A.L.: SelSus: Towards a reference architecture for diagnostics and predictive maintenance using smart manufacturing devices. In: 2015 IEEE 13th International Conference on Industrial Informatics (INDIN), Institute of Electrical and Electronics Engineers (IEEE) (2015); doi:10.1109/indin.2015.7281990

SOCRADES: Service-Oriented Cross-layer infRAstructure for Distributed smart Embedded devices (SOCRADES) project, European Commission, FP6 Programme. http://www.socrades.net (2016)

Stokic, D., Scholze, S., Barata, J.: Self-learning embedded services for integration of complex, flexible production systems. In: IECON 2011 – 37th Annual Conference of the IEEE Industrial Electronics Society (2011); doi:10.1109/iecon.2011.6119346

Taisch, M., Colombo, A.W., Karnouskos, S., Cannata, A.: SOCRADES roadmap: The future of SOA-based factory automation. Tech. rep., SOCRADES Project. http://www.socrades.net/Documents/objects/file1274836528.pdf (2009)

Webb, P., Asif, S.: Advanced flexible automation cell. In: 6th Innovation for Sustainable Aviation in a Global Environment (2011)

Wooldridge, M.: An Introduction to Multi-Agent Systems. Wiley, Harlow (2002)

Chapter 9
Engineering Workflow and Software Tool Chains of Automated Production Systems

Anton Strahilov and Holger Hämmerle

Abstract Application fields of automated production systems are varied, e.g. automotive, aerospace and food industry, just to name a few. The complexity of such production systems has significantly been increased in the last years, (Koren et al., CIRP Ann Manuf Technol 48(2):527–540, 1999). This increase was a result of the increased complexity and variance of products. As a result of this, the engineering workflow of automated production system has continuously been adapted to new requirements. In this regards, this chapter shows and describes the current engineering workflow of automated production systems based on experience in the field of production system for the automotive industry. The main focus of this description is set on the established tool-chains and used tools to create engineering information as well as data formats to save and exchange information between tools and involved personnel. In the introduction of this chapter, differences between an automated production system and a cyber-physical system are given. Current production systems could be named CPPS but this term is not popular in the field of production system builder as well as production owners. But in spite of that the end of this chapter gives an outlook of the future of automated production systems in direction of CPPS.

Keywords Tool chains • Exchange data formats • Automated production systems • Engineering data • Engineering workflow • Engineering tools • Automotive industry

List of Abbreviations

CPPS Cyber-physical production systems
CPS Cyber-physical production systems
OEM Original equipment manufacturer
OLP Offline robot program/offline robot programing

A. Strahilov (✉) • H. Hämmerle
EKS InTec GmbH, Heinrich-Hertz-Strasse 6, 88250, Weingarten, Germany
e-mail: anton.strahilov@eks-intec.de; holger.haemmerle@eks-intec.de

© Springer International Publishing AG 2017

S. Biffl et al. (eds.), *Multi-Disciplinary Engineering for Cyber-Physical Production Systems*, DOI 10.1007/978-3-319-56345-9_9

PLC Programable logical controller
PM Plant manufacturer/system manufacturer
VC Virtual commissioning
VE Virtual engineering

9.1 Introduction

In the past years, the term cyber-physical system has not been popular in the industrial area of production systems. A production system manufacturer was coping with mechatronic systems rather than *Cyber-Physical Production Systems* (CPPS). The reason for this is that, the term CPPS was not very popular and further the developed production system could not be referred to being a cyber-physical production system. The primary cause is the missing intelligence of implemented sub-components (e.g. electric drives, pneumatic cylinders, sensors, etc.) in the production system. Additionally, the information about the products, which have to be manufactured with the system, and the production processes are integrated into a higher-level *Programmable Logic Controller* (PLC) as a software/production program (Dilts et al. 1991). Consequently, sub-components could not react to product changes with modification of the production process by themselves. Furthermore, the communication of sub-components with each other in a production system was rarely used in the practice. The same is also true for the local communication of production systems of a production line between each other or with other systems via internet.

Unfortunately, production systems that are currently developed could not be labelled now as CPPS, because not all sub-components have built-in intelligence as well as using this intelligence within the production process. But the first step to use such CPPS in practice for automotive production has not been undertaken yet. Despite that significant changes of established tool chains for the design of production plants are not expected from the *production system builder*. Only changes or add-ons for functionalities of currently applied tools as well as on the engineering workflow are expected to be developed. In this regard, established tool chains in practice are presented in this chapter. To achieve a better understanding about the tool chains, a universal and established engineering workflow of production systems for automotive manufacturing will be presented.

The transition of a usual production system to a CPPS is the next step for each product system builder as well as system user. New innovative methods and tools will be required to design qualitative CPPS. Effective usage of these methods and tools during the design of the CPPS will be a significant success factor of acceptance by employees. During this, employers have to rethink established system design concepts of production systems. This rethink process will be taking more time. An adaptation of established development workflows will also be a necessary step within the rethink process. A similar rethink process also has to be performed by users of production system as well as by component manufacturers. Moreover, all three parties' production system builder, production system owner and component manufacturers have to work together to set a sound basis on which CPPS can be further designed.

The current chapter establishes an engineering workflow of production systems. The three main phases *System Development*, *Productive Use* and *Recycling/Re-Use* phases of the lifecycle of production systems are discussed. From those three phases, the state of the system development phase stand in focus of this chapter. In this regards, established tool chains in practice and common used exchange data formats are presented. Finally, a summary of the chapter is given which contains also a view of the further of automated production systems in direction of CPPS.

Across this chapter, some research questions described in Chap. 1 are taken into a count. Specifically, following research questions are from greatest interest for the current chapter: *information in and across value chains, quality assurance for information exchange* and *description of plug-and play capabilities and interfaces for engineering and run time* (see Chap. 1, RQI 1–RQ I3).

9.2 Engineering Workflow of Production System

The life cycle of production systems encompasses various phases which could be simplified as sequential process (Drescher et al. 2013). Thereby, the plant owner (*Original Equipment Manufacturer (OEM)*) authorizes a production system builder (*Plan Manufacture (PM)*) to develop and provide a production system regarding to its individual requirements (Li and Meerkov 2001). In the system development phase of the production system, mechanics, electronics and software of the system are designed (Groover 2007). Subsequently, the phase of productive utilization of the system starts and continues until the production system is used productively. With the ending of this phase, the production system will be recycled or modified, in such a way that the system could be used for a different production processes. In the following the development phase of production systems is presented in detail.

The production system's mechanics, electronics and software are designed and in a next step realized. Regarding to this, the production system development could be divided into the both phases system design and system realization (Fig. 9.1).

In system design, as its title suggests, the production system is about to be designed. Thereby, the three sub-phases *mechanical design*, *electrical design* and *software design* can be distinguished. In principle, those phases are executed in parallel and depend on each other. Additionally, system validation activities are carried out in parallel to the three phases. These activities can be divided into the phases *Virtual Engineering* (VE) and *Virtual Commissioning* (VC).

Into mechanical design, a detailed 3d geometric model of the production system is created. In this model, mechanical relations between product and production system are integrated together, e.g. position of welding points, gripper points, robots, etc. (see Chap. 1-RQ I1). Based on this geometric model, a 2d drawing of each part as well as aggregated assemblies is derived. These drawings are

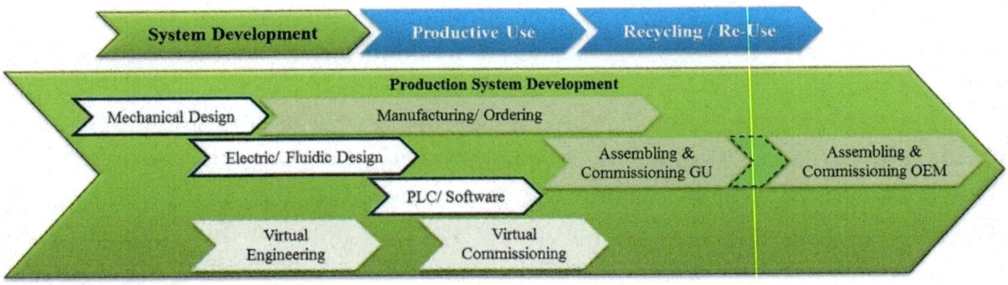

Fig. 9.1 Production system development process

required for the manufacturing of the system parts that are individually designed solely for the production system under development. For the remaining parts, a 2d drawing is required to become an overview about the used parts and the correct mounting position, e.g. screws, springs, nuts, flat washers, etc. In some cases, components provided by a manufacturer are used, e.g. electric drives, pneumatic drives, pneumatic valves, controllers, industrial robots, etc. Those components are assembled from a lot of parts and should not be produced by the system manufacturer. For such component's parts, the 2d drawing is not really useful and is not derived.

With the 3D geometric model of the production system, the layout of the system is defined as well. Thereby, the position of each part respectively component of the production system is determined. In this context, the position of product parts and external material flow of parts into the system is defined via the layout.

Another output of the mechanical design is the production process. This process strongly depends on the requirements defined by the product development engineers that intend the sequence of the assembly steps for his product. Also, the production process has a significant impact on the system's mechanical design respectively layout. In combination with the defining of the production process, offline robot programs (OLP) are also created. This output covers important relations between production system, products and production processes, e.g. processes performed via robots, sequence of movement of pneumatic/electric drives, change-over between product variant during production processes, etc. (see Chap. 1-RQ I1).

Regarding to the defined production system's parts and components of the mechanical design, a list of all installed parts is created. With this list, the quality of information exchange between mechanical engineer, pneumatic/electric engineer and purchasing can be ensured (see Chap. 1-RQ I2). Furthermore, the electrical engineer derives all parts and components that he requires as input for starting the design of the electric plan of the production system. In practice, the electric/fluidic design starts even before the final part list is determined. Regular changes on the electric plan regarding changes on the selected components are only one disadvantage of this procedure, e.g. changes of electrical drives, sensors, industrial robots, etc. But the premature beginning of the electric design phase is a consequence of time constraints since the development phase only accounts for

3–6 months in total. In some cases, these time period can be more than 6 months depending on the complexity of the production system (Drescher et al. 2013). Finally, the electrical plan of the entire production system is created. In this plan, all required data is summarized that is needed to connect components via electrical cables, e.g. cable types, port names, electrical sockets, requirements to the cable laying, additional components, etc.

Based on the created part list, pneumatic components are defined by the mechanical engineer together with the pneumatic engineer. Based on this list, a pneumatic plan of the production system is created by a pneumatic engineer. This pneumatic plan encompasses definitions regarding to use tubes, connections between the components, used valves to control cylinders, throttles to define the cylinder velocity, etc. Information about installation of the tubes is determined in the pneumatic plan in detail. In most cases, installation instructions are defined in the system realization phase and are not documented after that. The same is also valid for the installation of electrical cables. Despite that dependencies between electronics and pneumatics of the system are also added into both plans, e.g. connection between PLC signals and pneumatic valves as well as inductive sensors via electric cables, etc. Finally, production system's specific information about connections between mechanic, electric/pneumatic and PLC components are integrated with each other (see Chap. 1-RQ I1).

Parallel to the electrical/fluidic design, the validation of the mechanic design is carried out within the virtual engineering phase. In this phase, the collision free movement of each drive component (e.g. pneumatic cylinders, electric drives, robots, etc.) in the production system regarding to the production process is of major interest. Within virtual engineering the validation of production cycle time is also of paramount importance. Using this method, created outputs are collected and tested with each other (see Chap. 1-RQ I2).

With the ending of the electrical/fluidic design, the PLC/software design phase starts. In this phase, the software (control program) of the production system is created. The information contained in the electric plan and pneumatic plan as well as the knowledge about the production process of the production system is required. Furthermore, security-related aspects regarding to manufacturing personal and product will also be considered during the programming of the software. Finally, the complete software of the production system is created. Thereby, a testing of this program could be not done even before the real production system is built up.

In order to test and validate the production system's software the phase virtual commissioning is added as a sub-phase into the system design. In this phase, as its name suggests, the virtual commissioning method is used to test and validate the functionalities of the developed software. For this purpose, an extended 3D geometric model of the mechanics and a logic behaviour model of the production system are required to represent the real production system for the PLC that runs the developed software (Süß et al. 2015).

The manufacturing of the designed parts starts after completion of the parallel running phases after the mechanical design. Thereby, all parts developed for the current production system are produced and the remaining parts or components provided by suppliers are ordered. The duration of this phase depends on the complexity of the production system as well as on the ordered components and can account to one until 4 months. Regarding to this long period, this phase starts in some cases even before the mechanical design phase is completed.

In the automotive industry, the initial assembling and commissioning of the production systems is done by the production system builder. The system builder tests at his facilities the functionality of the system mechanic together with the interactions with the system electric, pneumatic and software. As soon as the automatic performing of the production process is tested, the system builder disassembled the system in separate modules and transports those modules to the plant owner (OEM's) shop floor. In the following step, these modules will be reassembled and commissioned again by the production system builder. Thereby, the production system is not tested again in detail, only minor adjustments to the local conditions are implemented.

9.3 Established Tool Chains in Practice

Regarding to the presented production system development process (Fig. 9.1), an overview about the results of each phase is displayed in Table 9.1. From this table, the relation between generated outputs of each phase and the processing of those outputs in other phases can be inferred. Thereby, it is evident from the overview that a lot of various outputs are generated during the production system development process. To generate those multiple outputs, a high number of various domain specific tools are used. Widely used domain specific tools are presented in this section and assigned to the established tool chains.

Furthermore, the significant data formats that serve to save and exchange outputs between the phases during the development process are listed. In this context, resulting difficulties during the exchange of outputs via data formats will also be considered. Regarding to this, the exchange data format AutomationML is introduced as possibility to achieve a qualitative information transfer between domain specific tools without information losses (see Chap. 1-RQ I2). Additionally, the utilization of AutomationML within the production system development process will be demonstrated via practical examples. A detailed description of the standard data exchange format AutomationML is given in Sect. 11.4.

In the following part of this section, the established tool chain for each phase of the production system development process is presented in separate sections. Based on this, a description of the output's content from each phase as well as of used data formats regarding to the used tools will be made in detail. Therefore, this section is divided into tool chain for mechanical design, electrical/fluidic design, PLC/software, virtual engineering and virtual commissioning.

Table 9.1 Overview of phase's outputs in the production system development process

Engineering phases	3d geometry	2d drawings	Layout plan	Process plan	Parts list	Offline robot programs	Documentation	Electric plan	Pneumatic plan	PLC/Software	Kinematic 3d simulation	Process simulation	Material flow simulation	Robot simulation	Behaviour simulation
Mechanical design	c	c	c	c	c	c	c								
Electric/fluid design							c	c	c						
PLC/Software				u			c	u	u	c					
Virtual engineering	u		u	u		u	c				cu	cu	cu	c	cu
Virtual commissioning	u		u	u			c	u	u	u	cu		cu	c	cu
Manufacturing/ ordering	u	u					c								
Assembling and commissioning	u	u	u	u		u	u	u	u	u					

c create; *u* use; *cu* create and use

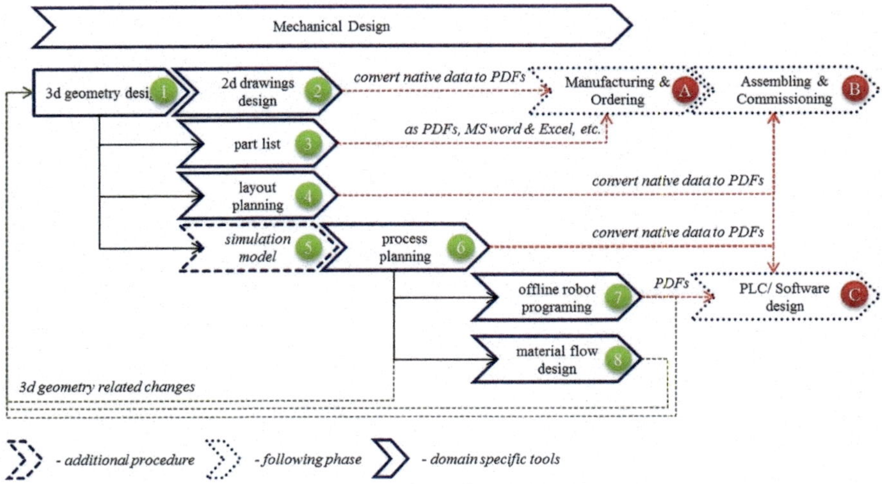

Fig. 9.2 Established tool chain of mechanical design of production systems

9.3.1 Tool Chain for Mechanical Design

Used tools during the mechanical design are extremely diverse and strongly depend on the data format that the PM has to deliver to the plant owner (OEM). A distinction between domain specific tools or rather between functionalities is made particularly in this contribution into Fig. 9.2. In this figure, connections between the various domain specific tools are added, too.

For better understanding, a tool for each output of the mechanical design is represented in the tool chain. It is necessary to keep in mind, that the tool chain doesn't show the sequence of using of each domain specific tool but only the relation between them. Those tools, respectively their outputs, are explained in the following paragraphs.

As well described in Sect. 9.1, the main output of the mechanical design is the 3d geometric model of the production system (Fig. 9.3). To create this model, various Computers-Aided Design (CAD) tools are already available in the market. Some of those widespread CAD tools are CATIA,[1] NX,[2] SolidWorks,[3] Creo,[4] AutoCAD.[5] In principle, these CAD tools provide the same functionalities while differing inter alia in terms of the various operations of the graphical user interface.

[1]http://www.3ds.com/products-services/catia

[2]http://www.plm.automation.siemens.com/en_us/products/nx/index.shtml

[3]http://www.solidworks.de/

[4]http://de.ptc.com/product/creo

[5]http://www.autodesk.de/products/autocad/overview

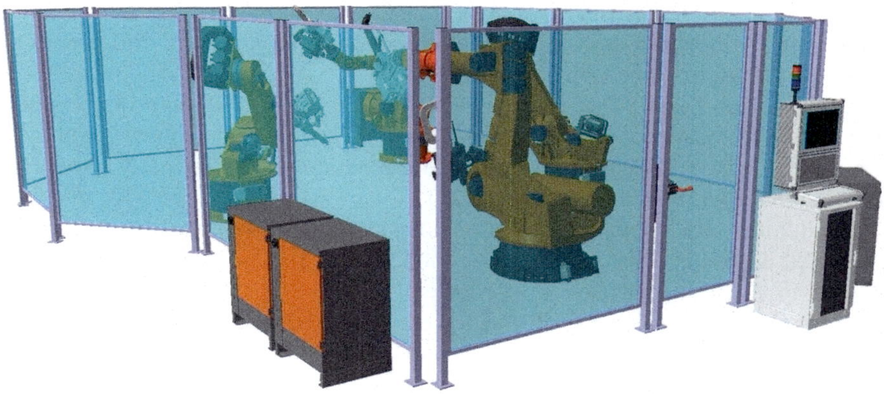

Fig. 9.3 3d geometric model of a production system

Table 9.2 Overview of used native data formats of some CAD tools

	Parts	Assemblies	2d drawings	Layout Plan	Process Plan
CATIA/DELMIA	*.CATParts	*.CATProduct	*.CATDrawing	*.CATProduct	–
NX/MCD			*.prt		
SolidWorks		*.sldprt	*.sldasm	*.slddrw	–
Creo	*.prt	*.asm	*.drw	*.lay	–
AutoCAD		*.dwg		–	

Usually, not every part of the production system has to be developed from scratch, because the production system manufacturer tend to use already existing parts or standard parts as much as possible, e.g. screws, springs, nuts, flat washers, etc. Such parts could be provided by the product system manufacturer or by third-party via standardized libraries. In this context, system components that are provided by a component manufacturer could be downloaded directly as 3d geometric model in various data formats after a detailed configuration. As add-on, some CAD tools provide libraries with standard parts as well as functionalities to create and manage user specific parts, e.g. CATIA V5 via workbench Catalog Editor.

To assemble various 3d geometric models to an entire assembly or module, various CAD tools provide a lot of functionalities that enable an exact positioning of parts or components relatively to each other as well as to define constraints between them (Fig. 9.2, 1). Normally, functionalities for design of 3d geometric models are separated from those functionalities that are used to create assembly models of the whole production system. In CATIA for example, both groups of functionalities are organized into different workbenches, i.e. Part Design and Assembly Design. A similar organization of those functionalities is made into other CAD tools as well, e.g. NX, SolidWorks, Creo, AutoCAD, etc. The essential notable difference between various CAD tools is the data format used to save the created 3d models. An overview of the various data formats is presented in Table 9.2.

This table is not an exhaustive list containing all available CAD tools and all native data formats that are supported by these tools. To find more information to this topic, please visit the web page.[6]

A significant benefit of using the same CAD tool for designing of parts and assembling of whole production systems is that no need for 3d geometry exchange amongst tools arises. Thus, integration of information related for parts and whole systems can be ensured (see Chap. 1-RQ I1).

In addition to the mentioned functionalities, CAD tools provide also features to derive 2d drawings from 3d geometric models of parts and entire assemblies as well as to create layout plans of production systems (Fig. 9.2, 2 and 4). During this, a qualitative exchange of information between mechanic and manufacturing or maintenance of production systems is guaranteed (in Chap. 1-RQ I2). Some data formats used by CAD tools are presented in Table 9.2. As we can see until here, CAD tools provide an extended spectrum of functionalities that allow use of solely one tool to create and prepare all required outputs from the mechanical design.

Normally, a rough concept of the production system is initially generated before the start of the mechanical design. Thereby, the mechanical engineer creates a rough plan of the layout and functionalities of the production system (Fig. 9.4). In this

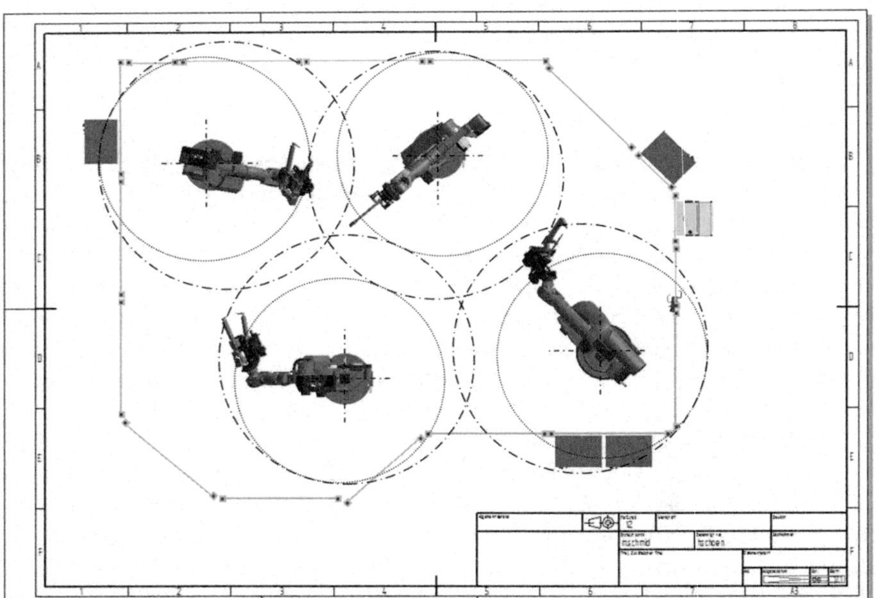

Fig. 9.4 Exemplary production system layout plan as drawing (see Fig. 9.3)

[6]https://de.wikipedia.org/wiki/Liste_von_CAD-Programmen

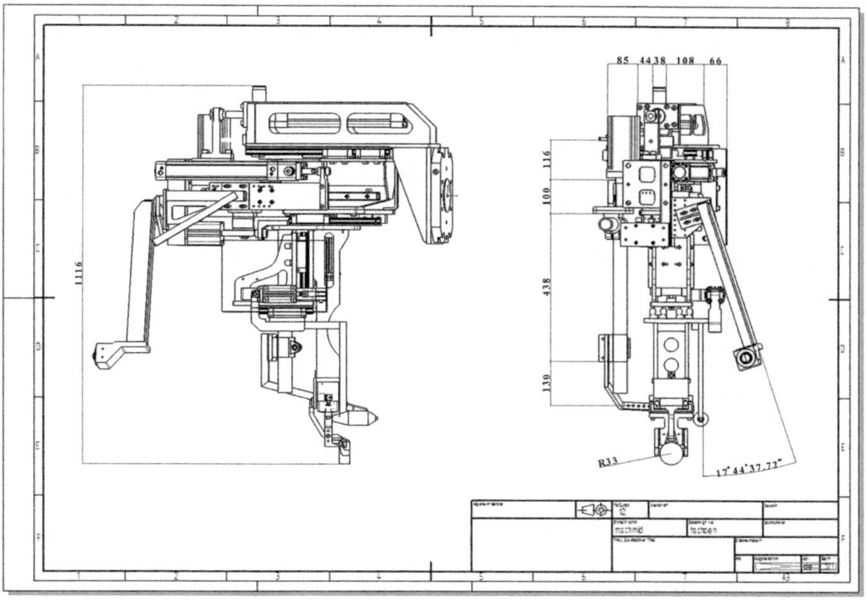

Fig. 9.5 Production process description as sequence diagram in MS excel

step, the same CAD tool is used that will also be used for the following detailed mechanical design. In this regard, a first draft of the production system process is also prepared (Fig. 9.5). Based on that, changes on the layout and the production process should be implemented in the 3d geometric design and considered for the subsequent steps of OLP and material flow design. Required modifications on the system's parts should be considered during creation of 2d drawing and following manufacturing as well.

Following, detailed 3d geometric models of the system's parts as well as assemblies are created in a CAD tool and saved in the CAD tools' native data formats. Based on these models, 2d drawings are derived and enhanced by specific manufacturing information that is required to produce every designed part, e.g. specific surface characteristics, part's dimensions, material information, etc. (Fig. 9.6). As a consequence, the final draft of the layout and production process together with the 3d geometric model of the system's mechanic is completed.

Subsequently, the manufacturing of system's parts and the ordering of customized components and standard parts respectively based on the 2d drawings is about to start (Fig. 9.2a). In practice, 2d drawings are converted to a neutral data format (e.g. *.dwg or *.dfx, *.) or saved as PDFs (Portable Document Format). Finally, a hard copy of each system's specific part is provided to the part manufacturer that initiates the parts production. 2d drawings are stored by plant owner (OEM) and serve as templates for the manufacturing of spare parts. Usually, those drawings are stored as PDFs and more rarely in native or neutral data formats.

		[sek.]	[sek.]	[sek.]	
	target process time		70,0		
1	Werkzeug mit Roboter an Werkstück, Position 1 schwenker	0,0	10,0	10,0	robot
2	cylinder 1 to work position	10,0	2,0	12,0	cylinder 1
3	cylinder 2 to work position	12,0	2,0	14,0	cylinder 2
4	cylinder 3 to work position	14,0	2,0	16,0	cylinder 3
5	cylinder 4 to work position	16,0	2,0	18,0	cylinder 4
6	cylinder 5 to work position	18,0	2,0	20,0	cylinder 5
7	cylinder 6 to work position	20,0	1,5	21,5	cylinder 6
8	cylinder 7 to work position	21,5	1,5	23,0	cylinder 7
9	electro drive 1 drive to 150mm	23,0	3,0	26,0	electric drive
10	tool on robot to product 1, position 2 move	26,0	10,0	36,0	robot
11	electro drive 1 drive to 40mm	36,0	3,0	39,0	electric drive
12	cylinder 7 to base position	39,0	1,5	40,5	cylinder 7
13	cylinder 6 to base position	40,5	1,5	42,0	cylinder 6
14	cylinder 5 to base position	42,0	2,0	44,0	cylinder 5
15	cylinder 4 to base position	44,0	2,0	46,0	cylinder 4
16	cylinder 3 to base position	46,0	2,0	48,0	cylinder 3
17	cylinder 2 to base position	48,0	2,0	50,0	cylinder 2
18	cylinder 1 to base position	50,0	2,0	52,0	cylinder 1
19	robot move to home position	52,0	10,0	62,0	robot

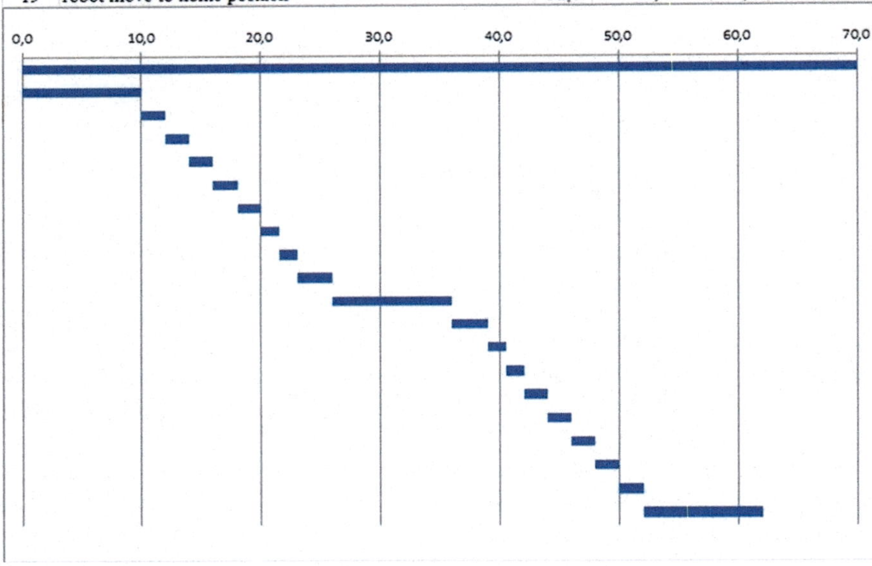

Fig. 9.6 2d drawing of a production system's assembly

Storage of 2d drawings in native data formats is applied by production system builder (PM) but generally not required by the OEM.

With the beginning of part manufacturing, the ordering process is started as well. For this purpose, a part list with all parts and components implemented into the production system is created (Fig. 9.7). To do this, some CAD tools provide functionalities to create a part list based on the 3d geometric model. This list could be stored in various data formats, e.g. text files, word documents, MS excel tables, PDF's, etc. In some cases, specific CAD tool add-ons are created by the PM that

piece	type	mm	name	man.numb.
2 x	profile_8_40x40	920	profile 8 40x40, natur	0.0.026.03
4 x	profile_8_40x40	2000	profile 8 40x40, natur	0.0.026.03
2 x	profile_8_40x40	519	profile 8 40x40, natur	0.0.026.03
4 x	profile_8_40x40	1306	profile 8 40x40, natur	0.0.026.03
4 x	profile_8_40x40	328	profile 8 40x40, natur	0.0.026.03
2x	profile_8_40x40	500	profile 8 40x40, natur	0.0.026.03
1 x	acrylglass_5mm_klar	541x862	acrylglass 5mm, klar	0.0.428.21
2 x	acrylglass_5mm_klar	300x862	acrylglass 5mm, klar	0.0.428.21
2 x	acrylglass_5mm_klar	328x862	acrylglass 5mm, klar	0.0.428.21
1 x	acrylglass_5mm_klar	1271x862	acrylglass 5mm, klar	0.0.428.21
28x	angle_8_40_black	-	angel 8 40, black	0.0.196.87
4x	tap_8_40_black		tap 8 40, black	0.0.196.86
60x	sliding_block_8_M8	-		
60x	hexagon_bolt_M8	-		
3x	hinge_8_PA_rechts_black	-	hinge 8 PA, right, black	0.0.026.12
2x	magnetic stop 8, black	-	magnetic stop 8, black	0.0.601.30
1x	hand_grip_PA_120_black	-	hand grip PA 120, black	0.0.391.35
20x	cap_8_40x40_black	-	cap 8 40x40, black	0.0.026.01
1x	mount part 6-8 for sefaty switch end position compact	-		0.0.473.23
1x	mount part 6-8 for sefaty switch end position & M12x1 code A	-		0.0.473.22
1x	safety switch end position			0.0.473.90
2x	safety switch M12x1 code A	-		0.0.473.25
1x	ADVU-50-250-A-P-A		pneumatic cylinder	
2x	ADVU-50-80-A-P-A		pneumatic cylinder	
1x	CPX-GE-EV-S-7/8-5POL			
2x	CPX-AB-4-M12X2-5POL			
2x	MTH-5/2-7,0-L-S-VI		velve	
2x	MTH-5/3B-7,0-L-S-VI		velve	
2x	cap profile 8, black	541	cap profile 8, black	0.0.422.26
4x	cap profile 8, black	300	cap profile 8, black	0.0.422.26
4x	cap profile 8, black	328	cap profile 8, black	0.0.422.26
2x	cap profile 8, black	1271	cap profile 8, black	0.0.422.26

Fig. 9.7 Part list of production system created manually in MS excel

export a part list from CAD tools based on the 3d geometric model of the entire production system. In doing so, the PM achieves a complete part list without any gaps.

As already explained, CAD tools provide an extended spectrum of functionalities to create a large part of mechanical design outputs. Only for process planning (Fig. 9.2, 6) the use of additional tools is recommended, e.g. design of processes via Process Simulate (Siemens), DELMIA (Dassault System), etc. Those tools used 3d geometric models as base to create 3d kinematic simulation models of the production system. In this regard, the offline robot programming and material flow design are supported by some process design tools as well (Fig. 9.2, 6 and 7). Here, offline robot programing could be performed based on the 3d geometry of the robot and all other production system's parts respectively. This generated OLP contains the information about motion trajectory and some process relevant signal references only (Fig. 9.8). Subsequently, OLP's detailing is performed during the PLC/software design of the production system since the information about the signal dependencies is already defined via the PLC/software (see Sect. 9.3.3). Ultimately, the production process of the system is provided to the assembling and

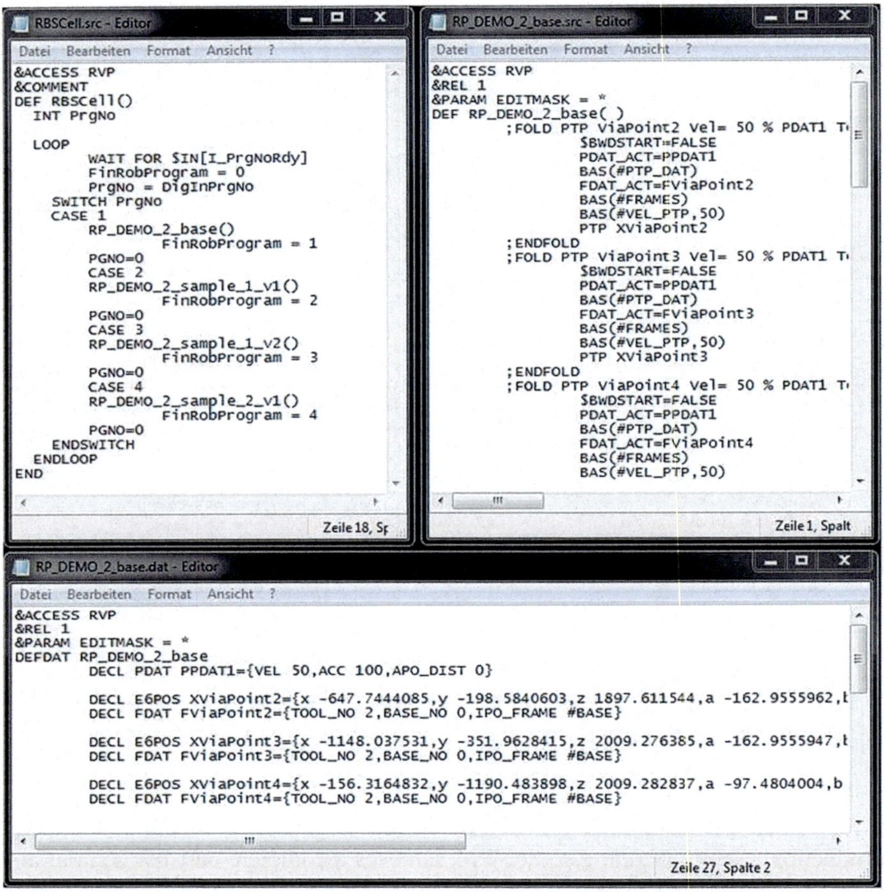

Fig. 9.8 Offline robot program of production system created via DELMIA V5

commissioning as well as to the PLC/software design phase as neutral data format, e.g. PDFs (Fig. 9.2b and c).

Based on practical experience, a simplified 3d model with additional kinematic information of the production system is required (Fig. 9.2, 5). As a consequence the enormous complexity of the design model containing each part as detailed 3d geometric model must be reduced for simulation purposes (Strahilov et al. 2012). For the execution of simulation, detailed models are superfluous and also deteriorate the performance of the simulation. Consequently, a separate simulation model has to be created for the purpose of process planning. The detailed description about this procedure is presented in the section of Virtual Engineering (Sect. 9.3.4).

9.3.2 Tool Chains of Electrical Design

Electrical/fluidic design is the second phase of the development process of a production system. In this phase, pneumatic plan and electrical plan of the production system are prepared and provided to the subsequent phases (Figs. 9.9 and 9.10). As purpose of these outputs, information about the connections between electric components respectively pneumatic components is described. Thereby, detailed requirements about used type of cables, in case of electric component, and tubes, in case of pneumatic components, are presented in those plans. Based on both outputs, Fig. 9.11 shows the established tool chain of electrical/fluidic design.

To design these plans, the part list created by mechanical design is required to identify electrical or pneumatic components implemented in the production system. Moreover, the final draft of the production process is required also to define additional electric or pneumatic components, e.g. to define the number of required valves to control pneumatic cylinders or throttles to set the velocity of cylinder's piston, electrical fuses, etc. Hence, additional system parts have to be added to the part's list and have to be taken into account for the ordering process. Finally, both designed plans are provided to the PLC/software design as input. Generally, electrical plans and pneumatic plans are provided to the following processes as neutral data format via a conversion to PDFs. Storage of those plans as tool specific formats is done by the PM and rarely provided to the OEM.

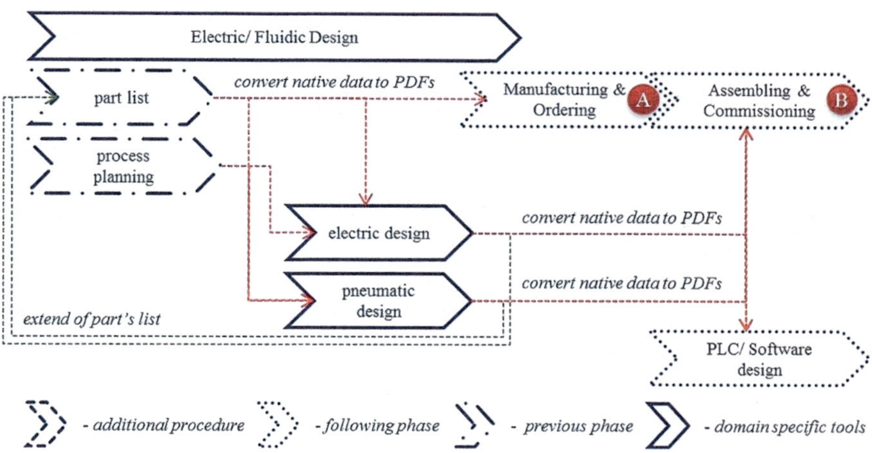

Fig. 9.9 Pneumatic plan of production systems

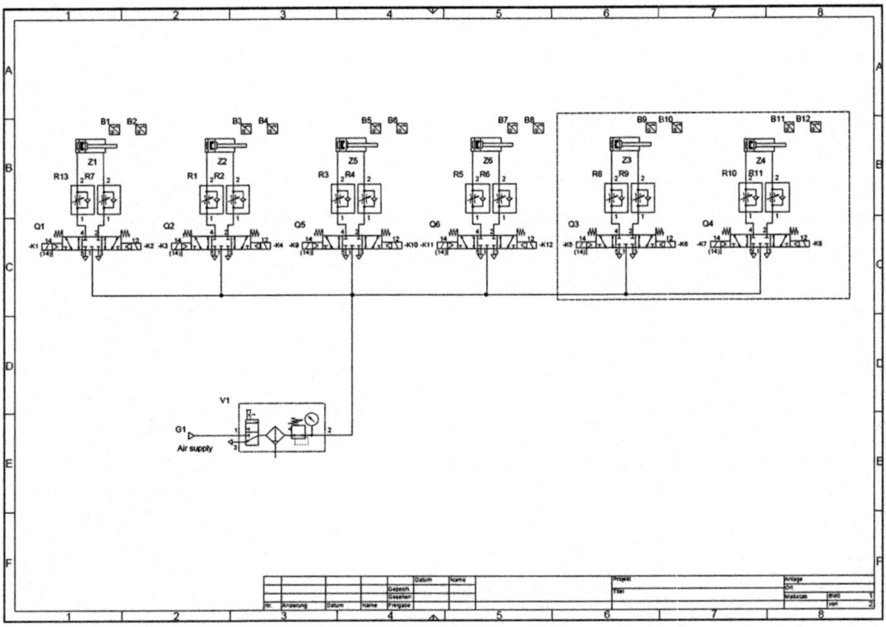

Fig. 9.10 Electric plan of production system

There are various electrical and fluidic design tools available in the market, e.g. EPLAN fluidic/electric,[7] FluidDraw,[8] E3.fluid,[9] elecworks™ Fluid,[10] DSHplus,[11] etc. Most of them provide the required functionalities to design electric and/or fluidic plans. Some of them can be used to perform simulations to test functionalities of the designed electrical or fluidic system as well, e.g. DSHplus. Other tools provide the option to jointly design electric and fluidic systems, e.g. FluidDraw, E3.fluid, elecworks™ Fluid, etc. Furthermore, some fluidic tools combine 2d drawings derived from 3d geometric models together with pneumatic plans or fluidic plans, e.g. FluidDraw. Aside from that, CAD tools used for mechanical design support additional functionalities to design electric and pneumatic plans but with focus on 3d cable laying, e.g. CATIA,[12] SolidWorks,[13] NX,[14] etc. For the design of

[7]https://www.eplan.de/en/solutions/electrical-engineering/

[8]https://www.festo.com/cms/nl-be_be/17099.htm

[9]http://www.zuken.com/en/products/electrical-wire-harness-design/e3-series/products/fluid

[10]https://www.eplan.de/en/solutions/electrical-engineering/

[11]https://www.fluidon.com/index.php/en/dshplus/

[12]http://www.3ds.com/products-services/catia

[13]http://www.solidworks.de/

[14]http://www.plm.automation.siemens.com/en_us/products/nx/index.shtml

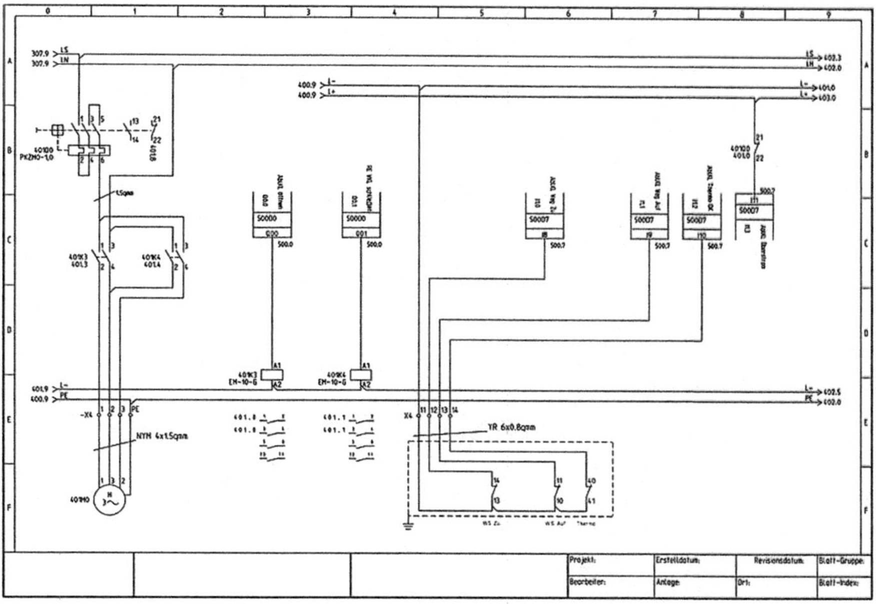

Fig. 9.11 Established tool chain of electric/fluidic design of production systems

electrical and pneumatic plans, such functionalities are not relevant since, 3d cable laying design is not possible in the very limited development time of production systems (see Sect. 9.2).

In the area of the development of production systems for car manufacturing, EPLAN is the most common design tool for electrical and fluidic design among prominent OEMs. In some cases, 2d drawing functionalities of CATIA are used to create electrical parts. In order to realize this, a 2d library with required symbols of pneumatic components is necessary.

As well as CAD tools, each electric and pneumatic design tool uses a specific data format to save designed plans (see Sect. 9.3.1). But in practice, the exchange of electric and pneumatic plans is carried out via data neutral formats, e.g. PDF. During a conversion to PDF, automated interpretation of content information in these plans cannot be realized without information losses. At end of this phase, these converted production system's plans are provided to the following PLC/software design phase.

9.3.3 *Tool Chain of PLC/Software Design*

At the end of electric/fluidic design, the PLC/software design is started (Fig. 9.12). In this phase, the control program of the production system is coded along with the human machine interfaces (HMI). To do this, production system's layout plan,

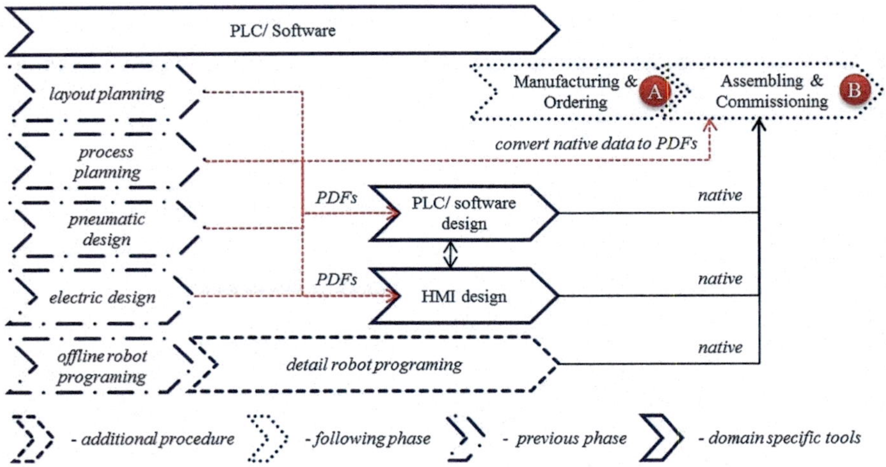

Fig. 9.12 Established tool chain of PLC/software design of production systems

process plan, pneumatic plan and electric plan that are created in the previous phase are taken as base. Thereby, these outputs are provided similar to both previous phases via PDFs.

Regarding to the electrical plan, hardware configuration of the production plan is created (cf. Fig. 9.13). With this configuration, all electrical components connected with the PLC via communication bus (e.g. PROFINET, PROFIBUS, etc.) are required for the PLC to define which components are expected using which bus address. In parallel, input and output signals of each component are defined also. Based on input and output signals, PLC software programming is initiated. Thereby, the production process and OLPs are taken into account and set as base to integrate all dependencies between the components into PLC's software. In this regards, the production process is mapped into the software also. In this phase OLPs must contain information about dependencies between the PLC's software and robot controllers on which these robot programs run. Parallel to this, detailing of robot programs has to be conducted.

As explained in Sect. 9.3.1, offline robot programs (OLP) are created based on the 3d simulation model of the production system. Such an OLP contains the information about motion trajectory and some process relevant signal references (Fig. 9.8). Based on that, a further detailing of these programs shall be done during the PLC/software design. The information about dependencies between system's components are for example dependencies between safety doors and components, release of extending or retracting of pneumatic cylinders, release of robot operating zones, etc. These dependencies are defined in the PLC's program and integrated into the robot programs via several signals, e.g. robots wait for release from PLC's software via Boolean signal to conduct an operation in a safety zone to prevent collision with other robots or system's component (Fig. 9.14).

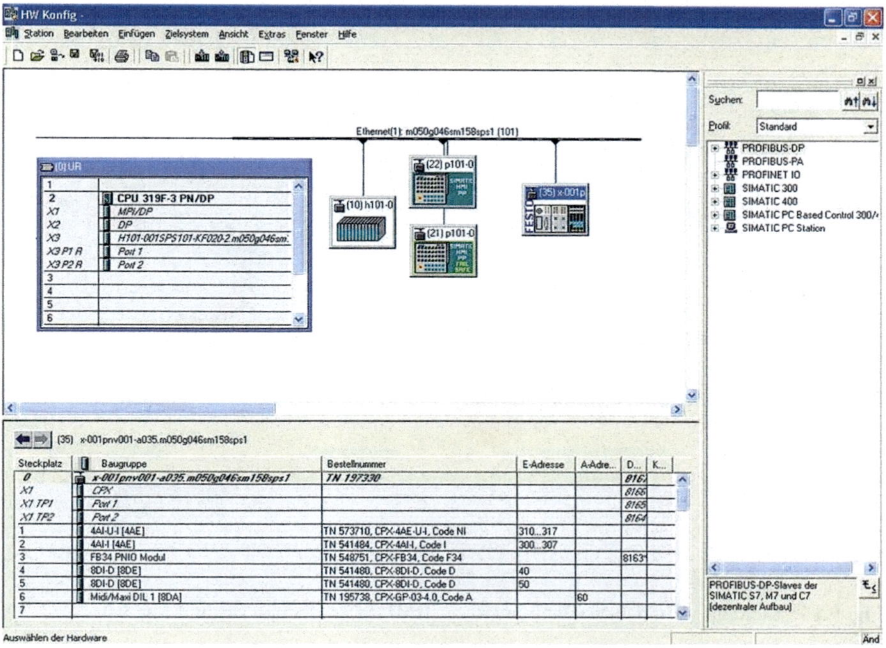

Fig. 9.13 Hardware configuration of a PLC's software via STEP7 (http://w3.siemens.com/mcms/automation/en/automation-systems/automation-software/Pages/Default.aspx)

Depending on the PLC software programming, HMI of the production system will be designed. To do this, the signals defined by PLC's software and layout plan of the production system are required. Based on both outputs, overview of the whole production system is created with additional information about signal status, e.g. end position of pneumatic cylinders, current position of electrical drives, released robot safe zones, etc. An HMI of a production system that is based on system's layout plan and extended by PLC signals is shown in Fig. 9.14.

For the PLC software design multiple tools are available that could be used to perform this design, e.g. STEP7,[15] TIA,[16] CODESYS,[17] PC WORX,[18] etc. Some of those tools can be used to create PLC software for various PLCs as well as for various embedded controllers, e.g. CODESYS. For this purpose, they could be extended via add-ons or additional tools that are from same or other

[15]http://w3.siemens.com/mcms/automation/en/automation-systems/automation-software/Pages/Default.aspx

[16]https://www.industry.siemens.com/topics/global/en/tia-portal/Pages/default.aspx

[17]https://www.codesys.com/#_

[18]https://www.phoenixcontact.com/online/portal/us?uri=pxc-oc-itemdetail:pid=2985259&library=usen&tab=1

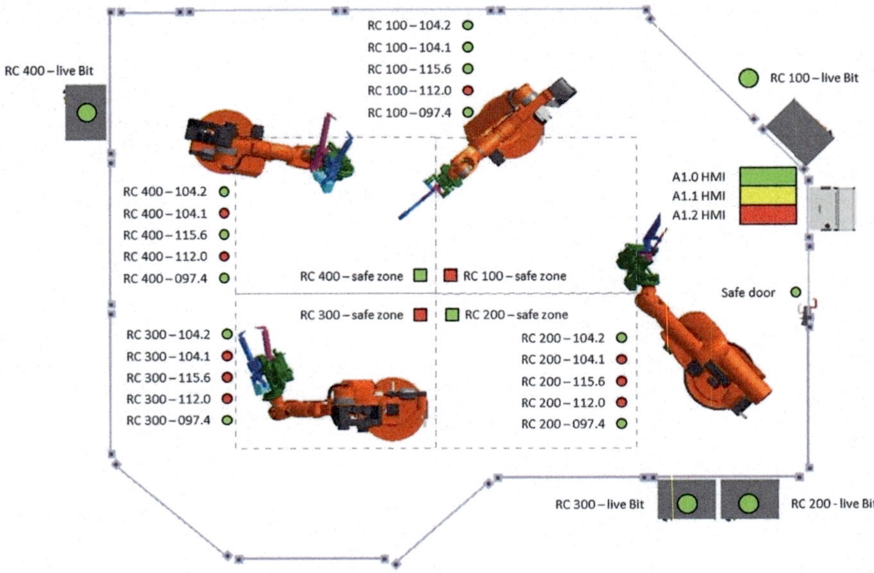

Fig. 9.14 Exemplary Human Machine Interface (HMI) of production system (Fig. 9.4)

tool provider. On the contrary, some tools can be used to design programs for specific PLCs, e.g. STEP7 and TIA. Independently from that, all tools support the same standard programming languages defined by IEC61131-3, e.g. Ladder Diagram (LD), Instruction List (IL), Function Block Diagram (FBD), Structured Text (ST) and Sequential Function Chart (SFC).[19] As expected, each of these tools use proprietary data formats to store native PLC software. Only some tools support the export and import of PLC software via a neutral data format, e.g. CODESYS via PLCOpen XML. By means of PLCOpen XML, exchange of PLC software code as well as HMI code is possible between various tools with marginal information losses (cf. Fig. 9.14). Furthermore, exchange of PLC software that is written in various programming languages defined by IEC61131-3 is possible by PLCOpen XML.

For the PLC software design multiple tools are available that could be used to perform this design, e.g. STEP7, TIA, CODESYS, PC WORX, etc. Some of those tools can be used to create PLC software for various PLCs as well as for various embedded controllers, e.g. CODESYS. At the end of this phase, the finalized detailed PLC software is provided to the assembling and commissioning phase of the production system (Fig. 9.15).

Even before commissioning starts, PLC software is tested via virtual commissioning (see Sect. 9.3.5). During system's commissioning on the shop-floor, each

[19]http://www.automation.com/pdf_articles/IEC_Programming_Thayer_L.pdf

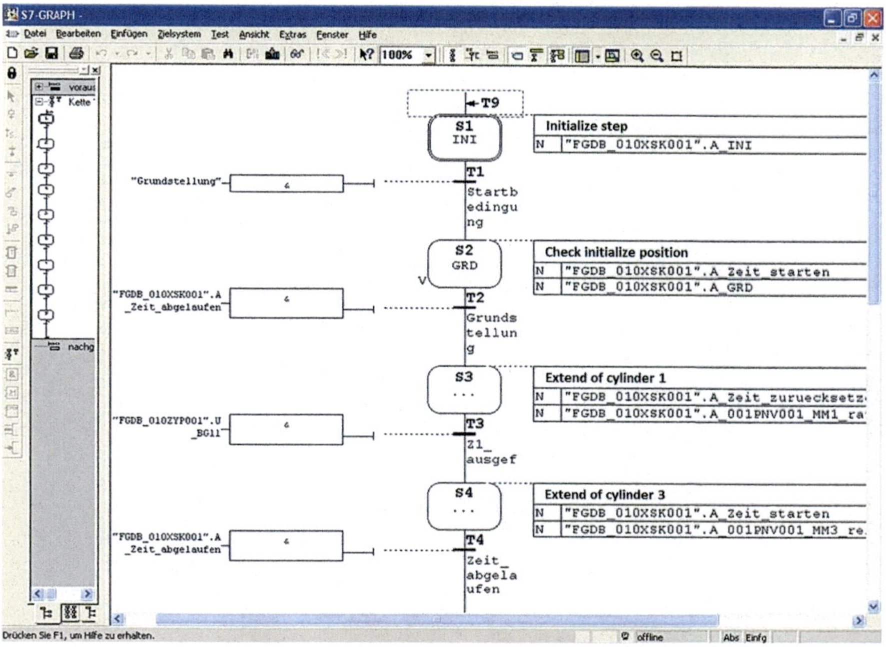

Fig. 9.15 Production process into PLC's software as step chain via STEP7

manual as well as automated function of the PLC software is tested in connection to the real hardware of the production system. e.g. manual extraction and retraction of each pneumatic cylinder via PLC, manual execution of robot movements, perform automated production processes with and without products, etc. In this context, fine adjustments on mechanic and electric components as well as on detailed robot programs is performed, e.g. adjustment of cylinder's velocity for extraction and retraction via throttle settings, teaching of robot target positions, minor modification of acceleration profiles of electric drives, etc.

9.3.4 *Tool Chain of Virtual Engineering*

Virtual Engineering (VE) is phase of the development process of production system that supports the system's mechanical design (Ovtcharova 2013). As base of this phase, a 3d simulation model of the production system is required. As explained already in Sect. 9.3.1, this 3d simulation model uses the 3d geometric model of the production system as base and integrates additional kinematic information of components into the production system, e.g. translation of pneumatic cylinders, rotation of electric drives, kinematic of robots, etc. (Fig. 9.16). Based on this model,

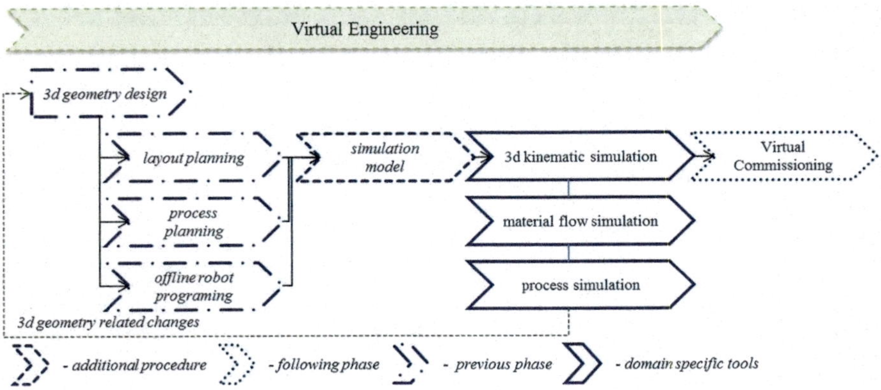

Fig. 9.16 Tool chain of virtual engineering (VC)

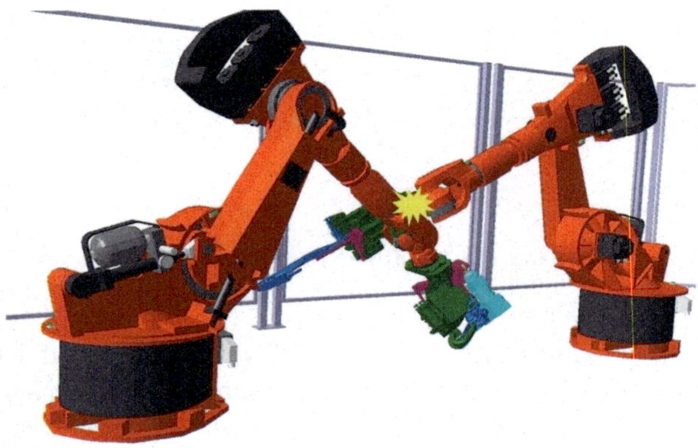

Fig. 9.17 Collision between robots during VE via 3d simulation model

collision free component's movement in connection with the system's mechanic have to be validated (Fig. 9.17). Furthermore, production processes of the system can be planned as well as tested with respect to process time and sequences via 3d visualization of movement. Of course, a first draft of the production process is required at the beginning of this step. In this context, OLPs of the production process are required also to animate the complex movement of robots. Thereby, OLPs represent a significant part of the production process and have a strong influence on process time and sequence. Visualization of movement of products as well as of system's own components (e.g. welding guns) is another important aspect within VE. Fort this purpose, the material flow model is needed that describes the position and orientation as well as dependencies between products or system's components during production process simulation, e.g. representation of gripping processes, conveyor transport processes, etc.

Based on practical experience in the area of production systems in car manufacturing, VE takes a major role in the development process as phase to achieve better quality of the mechanical design as well as to prevent errors based on the simulation (Ovtcharova 2013). In most cases, the modelling of the 3d simulation model is conducted with the same tool for performing the simulation. Typical simulation tools used in VE are DELMIA,[20] Process Simulate[21] and Mechatronic Concept Designer (MCD).[22] Some of those tools are integrated in a CAD tool that allows the use of the designed 3d geometric model without data transfer between the CAD and simulation tool, e.g. DELMIA into CATIA, MCD into NX, etc. Doing so, flawless data transfer can be ensured.

During checks of system's mechanic, difficulties of the mechanic should be identified and, required changes on system's mechanic have to be provided back to the mechanical design phase. After that, performed changes on the 3d geometric model of system's mechanic should be checked again based on the updated 3d simulation model. As soon as all mechanic checks are performed successfully, VE is completed. In most cases, the simulation model is stored by the PM in native tool specific data format and provided as same data forma to the OEMs, e.g. DELMIA as *.CATProcess, MCD as *.prt, etc. (see Table 9.2). Thereby, OEM requires the 3d simulation model by PM to use it for potential mechanical changes regarding integration of new or adapted products into the production system as well as to modify the same system for other production purposes in case of system reuse (see Sect. 9.2). The 3d simulation model is continuously used in virtual commissioning to visualize movements of the components (Sect. 9.3.5).

9.3.5 *Tool Chain of Virtual Commissioning*

The virtual commissioning can be observed as a separate phase of the development process of production systems (Fig. 9.18). In this phase, all functionalities of the PLC's software are tested based on a simulation model of the production system that doesn't exist in real hardware at this time (Süß et al. 2015; Damrath et al. 2015). For the purpose of VC, first draft of the PLC's software as well as working and tested simulation model of the production system must be available. This required simulation model, as well as VE's simulation model contain several sub-models, e.g. 3d simulation model, material flow and robot simulation (Sect. 9.3.4). Essential differences between VE's simulation and VC's simulation are on the one hand the additional behaviour model respectively behaviour simulation, that are needed to simulate logical behaviour of each component connected to the real PLC, and

[20]http://www.3ds.com/products-services/delmia/products/

[21]https://www.plm.automation.siemens.com/en/products/tecnomatix/manufacturing-simulation/assembly/process-simulate.shtml

[22]https://www.plm.automation.siemens.com/en/products/nx/for-design/mechatronics-design/

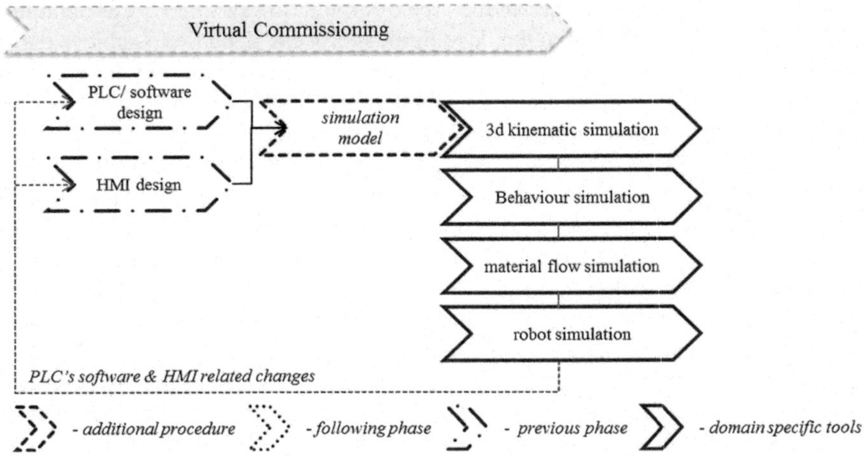

Fig. 9.18 Tool chain of virtual commissioning

on the other hand the missing process simulation, which is replaced by PLC's software. Furthermore, production system's 3d simulation model serves to visualize the movement of the system during the production process by the PLC via PLC's software but not to check system's mechanic such as VE.

Based on practical experience, the 3d simulation model created and used by VE is taken as base to use it for the purpose of VC. In the market, several tools provide required functionalities to prepare the 3d simulation model and to conduct VC, e.g. DELMIA,[23] Process Simulate,[24] NX-MCD,[25] RF::SGView and RF::SGEdit,[26] etc. In some of those tools, VE's simulation model can be taken without any further changes and modified for VC, e.g. DELMIA, NX-MCD, Process Simulate, etc. Based on practical experience, the complexity of a typical 3d simulation model is enormously high. As consequence of this, high computational performance is required to run the simulation in an approximate time step between 10 and 100 [ms]. Based on this, some VC's 3d simulation tools are adapted to deal with complex 3d simulation models, e.g. RF::SGView. Furthermore, the material flow defined by VE is used together with the 3d simulation model also. In most cases, both models are integrated in a single simulation model, e.g. DELMIA, Process Simulate, etc.

Additionally, the behaviour model of the production system is required by VC. This model represents the logical behaviour of each component connected to the PLC and communicates with the PLC via signals. For this purpose, the behaviour

[23]http://www.3ds.com/products-services/delmia/products/

[24]https://www.plm.automation.siemens.com/en/products/tecnomatix/manufacturing-simulation/assembly/process-simulate.shtml

[25]https://www.plm.automation.siemens.com/en/products/nx/for-design/mechatronics-design/

[26]http://www.rf-suite.de/en/products.html

model of each component installed in the production system has to be created. In practice, each OEM creates this model by himself or assigned a sub-contractor (Drescher et al. 2013). PMs receive these OEM's behaviour models and uses them to build up the entire behaviour model of the production system. In this context, each OEM stores behaviour models in tool specific data format. As a consequence, the PM is forced to use fixed tools to run the behaviour simulation and therefore also VC. Two widely used behaviour simulation tools in practice are WinMOD® and SIMIT.[27] Those tools provide several communication interfaces that allow exchange of simulation data in approximated real-time with other tools, e.g. WinMOD® with RF::SGView, SIMIT with NX-MCD, etc. Furthermore, additional functions can be used to automatically create whole behaviour models based on component's behaviour models, e.g. specific MS excel based assistance for WinMOD®.

Along with the behaviour simulation, robot program simulation is required as well. In this case, robot programs that are detailed during the PLC/software design are used and not OLPs (cf. Sects. 9.3.1 and 9.3.3). To run the simulation of those robot programs, specific tools are required that have to support specific robot programming languages, e.g. KUKA, ABB, Fanuc, Kawasaki, Mitsubishi, etc. For this purpose, the most robot manufactures provide a tool to create and simulate robot programs, e.g. RobotStudio,[28] KUKA.Sim,[29] etc. Such tools are standalone solutions and they expected the 3d geometric model of the production systems that have to be extended in a previous step by kinematic information. A distinction must be made between OLPs and detailed robot programs. In practice, OLPs are created by simulation tools that support process simulation and material flow as well as to export OLPs in several robot languages in parallel, e.g. export function of DELMIA as KUKA, ABB, Fanuc, Mitsubishi etc. Aside from that, RF::RobSim is a very common robot simulation tool that can process several real robot programs and simulate them in real time during VC (Süß et al. 2015). In practice, RF::RobSim is used in combination with WinMOD® and RF::SGView (Hämmerle and Drath 2014).

To achieve greater benefit of VC, modelling effort has to be kept to a minimum. For this reason, continuous use of VE's simulation model has to be done to prevent unnecessary repeated modelling time. In last years, the use of AutomationML[30] as neutral data format has proven itself to realize model exchange without data loss (Hämmerle and Drath 2014). Currently, several AutomationML export functions are developed to exchange 3d simulation models between process simulation tools (e.g. DELMIA, Process Simulation and NX-MCD, etc.) and VC's tools (e.g. RF::SGView and NX-MCD). Some renowned OEMs in the car manufacturing industry use AutomationML export functions of DELMIA and Process Simulate to transfer 3d simulation models to RF::SGView.

---

[27] http://w3.siemens.com/mcms/process-control-systems/en/distributed-control-system-simatic-pcs-7/simulation_training_systems/Pages/Default.aspx

[28] http://new.abb.com/products/robotics/robotstudio

[29] http://www.kuka-robotics.com/germany/de/downloads/software.html

[30] https://www.automationml.org/o.red.c/home.html

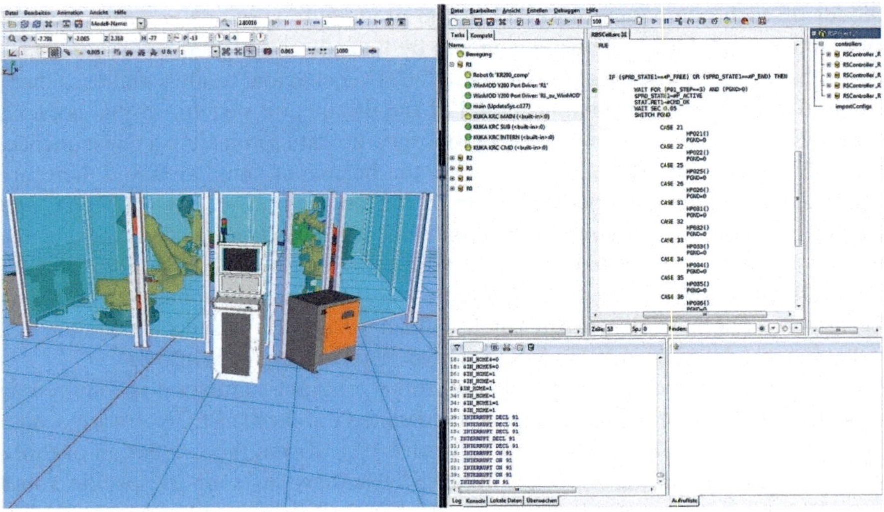

Fig. 9.19 Virtual Commissioning (VC) via RF::SGView and RF::RobSim

As soon as all functionalities of PLC's software are successfully tested, VC is completed. The result of this phase is the validated and optimized real PLC's software that runs on the target hardware PLC. As following step, commissioning of the real production systems can be initialized (Sect. 9.2) (Fig. 9.19).

9.4 Summary and Outlook

At the beginning of this chapter, production systems were not designated as CPPS. This statement was justified on several reasons. One of those reasons is the missing intelligence of components used today for production systems. Independently from that, first studies to use already existing smart components are successfully being performed, (VDE 2013). An engineering workflow of CPPS can be derived from the engineering workflow of production system, which was presented in this chapter. But, a detailed description of the engineering workflow without praxis experience will not be useful.

Despite this, significant differences between the engineering workflow of CPPS compared to the engineering workflow of current production systems are not expected. Only differences on used tools and exchange data formats are conceivable. Furthermore, additional or extended activities and tasks have to be integrated into the engineering workflow of CPPS, e.g. integration of component's and product's intelligence into PLC control programs.

Based on existing experience, the main difficulty into the engineering workflow of productions systems are the specific data formats of the used tools. Even worse is the use of standard data formats which do not support the exchange of data without losses, e.g. PDF. One exception is the use of the standard data format AutomationML to exchange simulation models between simulation tools and virtual commissioning tools (see Sect. 9.3.5). In this regards, a extend use of a standard data format into the whole workflow is an important step which has to be done during the design of the engineering workflow of CPPS.

From the point of view of the research question, the presented tool chains cover a great amount of the engineering information. This information is added by a user via several tools and it is stored in tool's specific data formats. Consequently, an integration of engineering information during the engineering workflow of a production system is not possible without additional efforts (Chap. 1-RQ I1). In case of a data exchange, tool neutral data formats are used to ensure a data exchange between different users during the engineering workflow in such a way that the content of the data can be read by a human, e.g. PDF, DWX, etc. Thus, the quality of exchanged data is assured but an automated reading of it via a computer is made difficult (Sect. 1-RQ I2). Checking of the quality of exchanged data is also performed during the virtual commissioning (Sect. 9.3.5).

Independently from all technical or managerial issues regarding to the engineering workflow of production process or CPPS, plant owner, plant builder and component manufacturer face their next challenge which could only be solved by joint action. In these activities, tool providers have to be also involved to design or adapt the used tools, e.g. new features, import/export functions, etc.

References

Damrath, F., Strahilov, A., Bär, T., Vielhaber, M..: Experimental validation of a physics-based simulation approach for pneumatic components for production systems in the automotive industry. In: 15th CIRP Conference on Modelling of Machining Operations. Elsevier (2015)

Dilts, D.M., Boyd, Whorms, H.H.: The evolution of control architectures for automated manufacturing systems. J. Manuf. Syst. **10**(1), 79–93 (1991)

Drescher, B., Stich, P., Kiefer, J., Bär, T., Strahilov, A., Reinhart, G.: Physikbasierte Simulation im Anlagenentstehungsprozess – Einsatzpotenziale bei der Entwicklung automatisierter Montageanlagen im Automobilbau. HNI-Verlagsschriftenreihe, Paderborn (2013)

Groover, P.: Automation, Production Systems, and Computer-Integrated Manufacturing. Prentice Hall Press, Upper Saddle River, NJ (2007)

Hämmerle, H., Drath, R.: Erfahrungen bei der virtuellen Inbetriebnahme. In: Tagungsband zur Automation (2014)

Koren, Y., Heisel, Y., Jovane, F., Moriwaki, T., Pritschow, G., Ulsoy, G., Van Brussel, H.: Reconfigurable manufacturing systems. CIRP Ann. Manuf. Technol. **48**(2), 527–540 (1999)

Li, J., Meerkov, M.: Production System Engineering. Springer, New York (2001)

Ovtcharova, J.: Virtual engineering: principles, methods and applications. In: International Design Conference – Designe 2010, 2013

Strahilov, A., Mrkonjić, M., Kiefer, J.: Development of 3D CAD simulation models for virtual commissioning. In: Proceedings of TMCE, 2012

Süß, S., Strahilov, A., Diedrich, C.: Behaviour simulation for virtual commissioning using co-simulation. In: 20th IEEE International Conference on Emerging Technologies and Factory Automation (2015). doi:10.1109/ETFA.2015.7301427

VDI/VDE-Gesellschaft Mess- und Automatisierungstechnik: Cyber-Physical Systems: Chancen und Nutzen aus Sicht der Automation Thesen und Handlungsfelde. Report of VDE. https://www.vdi.de/uploads/media/Stellungnahme_Cyber-Physical_Systems.pdf (2013). Accessed 03 Nov 2016

Chapter 10
Standardized Information Exchange Within Production System Engineering

Arndt Lüder, Nicole Schmidt, and Rainer Drath

Abstract Information exchange is one of the critical issues within the multi-disciplinary engineering chain of production system engineering. In the subsequent chapter the problem of identifying and standardizing an appropriate data exchange format for this field of application will be considered. It will be argued, why AutomationML can be an appropriate choice to fulfil current requirements.

Keywords Production system engineering chain • Data exchange • Standardized data exchange format • AutomationML

10.1 Introduction

The increasing global competition between companies from different global regions with completely different economic conditions forces European companies on the one hand to increase product variety, often until complete individualization to meet customer needs. In parallel, on the other hand, these companies are encouraged to increase production system flexibility regarding resource capabilities and quantities as well as regarding used production system technologies. Finally, they shall reduce the duration of both the product life cycle as well as the plant life cycle. But this results in an increased production system complexity which has to be handled within the entire production system life cycle adequately.

To tackle all these challenges different research and development programs are under progress with the German based Industrie 4.0 and the US based Industrial Internet Consortium as most prominent ones without neglecting the huge amount of worldwide initiatives. They intend to combine recent developments within information sciences like most advanced programming strategies with advances in

A. Lüder (✉) • N. Schmidt
Faculty Mechanical Engineering, Otto-von-Guericke University, Universitaetsplatz 2, 39106, Magdeburg, Germany
e-mail: arndt.lueder@ovgu.de; nicole.schmidt@ovgu.de

R. Drath
ABB AG Forschungszentrum, Wallstadter Straße 59, 68526, Ladenburg, Germany
e-mail: rainer.drath@de.abb.com

© Springer International Publishing AG 2017
S. Biffl et al. (eds.), *Multi-Disciplinary Engineering for Cyber-Physical Production Systems*, DOI 10.1007/978-3-319-56345-9_10

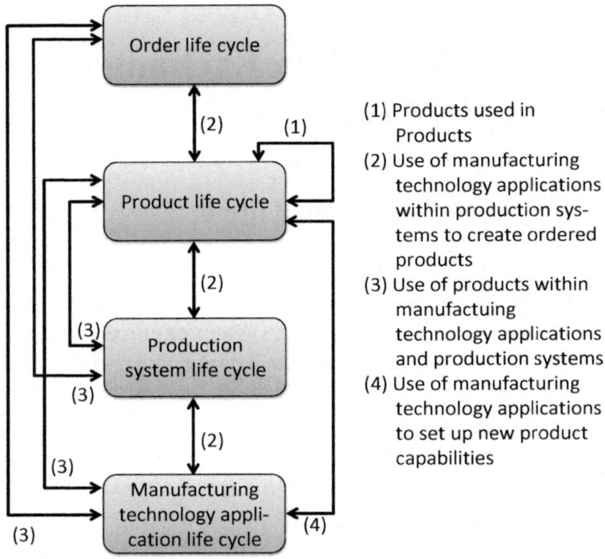

Fig. 10.1 Life cycles in the area of production systems

information processing hardware and communication system technologies as well as advanced system organization strategies (Kagermann et al. 2013; AGPI4.0 2015; IIC 2016). The intention is to reach information driven production system following the paradigm of Cyber Physical Production Systems (CPPS) including information driven production system development/engineering.

But an essential problem within such information driven systems is their inherent system complexity. To visualize the complexity here only two aspects will be reviewed.

The first aspect reflects the various life cycles ranging around production systems (Lüder et al. 2011; VDI 2014) (see Fig. 10.1). At first there is the production system life cycle covering engineering, commissioning, test, usage, maintenance, and disposal of a production system. It requires for the appropriate engineering the definition of the products to be produced as well as the definition of the manufacturing technologies to be applied for within production processes.

Thus, there is on the one hand a life cycle of application of manufacturing technologies covering design, engineering, commissioning, test, usage, support, and disposal of application systems of manufacturing technologies. On the other hand there is a product life cycle including design, engineering, manufacturing, sales, use, support, and disposal/recycling of products.

Beyond engineering the production system also depends on an order life cycle impacting the production system usage (as well as the manufacturing activity within the product life cycle). This order life cycle covers order creation, manufacturing, delivery, and post processing.

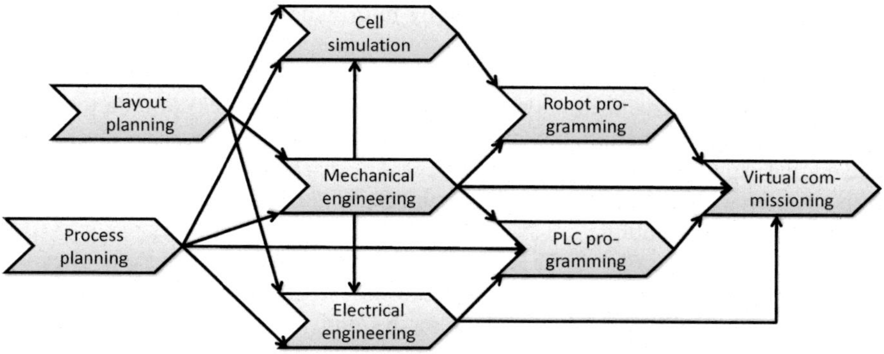

Fig. 10.2 Simplified engineering chain of production systems

Finally, the product life cycle is impacted by the manufacturing technology life cycle as it may enable new product features relevant in the design and engineering.

All these dependencies are depicted in Fig. 10.1. More details on the dependencies between product life cycle, production system life cycle, and order life cycle including their data dependencies can be found in Chap. 4.

It is easily visible that the dependency between these life cycles shall result in appropriate information exchanged between them.

The second aspect to be named is the complexity of engineering chains. Figure 10.2 provides a view on a simplified version of an engineering chain of a production system. It highlights that there are different engineering disciplines involved in the engineering chain executing different engineering activities depending on each other within a strongly coupled network. Just to give a size the number of different engineering activities only to within the process of engineering a body work production system within automotive industry covers more than 30 different engineering activities, the activities named in Fig. 10.2 among them. See also Chap. 9 for a practical engineering chain.

For each of these engineering activities optimized engineering tools have been developed enabling the executing engineers to concentrate on the relevant intellectual work trying to prevent them from stupid "clicking jobs". Hence, it is required to ensure the information exchange between the different involved engineering tools (Drath et al. 2011; Hundt and Lüder 2012; Schmidt et al. 2014) as also identified in research questions I1 and I2 in Chap. 1. This information exchange needs to cover all information relevant within at least two engineering tools or engineering activities.

Beyond the huge amount of information potentially to be exchanged within an information driven production system also the problem of the difference between data and information need to be reminded. Within engineering and use of production systems different data are created which will represent information. Data is usually seen as raw material in front of the information processing. They become information by assigning them a meaning in a way answering on "who", "what", "where" and "when" questions. If we apply information we can reach knowledge by answering "how" questions.

Summing this up, in information driven production systems it is required to define appropriate structuring (syntax) and meaning (semantics) of data to enable the appropriate information exchange among life cycles, engineering disciplines, and engineering activities as partially addressed by research question M2 in Chap. 1. This chapter will consider this problem and highlight possible stating points and approaches for its solution.

As the consideration of the complete field of production system goes far beyond this chapter, initially some relevant use cases of information exchange and application will be highlighted going beyond the current state of the art. They are succeeded by the consideration of information exchange technologies applicable in general and within these use cases especially mainly focusing on AutomationML as one example technology. Finally, the process of setting up a standardized information exchange structure is discussed.

10.2 Use Cases for Information Exchange

Within the Industrie 4.0 and Industrial Internet Consortium usually the horizontal and vertical integration of the different information processing units of production systems are addressed. Among the use cases relevant within the horizontal and vertical integration are

- the definition of appropriate production system hierarchies within the complete engineering chain,
- the integration of pre-developed production system units within larger production systems within engineering and commissioning,
- the exchange of control system engineering information,
- the consistent and up-to-date documentation of the production system as is, and
- the provision of engineering information at production system runtime and its combination with control information.

These use cases cover a representative set of requirements to data exchange in the field of production system engineering.

10.2.1 Use Case 1: Production System Hierarchies

As indicated in Fig. 10.2 and detailed in Chap. 9 of this book within the engineering of production systems different engineering activities/steps are executed. Let's just review the mechanical and electrical engineering steps.

Within the mechanical engineering the mechanical construction of the production system is defined covering the bottom up combination of materials to small components, small components to larger components, and finally to large system parts, all of them fulfilling a certain functionality within the production system. Some of the components are provided by suppliers and thereby not considered with internal details within the engineering, some other components are developed during the engineering based on purchasable materials like metal sheets, screws, beams, springs, pipes, etc. integration purchased sensors like proximity switches and actors like drives and valves. Usually within mechanical engineering the used engineering tools (CAD tools) enable the combination of elements to components in mostly user dependent and function and production process oriented hierarchy.

In contrast the electrical engineering will not care about a function or process oriented component hierarchy. It just will consider the sensors, actors, and control devices (including communication devices) involved within the production system and its relation to each other based on proper wiring. Thereby, the established hierarchy follows the structure of the electric circuits and the necessary communication signal exchange.

As easily visible both hierarchies will usually be different. In case of the a proper information exchange between mechanical and electrical engineering tools (and vice versa) it is required that imported information shall be mapped to the right hierarchy layer. Thus a definition of hierarchy layers as well as possible layer crossing groupings of elements as depicted in Fig. 10.3 need to be standardized and applied within the data exchange format (Grimm 2016).

Fig. 10.3 Examples of standardized hierarchy layers following ISA 95

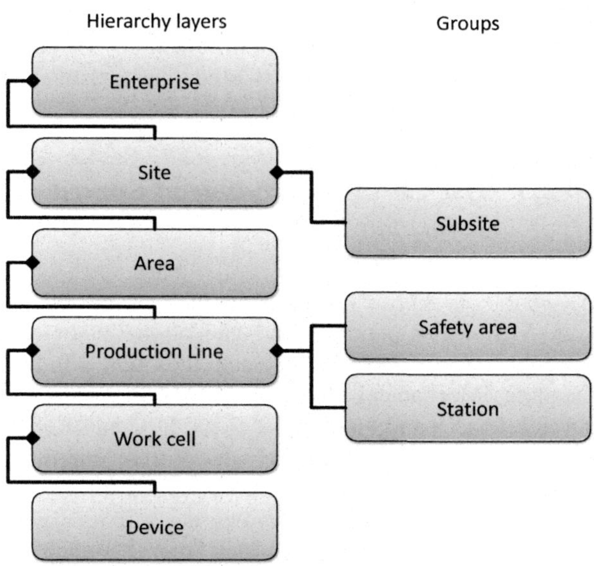

10.2.2 Use Case 2: Integration of Pre-developed Production System Units

The engineering especially of complex production systems is usually not done by a single person. It is also not always done again from scratch. On the contrary, it is good engineering practice to reuse pre-developed building blocks. These building blocks can be of different size and complexity ranging from standardized metal parts (plates, screws, etc.) up to complete technical systems like welding shops and combustion engines. This includes the integration of purchased components like drives, drive chains, or complete packing cells.

Theoretically each involved engineering discipline and each applied engineering tool can apply its own set of pre-developed components. But it is of much more value, if all engineering disciplines can apply a common library of pre-developed components (Wagner et al. 2010; Lüder et al. 2010).

The design of such a common pre-developed library requires two preconditions. The first is a common, all engineering disciplines spanning, system architecture (VDI 2009). This architecture shall comply with the different concepts applied within the production system engineering in the different engineering disciplines and harmonize them, i.e., a device shall be the same concept in each discipline. In addition a common conceptualization of the different system component properties and its consistent management is required. An example can be the component of a dive chain in a packing machine. It usually contains drive chain related attributes like the angular speed of a driven roll, the diameter of the roll, and the current and voltage assigned to the drive in a certain situation. All these attributes are functionally interrelated. These interrelations need to be representable in a component.

10.2.3 Use Case 3: Exchange of Control System Engineering Information

As visible in Fig. 10.2 an essential part of the engineering of a production system is the engineering and validation of control applications. Within these activities six of the engineering activities named in Fig. 10.2 are involved. Process planning, Mechanical engineering, and Electrical engineering define necessary control devices and their connections, PLC and Robot programming exploit them for control application design, and Virtual commissioning validate the control applications based on the defined device set.

To integrate the named engineering activities within an engineering chain it is required to have an appropriate representation of the common concepts of them. These common concepts at least contain the conceptualization of the control devices, their internal structure, and their physical and logical relations.

10.2.4 Use Case 4: Consistent and Up-To-Date Documentation

As named by Tauchnitz in (Tauchnitz 2016) in relation to the engineering of process control systems, during the life cycle of production systems different sources provide information about the structure of a control system. There are on the one hand the CA* tools for mechanical and electrical engineering providing first versions of the set of applied control devices and its interconnections. Currently they export created device lists (currently still often given as *.csv files) which subsequently are applied in PLC and Robot programming.

This methodology is applicable in case of an engineering process but it fails in case of production system runtime changes. If a device is changed or replaced during production system ramp-up of maintenance the application of device lists and the required search for the changed device within this list requires too much time to be meaningful.

The open question is the optimal way to ensure a proper way of change tracing at production system runtime. Here a detailed data exchange technology with a proper syntax and semantics related to control engineering is required.

10.2.5 Use Case 5: Combination of Engineering and Runtime Information

There is also another use case requiring the integration of engineering and runtime information of a production system. This use case emerges from the intention to use advanced production system maintenance strategies like condition monitoring.

The diagnosis and maintenance at production system runtime requires a detailed knowledge about the structure and behavior of the production system components including knowledge about the capabilities to identify component fault characteristics. Such knowledge can be extracted from the combination of production system engineering information, component documentations, and component state measurements. To enable this combination the named information need to be accessible at production system runtime.

To reach this accessibility it is necessary to store the engineering information as well as the component documentation (including documents like mounting guidelines and handbooks) in an accessible way and to integrate this information with access paths information usable to access control signals related to sensors, actuators, and component KPIs.

This storing and integration requires two essential prerequisites. At first there is a data exchange format required which can be integrated in a runtime system storage and accessed via clear access paths following syntax and semantics of the stored information. At second the same technology shall be applicable for the access to runtime control information enabling the definition of relations between runtime and engineering information.

10.2.6 Current Activities Related to Solution of the Use Cases

There are different developments on the way trying to provide solutions for the named use cases. The most appropriate are

- the development of modular and hierarchical production system architectures,
- the standardization of data exchange formats applicable during the complete life cycle of production systems, and
- the integration of engineering data representation and runtime communication system.

These activities will be considered in more detail.

10.2.6.1 Development of Modular and Hierarchical Production System Architectures

The most prominent development in the direction of a standardized architecture for the control of production systems is under development of the VDI/VDE-GMA working group "Industrie 4.0". It develops with the RAMI (VDI 2015) and the Industrie 4.0 component a reference structure to be applied in all Industrie 4.0 systems.

Less known are the work of the NAMUR working group "Automatisierung modularer Anlagen", the work of the AutomationML association, and the work of the VDA. The NAMUR made the decision to take up the results of the successful DIMA project (Obst et al. 2016) to develop a modular process industry system exploiting so called Module Type Packages. The AutomationML association currently develops a modelling strategy for production system components inspired from research and development activities within projects like SkillPro (Pfrommer et al. 2014), Avanti (Süß et al. 2015), and Conexing (Bartel et al. 2014). Finally, the VDA is developing a recommendation for the unique representation of production system hierarchies.

10.2.6.2 Standardization of Data Exchange Formats

With the aim to enable the implementation of standardized engineering tool interfaces currently different organizations develop data exchange formats. Following (Tauchnitz 2016) the most relevant criteria for the development of these data formats are the sequential standardization of data exchange structures for essential but manageable parts of the relevant information, the standardization of usually required information sets to stay with frequently used and lightweight formats, the enabling of user dependent extensions of the standardized data format, the compatibility with further developments on the same data format (even for further application cases), the application of existing standards, and the cooperative development integrating all relevant stakeholders from users to tool vendors.

Among others in this field the VDMA working group "Engineering", the NAMUR working group "PLS-Engineering", the VDA working groups "Logistics" and "Virtual commissioning", the GMA working group 6.16 "Integriertes Engineering in der Prozessleittechnik", the AutomationML association and the ProSTEP association are active.

10.2.6.3 Integration of Engineering Data Representations and Runtime Communication Systems

The accessibility of control information is considered by several organizations. A leading role are taking the communication system developing vendor organizations like ODVA, PNO, ETC, or Sercos International to name only a small subset. All of these organizations are developing communication systems defining among others profiles standardizing access paths to information within production system components (Reißenweber 2009). The main drawback of this approach is its dependency to special communication protocols.

A communication protocol independent access to control information is for example intended by the OPC Foundation with the development of the OPC UA technology family. For this technology family various other user organizations are trying to develop special profiles enabling the structured access to domain specific information.

Among these user organizations the AutomationML association has developed a methodology to represent engineering information by means of OPC UA. The resulting OPC UA profile is standardized within DIN SPEC 16592.

An overview of further research activities in the field of production system engineering is collected within (Ferscha 2016).

10.3 Information Exchange Technologies

Beyond the capability to cover the relevant information to enable the handling of the named use cases in (Lüder et al. 2014) a set of requirements to a data exchange format are concluded from a survey. These requirements are the following.

(A1) Information related to common concepts of all involved disciplines over the engineering process phases has to be exchanged in the same syntactically and semantically unique way.

(A2) Information about dependencies between different concepts of involved disciplines has to be exchanged.

(A3) Relevant dependencies between different concepts of involved disciplines have to be able to be parameterized and traced.

In addition the broad usability of exchange technologies requires a set of more technical requirements. From the viewpoint of the authors, the most relevant among them are the following.

(A4) The information exchange technology shall be XML based to enable a readability of the stored information for example for purposes of proper interface implementation. Using XML an implementer can validate the created files also manually to be more familiar with the development results.
(A5) The information exchange technology shall be applicable for free based on free available standards. Neither parts of the technology nor basic documentation is accessible only based on unreasonable access fees going beyond the membership fees of user organizations or the usual price of printed documents of publishers. If possible fees shall be avoided at all.
(A6) The information technology shall be applicable on different platforms including engineering stations as well as limited and embedded devices.
(A7) The information exchange technology shall enable the separated standardization of syntax and semantics of data objects. By this the extension of the standardized representation range shall be extendable on demand.
(A8) Beyond the standardized representation range users shall be enabled to integrate also not standardized information and represent it in a consistent and integrated way by the information exchange technology.

There are various data formats applicable to model production systems or parts of them. In the following a small set of such models is indicated and characterized representing classes of formats. These set has been selected to provide typical representatives of the data formats used in industrial practice. In some cases this might look artificial, but it reflects for example the case, that office tools (especially Excel) are the most frequent used tools in engineering.

The data exchange format ***STEP*** (STandard for the Exchange of Product model data) has been developed to cover product related information enabling the modelling of all information relevant within the product life cycle (PLM) (see also Chap. 4). It covers for example product geometry information, characterizing properties of products, and production process descriptions for the processes required for product creation (Xu and Nee 2009). It is standardized within ISO 10303. This standard series contains different profiles which are intended to be consistent among each other. Some of the defined profiles (especially the newer ones) are available as XML dialect. Hence, within production system engineering STEP is mostly applied for product and production process representation.

A possible substitute to STEP covering product information for PLM is PLM-XML, an XML based product data format specified and maintained by Siemens.

As STEP the data format Jupiter Tessellation (***JT***) is standardized as ISO standard within ISO 14306 (ISO 2012b). It has been developed to enable the vendor neutral representation of geometry information coming from CAD tools. It is intended to cover product models including the product structure, the geometry of structure elements, and structure element properties as modelled in collaborative product development processes as well as product lifecycle management processes (see also Chap. 4).

Substitutes to JT as geometry data formats are COLLADA, IGES, VDA-FS and DXF.

Within the office environment the data format Portable Document Format (***PDF***) has reached a wide acceptance. It is used for implementation technology independent text document content representation including the representation of graphical information on a virtual piece of paper by modelling geometry elements like vector graphics, text, etc. PDF is standardized in ISO 32000-1:2008. In production system engineering it is usually applied for manuals, layout plans, etc.

A substitute to PDF can be Postscript.

As in office environment also in the engineering chain of production systems, ***Microsoft Office tools*** are used to create various documents including documentations, descriptions, listings, etc. Widespread Microsoft Word and Microsoft Excel are applied and the reached results are stored in DOC, DOCX, XLS, and XLSX format. The applicability of these tools in the engineering chain of production systems is not really limited as they are very generic and can cover data from any engineering domain and phase.

Possible substitutes to office files are CSV files (for Excel) and data formats from tools like Open Office.

With XML Process Definition Language (***XPDL***) an XML based dialect has been standardized in Workflow Management Coalition (WfMC) (WfMC 2012) to enable business process modelling in a vendor neutral way. It enables the exchange of process definitions, following both graphics and semantics of business process workflows. Thus, it has to be regarded as the XML representation of the Business Process Modelling Notation (BPMN). In the engineering process of production systems XPDL can be applied for production processes representation.

A possible substitute to XPDL is BEPL.

A further XML based dialect is XML Metadata Interchange (***XMI***). It has been developed as meta format and is standardized by Object Management Group (OMG) (OMG 2015). Based on XMI domain specific XML based dialects can be developed like it is the case for the representation of UML project content.

In the engineering chain of production systems XMI is usually applied to represent production process information and production system information.

The ***AutomationML*** data exchange format is under development by AutomationML association (Drath 2010; Schmidt and Lüder 2015). It is standardized within IEC 62714 to provide a neutral, open, and XML based data exchange format. The intention of AutomationML is to enable consistent and lossless exchange of engineering information related to manufacturing system topology, geometry, kinematics, and control behavior. Therefore, AutomationML follows a modular structure integrating existing XML based data formats. Logically it is divided into the parts plant topology modelling (following CAEX 62424), geometry and kinematics modelling (following COLLADA), and control related logic data modelling (following PLCopen XML) where COLLADA and PLCopen files are referenced out of CAEX files.

Table 10.1 Modelling range of data formats

	AutomationML	STEP	JT	PDF	Office	XPDL	XMI
Production system topology/structure information	+	+	+	o	o	–	+
Mechanical engineering information	+	+	+	o	o	–	–
Electric and fluidic engineering information	+	+	o	o	o	–	o
Function describing information	+	+	–	o	o	+	–
Control engineering information	+	–	–	o	o	+	o
Further/generic engineering information	o	o	o	+	+	–	o
Dependencies between different disciplines	+	o	o	o	o	–	–

+ is good applicable
o is fair applicable
– is not applicable

In the engineering chain of production systems AutomationML is usually applied to represent production processes and production system information.

The different represented data formats have different modelling power related to the coverage of all information relevant for production system engineering. Table 10.1 subsumes the modelling range of the different selected data formats.

As the modelling range also the ability of the different data formats to fulfil the different requirements named above is not equal. Table 10.2 gives a summary about the capabilities of data formats to cope with requirements named above.

10.4 AutomationML

Within the above given tables it gets visible, that AutomationML is a proper candidate to be used within the use cases named above. Therefore, AutomationML will be described here in more detail and its main characteristics related to standardization of data exchange structures are given.

The AutomationML data format has been developed by AutomationML e.V. [see (AutomationML 2016)] as solution for the data exchange focusing on the engineering of production systems. It is an open, vendor neutral, XML-based, and free data exchange format which enables a domain and company spanning transfer of engineering data of production systems in a heterogeneous engineering tool landscape.

Table 10.2 Requirement fulfilment of data formats

	AutomationML	STEP	JT	PDF	Office	XPDL	XMI
(A1) Common concepts	+	+	o	+	+	o	+
(A2) Dependency types between concepts	+	+	–	o	o	–	+
(A3) Dependency properties/traceability	o	+	–	–	–	–	+
(A4) XML based	+	o	–	–	o	+	+
(A5) Free accessible	+	o	–	+	o	+	+
(A6) Platform independent	+	o	o	o	o	+	+
(A7) Separation of syntax and semantics	+	–	–	o	o	o	+
(A8) Integrate user dependent information	+	–	–	+	+	–	o

+ is good applicable
o is fair applicable
– is not applicable

AutomationML stores engineering information following the object oriented paradigm and allows the modelling of physical and logical plant components as data objects encapsulating different aspects. Objects may constitute a hierarchy, i.e. an object may consist of sub-objects and may itself be a part of a larger composition or aggregation. Additionally each object can contain information about object describing properties covering geometry, kinematics, and logic (sequencing, behavior, and control information) as well as further properties.

AutomationML follows a modular structure by integrating and enhancing/adapting different already existing XML-based data formats combined under one roof the so called top level format (see Fig. 10.4).

These data formats are used on an "as-is" basis within their own specifications and are not branched for AutomationML needs. Logically AutomationML is partitioned in:

- Description of the component topology and networking information including object properties expressed as a hierarchy of AutomationML objects and described by means of CAEX following IEC 62424 (IEC 2008),
- Description of geometry and kinematics of the different AutomationML objects represented by means of COLLADA 1.4.1 and 1.5.0 (ISO/PAS 17506:2012) (ISO 2012a),
- Description of control related logic data of the different AutomationML objects represented by means of PLCopen XML 2.0 and 2.0.1 (PLCopen 2012) and (partially) by means of MathML integrated in PLCopen XML and
- Description of relations among AutomationML objects and references to information that is stored in documents outside of the top level format using CAEX means.

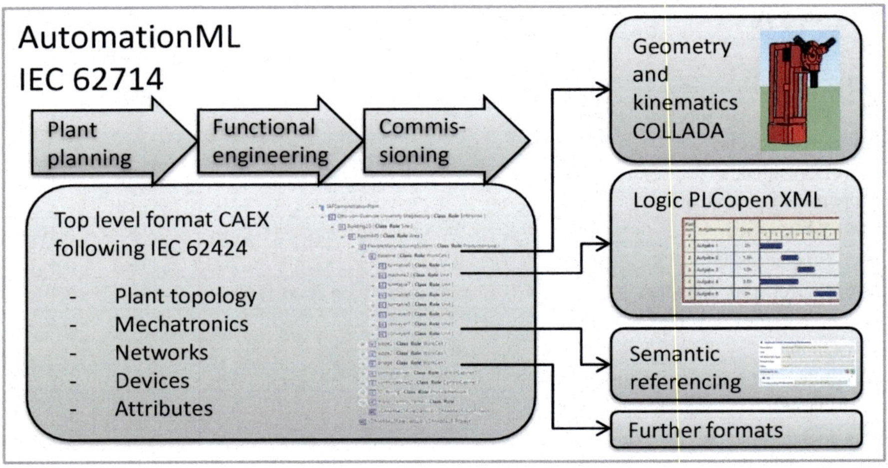

Fig. 10.4 Structure of AutomationML projects

AutomationML is currently standardized within the IEC standard series IEC 62714 (IEC 2014). For a more detailed description of AutomationML see (Drath 2010) and (Schmidt and Lüder 2015).

The foundation of AutomationML is the application of CAEX as top level format and the definition of an appropriate utilization fulfilling all relevant needs of AutomationML to model engineering information of production systems, to integrate the three named data formats CAEX, COLLADA, and PLCopen XML, and to enable an extension if necessary in the future.

As mentioned above, CAEX enables an object oriented approach (see Fig. 10.5). Thereby it enables the separation of syntax and semantics of the represented data objects. It is based on four main building blocks: role classes (RC), interface classes (IC), system unit classes (SUC), and internal elements (IE).

Role classes are intended to enable the definition of semantic of data objects in a kind of late semantic freeze by describing an abstract functionality of an object without defining the underlying technical implementation. Thus they serve for the formalization of the main concepts required in an application domain. Role classes can be for example the role classes motor and sensor indicating system structure semantics or LogisticalDevice and PhysicalDevice representing communication system semantics. Each role class is additionally equipped with attributes and external interfaces (EI) describing the role class in more detail representing role class properties and role class dependencies to other objects. For example the role class sensor may have an attributes to indicate the sensor vendor and the power consumption, an external interface (screwing) to indicate the mounting point of the sensor, and an external interface (power_connection_socket) to model the power supply of the sensor.

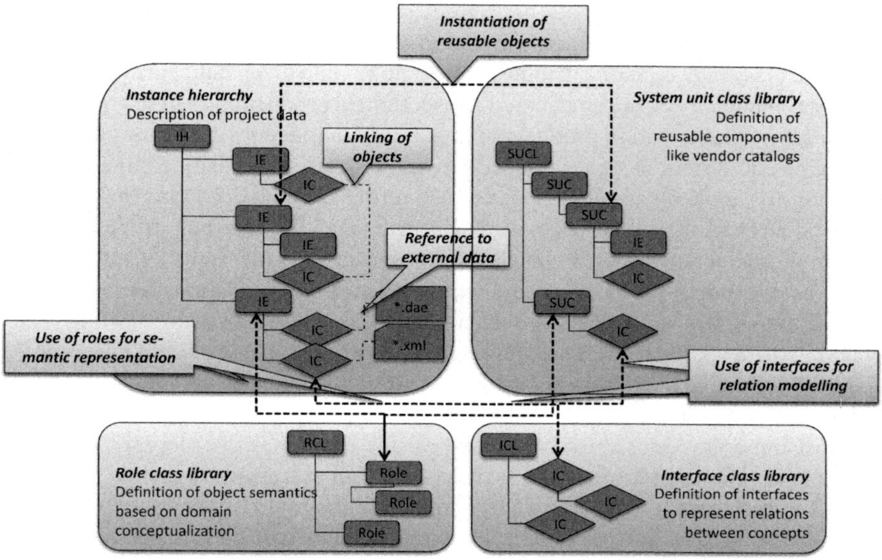

Fig. 10.5 AutomationML topology description architecture

Role classes are defined in role class libraries establishing a kind of tree structure of more detailed roles. AutomationML defines a set of basic role classes. Within Part 1 of the AutomationML standard (IEC 2014) the AutomationMLBaseRoleClassLib with fundamental role classes is defined. It contains with the AutomationMLBaseRole a role class each other AutomationML related role class has to be derived from. More detailed role classes are defined in the further parts of the AutomationML standard. For example Part 2 defines general role classes required for production system structuring and manufacturing process identification.

Interface serve as the model of the required relation between objects. They are derived from interface class describing an abstract relation an element can have to other elements or to information not covered within the CAEX based model (for example to geometry and kinematics modelling and behavior modelling). Thus they serve as formalization of relations between concepts. Examples of interface classes can be power supply and mounting position or the relation to a manual and a geometry description all relevant for a device.

Interface classes are defined in interface class libraries establishing a kind of tree structure of more detailed interfaces. AutomationML defines a set of basic interface classes. Within Part 1 of the AutomationML standard (IEC 2014) the AutomationMLInterfaceClassLib with fundamental role classes is defined. It contains with the AutomationMLBaseInterface an interface class each other AutomationML related interface class has to be derived from. More detailed interface classes are defined in the further parts of the AutomationML standard. For example Part 4 defines general interface classes required for control system structuring. Each interface class may have describing attributes. These attributes shall be used and filled with values in each occurrence of an instance of the interface class. For example a power supply interface class may contain attributes representing the voltage and current.

System unit classes are considered as ascertained production system components which are available within product catalogues of component vendors or component catalogues of system designer intended to enable reuse of already engineered (and successfully applied) system elements. Therefore they model more than just the concept. Instead they give a detailed description of the internal structure of the system component of interest. Therefore they contain a hierarchy of internal elements as structure representation and interfaces as possible relations representation. In addition system units classes, their internal elements, and their interfaces may contain attributes for property detailing. An example of a system unit class can be a special inductive approximate switch of a special vendor. System unit classes are defined and maintained within system unit class libraries.

Most important related to the content of this chapter is the property, that each system unit class shall reference at least one (but possibly more than one) role class it implements, i.e. which concepts are fulfilled by that system unit class. For example the system unit class sensor (and all system unit classes derived from it like inductive_sensor) fulfil the sensor role class. Thus it gets a semantics within the domain of interest. In addition, each interface integrated in the system unit class is derived (and by that assigned to) an interface class giving its semantics. Thus, system unit classes get their semantics by referencing role classes and interface classes.

The internal elements are the last important building block of CAEX. They arranged in a hierarchy named instance hierarchy and represent the individual engineering data to be exchanged. Each internal element represents a relevant data object including further internal elements as substructures, interfaces representing their relations to other internal elements, and attributes representing detailed object properties. An example of an internal element related to production system engineering can be the representation of a turntable with its internal sensors, actors, interfaces, and so on.

Again, the most important related to the content of this chapter is the property, that each internal element has to reference at least one (but possibly more than one) role class it implements, i.e. which concepts are fulfilled by that internal element. For example the internal element inductive_sensor_turntable fulfil the sensor role class and a role class related to the description of an inductive proximity switch within eCl@ss. Thus it gets a related semantics. Further, each interface integrated in the internal element is derived (and by that assigned to) an interface class giving its semantics. Thus, internal elements get their semantics by referencing role classes and interface classes. Beyond, internal elements can reference a system unit class acting as the template for the structure of the internal element providing an additional way of representing the object semantics.

Within Fig. 10.6 some of the named examples are represented. For details on the described structure and properties of CAEX and its use as roof format of AutomationML the authors refer to the different AutomationML whitepapers available at

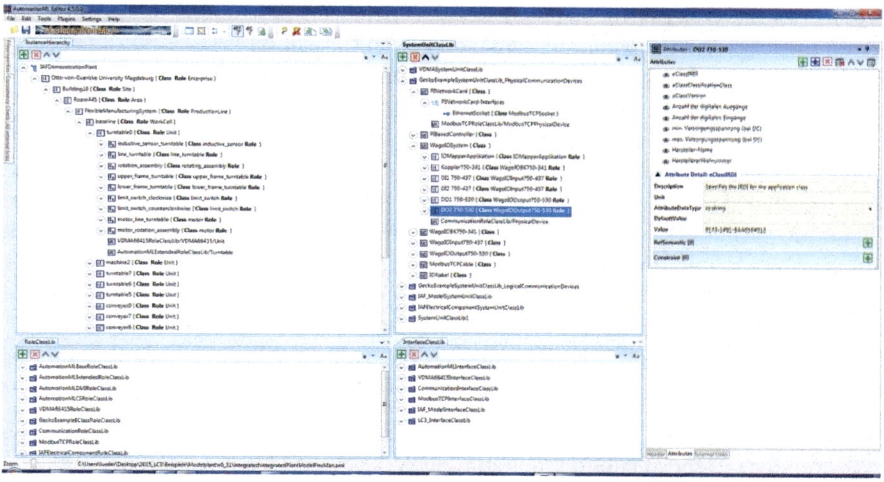

Fig. 10.6 AutomationML topology example

(AutomationML 2016) and to the AutomationML in a nutshell representation in (Schmidt and Lüder 2015).

10.5 Challenges Within Standardization of Information Exchange

As already previously mentioned above the setup of information driven production systems requires the definition of appropriate structuring (syntax) and meaning (semantics) of data to enable the appropriate information exchange among life cycles, engineering disciplines, and engineering activities.

This standardization is often trapped in a kind of standardization deadlock (Drath and Barth 2012). As the intention is to be applied in practical application cases the standardization process requires the involvement of practitioners. In addition the used information exchange technologies shall be most advanced and promising also for the future. Thus also researchers shall be integrated in the standardization process. Hence, a group of practitioners and researchers is formed developing the standard of interest.

But usually the developed standard will not fulfil all requirements from practical application cases as well as from the end users from the first beginning. By applying the standard in practice for example in engineering tools the quality of the standard is evaluated and new enhancement request will arise. These enhancement requests will be handled by the standardization group leading to a more mature version of the standard which can now be again applied and evaluated.

This cyclic process of standard development and application is an essential requirement to reach a high quality standard. But it can be broken by the need of implementation of engineering tools (and other information processing elements) applying this standard. Very often companies having an engineering tool (or another information processing element) as one of its products being relevant for the implementation of the standard clearly have limited interest in implementing an uncompleted standard just for evaluation. They cannot be sure on the technical and economic success of the standard. In addition, the sequential implementation of different versions of the standard may lead to non-satisfaction of their customers. Thus, these companies will refuse the implementation of intermediate versions of the standard braking up the cycle of standard development and evaluation.

To handle this challenge the information exchange standard shall fulfil the following properties:

(S1) The standardization shall follow a sequential process developing consistent versions of the standard integrating more advanced features step by step.
(S2) The standard shall enable the combination of standardized parts and non-standardized (proprietary) parts without loss of generality. If the proprietary parts become proven it shall be possible to take them over in the standard.
(S3) The standard shall enable the development of domain specific profiles. These profiles shall be designed in a way that they can be applied in combination and are, therefore, compatible to each other.
(S4) There shall be an integrated mechanism ensuring, that standardized parts, domain specific profiles, and proprietary parts will be compatible to each other.
(S5) The standard shall enable the easy adaptation of implementations of the standard within engineering tools and other information processing element by simply changing the configuration of the implementation without necessary changes within the implementation code.

In the following it will be discussed how the AutomationML technology and the standardization process of AutomationML are fulfilling the named requirements. But before this discussion will start, some essential properties of AutomationML are reminded.

(P1) AutomationML separates the standardization of syntax and semantics. The syntax is based on the applied XML format CAEX (and the other involved formats COLLADA, PLCopen XML, and MathML). It defined generic internal elements within an instance hierarchy being able to cover arbitrary physical and non-physical production system elements. The semantics of the data elements is defined by role classes and interface classes to be referenced by and integrated in the internal elements.
(P2) Internal elements can reference one or more role classes giving those different semantics in parallel. If the referenced role classes provide the internal element with attributes having the same name, there are means for attribute mapping enabling the clear distinction of role related attributes. In addition different

interfaces derived from different interface classes can be used in an internal element.

(P3) Role classes are specified in role class libraries and interface classes in interface class libraries which can be integrated in an AutomationML file following the needs of the application case. They are on the one hand independent from each other and on the other hand role classes are derived from each other building up a role class derivation tree with AutomationMLBaseRole as tree root and the takeover of all attributes and interfaces from the daughter role class from the mother role class. (A similar tree exists for interface classes with AutomationMLBaseInterface as tree root.)

The AutomationML standardization process is modular in three dimensions.

The first dimension is formed by the international standardization process. It starts with the discussion of new application cases and, therefore, parts within the AutomationML association. Based on this discussion working groups of the association will develop a first version of the part of the standard document of interest. This version is afterwards tested in test implementations by the AutomationML members. Based on the test results the standard document is improved until it gets stable. If the standard document is stable it is transferred to the German DKE (www.dke.de), a German national standardization organization. Here, in cooperation with further specialists the standard document is improved with respect to the requirements of an international standard and prepared for international standardization. If the standard document is prepared, it will be submitted to IEC for international standardization within IEC 62714.

The second dimension is formed by the stability cycles of the standard within the different involved organizations. Both, the AutomationML association as well as the IEC are reviewing and improving the standard documents in temporal cyclic way enabling the integration of intermediate improvement of the technologies applied as well as the standard itself.

The third dimension is formed by the parts of the AutomationML standard. Within the AutomationML association the AutomationML standard is codified by whitepapers, best practice recommendations and application recommendations. The whitepapers define the basements of AutomationML. For example Part 1 defines the CAEX profile AutomationML is based on, Part 3 specifies the use of COLLADA to represent geometry and kinematics, and Part 4 specifies the application of PLCopen to represent behavior models. In contrast Part 2 is not defining the use of an XML format but it defines basic, domain independent role class libraries. Application recommendations are defining domain specific profiles for the use of AutomationML. There are for example application recommendations related to the exchange of control system structures, transportation system models, and the use of OPC UA to represent an AutomationML project. The best practice recommendations are intended to reduce the interpretation range of the AutomationML standard and to provide domain independent features of the data format. For example, there are best practice recommendations related to the naming of AutomationML library versions, the modelling of reference designations, the modelling of parameter lists, or the structuring of AutomationML projects within an archive file.

Within the DKE and IEC set of documents is limited to parts. Up to now each IEC part is equivalent to a whitepaper. But following the IEC standard maintenance cycle the parts will combine whitepapers and best practice recommendations. Whether the application recommendations will be used to form IEC parts or not is currently not finally decided.

It is easily visible, that this standardization process fulfils the requirement (S1).

It is also easily visible that the different whitepapers and application recommendations will form domain independent and domain dependent profiles. They define the necessary semantics by giving appropriate role class libraries and interface class libraries covering all necessary concepts and relations between concepts of interest. As all these role class libraries and interface class libraries follow within its design (P3) and are applied for semantic representation following (P2) they can be applied independent and together depending on the application case, their application results in a consistent structure, and they are compatible to each other. Thus, requirement (S3) and (S4) are fulfilled.

The developed architecture of the AutomationML standard as well as the properties of CAEX enable users to define its own role class libraries and interface class libraries based on the predefined ones of AutomationML. They may extend existing semantics or define completely new semantics within these libraries and apply them within its own information exchange structures in extension to the specified AutomationML concepts. The AutomationML association has developed a procedure enabling users to disclose their internal structures to the members of the AutomationML for discussion. Within this discussion process the proprietary structures can be evaluated and finally be codified and standardized within an application recommendation or a best practice recommendation. Thereby requirement (S2) is fulfilled.

The capabilities of AutomationML to fulfil requirement (S5) is more hidden within the data format and its application on an importer or exporter of an engineering tool or another information processing element.

The implementation of exporters and importers requires a mapping of the tool internal information model to the information model of the data exchange format. This mapping needs to be defined and implemented depending on the used semantics (Hundt and Lüder 2012). As AutomationML defines the semantics by defining appropriate role classes and interfaces classes these classes can be applied to enable the configuration of the required mapping.

A possible solution can be the development of a mapping table between the applied role and interface classes and internal data elements and the implementation of a generic data access function reading an incoming or outgoing data element, looking in the table to find the related mapping and creating its appropriate representation. By this a general exporter and importer can be implemented tailored to the standard improvement by simply updating the configuration files fulfilling requirement (S5).

10.6 Summary

The intention of this chapter was to consider the problem of appropriate structuring (syntax) and meaning (semantics) definition for a file based data exchange technology applicable within information exchange among life cycles, engineering disciplines, and engineering activities of information driven production systems. Therefore, the chapter has discussed existing challenges and highlight possible stating points and approaches for its solution.

Initially, five use cases of information exchange and application within information driven production systems have been highlighted. They illustrate the strong potentials information driven production systems contain but also the challenges arise for its application. The use cases have been accompanied of current standardization activities undertaken to make the use cases possible.

As next step information exchange technologies have been discussed. Starting with requirements an information exchange technology has to fulfil in an information driven production system different - information exchange technologies have been reviewed and there modelling range and requirement fulfilment have been identified. This evaluation has highlighted AutomationML as one essential candidate to be used within the use cases of information driven production systems.

Following a deeper look in the structure and application of AutomationML and a clear description how AutomationML deals with the standardization of syntax and semantics five main challenges of the standardization of data exchange formats are named. It has been discussed how AutomationML deals with these challenges.

In general this chapter provides information about the power of AutomationML within the development of information driven production systems. It shall encourage users and developers, practitioners and researchers, and experts and novices to consider AutomationML, enhance its capabilities, and join the standardization efforts finally enabling information driven production systems.

Acknowledgement The authors of this paper intend to thank Björn Grimm (Daimler AG) for his valuable support and discussion of the content of this paper within the AutomationML association.

References

Arbeitsgruppen der Plattform Industrie 4.0: Umsetzungsstrategie Industrie 4.0 – Ergebnisbericht der Plattform Industrie 4.0. https://www.bmwi.de/BMWi/Redaktion/PDF/I/industrie-40-verbaendeplattform-bericht,property=pdf,bereich=bmwi2012,sprache=de,rwb=true.pdf (2015). Last Access 01 Feb 2016 (in German)

AutomationML e.V: AutomationML. www.automationml.org (2016). Last Access Sept 2016

Bartel, M., Schyja, A., Kuhlenkötter, B.: More than a Mockup – smart components: reusable fully functional virtual components from scratch. Prod. Eng. **8**(6), 727–735 (2014–12)

Drath, R. (ed.): Datenaustausch in der Anlagenplanung mit AutomationML. Springer, Berlin (2010)

Drath, R., Barth, M.: Concept for managing multiple semantics with AutomationML – maturity level concept of semantic standardization. In: 17th IEEE International Conference on Emerging Technologies & Factory Automation (ETFA 2012), Krakow, Poland, Proceedings, September 2012

Drath, R., Fay, A., Barth, M.: Interoperabilität von Engineering-Werkzeugen. Automatisierungstechnik. **59**(7), 451–460 (2011)

Ferscha, A.: Whitebook JKU Production Research, Johannes Keppler University Linz. http://www.pervasive.jku.at/download/JKU_PRODUCTION_WHITEBOOK.pdf (2016). Last Access Dec 2016

Grimm, B.: AutomationML in den Engineeringprozess einführen, ATP edition. Band **58**(Ausgabe 05), 42–51 (2016)

Hundt, L., Lüder, A.: Development of a method for the implementation of interoperable tool chains applying mechatronical thinking – Use case engineering of logic control, In: 17th IEEE International Conference on Emerging Technologies and Factory Automation (ETFA 2012), Krakow, Poland, Proceedings, September 2012

Industrial Internet Consortium (IIC): The industrial internet reference architecture technical paper. http://www.iiconsortium.org/IIRA.htm (2016)

International Electrotechnical Commission: IEC 62424: Representation of process control engineering – Requests in P&I diagrams and data exchange between P&ID tools and PCE-CAE tools. www.iec.ch (2008)

International Electrotechnical Commission: IEC 62714: Engineering data exchange format for use in industrial automation systems engineering – AutomationML. www.iec.ch (2014)

International Organization for Standardization: ISO/PAS 17506:2012 – Industrial automation systems and integration – COLLADA digital asset schema specification for 3D visualization of industrial data. www.iso.org (2012a)

International Organization for Standardization: ISO 14306:2012 – Industrial automation systems and integration – JT file format specification for 3D visualization. ISO Publisher, Research Triangle Park, NC (2012b)

Kagermann, H., Wahlster, W., Helbig, J. (eds.): Recommendations for implementing the strategic initiative INDUSTRIE 4.0 – Securing the future of German manufacturing industry, Final report of the Industrie 4.0 Working Group. http://www.acatech.de/fileadmin/user_upload/Baumstruktur_nach_Website/Acatech/root/de/Material_fuer_Sonderseiten/Industrie_4.0/Final_report__Industrie_4.0_accessible.pdf (2013) Last Access 01 Feb 2016

Lüder, A., Foehr, M., Hundt, L., Hoffmann, M., Langer, Y., Frank, St.: Aggregation of engineering processes regarding the mechatronic approach. In: 16th IEEE International Conference on Emerging Technologies and Factory Automation (ETFA 2011), Toulouse, France, Proceedings-CD, September 2011

Lüder, A., Hundt, L., Foehr, M., Wagner, T., Zaddach, J.-J.: Manufacturing system engineering with mechatronical units. In: 15th IEEE International Conference on Emerging Technologies and Factory Automation (ETFA 2010), Bilbao, Spain, Proceedings-CD, September 2010

Lüder, A., Schmidt, N., Rosendahl, R., John, M.: Integrating different information types within AutomationML. In: 19th IEEE International Conference on Emerging Technologies and Factory Automation (ETFA), Barcelona, Spain, Proceedings, September 2014

Object Management Group: XML Metadata Interchange (XMI) Version 2.5.1. www.omg.org/spec/XMI/ (2015)

Obst, M., Holm, T., Urbas, L., Fay, A., Kreft, S., Hempen, U., Albers, T.: Beschreibung von Prozessmodulen, ATP edition. Band **57**(Ausgabe 01–02), 48–59 (2016)

Pfrommer, J., Stogl, D., Aleksandrov, K., Schubert, V., Hein, B.: Modelling and orchestration of service-based production systems via skills. In: IEEE 19th Conference on Emerging Technologies and Factory Automation (ETFA 2014), Barcelona, Spain, Proceedings, September 2014

PLCopen association: PLCopen XML. www.plcopen.org (2012)

Reißenweber, B.: Feldbussysteme in der industriellen Praxis. Oldenbourg Industrieverlag, München (2009)

Schmidt, N., Lüder, A.: AutomationML in a nutshell, AutomationML consortium. www.automationml.org (2015)
Schmidt, N., Lüder, A., Steininger, H., Biffl, S.: Analyzing requirements on software tools according to the functional engineering phase in the technical systems engineering process. In: 19th IEEE International Conference on Emerging Technologies and Factory Automation (ETFA), Barcelona, Spain, Proceedings, September 2014
Süß, S., Strahilov, A., Diedrich, C.: Behaviour simulation for virtual commissioning using co-simulation. In: 20th IEEE Conference on Emerging Technologies & Factory Automation (ETFA), Luxembourg, Proceedings, 2015
Tauchnitz, T.: Engineering, Prozessdaten, Anlagendaten, Industrie 4.0 – alles wächst zusammen. Automation Kongress 2016, Baden Baden, Germany, Proceedings, June 2016
VDI/VDE: Industrie 4.0 – Wertschöpfungsketten, VDI/VDE Gesellschaft Mess- und Automatisierungstechnik. Status report (2014)
VDI/VDE-Gesellschaft Mess- und Automatisierungstechnik – Fachausschuss 7.21 Industrie 4.0: Status Report: Reference Architecture Model Industrie 4.0 (RAMI4.0). http://www.zvei.org/Publikationen/GMA-Status-Report-RAMI-40-July-2015.pdf (2015)
Verein Deutscher Ingenieure: VDI Guideline 3695 – Engineering von Anlagen – Evaluieren und optimieren des Engineerings. VDI Verlag, Düsseldorf (2009)
Wagner, T., Haußner, C., Elger, J., Löwen, U., Lüder, A.: Engineering processes for decentralized factory automation systems. In: Silvestre-Blanes, J. (ed.) Factory Automation, vol. 22. In-Tech, Austria. ISBN:978-953-7619-42-8. http://www.intechopen.com/articles/show/title/engineering-processes-for-decentralized-factory-automation-systems (2010)
Workflow Management Coalition: XML Process Definition Language (XPDL). www.xpdl.org (2012)
Xu, X., Nee, A.: Advanced Design and Manufacturing Based on STEP. Springer, New York (2009)

Part III
Information Modeling and Integration

Chapter 11
Model-Driven Systems Engineering: Principles and Application in the CPPS Domain

Luca Berardinelli, Alexandra Mazak, Oliver Alt, Manuel Wimmer, and Gerti Kappel

Abstract To engineer large, complex, and interdisciplinary systems, *modeling* is considered as the universal technique to understand and simplify reality through abstraction, and thus, *models* are in the center as the most important artifacts throughout interdisciplinary activities within model-driven engineering processes. *Model-Driven Systems Engineering (MDSE)* is a systems engineering paradigm that promotes the systematic adoption of models throughout the engineering process by identifying and integrating appropriate concepts, languages, techniques, and tools. This chapter discusses current advances as well as challenges towards the adoption of model-driven approaches in *cyber-physical production systems (CPPS)* engineering. In particular, we discuss how modeling standards, modeling languages, and model transformations are employed to support current systems engineering processes in the CPPS domain, and we show their integration and application based on a case study concerning a lab-sized production system. The major outcome of this case study is the realization of an automated engineering tool chain, including the languages SysML, AML, and PMIF, to perform early design and validation.

Keywords CPPS case study • Cyber-physical production systems • Model-driven systems engineering • Modeling standards • V-Model

11.1 Introduction

The increasing complexity of networked systems in the field of *Cyber-Physical Production Systems (CPPS)* (e.g., consider real-time control of wirelessly networked controller, sensors, and actuators from different vendors) demands a more compre-

L. Berardinelli (✉) • A. Mazak • M. Wimmer • G. Kappel
Business Informatics Group, Technische Universität Wien, Wien, Austria
e-mail: luca.berardinelli@tuwien.ac.at; mazak@big.tuwien.ac.at; wimmer@big.tuwien.ac.at; gerti@big.tuwien.ac.at

O. Alt
LieberLieber GmbH, Vienna, Austria
e-mail: oliver.alt@lieberlieber.com

© Springer International Publishing AG 2017

S. Biffl et al. (eds.), *Multi-Disciplinary Engineering for Cyber-Physical Production Systems*, DOI 10.1007/978-3-319-56345-9_11

hensive and systematized view of all aspects (e.g., physical, software, and network) in an engineering process. CPPS are being developed as part of a globally networked future world, in which *Products*, *Processes*, and *Resources (PPR)* interact with embedded hardware and software beyond the scope of single applications (Broy and Schmidt, 2014). CPPS engineering requires the integration of physical, software, network, and control aspects which are highly interwoven (Vangheluwe et al., 2016). Additionally, flexible control approaches are needed to adapt the systems' behavior to ever-changing requirements and tasks, unexpected conditions, as well as structural transformations (Lee, 2008).

The main requirements in the engineering of CPPS are (1) interoperability (i.e., the ability of CPPS and humans to connect and communicate), (2) virtualization (e.g., a virtual copy of the factory with sensed data), (3) decentralization (i.e., the ability of CPPS to make decisions on their own), (4) real-time capability (e.g., for supporting monitoring, analysis, planning, and execution (MAPE) cycles), (5) modularity (i.e., the flexible adaption of smart factories to changing requirements), and (6) cross-disciplinary methods to handle cross-cutting automation tasks (Kagermann et al., 2013). The realization of these aspects requires the synergistic integration of mechanical, electrical, network and software engineering, as well as the computer control of mechanical systems (Kyura and Oho, 1996). This again requires the integration of heterogeneous artifacts (e.g., design artifacts like piping and instrumentation diagrams, system control diagrams, etc.) together with their supporting tools, which are still often not well-integrated and typically not used in tandem during an engineering process (Jetley et al., 2013). However, this would be highly needed as, at the time of writing, the different engineering disciplines in isolation offer only partial solutions to meet the requirements of the envisioned engineering of CPPS and their combination is challenging as there exists a heterogeneous document/tool landscape in this domain (Vangheluwe et al., 2016).

It has to be also emphasized that appropriate methods of one engineering discipline are not necessarily applicable for another. For example, methods which enable software evolution like variability modeling or tracing are limited to the software domain (mostly dealing with requirements, software models, and program code). For the domain of mechanical engineering, e.g., in the field of automated production systems where naturally requirements will change over the system life-time (e.g., due to a changing product portfolio), methods of tracing need to be adapted and linked to well-established domain-specific methods (e.g., design structure matrix) (Vogel-Heuser et al., 2015). The need for explicit *modelling* is then rapidly arising (cf. Research Questions (RQs) of Chap. 1) requiring an appropriate set of tools and methodologies that meet various needs in the industrial automation domain (Jetley et al., 2013) throughout a multidisciplinary information flow.

Therefore, it is necessary to identify and implement a suitable subset of appropriate *models and standards* for guaranteeing the engineering quality in the field of CPPS (Broy and Schmidt, 2014). Models represent a system at different abstraction layers (e.g., requirements elicitation, analysis, design, implementation, validation, and verification), of different disciplines (e.g., process engineering, electrical engineering, mechanical engineering, software engineering), considering different aspects (security, performance, safety) and tasks (e.g., validation, verification, testing, optimization, design-space exploration) (Kagermann et al., 2013) (cf. Chap. 1, RQ M1).

Moreover, models can be used throughout the entire value chain, e.g., from product development through manufacturing engineering to production. Especially the production of the future requires models, e.g., for the virtual design, virtual planning, conceptualization, and simulation (Kagermann et al., 2013). The complex software required in CPPS is typically developed and refined iteratively in a model-driven way (Vogel-Heuser and Biffl, 2016). *Model-Driven engineering (MDE)* follows the principle that "everything is a model", which is also reflected in the systems engineering domain. However, MDE promotes the models to actually "drive" the engineering process by using generative and analytical techniques to automate the different tasks instead of using models solely as documentation or early sketches of the system (cf. Chap. 1, RQ I1).

Additionally to the general usage of "models", a set of appropriate standards (e.g., SysML (Friedenthal et al., 2014), MARTE (Object Management Group (OMG), 2016c), AML (IEC, 2014), etc.) is needed for integrating various engineering aspects, different stakeholder perspectives, tool-independent interoperability, as well as information that is needed to be exchanged at a specific engineering step. Moreover, appropriate modeling languages (providing a clear syntax as well as semantics) are essential in planning, designing, and realizing CPPS engineering. Currently, the interest in adopting system modeling languages is increasing in the industrial automation domain (Berardinelli et al., 2016). In this context, the main challenges (among others) are the *technical*, *syntactic*, and *semantic heterogeneity* as well as the *vertical integration* (i.e., among models required by domain/stakeholder-specific activities) and the *horizontal integration* (i.e., among models used in the same activity but performed by different stakeholders with their own, mainstream modeling languages/notations) (Mazak et al., 2016). One of the obstacles is a systematic adaption of models throughout the engineering process by identifying proper concepts, notations, techniques, tools, as well as their integration. Equally important are software tools to manage the complexity resulting from increasing functionality, customization, dynamics, and cooperation between different disciplines.

In this chapter, we present a set of appropriate MDSE standards, as used by authors in their research activities, for enabling the adaption of MDE principles in the CPPS domain. In particular, we focus on the role played by (1) the *System Modeling Language (SysML)* (Object Management Group (OMG), 2016b) as design notation for CPPS structural and behavioral modeling, (2) *AutomationML (AML)* as exchange standard for production system engineering tools, (3) *Performance Model Interchange Format (PMIF)* for performance modeling and analysis (e.g., via queuing networks to calculate performance indices like resource utilisation) (Cortellessa et al., 2011), and (4) model transformations as a strategic mechanism to integrate heterogeneous artifacts throughout the engineering of CPPS to deal with heterogeneities and to realize horizontal and vertical integration to ensure a holistic view on

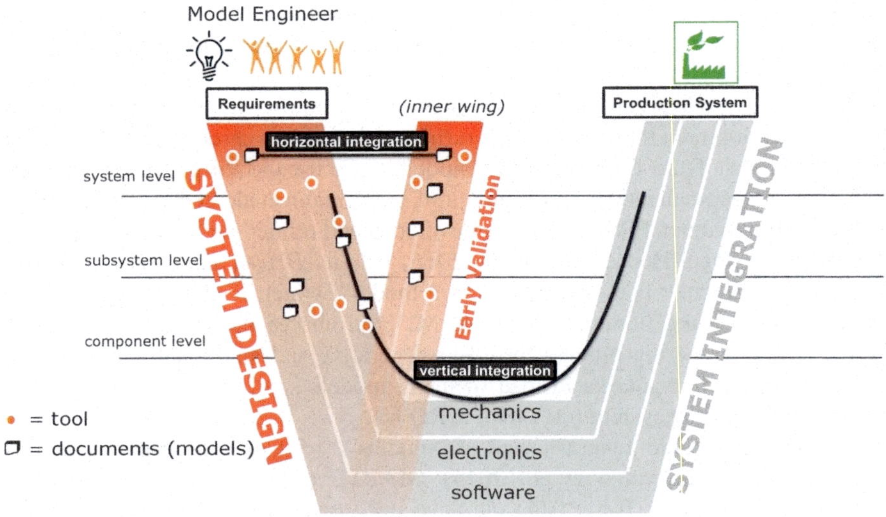

Fig. 11.1 A V-Model variant for CPPS engineering

the resulting technical system and to assemble different views[1] (cf. Chap. 1, RQ C2).

On top of these modeling and data exchange standards, model-based design and analysis methodologies can be devised in order to demonstrate the value of MDSE both in design and validation phases. For this purpose we use the V-Model,[2] presented in Fig. 11.1 as methodological foundation in order to express our framework in a technology neutral way.

Generally, the V-Model is a graphical representation of a system's development life-cycle specifying the sequence of steps during a generic engineering process. These steps include system design, domain specific engineering, system integration, as well as the verification and validation of system properties (Verein Deutscher Ingenieure (VDI), 2004). The V-Model has proven as structural approach for the development of interdisciplinary technical systems like CPPS (cf. Chap. 2 for more details). In particular, we propose in this book chapter an example of a model-based *Systems Engineering Technical Process*. We relate the MDSE standards to the inner wings of the V-Model to cover that phases of the V-Model needed to realize and analyze the engineering of CPPS and to support cross-disciplinary modeling built on MDE techniques. We utilize the synergies between MDSE and the V-Model to guide stakeholders to select and combine appropriate standards, languages, profiles, and formats to build their own MDE methodology (e.g., performance analysis

[1]Chapter 2 also discusses vertical and horizontal integration terms in the context of model-based engineering.

[2]See Chap. 2 for an introduction and comparison among engineering process models.

methodology) as needed for CPPS engineering on top of these standards (e.g., AML and PMIF).

The intention of this chapter is to provide an overview on how different standards for MDSE may be combined on the macro-perspective to cover different engineering phases and not to provide a complete treatment of each engineering phase in detail. For the latter, we provide pointers to existing literature.

The rest of this chapter is organized as follows. Section 11.2 provides a general introduction into model-driven engineering (MDE) and presents two major techniques of it: metamodeling and model transformations. Based on this foundation, we elaborate in Sect. 11.3 on a set of appropriate industry standards to apply MDSE techniques for CPPS engineering. In Sect. 11.4, we present examples based on a reference case study to show how the standards and techniques presented in the Sect. 11.3 are applied to realize CPPS and provide a critical discussion of the results. Finally, Sect. 11.5 concludes the chapter and outlines future challenges in MDSE.

11.2 Model-Driven Engineering in a Nutshell

Before we discuss in more detail how MDSE can be actually realized in the CPPS domain, we provide a general introduction into model-driven engineering (MDE). MDSE may be seen as a special interpretation and application of the general paradigm of MDE within the systems engineering domain.

In MDE, the abstraction power of models is applied to tackle the complexity of systems (Brambilla et al., 2012; Schmidt, 2006). MDE follows the principle "everything is a model" for driving the adoption and ensuring the coherence of model-driven techniques, in the same way as the principle "everything is an object" was helpful in driving the object-oriented techniques in the direction of simplicity, generality, and integration (Bezivin, 2005). Historically, MDE has been mainly applied in software engineering (Brambilla et al., 2012; Bezivin, 2005; Schmidt, 2006), but in recent years, the application of MDE has been increasing in the CPPS domain as well (Vyatkin, 2013; Hegny et al., 2010; Schütz et al., 2014).

A key principle of MDE is to address engineering with formal models, i.e., machine-readable and processable representations. Based on this foundation, modeling provides a set of advantages for driving the engineering process. The application of model validation, testing, verification, simulation, transformation, and execution enables the automation of engineering process steps and support the traceability of engineering artifacts to improve quality management to mention just a few benefits (Brambilla et al., 2012).

Furthermore, in MDE, models are considered to be connected (i.e., model elements may be linked beyond the boundary of one model) and dynamic (i.e., models may be analyzed and executed in some form) (Brambilla et al., 2012; Bezivin, 2005). Models can be (1) compared to reason about differences between model versions, (2) merged to unify different versions of a model, (3) aligned to create a global and integrated representation of the system from different viewpoints to reason about consistency, (4) optimized to improve their internal structure without changing their observable behavior, (5) refined to produce platform-specific models from platform-independent models, and (6) translated to other formalisms for code generation, verification, and simulation. Dedicated tool support for these tasks is available out-of-the-box in modeling environments, which can be customized for the modeling languages in use.

The two major MDE techniques are: (1) *metamodeling* for specifying modeling languages, i.e., the structure and content of valid models, and (2) *model transformations* to systematically manipulate models. In the following, we discuss these two techniques to provide the basis for subsequently showing how these techniques are used in MDSE.

11.2.1 Metamodeling

Metamodels play an important role in MDE (Kühne, 2006). They specify the abstract syntax of modeling languages (i.e., the language concepts and their relationships) that is the center of the modeling language definition and all other concerns such as concrete syntax (i.e., model notation) and semantics are defined based on these metamodels (Brambilla et al., 2012). MDE provides standardized metamodeling languages (also referred to as meta-metamodels) used for producing modeling environments such as the Meta Object Facility (MOF) (Object Management Group (OMG), 2003) for defining modeling languages that may be seen as pendant to Extended Backus-Naur Form (EBNF) (Wirth, 1977), which is the foundation for specifying textual languages.

MOF is based on a core subset of UML class diagrams, i.e., classes, attributes, and references. A metamodel gives the intentional description of all possible models within a given language. Practically, metamodels are instantiated to produce models which are in essence object graphs, i.e., they consist of objects (instances of classes) representing the modeling elements, object slots for storing values (instances of attributes), and links between the objects (instances of references), which have to conform to the UML class diagram describing the metamodel. Therefore, models are often represented in terms of UML object diagrams if their concrete syntax is neglected. This is especially true when models are automatically processed by the computer.

A model has to conform to its metamodel which is often indicated by the *conformsTo* relationship (cf. Fig. 11.2). In addition to the constraints defined by the metamodel, additional constraints may be defined based on the metamodel elements using a *constraint language*. The Object Constraint Language (OCL) (Object Management Group (OMG), 2010) is a standardized and formal language to describe expressions, constraints, and queries on models. As such, OCL is the language of choice for defining constraints going beyond simple multiplicity and type constraints defined by UML class diagrams and metamodels.

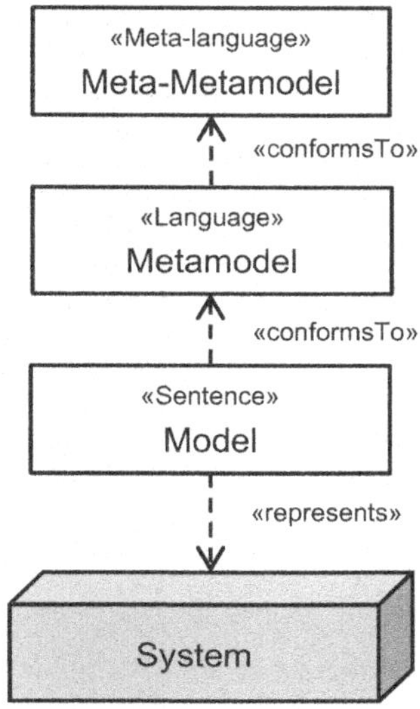

Fig. 11.2 Metamodeling pattern based on Kühne (2006)

Based on this meta-layer architecture, metamodeling environments allow generating modeling environments and providing generic tool support, which can be employed for all the modeling languages defined within a metamodeling environment. Thus, metamodeling environments empower the knowledgeable tool users to become tool developers, e.g., modeling languages may be easily extended with new modeling concepts or completely new languages may be developed without programming efforts.

11.2.2 Model Transformations

In a general sense, a model transformation is a program executed by a transformation engine which takes one or more models as input to produce one or more models as output as is illustrated by the model transformation pattern[3] (Czarnecki and Helsen, 2006) in Fig. 11.3. In MDE, model transformations are used to solve different tasks (Czarnecki and Helsen, 2006; Mens and Gorp, 2006; Lúcio et al., 2016) such as code generation, model refactoring, reverse engineering to name just a few. One important aspect is that model transformations are developed on the metamodel level, and thus, are reusable for all valid model instances.

In the MDE field, various model transformation kinds emerged in the last decade (Czarnecki and Helsen, 2006; Mens and Gorp, 2006; Lúcio et al., 2016) whereas two important kinds are differentiated in the following. A model transformation can be categorized as *out-place* if it creates new models from scratch, e.g.,

[3]Please note that the terms source/target models and input/output models are used synonymously.

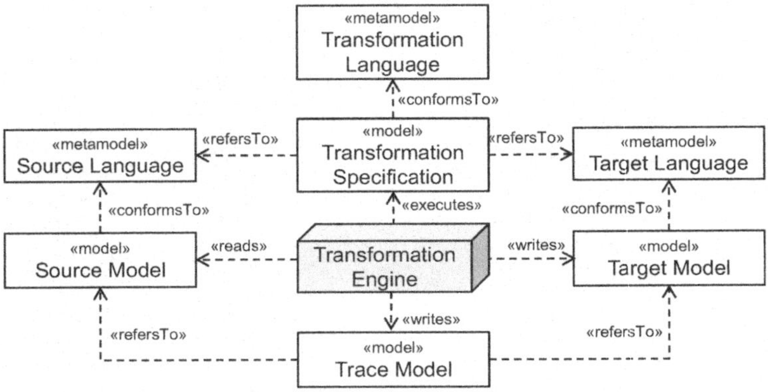

Fig. 11.3 Model transformation pattern based on Czarnecki and Helsen (2006)

producing an analysis model from a design model, or as *in-place* if it rewrites the input models until the output models are obtained, e.g., as it is the case in model execution or model optimization.

Several transformation languages emerged in the last decade which provide dedicated support for defining model transformations (Czarnecki and Helsen, 2006; Mens and Gorp, 2006). In this book chapter we are focussing on out-place transformations as we are interested in automating the transition from the early design steps to early validation steps as well as to the subsequent discipline-specific engineering tasks. Thus, our goal is to apply out-place transformations to produce from a design model other representations which can be used in discipline-specific tools as well as analysis tools.

For defining out-place transformations in this book chapter, we make use of the Query/View/Transformation (QVT) standard (Object Management Group (OMG), 2016a) which provides several languages to implement model transformations (Kurtev, 2007). For instance, by using the declarative QVT Relations language, transformation logic between two different metamodels is specified as a set of relations that must hold for the transformation to be considered successful. Relations contain a set of patterns used to match for existing source model elements in order to instantiate new target model elements or to modify existing ones. Since declarative approaches like QVT Relations allow for the specification of what has to be computed but not necessarily how, the transformations are defined in a very concise manner which allows to focus on the relations between different concepts instead of reasoning about how to encode them in imperative statements. Another benefit of using a declarative language such as QVT Relations is to allow for different application possibilities of the transformation specification. While the transformation can be executed in both directions, it is also possible to use the transformation to compare different models (if all relations are fulfilled by the source and target models) and in cases where differences exist, i.e., some relations are not completely fulfilled by the existing model elements, synchronization may be performed to restore the fulfillment of all specified relations.

In this book chapter we focus on the classical forward transformation capabilities of QVT Relations and refer the interested reader for other execution capabilities to Stevens (2010). However, we also see the comparison and synchronization capabilities of QVT Relations as interesting ingredients to further automate the system engineering process, e.g., consider change propagation and reverse engineering activities to mention just a few practical engineering tasks. Finally, by instantiating the relations on the model level, a trace model is produced for indicating which source elements have been transformed to which target elements by which relation (cf. bottom of Fig. 11.3).

After presenting the central techniques of MDE, we now proceed with their application in the domain of CPPS engineering. In particular, we elaborate on a selected set of standardized languages which are supported by metamodels in Sect. 11.3 and combine them in an automated engineering tool chain in Sect. 11.4 with the help of model transformations.

11.3 Selected MDSE Standards for CPPS Engineering

In this section, we elaborate on a selected set of industry standards to introduce MDSE in the CPPS domain. This selection is far from being exhaustive, rather it aims at providing a coherent subset of modeling standards, proposed by the experience of the authors. These standards can be suitably integrated in model-driven methodologies for CPPS design activities as well as for the early verification and validation activities.

In particular, we selected (1) SysML, a general purpose graphical modeling language explicitly devised for requirements specification and system design, (2) AML for realizing data exchange among production system engineering tools, and (3) PMIF for supporting performance modeling via the queuing network (QN) notation and as standardized data exchange format among QN solvers. By using these proposed standards, we can take full advantage of those MDE techniques presented in the previous section to automate a particular *engineering tool chain framework* which focusses on design, data exchange, and analysis. This framework is not limited to the presented set of standards as also other standards may be used within this framework.

In the following, we describe each of the selected standards at a glance and give pointers to external resources for the interested reader.

11.3.1 Systems Modeling Language (SysML)

SysML is a graphical modeling language standardized by the Object Management Group (OMG) for the development of large-scale, complex, and multi-disciplinary systems (Object Management Group (OMG), 2016b). SysML is derived and extended from the Unified Modeling Language (UML), a graphical, general-purpose software modeling language that is currently the most adopted one in model-based software engineering (Hutchinson et al., 2011). SysML reuses a subset of UML elements and introduces new elements, e.g., for requirements modeling. SysML is defined as a *UML profile*. In addition to SysML, several other UML profiles have been standardized by the OMG such as the MARTE profile (Object Management Group (OMG), 2016c) which we will discuss later.

To better understand the relationship between UML and SysML, we will explain it in more detail. With UML profiles, UML provides a language-inherent extension mechanism for customizing UML concepts for particular domains and platforms. This mechanism allows engineers to extend UML to create new modeling concepts (derived from existing ones) comprising specific properties that are suitable for the domain of interest. A *stereotype*, denoted by the keyword `«stereotype»`, is one of three types of extensibility mechanisms, the other two are *tagged values* and *constraints* (Booch et al., 2005). By using stereotypes any UML metaclass can be extended with additional meta-attributes (i.e., tagged values) and additional modeling constraints. The profiling mechanism has been extensively adopted by the UML community to broaden the adaption of UML as design modeling notation in several domains. For more details, we refer the interested reader to Booch et al. (2005) and Seidl et al. (2012).

SysML provides modeling concepts and diagrams for representing *requirements*, *structure*, *behavior*, and *parametrics* (i.e., mathematical constraints) of a system which are linkable to trace requirements and to connect structure with behaviour (cf. Fig. 11.4). We now present only a small subset of SysML which we use later on in Sect. 11.4.2.1.

To represent the *structure of systems*, the UML class diagram and composite structure diagram are adapted and renamed into *block definition diagram (BDD)* and *internal block diagram (IBD)* by SysML. By using BDD, the structural

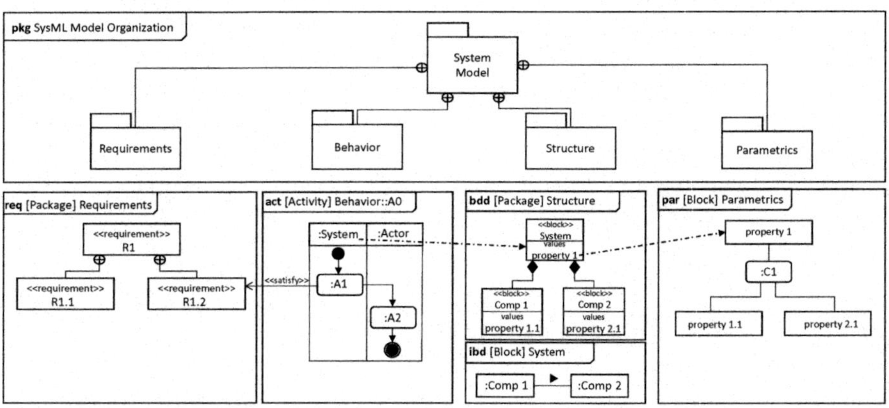

Fig. 11.4 Overview on SysML diagram types and their connections (Friedenthal et al., 2014)

decomposition of a system into so-called *blocks* can be defined. A block represents a modular component of a system. Its definition comprises the component's properties and relationships to other components. Important relationships between blocks include:

- *Composition relationships* representing the decomposition of a block into sub-blocks (called *parts*);
- *Reference associations* representing logical references between blocks which are parts of different composite blocks;
- *Dependency relationships* denoting that a change on one block may cause a change on other blocks;
- *Generalization* relationships representing classifications of blocks.

By using IBD, the connections between parts of a compound block can be defined. In particular, they include:

- *Ports* to define connection points between blocks;
- *Connectors* to connect blocks via ports and enable interaction.

SysML provides an integration framework for discipline-specific design models, e.g., mechanical, electrical, and software models. The system model captures the overall design of a system on a high-level of abstraction and traces this design to discipline-specific models. For this reason it fits to the multi-disciplinary engineering process required for realizing CPPS (cf. Fig. 11.1). SysML has been already adopted in several domains[4] for developing complex and multi-disciplinary systems, e.g., in the aerospace and defense industry. SysML is now also emerging in the automation domain (Feldmann et al., 2014). As SysML is a profile extension of UML, model-driven techniques and tools are directly applicable to this standard. For more information on SysML, we kindly refer the interested reader to Alt (2012).

11.3.2 Modeling and Analysis of Real-Time Embedded System Profile (MARTE)

MARTE (Object Management Group (OMG), 2016c) is another standard introduced by the OMG. MARTE's target modeling domain includes reactive systems interacting with the external environment through sensors and actuators, e.g., consider transportation, factory automation, hardware/software controllers, and various embedded electronic appliances also including mobile communications. As MARTE is designed as a UML profile, it is applicable to UML, and by this, also to all other UML profiles such as SysML.

MARTE includes many sub-profiles structured around two main concerns (1) modeling software and hardware structures, e.g., by the *software resource model*

[4]See http://www.omgsysml.org for an overview.

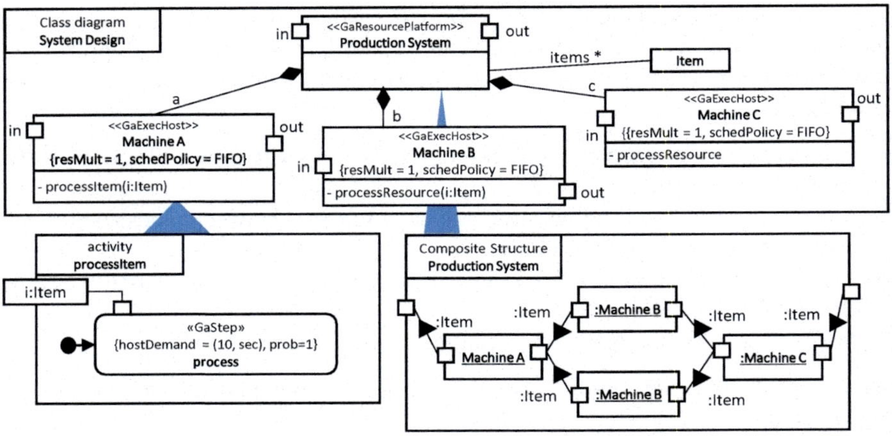

Fig. 11.5 A production system modeled in UML with MARTE annotations

profile and *hardware resource model profile*, and (2) enriching design models to obtain analysis models with additional parameters required by model-based methodologies for performance and schedulability analysis (e.g., response time of software/hardware execution hosts or scheduling policies of tasks).

Figure 11.5 shows a simple example of a production system modeled in terms of a UML class diagram with annotations from a subset of the MARTE profile. By using MARTE, the role of different classes can be identified. A production system can be considered as a collection of active resources, e.g., machines. As shown in Fig. 11.5, the production system is composed by three different kinds of machines, `Machine A`, `Machine B`, and `Machine C`. The production system's layout is presented by a *composite structure diagram* (cf. Fig. 11.5, on the bottom right). It includes 4 machines in total (one of type `A` and one of type `C`, both, connected by two machines of type `B`). In our example, these machines provide processing services for items entering the system to be processed by `Machine A`, flowing through `Machine B`, and then leaving the system after the last processing step carried out on `Machine C`. This flow is indicated by the black triangles showed in the composite structure diagram of the production system. Thereby, each item is processed following a *first-in-first-out (FIFO)* scheduling policy. Additionally, quantitative information can be annotated on structural (e.g., the resources' multiplicities) as well as on behavioral specifications. For example, *timed properties* can be assigned to actions as shown in the `processItem` activity diagram. In this diagram the modeling behavior of `processItem()` operation of `Machine A` is described. Each execution of the `processItem()` operation lasts (exactly) 10 s, seizing its execution host, the `Machine A` instance shown on the composite structure diagram.

MARTE provides concepts required for model-driven design and analysis of systems, but it is independent of any design and analysis methodology. Moreover, the general framework for quantitative analysis provided by MARTE is intended to be refined/specialized for the specific methodology of choice (e.g., dependability analysis; Bernardi et al. 2011). As for SysML, MARTE is a profile extension of UML, thus model-driven techniques and tools are directly applicable to this standard. For more information on MARTE, we kindly refer the interested reader to Object Management Group (OMG) (2016c) and Selic and Gérard (2013).

11.3.3 Performance Modeling Interchange Format (PMIF)

PMIF is conceived as a common representation for system performance model data that could be used to move queuing network models between analysis and simulation tools (Smith and Williams, 1999). Its creators were interested in tool interoperability for performance engineering (Smith, 1990). The first version of PMIF (Smith and Williams, 1999) addresses a specific type of performance model, i.e., queuing network models that may be solved using analytical solution algorithms.

Providing enhanced support for extra-functional modeling and analyses of CPPS is in particular of high importance in the early design and validation steps (Malavolta et al., 2013). In this regard, *queuing network (QN)* models provide a powerful notation widely used to represent and analyze resource sharing systems like CPPS (Schleipen and Drath, 2009). As summarized in Cortellessa et al. (2011):

> Informally, a QN is a collection of interacting *service centers* representing system resources and a set of *customers* representing the users that share the resources. It can be represented as a direct graph whose nodes are service centers and edges represent the potential paths of customers' service requests. Several different classes of customers can circulate over the network at the same time, each class representing a set of customers with homogeneous behavior (i.e., paths and amounts of service requests).

The construction of a QN can be split in two main steps, (1) definition of service centers, their number, and the interconnections, and (2) parameterization of the arrival processes, i.e., the definition of job classes, the service rates, scheduling policies and the routing probability among servers. Figure 11.6 gives a graphical overview of a generic QN model, its typical modeling elements, and their relationships as already defined for the production system modeled with UML/MARTE in Sect. 11.3.2. Each machine instance is represented as a server. The produced items

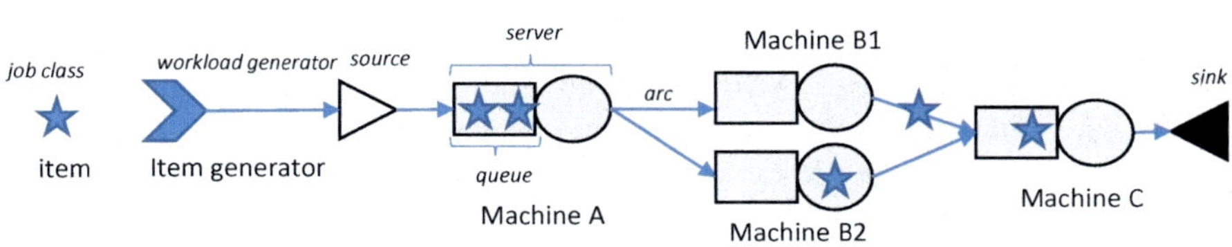

Fig. 11.6 A production system modeled as an open queuing network

are instances of the same job class. Jobs enter the system at a certain arrival rate and flow to servers, waiting for being served, seizing the server for a certain amount of time (a.k.a., service time) then proceed to the next server connected with arcs until they leave the system through sink nodes.

PMIF provides a common serialization format for this kind of models. In this paper, we adopt a variant of the PMIF metamodel presented in Troya and Vallecillo (2014), suitably updated for being easily integrated in model-driven performance analysis methodologies. Following this variant, a QN model is a graph with **Node**s connected by **Arc**s. There are two types of nodes, **Server**s and **NonServer**s. The former provide a processing service while the latter represent origin (*SourceNode*) and exit (*SinkNode*) for entities flowing through the QN. These entities are referred as *customers* or *jobs*. Different job classes are defined through workloads (**Workload**). A **Workload** can be open or closed depending on the capability of jobs to enter/leave the **QNM** (i.e., open) or not (i.e., closed with a fixed *job number*). Any **Workload** specification includes the sequence of *transits to* different **Server** where jobs can ask for one or more *service request*s (**ServiceReq**). At each **Server**, the next job to be processed is decided by a scheduling policy (e.g., first-come-first-served).

Finally, *timed properties* like job inter-arrival time (for open workload), and service times (for servers) are usually given as part of a performance analysis scenario (e.g., arrival probability distribution). Other timed properties (e.g., the waiting and completion time for single jobs or for the whole workload) are obtained as part of analysis results and used by tools to compute performance indices.

There are currently several tools for solving QN models providing their own model representations. In Troya and Vallecillo (2014), the reader can find a recent list of QN solvers together with their evaluation techniques (e.g., analysis or discrete event simulation), model format, allowed probability distributions (e.g., for generating inter-arrival times and service times) and supported QN types (e.g., open/closed). As we build on the existing metamodel for PMIF (Troya and Vallecillo, 2014), model-driven techniques and tools are directly applicable to this standard as well. For more information on QNs and PMIF, we kindly refer the interested reader to Smith and Williams (1999), Smith et al. (2010), and Troya and Vallecillo (2014).

11.3.4 AutomationML

AML is a neutral, free, open, XML-based, and standardized data exchange format, which aims for data exchange within the engineering process of production systems (IEC, 2014). We present an overview on AML in Fig. 11.7. In particular, typical elements in an AML production system model comprise: (1) the *plant structure* including devices and communication structures, expressed as a hierarchy of AML objects and described by means of CAEX which follows the standard IEC 62424 (Schleipen et al., 2008), (2) *geometry and kinematics* represented

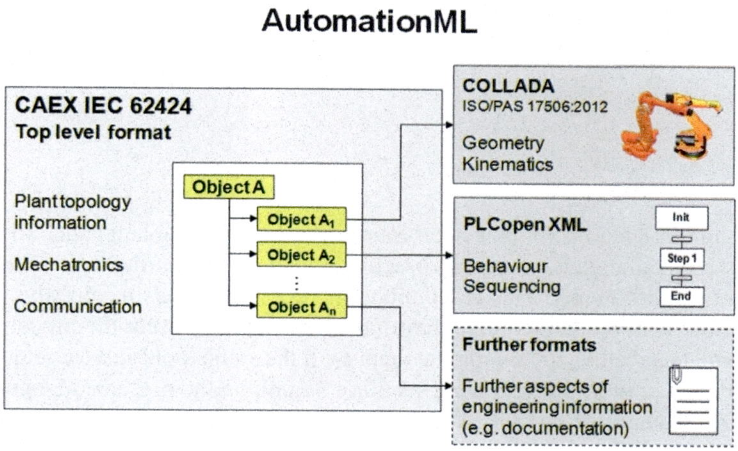

Fig. 11.7 AML overview taken from IEC (2014)

by COLLADA 1.4.1 and 1.5.0 (ISO/PAS 17506:2012) (ISO/PAS, 2012), and (3) *control related logic data* (i.e., PLCopen XML 2.0 and 2.0.1; PLCopen 2011). Since the foundation of AML is the application of CAEX as top level format, we focus on this part of AML in this section and the following one.

CAEX stores engineering information following a prototype-oriented paradigm. It allows the modeling of physical and logical system components as data objects encapsulating different aspects. CAEX objects (namely, internal elements **IE**) and their interfaces (namely, external interfaces **ExtI**) can be specified from scratch or suitably instantiated (that means *cloned*) from existing prototypical classes (namely, **SUC**), defined and collected in system unit class libraries (**SUClib**). Both CAEX objects and classes may consist of other sub-elements, and may themselves be part of a larger composition or aggregation.

Both classes, objects, and their interfaces are semantics agnostic. CAEX provides role classes (**RC**) and interface classes (**IC**) to assign semantics to **IE**s and **ExtI**, respectively. Both **RC** and **IC** are defined and collected in libraries (**RClib** for **RC** and **IClib** for **IC**). AML objects and classes can then support and/or require such roles.

Finally, individual objects are modeled in an instance hierarchy (**IH**) which, in turn, may contain internal elements (**IE**). **IE**s can be instantiated from **SUC**s and arranged in accordance with the supported and required roles. External interfaces (**ExtI**), instantiated from **IC**s, are used to interlink objects (**IL**) among each other or with externally stored information (e.g., COLLADA or PLCopen XML files).

In previous work (Biffl et al., 2015), we have defined a metamodel for AML considering the CAEX part. Using this metamodel, we can represent AML data as models and apply model-driven techniques and tools for AML as well. This allows, e.g., to transform system models defined in SysML to AML and vice versa.

For more details about AML, we kindly refer the interested reader to the different AML whitepapers available at IEC (2014). In addition, Chap. 9 of this book focuses on the role of AML as a potential standardized data format and it is exemplarily presented for the case of virtual commissioning of a production system.

11.3.5 Synopsis

To sum up this section, we presented four different modeling languages which are all based on metamodels which are directly defined by standardization bodies or by the scientific community. This is an important requirement as this already resolves the potential data model heterogeneities and allows to represent the models in the same format as well as to manipulate them with the same tools and techniques. We exploit this feature in the following section to combine the presented languages in an automated engineering tool chain framework.

11.4 MDSE of CPPS in Action

In this section, we show how the presented languages of Sect. 11.3 can be combined to support the envisioned CPPS engineering process. Figure 11.8 depicts how these languages are aligned with the engineering activities of the V-Model:

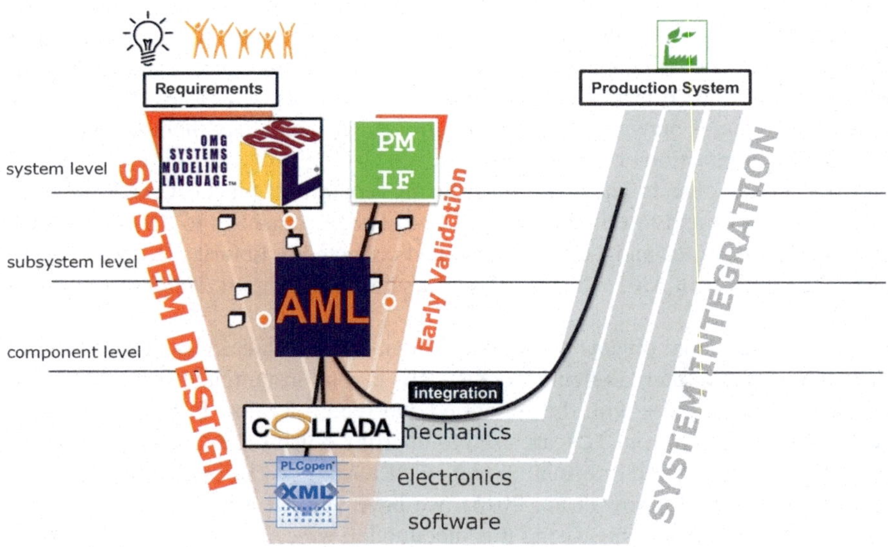

Fig. 11.8 Populating the V-Model with concrete languages for requirements and design, data exchange, and analysis

- **Requirements and design language**: We use SysML as a front-end requirements and design notation. In particular, we model the requirements of the system as well as its design covering its structure and behavior. Furthermore, we show how MARTE stereotypes can be used to provide the necessary information to perform early performance analysis.
- **Data exchange language**: AML is primarily devised as a data exchange among automation tools in our engineering tool chain. It eases vertical integration with other domain-specific activities (e.g., virtual commissioning with COLLADA-based tools and PLC programming; Lüder et al. 2015). In particular, we show how the information defined in the SysML model can be exchanged on the basis of CAEX.
- **Analysis language**: PMIF is employed to share models with QN solvers to compute properties of interest for early design validation. In order to analyse the performance of a design given in SysML and MARTE, we compute the necessary properties to validate if the stated requirements are fulfilled or not with the help of an existing QN solver.

As we will see later, the transitions between the different engineering activities are automated by model transformations. This allows to exchange the design models between engineering tools used in the discipline-specific engineering activities, as well as to perform early validation and verification of the design models before the discipline-specific engineering activities start. Furthermore, we would like to highlight that based on the trace model generation by model transformation engines, traceability between the design models, data exchange models, as well as analysis models is provided automatically.

The rest of this section is organized as follows. We first provide the description of a reference case study, a lab-sized production system hosted at IAF of the Otto-v.-Guericke University Magdeburg (*Equipment Center for Distributed Systems*, 2016) which is subsequently used as a running example to exemplify the integration of SysML, PMIF, and AML as well as to discuss its benefits and challenges. First, we provide three different models of this given production system using SysML, AML, and PMIF as modeling languages for the sake of CPPS design (via SysML), data exchange, (via AML) and early model-based validation activities (e.g., performance analysis through queuing network represented by PMIF). Second, we describe how the integration between (1) SysML and AML and (2) AML and PMIF has been realized to automate the validation of requirements on SysML design models.

11.4.1 Case Study

The IAF production system (cf. Fig. 11.9) consists of a transportation line made of sets of turntables, conveyors, and multi-purpose machines. Each turntable is equipped with an inductive proximity sensor for material detection and a motor for

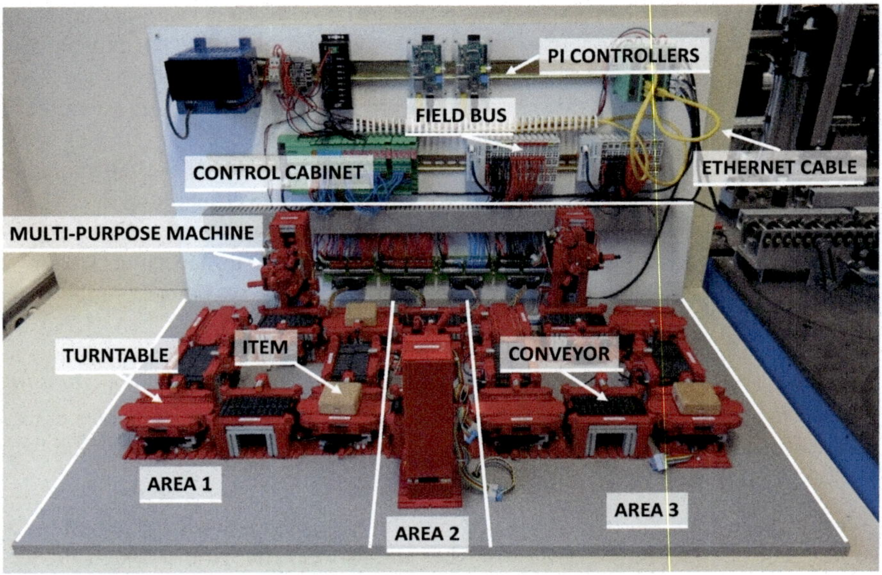

Fig. 11.9 The lab-sized production system hosted at IAF of the Otto-v.-Guericke University Magdeburg (*Equipment Center for Distributed Systems*, 2016)

table rotation. The transportation line is wired to a modular fieldbus I/O system, which, in turn, communicates with Raspberry Pi based controllers by Ethernet. The Raspberry Pi based controller is running a Programmable Logic Controller (PLC) program (PLCopen, 2011) governing the transportation line. Such programs logically divide the transportation line in three areas as depicted in Fig. 11.9. The production plant is supposed to continuously processes items via its multi-purpose machines located in different areas. Turntables and conveyors are in charge of moving such items to these machines following a predefined process.[5]

11.4.2 CPPS Modeling

Systems design is the process of defining (1) the architecture, in terms of components, modules, interfaces, and (2) the functionalities that such a system should provide in order to satisfy the specified requirements. In the following, we design a CPPS virtual prototype of the reference case study using SysML and AML. Then we build a QN performance model based on PMIF to validate the virtual prototype.

[5]Models realized for this case study can be downloaded from our companion web site at the following address http://www.sysml4industry.org/wp-content/uploads/2016/08/models-1.zip

We represent the overall structure and behavior of the transportation line that is further decomposed in three groups of resources, called areas. We then provide a detailed design of `Area1`, the leftmost logical area in Fig. 11.9, together with its internal resources (i.e., four `turntables`, four `conveyors`, and one multipurpose `machine`) as well as resource-specific behaviors.

Of course, the proposed models can be built from scratch in the different stages of the engineering process and they can be used isolated for documentation purposes (e.g., by SysML diagrams) or for analysis purposes (e.g., calculating the minimum, maximum or average processing time for items). However, in order to realize an efficient engineering process realizing the potential benefits of formal models, repetitive manual tasks should be eliminated as much as possible. Instead model transformations should be employed for automating these tasks as much as possible.

11.4.2.1 Modeling in SysML

We employ SysML (Object Management Group (OMG), 2016b) to support requirements specification and system design of the CPPS engineering process as sketched in Fig. 11.8. Figures 11.10 and 11.11 show parts of the IAF plant SysML model which we discuss in the following.

SysML provides modeling concepts and a diagram type to represent text-based requirements and their relationships to other SysML modeling elements, as discussed in Sect. 11.3. The requirements diagram in Fig. 11.10 depicts functional (`FR`) and non-functional (`NFR`) requirements for the `IAF Plant`.

The `FR` prescribes a constraint on the logical architecture of the transportation line, which has to be divided in three distinct areas of turntables, conveyors, and machines. This requirement will probably affect the logical representation of the transportation line as programmed in the PLC code deployed on Raspberry PI controllers (cf. Fig. 11.9).

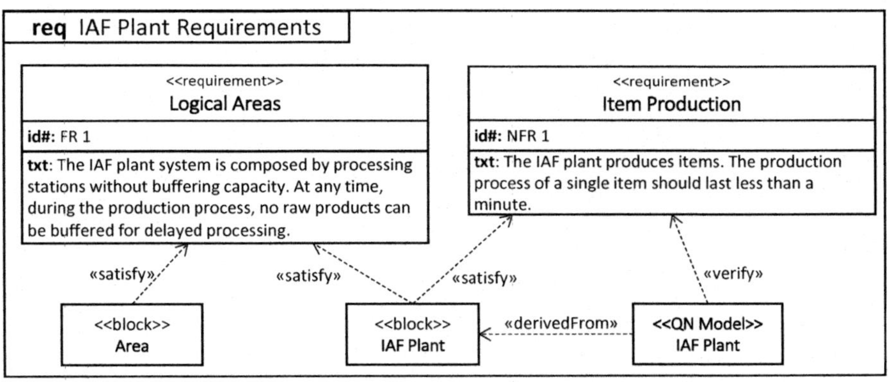

Fig. 11.10 Excerpt of the SysML requirement model of the IAF Plant

The NFR states a production constraint setting a minimum production threshold. This is a typical performance requirement that can be validated through analysis of a queuing network representation of the IAF Plant. For this reason, we extend the SysML design of the IAF Plant with MARTE annotations and derive a PMIF-based queuing network for the sake of requirements validation.

Both requirements describe system-level properties, which have to be satisfied by the whole IAF plant. For this reason, a satisfy relationship is depicted from the top-level structural SysML block representing the whole IAF plant to these requirements.

Concerning the design of the IAF Plant, the structural elements and their relationships are defined via blocks and containment relationships in the BDD in Fig. 11.11a. A top level IAF Plant block includes an Area block, which in turn has Turntable, Conveyor, and Machine blocks. The ports depicted on the blocks' borders represent the potential input (e.g., inA, for receiving Areas), output (e.g., outT from Conveyors to Turntables), and bidirectional (e.g., in_outC, for Machines interacting with Conveyors) interaction points among these structural elements. Finally, blocks can be used to define the product types, e.g., the generic item shown as wooden piece in Fig. 11.9.

The IBD depicted in Fig. 11.11c defines the structure of the transportation line with three connected areas. We then detail the internal structure of the Area 1 as a graph of parts, ports, and connectors. Parts represent tX and cX properties in the BDD with the following naming convention: *'part name': 'type name'*, where *'type name'* refers to the corresponding block defined in the IAF Plant Design BDD (e.g., t1:Turntable). Connectors have both structural and behavioral functions, specifying both (1) links between parts via ports, and (2) item flows between parts depicted with black directed arrows.

Once the system architecture has been defined, we proceed with the behavioral specification[6] of processing resources, i.e., Turntable, Conveyor, and Machine blocks by defining the resource-specific operations turn(), transfer(), and process(), respectively, applied on items. The detailed behaviors are modeled through Activity Diagrams in Fig. 11.11b. Each activity is realized with a single action receiving the parameter Item *i* (i.e., the box at the border of the activity diagram).

In Fig. 11.11d such operation-specific activities are combined in a system-level activity where a workflow of call operation actions correspond to the processing steps applied on items flowing through Turntables, Conveyors, and Machines placed in the Area1 of the IAF Plant. We assume that the processing starts at turntable T0 and proceed up to turntable T2. Here the item can leave Area1 to be processed by resources placed in Area2, or continue to conveyor C2. A similar alternative choice happens in turntable T3, where the item can be

[6]It is worth noting that SysML provides different behavioral notations and diagrams such as State Machines and Interactions (a.k.a., Sequence Diagrams). It is up to the modeler to choose the notation that better fits with her needs.

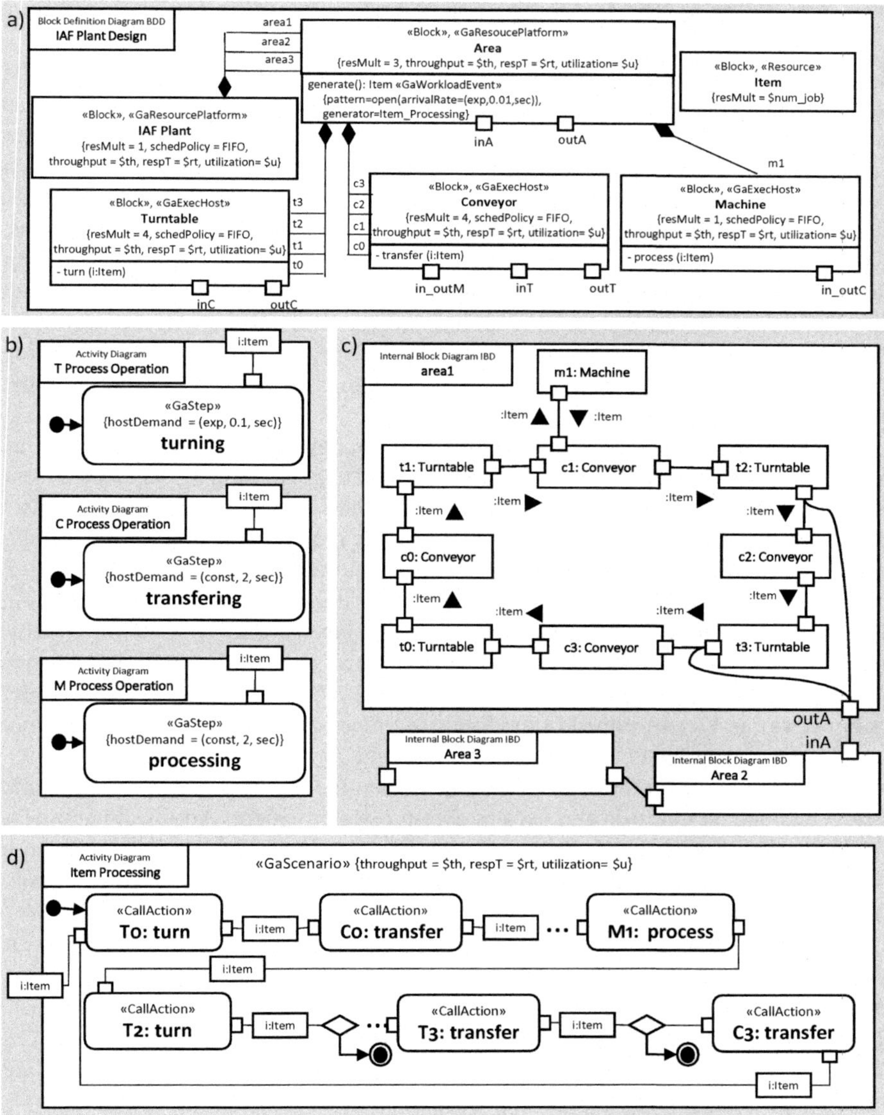

Fig. 11.11 Excerpt of the SysML design model of the IAF plant through block definition diagrams (**a**, **d**), Activity diagrams (**b**), Internal block diagrams (**c**)

moved to `Area2` or to conveyor `C3` where the whole is supposed to restart from the beginning at turntable `T0`. This processing scenario is realized on top of the `IAF Plant` layout shown in the IBD in Fig. 11.11c.

We now proceed with the description of how MARTE is used to analyze the given design model with respect to the requirements model.

11.4.2.2 Profiling SysML Models with MARTE

In MARTE, a system is a platform where resources provide services that can be acquired and released. In this respect, both, the whole IAF plant and its areas are identified as platforms (`GaResourcePlatform`). The structural constraint imposed by the `FR` requirement in Fig. 11.10 suggests to consider also the Area block as a logical sub-collection of resources as depicted in Fig. 11.9. MARTE allows the distinction of different kinds of resources depending on specific purposes, i.e., processing, storage, and communication. In particular, in performance modeling a `GaExecHost` can be any resource that executes behavior. Therefore we applied this stereotype to SysML blocks modeling the resources of the IAF transportation line, i.e., `Turntable`, `Conveyor`, and `Machine`, which host and provide `turn()`, `tranfer()`, and `process()` operations as services to accomplish the production of items depicted in Fig. 11.11d.

The aforementioned stereotypes come with predefined properties which act as placeholders (1) to store input parameters required to generate analysis models from SysML ones and (2) to store output results generated by the chosen analysis tools.

In order to enable the validation of the `NFR` requirement imposing a timing constraint on the item production process, we require the model-driven generation of a queuing network model representing the Area 1 of the IAF transportation line. MARTE explicitly supports performance analysis with its Generic Quantitative Analysis Model (GQAM) and Performance Analysis Model (PAM) sub-profiles (Object Management Group (OMG), 2016c). The performance model parameters can be partitioned in the following main categories: (1) the operational profile, (2) the workload, (3) the resource demands.

The operational profile is a collection of data that stochastically represent the usage that agents make of a system in a certain environment. A typical parameter is the probability to invoke a certain service. In this example, we consider a unique system-level service (i.e., item processing) then we omit this parameter (i.e., probability equal to one).

The workload represents the intensity of system services requests from agents. It is annotated by the `GaWorkloadEvent` applied on the `generate()` operation on the `Area` block that collect all the resources involved in the production process. `Items` enter `Area1` from turntable `T0` and leave it from turntables `T2` and `T3`, therefore we considered an open workload of items with a mean inter-arrival time of 10 s (expressed as a Poisson distribution random variable).[7]

Finally, resource multiplicities (resMult) and demands are annotated on `GaExecHosts` block and operations' Actions in terms of execution times obtained, as for arrival pattern specification, random variables obtained from exponential (exp) and constant (const) probability distributions.

[7] A closed pattern would be considered in a system-level production process involving all the resources of three areas. In that case a similar annotation would be applied on the IAF Plant block.

The output of the analysis may also be stored in properties of MARTE. Common performance indices are throughput, response time, utilization but the possible performance indices and the granularity of the results depend on the capability of the chosen analysis tool.[8] Indices can be calculated both at system-level or for single resources. We refer the reader to Sect. 11.4.2.3 for their description in context of the given case study. In Fig. 11.11 we store the output results that refer to a particular behavioral modeling element, e.g., the Item Processing activity (`GaScenario`, `throughput`, `respT`, `utilization` properties), or to a particular structural one (`GaResourcePlatform`, `throughput`, `respT`, `utilization` properties). In MARTE, variables, shown as $-prefixed Strings in Fig. 11.11, can be used as placeholder to be replaced with actual results.

It is worth noting that the annotation of SysML models should be considered as an integral part of model-driven methodology for early validation of systems adopting SysML as the main modeling notation. In this respect, MARTE does not prescribe the adoption of any annotation strategy or model-driven methodology. The interested reader is kindly referred to Cortellessa et al. (2011) for a general overview on model-driven performance engineering and analysis methodologies.

11.4.2.3 Modeling in PMIF

An excerpt of the `IAF Plant` modeled in PMIF is illustrated in Fig. 11.12. Such as AML, PMIF is primarily devised as an exchange standard for interoperability among queuing network (QN) solvers. Moreover, as discussed for AML, PMIF does not provide a standard concrete notation. Therefore we provide here an ad-hoc notation for the sake of explanation aided by a graph-oriented diagram. The legend on the right side of the diagrammatic representation relates this notation with PMIF concepts.

The so-called queuing network topology (i.e., the graph of servers and arcs) of the IAF Plant QN model includes a distinct server for each turntable (`Tx`), conveyor (`Cx`), and machine (`Mx`), connected through arcs. A source node is connected to the turntable `T0` and two sink nodes can be reached from turntable `T2` and `T3`, where the item processed by the plant can leave the QN, i.e., the `Area1` of the IAF Plant to enter `Area2`. The produced items that circulate among servers are referred as customer or job in the QN jargon. Jobs are generated and injected in the QN via `T1` by a workload generator.

The arrival process of item jobs is determined by a probability distribution that generate the inter-arrival times between consecutive arrivals. In this example, we considered a Poisson arrival process of items always (then probability set to 1) to `T0`. We then assume mutually independent arrivals of items every 100 s, on average (obtained from the reciprocal of arrival parameter 0.01, a.k.a. lambda parameter).

[8] For an example, the reader can consolidate the manual of the JMT queuing network solver (Casale and Serazzi, 2011).

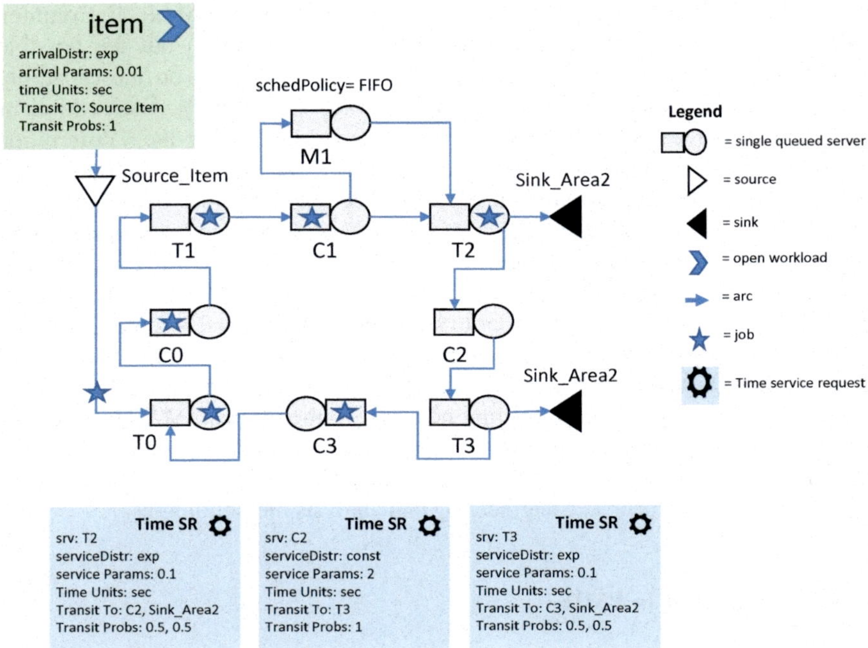

Fig. 11.12 Excerpt of IAF Plant modeled in PMIF

The generated item jobs can potentially visit all the servers reachable by the QN topology. However, the possible paths of such jobs are determined by a collection of service requests (SR) that Item jobs require to the visited server. A subset of these SRs required by Items to `T2`, `C2`, and `T3` servers (`srv`) are shown at the bottom of Fig. 11.12. They are expressed in term of service time (Time SR) during which the server is seized to process the current job before releasing to the next. Again, the service time is modeled as a stochastic variable governed by an exponential probability distribution whose average value is obtained as a reciprocal of the lambda parameter (e.g., 10 s for `T3`). It is worth noting how the `transitTo` and `transitProb` attributes of `T2` and `T3` SRs contains two ordered values, one for each arc outgoing the server. The next server is then determined by pairing the value of these attributes.

The resulting QN model can be directly modeled in PMIF using the xQNM tool presented in Troya and Vallecillo (2014) or in JMT (Casale and Serazzi, 2011). Figure 11.13 shows some simulation results generated by JMT and annotated back on the IAF Plant SysML model by replacing the MARTE variables previously assigned to properties of MARTE stereotypes (cf. the $-prefixed terms in Fig. 11.11).

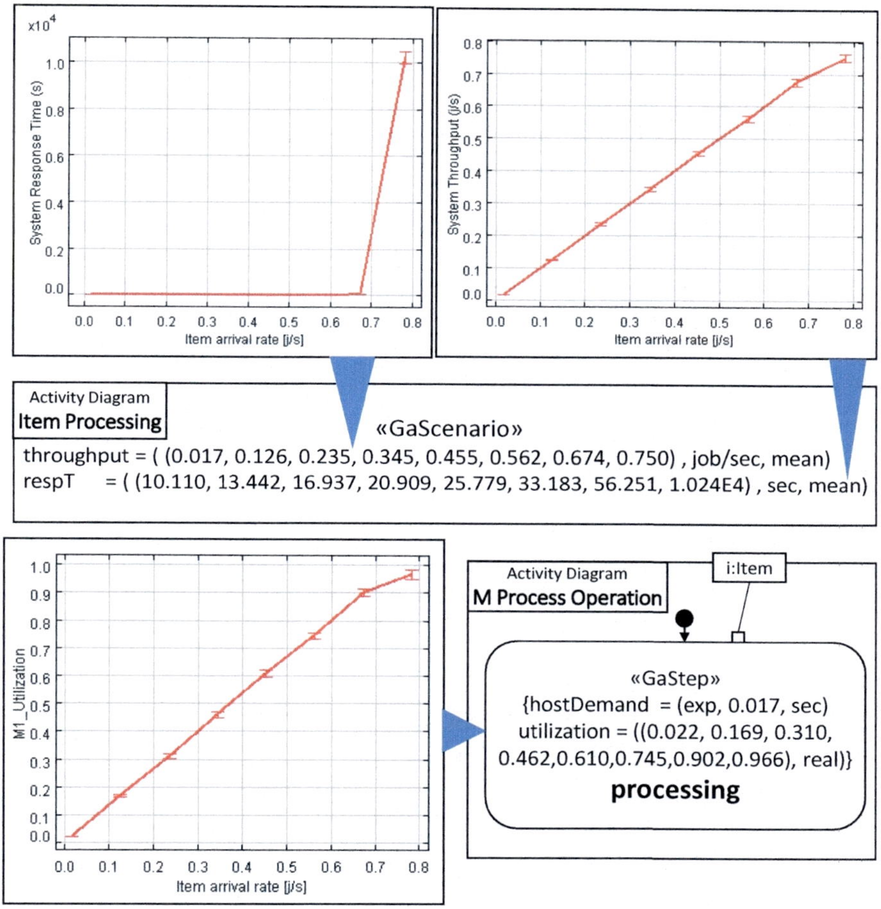

Fig. 11.13 Analysis results and their annotation on IAF Plant SysML model with stereotypes and properties of the MARTE profile

We calculated the system response time and system throughput[9] for the item processing scenario depicted in the activity of Fig. 11.11d. We conducted an early what-if analysis by increasing the arrival rate on items from 1 item per minute up to 1 item per second to test production capacity of that area.

`Area1` can manage up to 40 item arrivals per minute (0.674 item/s) with an average response time (i.e., the time required to execute the activity `Item Processing`) of 56 s per item. This processing time for items does not violate

[9]The QN model represents only Area1's process and resources, therefore it corresponds to the system under analysis.

the maximum processing time of 60 s allowed for the item production process (cf. the `NFR` in Fig. 11.10).

However, the given response time is an average response time and the utilization of `Area1`'s resources rapidly grows to the critical 0.9 threshold (i.e., the resource is busy for the 90% of its time) as shown for machine `M1`. Therefore the production capacity of `Area1` is currently limited to 40 items per minute and any increment in the production rate (e.g., to 45 item per minute, 0.750 item/s) would cause the violation of the given requirement due to the increasing queue of items waiting for `M1` processing.

Thanks to these early analysis results, the modeler/analyst can then decide (1) to keep or reduce the current production rate of `Area1`, (2) to increase the processing power of the multi-purpose machine `M1`, or (3) to decrease the arrival rate of items, and then the production rate of `Area1`.

11.4.2.4 Modeling in AML

An excerpt of the IAF Plant modeled in AML is shown in Fig. 11.14. It is worth noting that AML does not provide a standard concrete notation (such as SysML does). Therefore we provide here an ad-hoc notation for the sake of explanation in terms

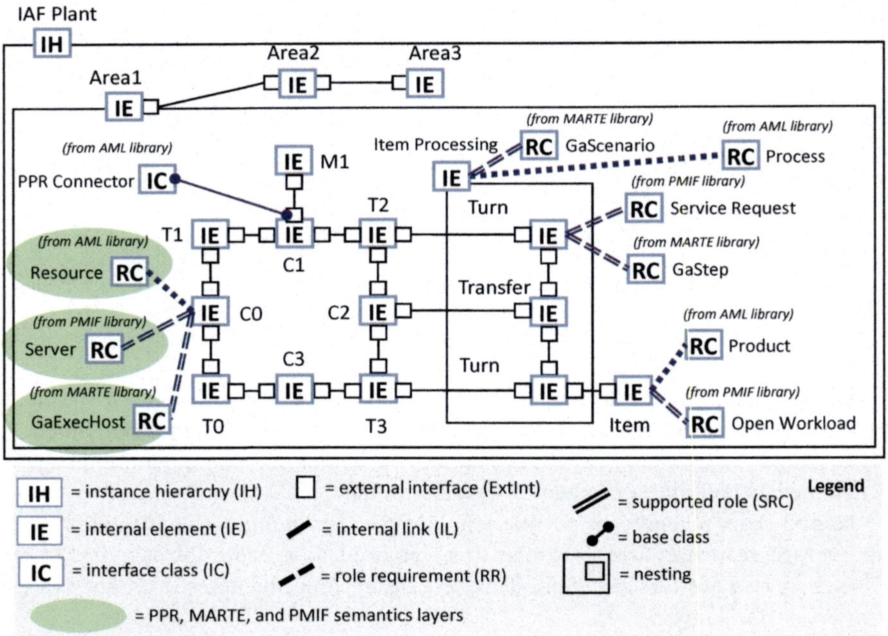

Fig. 11.14 Excerpt of IAF Plant modeled in AML

of a graph-oriented diagram. The legend below the diagrammatic representation of IAF Plant explains this notation with AML concepts.

Following the requirements in Fig. 11.10, also the AML representation of the IAF Plant's transportation line is divided in three areas (`Area1`, `Area2`, and `Area3`), each composed of turntables, conveyors, and machines. The root node is the `IAF Plant` instance hierarchy (**`IH`**). Its nested structure includes three internal elements (**`IE`**), one for each area that logically divide the transportation line. Then, the detailed architecture of `Area1` is shown as a graph of **`IE`**s (Lüder et al., 2013). Turntables (`Tx` **`IE`**s) and conveyors (`Cx` **`IE`**s) are nodes alternating each other in closed paths. An additional node for a machine `M1` is linked to the `C1` **`IE`**. Each **`IE`** has a predefined set of interaction points modeled as external interfaces (**`ExtI`**) connected via internal links (**`IL`**). According to Schleipen and Drath (2009), such **`IE`**s can be classified in (1) products, (2) processes, and (3) resources (PPR) (Schleipen and Drath, 2009).

Resources are entities involved in the production that process and handle products. Depending on the reference engineering discipline, software, mechanical, and electronic elements can all be referred as resources collaborating to product handling.

Products are the produced goods processed by the IAF plant through its resources. The IAF plant is supposed to produce a generic item that is moved by turntables and conveyors towards multi-purpose machines.

Processes consist of processing steps realized by the IAF plant resources. The IAF plant provides a production process for items determining their movements among resources.

Products, processes, and resources are all modeled as **`IE`**s as done for turntable `T3`, the processing step `Turn` realized by `T3`, and the product `Item`.

The instance hierarchy with its internal elements, external interfaces, and internal links is semantic-agnostic, i.e., there is no distinction between nodes and arcs building such a graph-like structure. For this purpose, AML provides the role class libraries (**`RClib`**) and role classes (**`RC`**) concepts for nodes and interface class library and interface classes for external interfaces. In particular, AML provides several standard role class library (**`RClib`**) and interface class library libraries (**`IClib`**). The AML **`RClib`** defines `Product`, `Process` and `Resource` **`RC`**. We then build a PPR semantics layer in the `IAF Plant` **`IH`** structure by assigning `Product`, `Process`, and `Resource` role classes to the corresponding **`IE`**s of the `IAF Plant` instance hierarchy via role requirement relationships (**`RR`**) as shown in Fig. 11.14 for `Item`, `Item Processing`, and `C0`, respectively.

For the sake of the envisioned integration with SysML/MARTE and PMIF, we further enrich the AML model with two additional semantics layers. We created two new role class libraries, `PMIF` **`RClib`** and `SysML/MARTE` **`RClib`** that define the corresponding language-specific concepts by role classes of the same name.

The PMIF **`RClib`** introduces in AML role classes like `Server`, `Service Request`, and `Open Workload` role classes. A PMIF semantics layer is then realized on top of the same `IAF Plant` **`IH`** by assigning these role classes to resources (e.g., `C0`), processing steps (e.g., `Turn`), and products (e.g., `Open Worload`) to facilitate the integration with PMIF as further discussed in Sect. 11.4.3.2.

The `SysML/MARTE` **`RClib`** supports the modeling of a AML semantics layer for these two OMG standards. We assigned the AML **`RC`** counterparts of MARTE stereotypes (`GaExecHost`, `GaScenario`, and `GaScenario`) to plant resources and processes to facilitate the integration of AML with SysML models annotated with MARTE, as discussed in Sect. 11.4.3.2.

11.4.3 CPPS Engineering Chain Automation

The proposed engineering chain automation approach aims at bridging system design with early validation of system performance. As we discussed for the used modeling languages in the engineering chain, all of them are supported by metamodels which define the languages explicitly and also define the possible structures and content of the models. This allows model transformations to be formulated on the metamodel level which ensures their executability for any valid instance. However, to realize a valid transformation with a specific purpose, also the models have to be rich enough and systematically built by following modeling rules to be transformed to the target language.

Figure 11.15 gives a graphical overview on the engineering chain automation. In particular, we use a transformation chain to transform SysML models to AML ones with a first transformation and AML models into PMIF ones with a second transformation. In both cases, the integration of the languages is carried over in two consecutive steps:

1. Modeling rules to be applied on models;
2. Model transformation rules specification and execution.

The first step suitably extends the source model (SysML in SysML/AML, AML in AML/PMIF) with information required by the second step, i.e., model transformations, to obtain a complete and useful target model.

11.4.3.1 Integrating SysML and AML

We presented for the first time the SysML/AML integration in Berardinelli et al. (2016). Consequently, we give here only an overview of the previously presented approach.

Modeling Rules for SysML The first step of the SysML/AML integration is realized through the UML profiling mechanism which allows to add further information to the SysML models needed for the production of AML models. An excerpt of the AML4SysML profile is shown in Fig. 11.15a. The AML4SysML

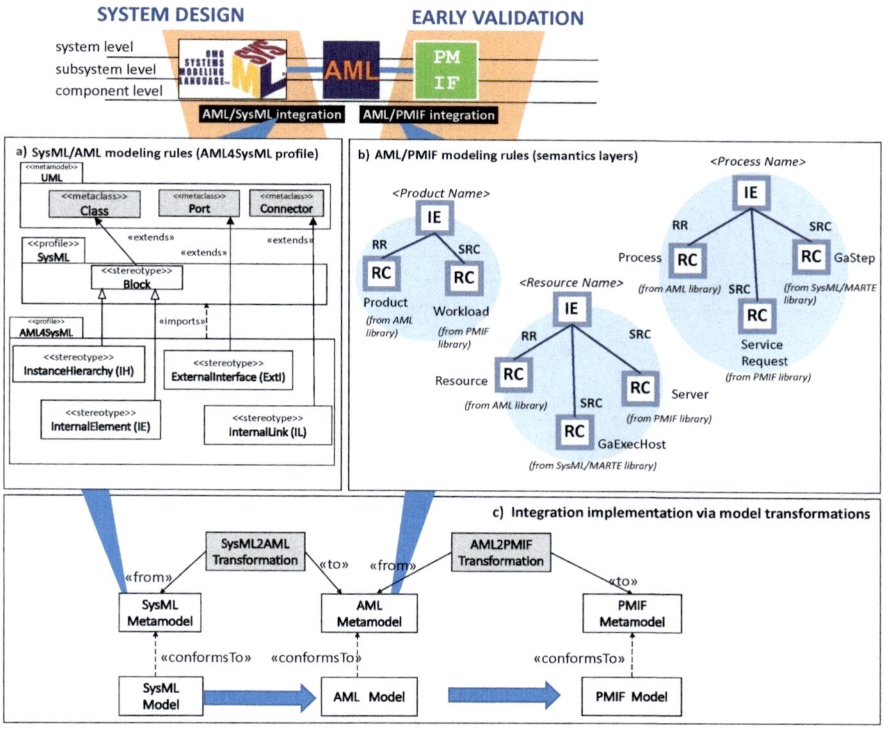

Fig. 11.15 SysML, AML, and PMIF transformation chain (**c**) involving the AML4SysML profile (**a**) and semantics layers (**b**)

profile maps AML concepts to structural concepts from SysML such as `Block`, `Port`, and `Connector`, i.e., it addresses the structural modeling of CPPS in SysML. As a consequence, the selection of SysML diagram types are restricted to structural ones, i.e., BDDs and IBDs.

A choice behind the design of the AML4SysML profile is the coupling of our AML4SysML profile with the SysML profile so that the application of the former always requires the import of the latter. Consequently, any AML model annotated with the AML4SysML profile is always managed as a SysML model. For example, the stereotypes for **`IH`** and **`IE`** AML concepts are specialization of the SysML block stereotype (cf. Fig. 11.15a).

SysML to AML Model Transformations For automating the transition from SysML to AML, we employ model transformations. As we use profiles in addition to metamodels in our integration chain, it is necessary to use transformation technologies, which are able to work with profiles and their applications on models such as it is possible in QVT.

Figure 11.16 provides an excerpt of the transformations needed for our setting. In particular, it is showing a graphical visualization of a QVT transformation excerpt

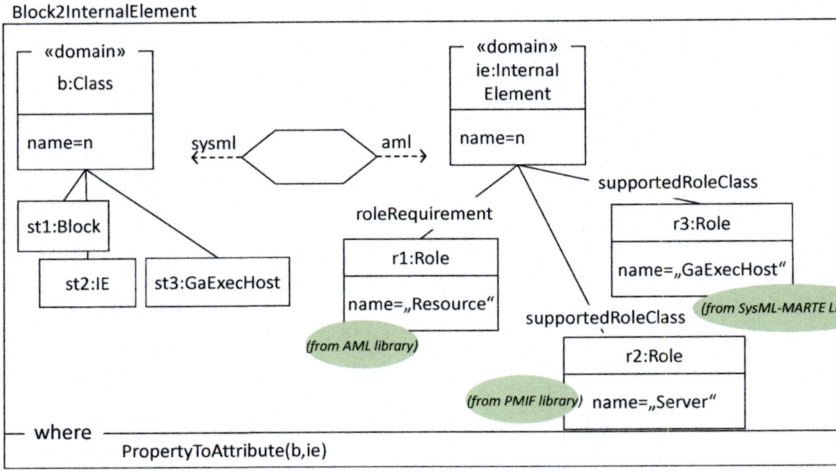

Fig. 11.16 Excerpt of the *SySML-to-AML* transformation in QVT

realizing one of the mappings between SysML and AML language concepts, namely the mapping between the SysML *Block* concept, which is realized by instantiating the metaclass `Class` of the UML metamodel (as it is the base for the `Block` concept in SysML) and the **`IE`** metaclass from the AML metamodel. The left hand side of the transformation matches UML Classes annotated with three stereotypes from three different profiles, i.e., `Block` (from SysML), `IE` (from AML4SysML), and `GaExecHost` (from MARTE).

The corresponding modeling pattern on AML models is depicted on the right hand side of the transformation: an internal element with three distinct assigned roles (via role requirement and supported role class relationships), i.e., `Resource` (from the standard AML library), `Server` (from the PMIF library), `GaExecHost` (from the SysML/MARTE library).

For realizing this mapping, QVT provides the possibility to define a relation, which is matched by QVT engines executed in the forward transformation mode on the input model elements (in our case, the SysML model elements) to produce the output model elements and to calculate their feature values (in our case, the AML model elements). Similar relations are possible to develop for all the mappings reported in Berardinelli et al. (2016) to obtain an executable specification of the proposed mappings.

11.4.3.2 Integrating AML and PMIF

In Berardinelli et al. (2016), we proposed a model-driven performance engineering approach for CPPS through the combination of AML with PMIF discussing three possible integration strategies based on the native AML integration mechanism (IEC, 2014). The one which applies for our given engineering chain automation is via model transformation which we summarize in the following.

Modeling Rules for AML Figure 11.15b depicts the modeling rules applied on AML models to close the semantics gap with the target model, i.e., PMIF. The goal promoted by the proposed modeling rules is creating a domain-specific semantics layer for PMIF in AML. A semantic layer is a collection **RC**s that is applied on a semantics-agnostic graph represented by an **IH** (e.g., `Area1` in Fig. 11.14). These **RC**s are suitably collected in AML **RClib**s and applied on purpose via **RR** and **SRC** relationships.

The PMIF semantics layer includes a PMIFRoleClassLib **RClib**. In particular, a new **RC** is created for each metaclass of the PMIF metamodel and inherits the name from the corresponding metaclass.

The domain-specific semantics layering for PMIF is realized by assigning the PMIF-specific **RC**s to **IE**s. In order to apply such a new PMIF domain-specific semantics layer, **RR** and **SRC** relationships[10] are required that connect **IE**s to **RC**s, as shown in Fig. 11.15b.

For this purpose we add the new PMIF domain-specific layer on top of the PPR concepts, in accordance with the following three modeling rules:

Resource ⇝ Server: A resource **RC** is an entity involved in production and executes processes and handle products (Schleipen and Drath, 2009). Similarly, a **Server** represents a component of the execution environment that provides some processing service (Smith and Williams, 1999).

Product ⇝ Workload: A product **RC** depicts a produced good, processed by resources (e.g., material handling, creation of intermediate products) (Schleipen and Drath, 2009). In this respect, a **Workload** represents a collection of jobs, i.e., characterizes a product type, that make similar **ServiceReq**s to **Server**s (i.e., resources).

Process ⇝ ServiceRequests: A process **RC** represents a production process including sub-processes, process parameters and the process chain, modifies products and produces final products out of different sub-products (Schleipen and Drath, 2009). Similarly, a **ServiceReq** associates the **Workload** (i.e., product types) with a **Server** (i.e., resources) (Smith et al., 2010). Therefore, an ordered set of **ServiceReq**s (Smith et al., 2010) builds up a process (Schleipen and Drath, 2009).

AML to PMIF Model Transformation For this transformation, we do not require the usage of profiles as for the SysML to AML transformation, but we have to make use of the AML libraries as well as their applications on the input models. Figure 11.17 shows an excerpt of a model transformation from AML to PMIF implemented with QVT relations exploiting the PPR, PMIF, and SysML-MARTE semantics layers.

[10] **RR**s assign **RC**s of a mandatory primary semantics layer, while additional ones can be assigned via **SRC**.

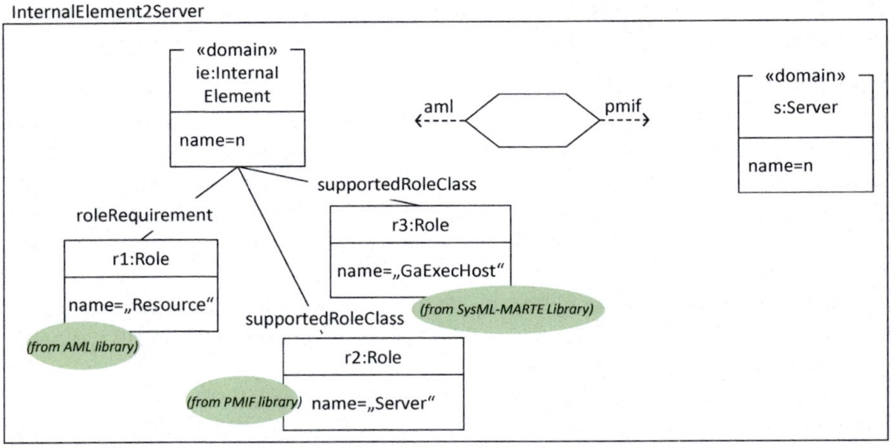

Fig. 11.17 Excerpt of the *AML-to-PMIF* transformation in QVT

In particular, it shows a relation that generates a PMIF **Server** from an AML **IE** playing the role of PMIF **Server** as modeled by the **SRC** relationships from such **IE** to **RC**s of the PMIF **RClib**. Similar relations can be implemented to transform AML **IE**s, **ExtI**s, and **IL** into PMIF **Server**s and **Arc**s as has been done in Berardinelli et al. (2016).

It is worth noting that the left hand side of the QVT transformation in Fig. 11.16 corresponds to the right hand side of the QVT transformation in Fig. 11.17. Once combined, these two QVT transformations realize the end-to-end integration among SysML and PMIF adopting AML as an intermediate model, as depicted Fig. 11.15c.

11.4.4 Synopsis

The engineering chain shown in our case study relies on language-specific mechanisms, i.e., (1) the profiling mechanism for OMG standard languages like UML and SysML, and (2) the specification of **RClib** libraries in AML.

Both mechanisms create a set of additional information (referred as semantics layer for AML) attached to source model elements (e.g., a SysML block or an AML **IE**) via ad-hoc relationships (e.g., **RR**s and **SRC**). These layers are then accessed by model transformation to generate new target models. The extent of mappings between source and target modeling languages like SysML and AML, then depends on (1) the completeness of profiles and **RClib**s, and, of course, (2) on the requirements of the adopted model-driven methodology.

In Berardinelli et al. (2016), we already evaluated the current status of the AML4SysML profile that currently maps a subset of AML concept to SysML model elements.[11] In this chapter, we introduced for the first time a role class library for SysML and MARTE. Currently it includes the small subset of the MARTE stereotypes used in the proposed case study and its extension is left as future work.

Furthermore, we presented a model-driven performance methodology based on the combination of SysML, AML, and PMIF. In particular, PMIF does not provide extension mechanisms and does not allow the modeling of requirements as available is SysML. For this reason, we deliberately ignore the mappings of SysML requirements to AML and then their translation to PMIF. However, the AML native extension mechanisms based on role classes can be also used to map SysML requirements in AML as additional internal elements and relationships to an extended version of the SysML/MARTE role class library.

Finally, we can state that for the engineers, the knowledge on a subset of SysML and MARTE is powerful enough to perform early design and validation of systems such as the IAF plant which also has to meet certain performance indicators. In future work, we will investigate how general the proposed solution is in the context of CPPS.

11.4.5 Critical Discussion

In this section we proposed an *engineering chain* or a *technical process* (cf. Chap. 2) for the architectural design and performance analysis of CPPS, leveraging model-driven methodologies, modeling languages and tools from the authors' background.

An appropriate selection of methodologies, languages and tools strictly depends on the chosen *views* on the CPPS under study. In the proposed example we focused on three main tasks, i.e., requirement specification, system design and performance analysis, adopting two main models and modeling languages, namely SysML for requirement and architectural specification tasks and PMIF for performance analysis. Model transformation are then required to support the *information flow* (see Chap. 2) between system models and analysis models.

Of course, the proposed technical process is limited in terms of the supported tasks, we deliberately choose a general purpose modeling language such as SysML, and two free, open, XML-based data exchange file formats, namely AML and PMIF. We see the following two advantages of our design rationale:

Openness for other modeling concerns: The proposed technical process remain open to additional process tasks e.g., by extending the CPPS SysML-based design by UML profiles that increase the informative content of the information flow among the planned steps.

Openness for other target tools: Additional target models and tools may be attached to the current information flow through the adopted data exchange

[11] An up-to-date AML/SysML concepts mapping table is available at http://www.sysml4industry.org/?page_id=299

formats, i.e., AML and PMIF. For example, WEASEL[12] is Web service that allows the user to send a PMIF-based QN model to the server, remotely execute model transformations to tool-specific file formats, and receive back performance indices calculated on different solvers. In addition, also other tools building on AML may be used in the technical process such as for deriving implementation artefacts, e.g., OPC UA specifications (Schleipen et al., 2015), which may be employed in later engineering phases.

Of course, the implementation of a model-driven, customized technical process poses both methodological and technical challenges. On the one hand, new views can be added to support additional methodologies with native language extension mechanisms such as UML profiling (e.g., combination of performance with reliability and availability analyses (Berardinelli et al., 2009) or variability modeling (Weilkiens, 2011)). On the other hand, new technical challenges may arise requiring the implementation of additional, possibly bi-directional transformations among new source/target models which is a challenge on its own (France and Rumpe, 2007).

11.5 Conclusion and Future Challenges

In this book chapter, we have outlined how systems engineering processes can be supported by novel model-driven engineering techniques to realize MDSE. In particular, we have shown (1) how virtual prototypes are produced by using SysML, (2) how this information may be exchanged on basis of AML to provide vertical as well as horizontal engineering tool integration in the context of the V-Model, and (3) how virtual prototypes are validated based on formal analysis methods such as queuing networks. All of this is facilitated by a set of machine-readable models and well defined model transformations between them.

The reference model we presented in this book chapter may be further exploited by new language combinations as well as extended for other scenarios. In particular, we see research required to classify existing MDSE approaches with respect to the tasks which are automated. In addition, empirical user studies are required to evaluate how well MDE approaches are received by practitioners. Furthermore, the following research lines are of major interest to further develop the area of MDSE.

Requirements modeling and their validation and verification. It starts already by the requirements specification as different properties may be expected from the system. As we focussed in this book chapter on non-functional properties such as performance attributes of the system, also functional properties such as temporal properties ensuring safety requirements may be needed as well. In previous work, we have provided a method to construct property languages which allow

[12]WEASEL Web Service http://sealabtools.di.univaq.it/tools.php

to formulate temporal properties of interest within the domain-specific language and to use existing model checkers to verify them (Meyers et al., 2014). Utilizing this approach for systems engineering is an interesting future research line, e.g., to generate a property language for SysML which would require to handle parametrics diagrams as well during the modeling checking process.

Synthesis of discipline-specific models. Another research line is the investigation of transferring system design models established in the early phases to discipline-specific models (Kernschmidt et al., 2013). Here a dedicated step to refine the early systems designs into discipline-specific solutions is necessary, first of all to clarify which elements are realized with mechanical, electrical, and software components. SysML4Mechatronics (Kernschmidt and Vogel-Heuser, 2013) seems a very promising candidate to perform this refinement step. Subsequently, dedicated transformations are needed to derive different models for the specific disciplines starting from mCAD over eCAD to software models. Providing full traceability from the systems engineering models to the domain-specific models is a major requirement to deal with evolution concerns as well as with verification and validation concerns.

Model-driven design-space exploration. One major benefit of systems engineering is to explore the design-space before one design is chosen to be realized. For performing design-space exploration efficient search techniques are needed as exhaustive search, i.e., enumerating each possible design solution, is for most cases not feasible due to the combinatorial explosion. A very recent emerging model transformation approach is search-based model transformation (Fleck et al., 2016) which combines the power of model transformations to systematically manipulate models and the highly-scalable search capabilities of meta-heuristic search algorithms. In particular, by formulating objectives for the design models (also including multi-objectives) in terms of fitness functions, the search algorithms are guiding the transformation rule applications to find good design models.

Model-driven product lines. In this book chapter we considered the creation of models from scratch except the availability of a model library providing already existing building blocks which have to be combined. An alternative approach is to build a family of systems which allows to derive different concrete realizations. This approach is often referred to as product-line engineering. A product-line provides a description of the commonalities as well as differences of the concrete system realizations covered by the product-line. The design process is then reduced to configuring a concrete realization, e.g., by using a feature model to select certain features and using model completion to derive a valid realization in cases where only a partial selection is performed. The combination of product-line engineering and SysML has been already discussed in the literature (Papakonstantinou and Sierla, 2013; Maga and Jazdi, 2010) and different efforts are ongoing to further support product-line engineering with SysML especially in multi-disciplinary settings (Vogel-Heuser et al., 2015).

Model profiling. We consider this research field as the natural continuation and unification of different already existing techniques with respect to the usage of models in the context of MDE. Model profiling continues the research lines of (1) process mining (Leemans and van der Aalst, 2015), (2) specification mining (Dallmeier et al., 2012), (3) FSA learning (Giles et al., 1992), (4) data profiling (Abedjan et al., 2015), (5) program profiling (Graham et al., 1982), and (6) data mining as well as (7) data analytics (Fayyad et al., 1996). All these techniques aim at better understanding the concrete data and events used in or by a system and by focusing on particular aspects of it. Therefore, we consider model profiling as a very promising field to bridge the gap between the design time and runtime phases in the current state-of-the-art in MDE. In particular, the automated information upstream from operations to the design is highly needed to improve the design of a system continuously with additional knowledge from operations.

Acknowledgements This work has been supported by the Christian Doppler Forschungsgesellschaft and the BMWFW, Austria and by the Austrian Research Promotion Agency (FFG) within the project "InteGra 4.0 - Horizontal and Vertical Interface Integration 4.0".

References

Abedjan, Z., Golab, L. and Naumann, F.: Profiling relational data: a survey. VLDB J. **24**(4), 557–581 (2015)

Alt, O.: Modellbasierte Systementwicklung mit SysML. Carl Hanser Verlag, Munich (2012)

Berardinelli, L., Bernardi, S., Cortellessa, V., Merseguer, J.: UML profiles for non-functional properties at work: analyzing reliability, availability and performance. In: 'Proceedings of NFPinDSML Workshop @ MoDELS' (2009)

Berardinelli, L., Biffl, S., Lüder, A., Mätzler, E., Mayerhofer, T., Wimmer, M., Wolny, S.: Cross-disciplinary engineering with AutomationML and SysML. Automatisierungstechnik **64**(4), 253–269 (2016)

Berardinelli, L., Mätzler, E., Mayerhofer, T., Wimmer, M.: Integrating performance modeling in industrial automation through AutomationML and PMIF. In: Proceedings of the IEEE International Conference on Industrial Informatics (INDIN), pp. 1–6 (2016)

Bernardi, S., Merseguer, J., Petriu, D.C.: A dependability profile within MARTE. Softw. Syst. Model. **10**(3), 313–336 (2011)

Bezivin, J.: On the unification power of models. Softw. Syst. Model. **4**(2), 171–188 (2005)

Biffl, S., Lüder, A., Mätzler, E., Schmidt, N., Wimmer, M.: Linking and versioning support for AutomationML: a model-driven engineering perspective. In: Proceedings of 2015 IEEE International Conference on Industrial Informatics (INDIN), pp. 499–506 (2015)

Booch, G., Rumbaugh, J., Jacobson, I.: The Unified Modeling Language User Guide, 2nd edn. Addison-Wesley, Reading, MA (2005)

Brambilla, M., Cabot, J., Wimmer, M.: Model-Driven Software Engineering in Practice. Morgan and Claypool, San Rafael (2012)

Broy, M., Schmidt, A.: Challenges in engineering Cyber-Physical Systems. Computer **47**(2), 70–72 (2014)

Casale, G., Serazzi, G.: Quantitative system evaluation with Java modeling tools. In: Proceedings of the 2nd ACM/SPEC International Conference on Performance Engineering (ICPE), pp. 449–454 (2011)

Cortellessa, V., Di Marco, A., Inverardi, P.: Model-Based Software Performance Analysis. Springer, Berlin (2011)

Czarnecki, K., Helsen, S.: Feature-based survey of model transformation approaches. IBM Syst. J. **45**(3), 621–645 (2006)

Dallmeier, V., Knopp, N., Mallon, C., Fraser, G., Hack, S., Zeller, A.: Automatically generating test cases for specification mining. IEEE Trans. Softw. Eng. **38**(2), 243–257 (2012)

Equipment Center for Distributed Systems: http://www.iaf-bg.ovgu.de/en/technische_ausstattung_cvs.html (2016). [Online; accessed 30 Oct 2016]

Fayyad, U.M., Piatetsky-Shapiro, G., Smyth, P.: From data mining to knowledge discovery: an overview. In: Fayyad, U.M., Piatetsky-Shapiro, G., Smyth, P., Uthurusamy, R. (eds.) Advances in Knowledge Discovery and Data Mining. American Association for Artificial Intelligence, pp. 1–34. AAAI Press, Menlo Park, CA (1996)

Feldmann, S., Kernschmidt, K., Vogel-Heuser, B.: Combining a SysML-based modeling approach and semantic technologies for analyzing change influences in manufacturing plant models. In: Proceedings of the 47th CIRP Conference on Manufacturing Systems (CMS) (2014)

Fleck, M., Troya, J., Wimmer, M.: Search-based model transformations with MOMoT. In: Proceedings of the 9th International Conference on Theory and Practice of Model Transformations (ICMT), pp. 79–87 (2016)

France, R.B., Rumpe, B.: Model-driven development of complex software: a research roadmap. In: Proceedings of the International Conference on Software Engineering (ICSE), pp. 37–54 (2007)

Friedenthal, S., Moore, A., Steiner, R.: A Practical Guide to SysML: the Systems Modeling Language. Morgan Kaufmann, Amsterdam (2014)

Giles, C.L., Miller, C.B., Chen, D., Chen, H.-H., Sun, G.-Z., Lee, Y.-C.: Learning and extracting finite state automata with second-order recurrent neural networks. Neural Comput. **4**(3), 393–405 (1992)

Graham, S.L., Kessler, P.B., Mckusick, M.K.: Gprof: a call graph execution profiler. SIGPLAN Not. **17**(6), 120–126 (1982)

Hegny, I., Wenger, M., Zoitl, A.: IEC 61499 based simulation framework for model-driven production systems development. In: Proceedings of the IEEE Conference on Emerging Technologies and Factory Automation (ETFA), pp. 1–8 (2010)

Hutchinson, J., Whittle, J., Rouncefield, M., Kristoffersen, S.: Empirical assessment of MDE in industry. In: Proceedings of the 33rd International Conference on Software Engineering (ICSE), pp. 471–480 (2011)

IEC: IEC 62714 – Engineering data exchange format for use in industrial automation systems engineering – AutomationML. http://www.iec.ch (2014)

ISO/PAS: ISO/PAS 17506:2012 Industrial automation systems and integration – COLLADA digital asset schema specification for 3D visualization of industrial data. http://www.iso.org (2012)

Jetley, R., Nair, A., Chandrasekaran, P., Dubey, A.: Applying software engineering practices for development of industrial automation applications. In: Proceedings of the 11th IEEE International Conference on Industrial Informatics (INDIN), pp. 558–563 (2013)

Kagermann, H., Wahlster, W., Helbig, J.: Recommendations for implementing the strategic initiative INDUSTRIE 4.0 – securing the future of German manufacturing industry. Final Report of the Industrie 4.0 Working Group, Forschungsunion im Stifterverband für die Deutsche Wirtschaft e. V. (2013)

Kernschmidt, K., Vogel-Heuser, B.: An interdisciplinary SysML based modeling approach for analyzing change influences in production plants to support the engineering. In: Proceedings of the IEEE International Conference on Automation Science and Engineering (CASE), pp. 1113–1118 (2013)

Kernschmidt, K., Barbieri, G., Fantuzzi, C., Vogel-Heuser, B.: Possibilities and challenges of an integrated development using a combined SysML-model and corresponding domain specific models. In: Proceedings of the 7th IFAC Conference on Manufacturing Modelling, Management, and Control (MIM), pp. 1465–1470 (2013)

Kühne, T.: Matters of (Meta-)modeling. Softw. Syst. Model. **5**(4), 369–385 (2006)

Kurtev, I.: State of the art of QVT: A model transformation language standard. In: Proceedings of the International Symposium on Applications of Graph Transformations with Industrial Relevance (AGTIVE), pp. 377–393 (2007)

Kyura, N., Oho, H.: Mechatronics–an industrial perspective. IEEE/ASME Trans. Mechatron. **1**(1), 10–15 (1996)

Lee, E.A.: Cyber physical systems: design challenges. In: Proceedings of the 11th IEEE International Symposium on Object-Oriented Real-Time Distributed Computing (ISORC), pp. 363–369 (2008)

Leemans, M., van der Aalst, W.M.P.: Process mining in software systems: discovering real-life business transactions and process models from distributed systems. In: Proceedings of the 18th International Conference on Model Driven Engineering Languages and Systems (MoDELS), pp. 44–53 (2015)

Lúcio, L., Amrani, M., Dingel, J., Lambers, L., Salay, R., Selim, G. M.K., Syriani, E., Wimmer, M.: Model transformation intents and their properties. Softw. Syst. Model. **15**(3), 647–684 (2016)

Lüder, A., Schmidt, N., Helgermann, S.: Lossless exchange of graph based structure information of production systems by AutomationML. In: Proceedings of IEEE 18th Conference on Emerging Technologies and Factory Automation (ETFA), pp. 1–4 (2013)

Lüder, A., Schmidt, N., Rosendahl, R.: Data exchange toward PLC programming and virtual commissioning: is AutomationML an appropriate data exchange format? In: Proceedings of the IEEE 13th International Conference on Industrial Informatics (INDIN), pp. 492–498 (2015)

Maga, C.R., Jazdi, N.: An approach for modeling variants of industrial automation systems. In: Proceedings of the IEEE International Conference on Automation Quality and Testing Robotics (AQTR), pp. 1–6 (2010)

Malavolta, I., Lago, P., Muccini, H., Pelliccione, P., Tang, A.: What industry needs from architectural languages: a survey. IEEE Trans. Softw. Eng. **39**(6), 869–891 (2013)

Mazak, A., Wimmer, M., Huemer, C., Kappel, G., Kastner, W.: Rahmenwerk zur modellbasierten horizontalen und vertikalen Integration von Standards für Industrie 4.0. In: Vogel-Heuser, B. et al. (eds.) Handbuch Industrie 4.0. Springer, Berlin (2016)

Mens, T., Gorp, P.V.: A taxonomy of model transformation. Electron. Notes Theor. Comput. Sci. **152**, 125–142 (2006)

Meyers, B., Deshayes, R., Lucio, L., Syriani, E., Vangheluwe, H., Wimmer, M.: Promobox: a framework for generating domain-specific property languages. In: Proceedings of the 7th International Conference on Software Language Engineering (SLE), pp. 1–20 (2014)

Object Management Group (OMG): Meta Object Facility (MOF) 2.0 Core Specification. OMG Document ptc/03-10-04 (2003)

Object Management Group (OMG): Object Constraint Language (OCL) Specification. Version 2.2. OMG Document formal/2010-02-01 (2010)

Object Management Group (OMG): Meta Object Facility (MOF) 2.0 Query/View/Transformation (QVT). OMG Document formal/2016-06-03 (2016a)

Object Management Group (OMG): OMG Systems Modeling Language (OMG SysML). http://www.omg.org/spec/SysML/1.4/ (2016b)

Object Management Group (OMG): UML Profile for MARTE. Version 1.1. http://www.omg.org/spec/MARTE/1.1/PDF (2016c)

Papakonstantinou, N., Sierla, S.: Generating an Object Oriented IEC 61131-3 software product line architecture from SysML. In: Proceedings of the IEEE 18th Conference on Emerging Technologies and Factory Automation (ETFA), pp. 1–8 (2013)

PLCopen: PLCopen. http://www.plcopen.org (2011)

Schleipen, M., Drath, R.: Three-view-concept for modeling process or manufacturing plants with AutomationML. In: Proceedings of the IEEE Conference on Emerging Technologies Factory Automation (ETFA), pp. 1–4 (2009)

Schleipen, M., Drath, R., Sauer, O.: The system-independent data exchange format CAEX for supporting an automatic configuration of a production monitoring and control system. In: Proceedings of the IEEE International Symposium on Industrial Electronics (ISIE), pp. 1786–1791 (2008)

Schleipen, M., Selyansky, E., Henssen, R., Bischoff, T.: Multi-level user and role concept for a secure plug-and-work based on OPC UA and AutomationML. In: Proceedings of the 20th IEEE Conference on Emerging Technologies and Factory Automation (ETFA), pp. 1–4 (2015)

Schmidt, D.: Guest Editor's Introduction: Model-Driven Engineering. Computer **39**(2), 25–31 (2006)
Schütz, D., Legat, C., Vogel-Heuser, B.: MDE of manufacturing automation software – integrating SysML and standard development tools. In: Proceedings of the 12th IEEE International Conference on Industrial Informatics (INDIN), pp. 267–273 (2014)
Seidl, M., Scholz, M., Huemer, C., Kappel, G.: UML@Classroom. Springer, New York (2012)
Selic, B., Gérard, S.: Modeling and Analysis of Real-Time and Embedded Systems with UML and MARTE: Developing Cyber-Physical Systems. Elsevier, Heidelberg (2013)
Smith, C.U.: Performance Engineering of Software Systems. Addison-Wesley Longman Publishing Co., Inc., Reading, MA (1990)
Smith, C.U., Llado, C.M., Puigjaner, R.: Performance Model Interchange Format (PMIF 2): a comprehensive approach to Queueing Network Model interoperability. Perform. Eval. **67**(7), 548–568 (2010)
Smith, C.U., Williams, L.G.: A performance model interchange format. J. Syst. Softw. **49**(1), 63–80 (1999)
Stevens, P.: Bidirectional model transformations in QVT: semantic issues and open questions. Softw. Syst. Model. **9**(1), 7–20 (2010)
Troya, J., Vallecillo, A.: Specification and simulation of queuing network models using domain-specific languages. Comput. Stand. Interfaces **36**(5), 863–879 (2014)
Vangheluwe, H., Amaral, V., Giese, H., Broenink, J., Schätz, B., Norta, A., Carreira, P., Lukovic, I., Mayerhofer, T., Wimmer, M., Vallecillo, A.: MPM4CPS: multi-paradigm modelling for Cyber-Physical Systems. In: Proceedings of the Project Showcase @ STAF 2015, pp. 1–10 (2016)
Verein Deutscher Ingenieure (VDI): Design methodology for mechatronic system–VDI 2206 (2004)
Vogel-Heuser, B., Biffl, S.: Cross-discipline modeling and its contribution to automation. Automatisierungstechnik **64**(3), 165–167 (2016)
Vogel-Heuser, B., Fay, A., Schaefer, I., Tichy, M.: Evolution of software in automated production systems: challenges and research directions. J. Syst. Softw. **110**, 54–84 (2015)
Vogel-Heuser, B., Fuchs, J., Feldmann, S., Legat, C.: Interdisziplinärer Produktlinienansatz zur Steigerung der Wiederverwendung. Automatisierungstechnik **63**(2), 99–110 (2015)
Vyatkin, V.: Software engineering in industrial automation: state-of-the-art review. IEEE Trans. Ind. Inf. **9**(3), 1234–1249 (2013)
Weilkiens, T.: Systems Engineering with SysML/UML: Modeling, Analysis, Design. Morgan Kaufmann, Waltham (2011)
Wirth, N.: What can we do about the unnecessary diversity of notation for syntactic definitions? Commun. ACM **20**(11), 822–823 (1977)

Chapter 12
Semantic Web Technologies for Data Integration in Multi-Disciplinary Engineering

Marta Sabou, Fajar J. Ekaputra, and Stefan Biffl

Abstract A key requirement in supporting the work of engineers involved in the design of *Cyber-Physical Production Systems* (CPPS) is offering tools that can deal with engineering data produced across the various involved engineering disciplines. Such data is created by different discipline-specific tools and is represented in tool-specific data models. Therefore, due to this data heterogeneity, it is challenging to coordinate activities that require project-level data access. *Semantic Web technologies* (SWTs) provide solutions for integrating and making sense of heterogeneous data sets and as such are a good solution candidate for solving data integration challenges in multi-disciplinary engineering (MDE) processes specific for the engineering of cyber-physical as well as traditional production systems. In this chapter, we investigate how SWTs can support multi-disciplinary engineering processes in CPPS. Based on CPPS engineering use cases, we discuss typical needs for intelligent data integration and access, and show how these needs can be addressed by SWTs and tools. For this, we draw on our own experiences in building Semantic Web solutions in engineering environments.

Keywords Multi-disciplinary engineering • Data integration • Semantic Web • Linked data • Ontology-based information integration

12.1 Introduction

The Industrie 4.0 initiative considers added-value processes that rely on strong data integration across various stakeholders, engineering disciplines, and engineering and operation phases; examples are the use cases and application examples in VDI/VDE (2014) and Industrie 4.0 WG (2013). The shorter lifecycles and higher

M. Sabou (✉) • F.J. Ekaputra
Technische Universität Wien, Favoritenstrasse 11, Wien, Austria
e-mail: Marta.Sabou@ifs.tuwien.ac.at; Fajar.Ekaputra@tuwien.ac.at

S. Biffl
Technische Universität Wien, Wien, Austria
e-mail: Stefan.Biffl@tuwien.ac.at

© Springer International Publishing AG 2017

S. Biffl et al. (eds.), *Multi-Disciplinary Engineering for Cyber-Physical Production Systems*, DOI 10.1007/978-3-319-56345-9_12

variation of products require better integration (a) between the *life cycles* of products and the associated production systems and (b) between the *engineering and operation phases* of these production systems. These issues have been discussed in detail during the introduction to *multi-disciplinary engineering* (MDE) for CPPS in Chap. 1. The more complex design-time and run-time variation points in the design of products and CPPS considerably enlarge the state space of these systems compared to traditional products and production systems. Therefore, MDE for CPPS has stronger data integration needs compared to the MDE of traditional, more or less fixed, production systems.

Semantic Web technologies (SWTs) are a family of knowledge-based approaches suitable to deal with the data heterogeneity aspects of CPPS and to enable advanced capabilities of such systems that inherently rely on data integration (e.g., handling disturbances, adapting to new business requirements). Unlike traditional knowledge-based approaches (Legat et al. 2013), SWTs aim to address data heterogeneity in Web-scale settings thus tackling challenges in terms of data size, heterogeneity, and level of distribution (Berners-Lee et al. 2001). SWTs enable a wide range of advanced applications (Shadbolt et al. 2006) and they have been successfully employed in various areas, ranging from pharmacology (Gray et al. 2014) to cultural heritage (Hyvönen 2012) and e-business (Hepp 2008).

In this chapter, we investigate Semantic Web-based data integration as an approach to cater for addressing heterogeneity needs in MDE scenarios specific to CPPS. For that, we draw on several years of experience in using SWTs for creating flexible automation systems with industry partners as part of the Christian Doppler Laboratory "Software Engineering Integration for Flexible Automation Systems" (CDL-Flex)[1].

Concretely, we aim to address two of the over-arching research questions that are core to this book and were specified in Chap. 1. First, we engage in "*the analysis of typical requirements for the integration of engineering project data coming from heterogeneous data sources*" (RQ I2.a). For this, we identify in Sect. 12.2 concrete needs for semantic integration in CPPS settings and validate those needs with eight use cases and application examples introduced in VDI/VDE (2014) and Industrie 4.0 WG (2013).

Second, with respect to RQ I1 from Chap. 1, we introduce SWTs as examples of *methods and technologies that support the integration on information within and across value chains of products, production systems, and production technologies*, with a particular focus on horizontal data integration across MDE teams during the engineering of CPPS. To that end, we introduce in Sect. 12.3 the basics of SWTs with a special focus on their use for data integration and explain how core SWT capabilities address the industry needs detailed in Sect. 12.2. In Sect. 12.4, we sketch the current uptake of SWTs in representative approaches for CPPS engineering. Finally, we provide an example use case where SWTs were used to support data integration during the multi-disciplinary engineering phase of a production system, namely a hydro power plant (Sects. 12.5 and 12.6). We conclude in Sect. 12.7.

[1]CDL-Flex: http://cdl.ifs.tuwien.ac.at/

In this book, several chapters relate to the topics discussed in this chapter. Chapter 4 on Product Lifecycle Management Systems discusses data and information management issues arising from the advanced use of Model-Based Systems Engineering (MBSE) methods, which need to deal with data interfaces to several systems engineering disciplines. Section 4.2.2 on Model Representation summarizes standards for multi-disciplinary product engineering, such as STEP, *product manufacturing information,* JT, and SysML. Section 4.2.3 on Information management and integration discusses the role of semantic technologies as a neutral or intermediate layer between different areas of the development process and as foundation for data analytics to integrate product design data, production process data, and quality measurements.

Chapter 9 on Engineering Software Tool Chains discusses engineering data formats for the exchange of data between mechanical design, electrical design, and software design. Chapter 10 on the problem of standardized information exchange within production system engineering presents requirements for information exchange technologies for multi-disciplinary engineering settings. Chapter 15 on setting up a Multi-Disciplinary Knowledge Base for Deterministic Product Ramp-Up Processes reports on the application of Semantic Web technologies for mapping descriptions of device-independent production processes to descriptions of device-dependent production system capabilities as foundation for deciding on the feasibility to run a specified production process in a given production context. Chapter 16 on Model Quality Assurance for Multi-Disciplinary Engineering presents a change review process for multi-disciplinary models, which can also be applied to engineering models expressed with Semantic Web technologies.

12.2 Industry Needs for Semantic Web Technologies

Semantic integration is a foundation for engineering tool support across disciplines, based on isolated engineering tools and their data models. Biffl et al. (2016) have considered the research question on needs for semantic support in multi-disciplinary production system engineering.

In the following, we summarize four usage scenarios that illustrate important needs for semantic data integration, which pose challenges to the domain experts in their daily work. Then we describe these needs for semantic data integration. A more detailed description of these usage scenarios and needs can be found in Biffl et al. (2016). We validate the needs for semantic data integration with eight use cases and application examples introduced in VDI/VDE (2014) and Industrie 4.0 WG (2013).

Scenario 1, called *Discipline-Crossing Engineering Tool Networks*, focuses on the correct and consistent propagation of engineering data in a MDE context between engineering activities, engineers, and tools. A semantic challenge is the heterogeneous modeling of the discipline-specific views on the same objects (e.g., plant sensors, drives, and controllers). Currently, these common concepts, which appear in different forms in several disciplines, are not explicitly represented in discipline-specific and isolated tools, which makes data analysis across disciplines more difficult and dependent on the interpretation by human experts. Adequate modeling and integration of engineering knowledge would provide the foundation for better production systems quality and for better engineering tool networks. For this scenario, the needs N1 to N4 and N6 (see Table 12.1) for engineering knowledge modeling and integration capabilities are relevant.

Scenario 2, called *Use of existing Artifacts for Plant Engineering*, concerns the reuse and protection of knowledge represented in the "digital shadows" of components in technical systems during engineering. Key issues concern the description of the reuse requirements for production systems and the capabilities of reusable devices and components for the effective and efficient identification and preparation of reusable production system components. A mapping method for the evaluation of component models to decide about the potential usability of the component within a production system requires well-defined semantics for production system components in the production system hierarchy to improve the quality and efficiency of production systems engineering. For this scenario, the needs N1 to N6 (see Table 12.1) for engineering knowledge modeling and integration capabilities are relevant.

Scenario 3, called *Flexible Production System Organization*, concerns the Industrie 4.0 value chain processes "Commissioning" and "Use for production", in particular the run-time flexibility of production systems during the operation phase. The dynamic integration or change of components within the production system at run time requires well-defined semantics for describing the capabilities, access paths, and control-related features of these components. In addition, the CPPS needs the capability to reason about the information provided by the component and to integrate the component at run time into the target production processes. This capability requires the integration of advanced knowledge about the production system and the product within the production system control system at production system run time. For this scenario, the needs N1 to N3, N5, and N6 (see Table 12.1) for engineering knowledge modeling and integration capabilities are relevant.

Scenario 4, called *Maintenance and Replacement Engineering*, focuses on the "Maintenance and decomposition planning" and "Maintenance" phases of the Industrie 4.0 value chains. Automation support for the assessment of the impact of changes to selected plant components or devices requires strong integration of the diverse kinds of knowledge coming from several roles in the engineering process with the maintenance knowledge during production system operation. Maintenance and repair strategies require the combination of engineering and run-time information of a production system to achieve improved maintenance capabilities of production system components. Such scenarios require a common semantics of engineering and run-time information related to system components and devices. General behavior models of components are required, which exploit engineering information and specific system knowledge, and can be combined with run-time information coming from the production system. For this scenario, all needs N1 to N7 (see Table 12.1) for engineering knowledge modeling and integration capabilities are relevant.

Table 12.1 Production system engineering use cases (Vx) and application examples (Ix) and needs for semantic engineering knowledge modeling and integration capabilities (Nx)

Production system engineering needs and Use cases/application examples		V1	V2	V3	I1	I2	I3	I4	I5
N1	Explicit engineering knowledge representation	++	++	++	++	++	++	++	++
N2	Engineering data integration	++	++	++	++	++	++	++	++
N3	Engineering knowledge access and analytics	++	++	++	++	++	++	++	++
N4	Efficient access to semi-structured data in the organization and on the Web	+	++	++		++	+	+	++
N5	Flexible and intelligent engineering applications	++	++	++	++	++	++	++	++
N6	Support for multi-disciplinary engineering process knowledge		+	++		++	++		
N7	Provisioning of integrated engineering knowledge at production-system run time	++	++	++	++	++	++	++	++

++ ... strong need; + ... considerable need; empty cell ... no significant need

The four scenarios assume strong integration of knowledge coming from several roles in the engineering and maintenance process based on the well-defined semantics for a set of shared common concepts allowing to combine engineering views, such as the production plant topology, mechanical construction geometry and kinematics, electrical communication and wiring, behavior information on processes and production resources, and PLC control code.

From these four scenarios, Biffl et al. (2016) derived the following seven needs (Nx) for semantic engineering knowledge modeling and integration capabilities.

N1—Explicit Engineering Knowledge Representation The expressiveness of the modeling approaches currently in use is not sufficient for the expression of knowledge needed to automate engineering processes for CPPS engineering. A key requirement of all scenarios is well-defined semantics of engineering and operation knowledge coming from several disciplines and tools for explicit knowledge representation and design.

N2—Engineering Data Integration In MDE, the heterogeneous data models in typical engineering tools use different terms and data formats for similar concepts, which makes the parallel engineering of CPPS more difficult, costly, and risky. A key requirement of all scenarios is the semantic integration of heterogeneous engineering and operation knowledge coming from several disciplines and tools.

N3—Engineering Knowledge Access and Analytics In the MDE of CPPS, domain experts in engineering and operation need to access and analyze the integrated data model based on the capabilities to provide formally represented and integrated engineering. A requirement of all scenarios are effective and efficient mechanisms for querying engineering model versions and for defining and evaluating engineering model constraints and rules across MDE views.

N4—Efficient Access to Semi-structured Data in the Organization and on the Web MDE domain experts use many data sources, while engineering process automation mostly uses structured data that follow a metamodel. A requirement of most scenarios is efficient automated access to semi-structured data, such as technical fact sheets that include natural language text, or linked data, such as component information from a provider or in the organization to automate support for reuse processes.

N5—Flexible and Intelligent Engineering Applications (for the Automation of Engineering) Intelligent engineering applications, such as defect detection and constraint checking, can be designed based on the capability of knowledge access and analytics on an integrated production system plant model. A requirement of some scenarios are flexible engineering applications that are driven by the description of the production system and therefore can adapt to changes in a CPPS both at design time and at run time.

N6—Support for Multi-Disciplinary Engineering Process Knowledge In MDE, a goal is to increase the quality and efficiency of the MDE process by representing engineering process responsibilities and process states linked to the production system plant model. A requirement of all scenarios are an extension of the description of the CPPS with model versions and knowledge on process attributes for analysis and improvement, such as tentative changes.

N7—Provisioning of Integrated Engineering Knowledge at System Run Time In a CPPS context, domain experts and CPPS mechanisms need engineering knowledge at run time to assess in a situation, which needs changing the system, the set of options for a successful change. A requirement of some scenarios is timely access to integrated engineering knowledge at system run time to support applications that depend on reacting to real-time processes.

Table 12.1 provides an overview rating to what extent the eight Industrie 4.0 use cases and application examples on CPPS engineering from VDI/VDE (2014) and Industrie 4.0 WG (2013) require the needs for semantic engineering knowledge modeling and integration capabilities, N1 to N7. In the following, we will discuss for each use case and application example the rationale for the ratings given in Table 12.1.

V1. ***VDI Use Case 1*** *Optimization of batch processes in a mid-sized company* (VDI/VDE 2014). The core goal of this use case is to improve the optimization of the adaptation of batch process recipes for the production in a specific industrial plant by using external computation power.

The solution approach builds on expertise from the disciplines of production process engineering and simulation as well as software service engineering to enable the description of recipes and plants as input to a simulation service that allows adapting the recipe to the needs of a local production engineer.

In this use case, explicit engineering knowledge representation (N1) and engineering data integration (N2) on the production process and on the plant characteristics are strongly relevant to enable engineering knowledge access and analytics (N3) for the optimization of the production process settings. In the description of the use case, the need for efficient access to semi-structured data in the organization and on the Web (N4) is not explicitly stated but probably useful, e.g., to interpret data from technical fact sheets coming from technology providers. The envisioned system is a flexible and intelligent engineering application (N5), which provides integrated engineering knowledge at production-system run time (N7).

V2. ***VDI Use Case 2*** *Plug-and-produce in modular industrial plants* (VDI/VDE 2014). Core goal of this use case is to reduce the engineering effort needed during the operation phase to exchange a processing module against a functionally equivalent new module by using external computing and simulation power.

The solution approach builds on expertise from the disciplines of CPPS engineering, including the coordination of modules, module control, and visualization to enable the description of requirements for and capabilities of modules as input to automatically linking the module interfaces to the interfaces of the production system and adapting the control, communication, and visualization as needed.

In this use case, the needs N1 to N5 and N7 are strongly relevant to support the formal description and model analysis of a module to enable automating engineering processes during the operation phase that currently depend strongly on human engineering and configuration activities. For describing the tentative use of processing modules in a solution option, N6 is relevant for the extension of the plant structure with engineering process elements.

V3. ***VDI Use Case 3*** *Self-correction of a discrete manufacturing process* (VDI/VDE 2014). Core goal of this use case is to reduce the downtime of production by enhancing a machine for punching and bending with measurement technology for the self-correction of tool parameters to counter changes in the production environment.

The solution approach builds on expertise from the disciplines of the process engineer, the tool vendor, and the CPPS engineer to define system goals for the process, such as minimal process duration or minimal energy consumption, to describe how changes in tool parameters may have an impact on achieving these system goals, and to design a distributed system to collect the data for the optimization from heterogeneous sources.

In this use case, the needs N1 to N7 are strongly relevant to enable the description of system goals, the means to achieve these system goals, the mechanism for reasoning on tentative solution options, including access to external data sources and services for the optimization.

I1. ***Industrie 4.0 WG Example Application 1*** *Energy consumption by a vehicle body assembly line* (Industrie 4.0 WG 2013). Core goal in this example application is to reduce the energy consumed by a vehicle body assembly line while the line is not in use by envisioning new versions of the involved machines that allow their systematic control for going into an energy-saving mode.

The solution approach builds on expertise from the disciplines of machine tool vendors, process engineers, and CPPS engineers to provide advanced machines that can be systematically powered down during breaks in production for better energy efficiency while keeping the system ready to restart production.

In this example application, the needs N1 to N3 are strongly relevant to enable the description of production system goals, machine capabilities and their control, and a mechanism to optimize and coordinate a set of machines at run time. The needs N5 and N7 concern the design of an engineering application, which will provide integrated engineering knowledge at run time as foundation for data-based optimization at run time based on sensor data.

I2. ***Industrie 4.0 WG Example Application 2*** *End-to-end system engineering across the entire value chain* (Industrie 4.0 WG 2013). Core goal in this visionary example application is to allow manufacturing individual products by adapting the IT support for systems engineering to enable a global overview from the perspective of the product that is manufactured.

The solution approach builds on expertise from the disciplines of software tool vendors and CPPS engineers to evolve the IT landscape from the current state of IT systems with a variety of interfaces that are hard to change and maintain towards the vision of end-to-end systems engineering, in which software tools from several vendors work together seamlessly.

In this example application, the needs N1 to N3 are strongly relevant to enable the integrated description of the production system and alignment with the interfaces of the relevant IT tools used. Some of these IT support systems will be intelligent engineering applications (N5), which need access to the engineering process knowledge (N6), probably also to semi-structured data in the organization and on the Web (N4). To support the complete value chain, the engineering knowledge has to be provided at run time (N7).

I3. ***Industrie 4.0 WG Example Application 3*** *Supporting custom manufacturing: how an individual customer's requirements can be met* (Industrie 4.0 WG 2013). Core goal in this example application is to enable a producer for reacting to last-minute requests for changes prior or during production based on the customer- and product-specific coordination of MDE with business value chains.

The solution approach builds on expertise from the disciplines of *manufacturing execution system* (MES) tool vendors and CPPS engineers to evolve product line engineering and manufacturing support from a solution tightly coupled with the production line hardware to a flexible product line that allows mixing and matching the equipment and production resources from several parts of the product line.

In this example application, the needs N1 to N3 are strongly relevant to enable the integrated description of the CPPS and the product lines. The flexible MES will be an intelligent engineering application (N5), which needs access to the engineering process knowledge (N6), such as changes to the product line, probably also to semi-structured data in the organization and on the Web (N4). To support the changes during production, the engineering knowledge has to be provided at run time (N7).

I4. ***Industrie 4.0 WG Example Application 4*** *Telepresence for manufacturing system diagnosis and maintenance* (Industrie 4.0 WG 2013). Core goal in this example application is to make manufacturing system maintenance more effective and more efficient by introducing a telepresence platform that provides advanced services to support manufacturing systems in finding appropriate experts for their diagnosis and maintenance.

The solution approach builds on expertise from the disciplines of service providers, manufacturing system tool vendors, and CPPS engineers to evolve the current business process of servicing the machines of one vendor to the coordinated data-driven servicing of a manufacturing system, probably consisting of machines coming from several vendors.

In this example application, the needs N1 to N3 are strongly relevant to enable the integrated description of the manufacturing system, a CPPS, and the servicing functionality for machines coming from several vendors. The telepresence platform will include intelligent engineering applications (N5), which probably will need access to semi-structured data in the organization and on the Web (N4). To support changes to the manufacturing system during production, the engineering knowledge has to be provided at run time (N7).

I5. ***Industrie 4.0 WG Example Application 5*** *Sudden change of supplier during production due to a crisis beyond the manufacturer's control* (Industrie 4.0 WG

2013). Core goal of this example application is to improve the fallback planning of a sudden change of supplier during production by using external simulation power.

The solution approach builds on expertise from the disciplines of production process engineering and simulation as well as software service engineering to enable the description of production processes and related value chain process steps as input to a simulation service, which allows adapting the production process and logistics to the needs of a production engineer facing the sudden change of supplier during production.

In this example application, explicit engineering knowledge representation (N1) and engineering data integration (N2) on the production process and on related value chain process steps are strongly relevant to enable engineering knowledge access and analytics (N3) for the simulation of the change impacts. In the description of the example application, the need for efficient access to semi-structured data in the organization and on the Web (N4) is not explicitly stated but probably very useful, e.g., to interpret data from technical process descriptions. The envisioned simulation system is a flexible and intelligent engineering application (N5), which provides integrated engineering knowledge at production-system run time (N7).

In summary, Table 12.1 indicates that the use cases and application examples in the Industrie 4.0 research roadmaps (VDI/VDE 2014) and (Industrie 4.0 Working Group 2013) show significant relevance of the needs identified in Biffl et al. (2016). All use cases and application examples strongly require N1 to N3, N5, and N7. Some use cases and application examples strongly require N4 and N6. In the next section we investigate to what extent Semantic Web technologies provide capabilities that can address the needs identified in this section. To that end, we introduce the basics of these technologies first and then discuss how their core capabilities address the various needs N1–N7.

12.3 Semantic Web Technologies: Key Concepts and Capabilities

The core motivation behind Semantic Web technologies is to improve information access on the Web. For example, the large size and diversity of Web data are key challenges for finding complex information with high precision by using simply keyword-based search mechanisms. SWTs aim to augment the traditional Web consisting of textual Web pages, with a semantic layer (Berners-Lee et al. 2001). This semantic layer contains a description of the Web data in a format that is easier to read and interpret for computer programs than textual information. As a result, this semantic layer has the potential to enable advanced information access tasks on the Web. For example, complex semantic search algorithms can be realized which handle queries that are more complex and lead to more precise results than keyword-based search on textual data.

In this section, we introduce the key elements of SWTs (Sect. 12.3.1) and focus on explaining the technological aspects of data integration with SWTs (Sect. 12.3.2). We conclude this chapter by enumerating a set of SWT capabilities relevant for addressing the CPPS engineering specific needs for engineering knowledge modeling and integration derived in Sect. 12.2 (Sect. 12.3.3).

12.3.1 Key Elements of Semantic Web Technologies

Semantic Web and Linked Data technologies aim (1) to enrich data with semantic information in a format that machines can process, (2) to publish it using Web based languages, and (3) to provide advanced data analytics capabilities that rely on reasoning capabilities (Shadbolt et al. 2006). As such, these technologies are highly suitable for large-scale data integration and analysis of heterogeneous and distributed datasets. Such distributed and heterogeneous datasets are often used for storing data resulting from production systems engineering processes. This section provides an overview of the core elements of Semantic Web technologies.

Expressing and Encoding Meaning with Ontologies A core element of SWTs are ontologies (Gruber 1993), formal domain models describing concepts in a domain and their relationships using logics based formalisms so that computer programs can process and reason with these descriptions. For example, a mechanical engineering ontology, such as depicted in Fig. 12.1, could describe concepts such as Conveyer or Turntable. Data items (e.g., a specific belt conveyor referred to as Conv1) are then described in terms of ontology concepts (e.g., by associating Conv1to the concept BeltConveyor by means of the instanceOf relation).

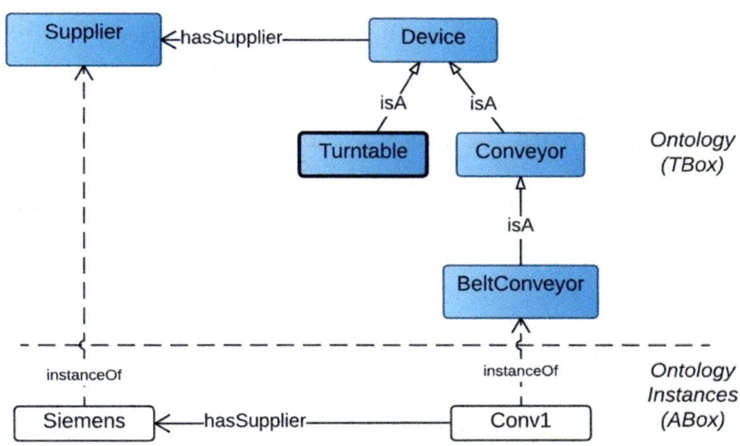

Fig. 12.1 Example ontology in the mechanical domain

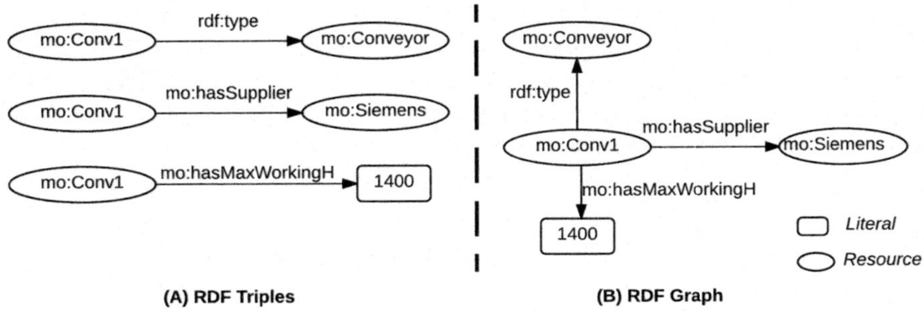

Fig. 12.2 Example RDF Triples (**a**) and their integration in an RDF Graph (**b**)

Use of Global Identifiers Since the goal of SWTs is to make data public on the Web, ontology elements as well as each data element to be described in the Semantic Web are assigned a unique web URL, for example, http://data.example.eu/dataset/Conv1, for the data element Conv1 or http://data.example.eu/ontology/BeltConveyor for the BeltConveyor concept of the ontology. The structure of the URLs usually indicates the name of the dataset (in this case example) as well as the type of the entity, which can be either part of the abstract data model (i.e., the ontology) or of the dataset.

Semantic Web Knowledge Representation Languages To represent Semantic Web specific data, a set of languages have been developed, most notably RDF[2] (Resource Description Format), RDF(S)[3] (RDF Schema) and OWL[4] (Web Ontology Language). OWL builds on RDF(S) but allows expressing more complex semantics than RDF(S). While relational databases rely on a relational (i.e., table like) data model, Semantic Web specific languages adopt a triple (or graph based) model with data being represented as triples. For example, to declare that Conv1 is a BeltConveyor, a triple is created stating that $<$ Conv1, isA, BeltConveyor $>$. Figure 12.2 illustrates triples that refer to the Conv1 resource (part A) and show how these are combined into an equivalent graph based structure (part B).

Formality and Reasoning All Semantic Web specific languages are based on formal logics and possess an associated semantics that enables performing reasoning activities. For example, OWL is based on Description Logics (Baader et al. 2003) and has a model-theoretic semantics. This enables the following reasoning tasks: subsumption checking (e.g., to deduce super- or sub-class relations between ontology classes based on their definitions); consistency checking (i.e., to detect logical

[2]RDF: https://www.w3.org/RDF/

[3]RDF(S): https://www.w3.org/TR/rdf-schema/

[4]OWL: https://www.w3.org/TR/owl2-overview/

contradictions within an ontology); or instance classification (i.e., identifying the most appropriate ontology class for a specific instance).

For a more in-depth presentation of SWTs, we refer the interested reader to (Sabou 2016).

12.3.2 Data Integration with Semantic Web Technologies

Semantic Web technologies are well suited to support large-scale data integration scenarios (Wache et al. 2001; Noy 2004). Ontologies can be used to provide a semantic bridge for information integration. Concretely, ontologies are often developed with the goal to support data integration (Noy 2004). For example, developers of several applications can agree on a general ontology and then extend this ontology with concepts and properties specific to their own applications. Since individual applications share a common semantic ground, this enables easily finding correspondences between them and therefore integrating their data. A set of high level ontologies such as SUMO (Niles and Pease 2001) and DOLCE (Gangemi et al. 2003) have been developed specifically for supporting data integration scenarios.

Wache et al. (2001) identify three typical approaches of ontology-based data integration. First, in the *single ontology* approach, one global ontology is used as a reference model for specifying the semantics of various data sources that need to be integrated. This approach is best suited when the integrated datasets are semantically close and it is feasible to define a shared vocabulary for their integration. Second, if the semantic gap between the various datasets is too broad to grant the creation of a single over-arching ontology, a *multiple ontology* approach can be followed. In this case, each source will be semantically described by its own local ontology and then mappings will be declared between these local ontologies. Finally, the *hybrid ontology* approach combines the previous two approaches: local ontologies are defined for each data source and then integrated through a shared, global ontology. In this case, mappings are established between the local and global ontologies.

The ability to define links and transformations between ontologies is a key enabler for data integration. *Ontology matching* techniques (Euzenat and Shvaiko 2013) are examples of such mechanisms for defining correspondences and links between ontologies. Indeed, when combining data sources that are described according to different ontologies, a set of mappings can be defined between ontology elements (Kovalenko et al. 2013). An ontology mapping specifies how elements of two ontologies relate, for example, that concept Weight in a mechanical engineering ontology has the same meaning as concept Mass in an electric engineering ontology. Through such mappings, joint terminologies can be established between diverse disciplines both for (1) improving the communication of experts but also (2) thanks to the formal nature of ontologies, for automatically integrating engineering discipline-specific tool data. Furthermore, thanks to their explicit nature (i.e., being declared as opposed to hard-coded), ontology models and mappings can be reused from one project to another.

Additionally to ontology matching, *Linked Data* technologies support data integration at Web-scale. To that end, they rely on a stack of technical standards for publishing, querying, and annotating ontological information on the Web e.g., RDF(S), OWL, SPARQL (SPARQL Query Language for RDF)[5]. Since with Linked Data technologies datasets are made available online through Web based standards, links can be explicitly specified and recorded between the elements of these datasets (similarly to hyperlinks in HTML). Most often, an owl : sameAs link is created between URLs from different datasets that represent the same real-life entity. These links enable computer programs to understand that two syntactically different terms refer to the same entity, thus again facilitating data integration. Due to the formal semantics of the mapping constructs, reasoning mechanisms can exploit these cross-model links to discover new knowledge that is only implicitly represented. For example, if the mo : Device concept of the example ontology is declared equivalent to the Component concept of another ontology, by virtue of reasoning, it can be deduced that any subconcept of mo : Device (e.g., mo : Turntable, mo : Conveyer) is also a subconcept of Component. This ability to reason about cross-dataset links enables data integration applications.

12.3.3 Semantic Web Capabilities

Taking into consideration the technology details described in Sect. 12.3.1, in (Sabou 2016) we introduced a set of SWT capabilities that are important for addressing the aspects of heterogeneity in CPPS engineering (Sabou 2016). To that end, we took the set of ontology-specific technology features identified by Oberle (2014) as a starting point for defining these capabilities and revised those from the perspective of the engineers' needs. Therefore, they are not always purely technical capabilities but rather useful functionalities that can support the various needs of engineering scenarios introduced in Sect. 12.2. We now discuss these Semantic Web capabilities and how they support typical industry needs (from Sect. 12.1). An overview of our analysis is depicted in Table 12.2.

C1—Formal and Flexible Semantic Modelling Semantic (or conceptual) modelling is achieved with *ontologies*, which facilitate capturing a universe of discourse with their modelling primitives (classes, properties, and instances). Ontology models are represented by means of formal, logics-based knowledge representation languages that assign unambiguous meaning to modelling constructs. By formally explaining the meaning of data, ontologies make data easier to understand to a wider range of users, both humans and machines.

[5]SPARQL: https://www.w3.org/TR/rdf-sparql-query/

Table 12.2 Needs for engineering knowledge modeling and integration (Nx) vs. Semantic Web technology capabilities (Cx)

Needs vs. Semantic Web capabilities		C1—Formal and flexible semantic modeling	C2—Intelligent, web-scale knowledge integration	C3—Browsing and exploration of distributed data sets	C4—Quality assurance of knowledge with reasoning	C5—Knowledge reuse
N1	Explicit engineering knowledge representation	++	+			+
N2	Engineering data integration	+	++			+
N3	Engineering knowledge access and analytics	++	++	+		++
N4	Efficient access to semi-structured data in the organization and on the Web	+	+	++		
N5	Flexible and intelligent engineering applications	+	++	+	++	
N6	Support for multi-disciplinary engineering process knowledge	+	++	+	++	++
N7	Provisioning of integrated engineering knowledge at production-system run time	+		+	++	+

Strong (++) and moderate (+) support of a need by a capability

C1 addresses the need for explicit knowledge representation (N1) and therefore it is marked with "++" in Table 12.2. Semantic models (1) enable reasoning and querying functionalities which provide access to and analytics on engineering knowledge (N3) and (2) facilitate data integration (N2) because meaningful relations between datasets can be clearly specified. Data integration can also be achieved with semi-structured data (e.g., technical fact-sheets including natural language) both within the organization and on the Web (N4) by specifying links to this data. Semantic engineering models support the creation of intelligent engineering applications (N5) and are a pre-requisite for addressing the need of providing integrated engineering knowledge at system run time (N7). Semantic modelling technique can help explicitly represent engineering process knowledge, such as engineering process responsibilities and process states (N6).

C2—Intelligent, Web-Scale Knowledge Integration This capability refers to the possibility of using SWTs to tackle data heterogeneity by automatically and flexibly solving complex data integration problems on a large scale. Examples from other domains include advanced applications that integrate many millions of data elements in pharmacology (Groth et al. 2014) or media publishing (Kobilarov et al. 2009). C2 further enhances the knowledge representation capabilities of SWTs as it allows specifying links between models (N1). C2 enables engineering data integration (N2) and knowledge access and analytics on this integrated data (N3) which are often used as a key ingredient of flexible and intelligent engineering applications (N5). Linked data facilitates access to semi-structured data in the organization and on the Web by providing mechanisms to interlink these resources (N4). Finally, C2 provides support for multi-disciplinary engineering process knowledge which is distributed and must be integrated for a meaningful insight onto project-wide processes (N6).

C3—Browsing and Exploration of Distributed Data Sets Linked Data technologies enable user-friendly browsing and exploration of distributed data sets (Hausenblas et al. 2014). In the context of engineering applications, this capability can be used to browse and explore both engineering models internal to an organization and external data sources, such as Web resources coming from third-party providers. C3 is a core requirement for efficient access to semi-structured data in the organization and on the Web (N4) and for engineering knowledge access through browsing, exploration and navigation. Browsing capabilities are important features of flexible and intelligent engineering applications (N5). By supporting sense-making and following links across engineering disciplines, capability C3 supports the increased productivity of multi-disciplinary engineering processes (N6). Finally, navigational data access interfaces can provide access to integrated engineering knowledge during the production-system run time (N7).

C4—Quality Assurance of Knowledge with Reasoning Given the mission-critical character of engineering projects, inconsistencies, defects and faults among diverse engineering models should be discovered as early as possible. Therefore, the quality assurance of engineering knowledge with advanced checks is highly relevant. SWTs address this requirement by formally representing (C1) and interlinking (C2) engineering knowledge and then automating a variety of quality assurance tasks through reasoning mechanisms. Semantic Web reasoning facilities can be used for supporting a wide range of tasks, but the ability to ensure quality (e.g., through consistency checks) is particularly important for the engineering domain.

Capability C4 allows detecting defects, faults and inconsistencies and as such it increases the productivity of multi-disciplinary engineering process (N6) while ensuring that high-quality engineering knowledge is provided at production-system run time (N7), often as part of intelligent engineering applications that both act intelligently (i.e., deduce new information from existing information through reasoning) and reliably with respect to quality assurance (N5).

C5—Knowledge Reuse One implication of the declarative nature of ontologies is that knowledge represented in ontologies can be easily reused among different application use cases. This reusability of knowledge is one of the fundamental concepts underlying SWTs (Simperl 2009), (Arp et al. 2015).

Knowledge reuse activities are often integral part of semantic modelling processes by reusing existing models (N1) and of data integration by reusing already specified mappings between semantic models (N2) thus increasing the productivity of multi-disciplinary engineering processes by avoiding creating this knowledge from scratch (N6). C5 strongly supports the need for engineering knowledge access and analytics (N3). Lastly, the provision of integrated engineering knowledge at production system run time can leverage on knowledge reuse (N7).

12.4 Adoption of Semantic Web Technologies in Multi-Disciplinary Engineering Settings

In previous work Sabou and Biffl (2016) and Sabou et al. (2016), we have found that *SWTs were successfully used to support various aspects of production system's engineering*, including requirements management (Feldmann et al. 2014), engineering artifact design optimization (Tudorache and Alani 2016), consistency management across diverse engineering models (Tudorache and Alani 2016; Steyskal and Wimmer 2016; Feldmann et al. 2016), creation of control system setup to enable product ramp-up (Willmann and Kastner 2016), simulation design and integration (Novák and Šindelár 2016) and project management (Grünwald et al. 2014). These diverse use cases from the various life cycle stages of production systems are enabled at a technical level by the following three individual tasks: model consistency checking, flexible comparison and model integration. We discuss these tasks in what follows.

Model consistency management is the task of detecting defects and inconsistencies in engineering models from individual disciplines as well as across inter-related models from diverse engineering disciplines. This task is particularly relevant in multi-disciplinary engineering projects to avoid that defects in artifacts of individual disciplines are propagated to related artifacts in other disciplines. For example, a sensor type specified in the physical topology model (mechanical engineering) must match the information in the corresponding electrical plan (electrical engineering) and the value range for control variables (software engineering). Defects may also arise from inconsistencies between disciplines without being defects in any of the single discipline views. Because these interdisciplinary relations are not represented in a machine-understandable way, they cannot be checked and managed easily with standard tool support. For example, Steyskal and Wimmer (2016) focus on consistency management among different, overlapping views (or models) of the same complex systems. The proposed technical solution relies on RDF to encode different system views in a uniform manner (making use of the C1 capability) and the emerging *Shapes Constraint Language* (SHACL)[6] to define the inter-viewpoint dependencies. By relying on *Reasoning* SWT capabilities (C4), these dependencies can be automatically checked during modeling time to uncover potential inconsistencies between the various models.

Flexible comparison refers to performing comparison among descriptions of engineering objects. In engineering settings, such comparisons are often performed between engineering objects that should be replaced or interchanged, e.g., a comparison between the capabilities of an engineering unit to be replaced (e.g., a device) and a new unit. For example, Feldmann et al. (2016) focus on ensuring compatibility between mechatronic modules that need to be replaced in a given system configuration. This use case requires means for (1) identifying modules compatible with a module that needs to be replaced and (2) identifying and resolving conflicts in a given system configuration as a follow up of a module change. For this, the authors propose using an ontology for representing compatibility information and encoding and checking compatibility through SPARQL queries (thus relying on capabilities C1 and C4). Similarly, flexible comparison enables recommending a control system setup for efficient product ramp-up processes (Willmann and Kastner 2016). This task requires assembling a production plan for the target production system by modifying and adjusting the production system of the source system in a way that it flexibly reuses artifacts (e.g., devices, configurations, raw materials) from the target site.

Model integration aims to bridge semantic gaps in engineering environments between engineering models. These models are often created by diverse engineering disciplines who use different terminologies that need to be semantically aligned. For example, a shared concern in the various works described above which aim to solve model consistency checking (Tudorache and Alani 2016; Steyskal and Wimmer 2016; Feldmann et al. 2016) is that solving data integration is a prerequisite for reaching their goal. Most authors chose an ontology-based data integration approach in line with Wache et al.'s *hybrid ontology model* (2001). Concretely, an

[6]SHACL: https://www.w3.org/TR/shacl/

ontology is built that captures the common concepts among engineering disciplines and plays the role of a semantic bridge to integrate data described in terms of discipline-specific local ontologies (C1, C2). For example, in their work to *detect inconsistencies* between different engineering models of the same system, Feldmann et al. (2016) observe that the engineering models created to describe the same system overlap to some extent. The overlaps between models should be kept consistent by defining which parts correspond to each other as a basis for compatibility checks. The proposed solution includes (1) defining a base vocabulary that contains the common concepts used by the various models and (2) using a common data representation language (namely RDF) to encode the various models in a uniform way and describe equivalent *mappings* between their corresponding elements.

To conclude, we observe that most approaches that use SWTs in various engineering settings rely on **data integration** as a pre-requisite for providing more advanced functionalities (e.g., consistency management). Therefore, in the rest of this chapter we focus on discussing data integration approaches with SWTs. In Sect. 12.5 we provide a use case that illustrates data integration needs in multi-disciplinary engineering settings and then illustrate the concrete SWT solution developed for this use case in Sect. 12.6.

12.5 Use Case: Engineering Data Integration in a Multi-Disciplinary Engineering Setting

To exemplify the use of SWTs for data integration, we consider a use case related to the development of a hydro power plant. Although a hydro power plant is not a CPPS itself, its development process provides a good example for a multi-disciplinary engineering setting which also characterizes CPPS. Engineers from different engineering disciplines work on their own part of the system but rely on data exchange to coordinate their work with other engineering teams. In particular, in our use case, engineers exchange data about *signal* information. Signals are one of the core information artifacts in the course of developing power plants and consist of structured key value pairs representing communication links between different power plant components. Depending on their size, hydro power plants manage 40,000–80,000 signals. Signal information is typically exported from the discipline-specific tools in machine-readable formats, such as Comma Separated Values (CSV) or eXtensible Markup Language (XML).

There is a strong need for data integration to alleviate the effects of the heterogeneity of terms used for the same concept across engineering disciplines and tools. For example, information about a CPU can be stored as part of the composite programmable logic controller (PLC) address in EPLAN[7] (electrical engineering

[7]EPLAN: http://www.eplan.at/

tool) data or as property LK_BSE in OPM (Object Process Methodology, typically used in mechanical engineering) data. Such heterogeneous representations of the same engineering artifact within diverse engineering models raise the need for propagating changes across data from different engineering disciplines and for detecting potential defects across engineering models. In this particular example, changes to the CPU at the mechanical level (e.g., replacement with a new version) must be communicated with the electrical engineers to update their models accordingly. At the same time inconsistencies in representing CPU data in diverse engineering models must be identified. However, a pre-requisite of change and inconsistency management is the meaningful integration of engineering data originating from different disciplines.

In our use case, there are three different **data sources** that need to be integrated for purposes of change and inconsistency management:

- Mechanical Engineering Data (OPM). The OPM tool is used in mechanical engineering to develop the plant topology and its components. OPM exports data as CSV files representing the design of the overall structure of the mechanical components. The exported file can also contain information about software components.
- Electrical Engineering Data (EPL). Electrical engineers use the EPLAN tool to develop the electrical component of the power plants. EPLAN also provides a CSV file export containing information about the electrical setup and its link to the mechanical components. It also contains specific information about electrical components, which may or may not be useful to stakeholders from other engineering domains.
- Project Management Data (PMD). Additionally, in many use cases it is important to consider project management data such as the engineering project, customer, and engineering activities, typically available as spreadsheet files.

In Sect. 12.6 we describe the SWT based data integration approach used in this use case.

12.6 A SWT-Based Solution for Data Integration

Our data integration approach in this use case was based on Wache et al.'s *hybrid ontology model* (2001). Concretely, the semantics of each individual data source is described by its own ontology (called *local ontologies*) and these local ontologies are mapped to a common ontology, which acts as a shared vocabulary across the terminologies of various disciplines. Our generic solution approach is depicted in Fig. 12.3. We will now discuss the main technical elements to realize this approach.

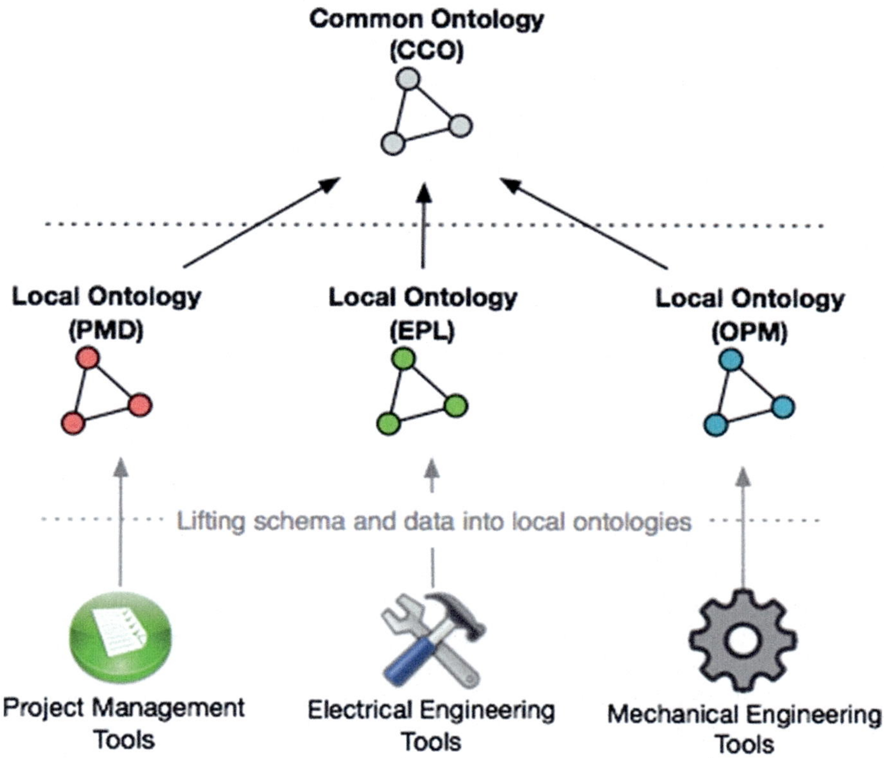

Fig. 12.3 Conceptual data integration solution

12.6.1 Ontologies Used for Data Integration

Local ontologies are obtained through an automatic transformation from the input CSV files to RDF documents. Currently, the resulting ontologies are semantically lightweight and primarily serve to enable easier data transformation between local data and common ontology.

The Hydro Power Plant *common concept ontology (CCO*[8]*)* represents the common information relevant for different engineering disciplines involved in the use case. As detailed in (Ekaputra et al. 2016), the CCO consists of two major parts. First, it contains concepts that describe organizational level aspects, including the *Project*, the *Customer* for whom the project is performed, as well as the *Engineers* (and *Engineering Roles*) necessary to realise the project. Engineers conduct *Engineering Activities,* which take as input and create as their output various *Engineering Documents* (e.g., signal lists, design documents). Engineering

[8]CCO: http://data.ifs.tuwien.ac.at/engineering/cco

documents are versioned and reviewed by the customer, thus constituting an important exchange medium between the customer, who requested a project, and the engineering team executing that project. Second, the CCO describes various *Engineering Objects* created during the engineering project, such as *Software Objects*, *Mechatronic Objects*, and *Electrical Objects*. *Physical Signals* and *Logical Signals* represent the links between engineering objects created by different engineering disciplines and how these diverse components can command or exchange data with each other.

12.6.2 Mappings Across Local and Common Ontologies

Mappings between the common and local ontologies ensure integrated access to local data through the common ontology. As detailed by Kovalenko and Euzenat (2016), diverse mapping scenarios arise between engineering data structures such as local and common ontologies, including: mapping between a property in one ontology and a (mathematical) combination of properties in other ontologies; or mappings between structures where different conceptualizations are used (e.g., mapping a concept in one ontology to a property in another). Kovalenko and Euzenat (2016) have distilled seven such transformation scenarios (and their variants) and showed that, among available technologies, SPARQL Construct, EDOAL[9] and SPIN[10] have the expressivity to express all these complex transformations. We chose SPARQL Construct because it as a W3C standard and many other languages (e.g., SPIN, or the upcoming SHACL language) use it as backend implementation. An excerpt of mappings between local ontologies (i.e., local electrical ontology and local mechanical ontology) and common concept ontology is shown in Fig. 12.4. Classes are defined in bold while properties are preceded by "-".

Kovalenko and Euzenat (2016) identified several types of ontology mappings that can exist between engineering data models. Figure 12.4 exemplifies a few types of these mappings, including:

(1) Direct mapping, e.g., Function_Text in the Electrical Ontology has the same meaning as longText in the Mechanical Ontology,
(2) Structural Granularity arises when different modeling elements need to be mapped to each other. For example, we mapped PLC_address, a property in the Electrical Ontology to the concepts of Rack, Channel, and Position in the CCO, and
(3) Datatype transformations specify some transformations between data values through mathematical or other custom-defined functions, e.g., transforming the

[9]EDOAL: https://ns.inria.fr/edoal/1.0/
[10]SPIN: https://www.w3.org/Submission/spin-overview/

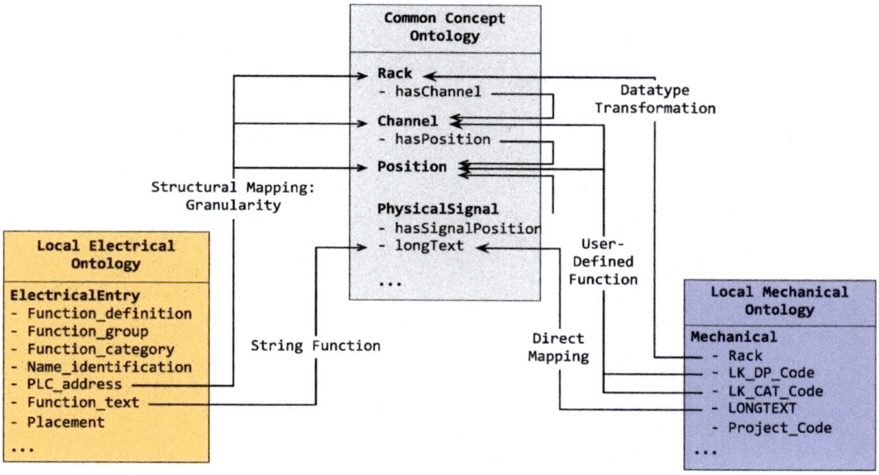

Fig. 12.4 An example mapping of local ontologies to common concepts

Table 12.3 Example data from local mechanical ontology

Mech. entry	Project code	Rack	LK_CAT code	LK_DP code	Long text
M1	KOM	2	5702	112	6 kV GENERAL DISTRIBUTION METERING PT FUSE TRIPPED
M2	KOM	2	5702	135	6 kV GENERAL DISTRIBUTION COUPLING CB OPENED

value of the Rack class in the CCO, which uses a certain encoding standard, to a value using another coding standard for the property Rack in the Mechanical Ontology.

As an example of data transformation based on this mapping, we show sample data from both local mechanical and electrical ontologies in Tables 12.3 and 12.4 respectively. From the electrical data and the mapping, we can derive the information about contained *Rack*, *Channel*, and *Position*. The *longText* property of *PhysicalSignal* can also be derived using a split string function in the *Function Text* field by the "\n" character.

From the mechanical data we can derive the information about *Rack* and *longText*. Additionally, we can derive the information about *Channel* and *Position* using a user-defined function. The transformation result from these two local ontologies using the mappings from Fig. 12.4 is shown in Fig. 12.5.

Table 12.4 Example data from local electrical ontology

Elec. entry	Name identification	PLC address	Function text
E1	=0CKN10GH001+0CKN01-A26:1	050.04.02.4.00	6 kV GENERAL DISTRIBUTION METERING PT FUSE TRIPPED\nKOM.0.BBA00. ED010.XM01
E2	=0CKN10GH001+0CKN01-A27:8	050.04.02.5.07	6 kV GENERAL DISTRIBUTION COUPLING CB OPENED\nKOM.0.BBA00. GS104.XB01

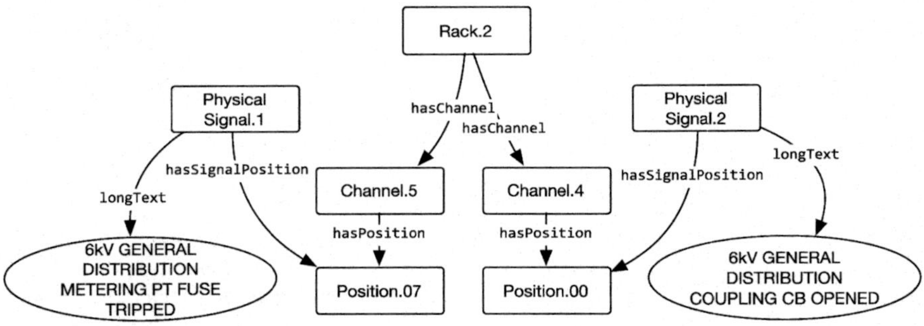

Fig. 12.5 Common concept ontology instances from two local ontology instances in Tables 12.3 and 12.4

12.6.3 Implementation Details and Functionality

In the implementation, we derived the process steps for engineering data integration by adapting the ontology based information integration approach (Wache et al. 2001, Calvanese et al. 2001). We utilize an IDEF-0[11] style diagram to structure the proposed approach, in which processes are shown as boxes and resources are shown as directed arrows (see Fig. 12.6). Input is shown as incoming arrows from the left hand side of the box, output is drawn as outgoing arrows to the right hand side of the box, consumable resources and stakeholders are depicted as input arrows from the bottom of the box and standards are indicated by incoming arrows from the top of the box used in the reference process.

[11]IDEF-0: http://www.idef.com/idefo-function_modeling_method/

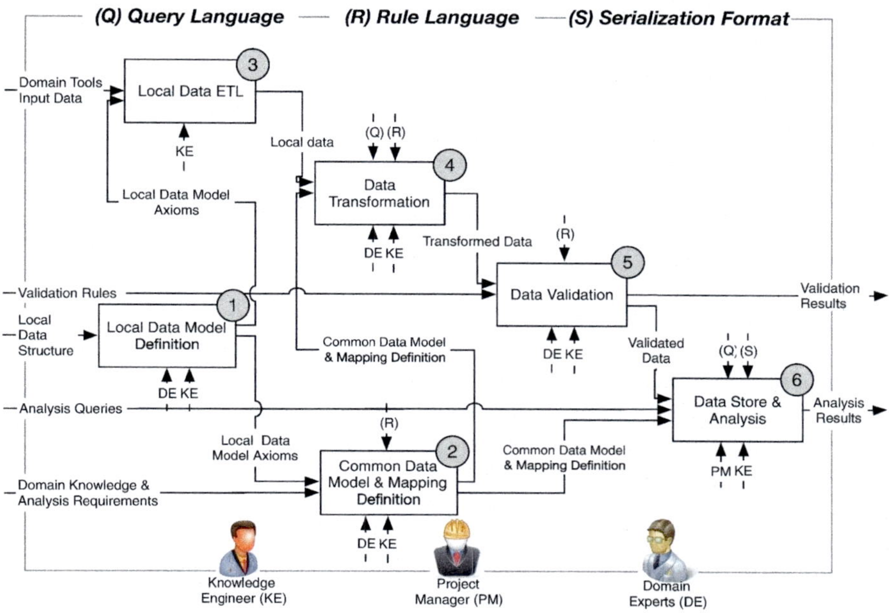

Fig. 12.6 An adapted OBII approach for Engineering Data Integration

We now explain each process step and its implementation within our prototype:

1. **Local Data Model Definition**
   This step encapsulates the process of defining local ontologies. These ontologies represent data coming from local tools (e.g., Eplan CSV data from electrical engineering). Due to the table-like structure of spreadsheet files exported from the local tools, we develop lightweight ontologies with only one class for each tool as this is sufficient to represent the data in this particular use case. Other use cases might require a richer local ontology for representing local data.
2. **Common Data Model & Mapping Definition**
   This step handles the process of developing a common data model and its mappings to the local data models. To support this goal, vocabularies and standards are required to formalize the data model and mappings. More details about the Common Concept Ontology used as the common data model and the mapping technology used for our prototype is available in Sects. 12.6.1 and 12.6.2.
3. **Local Data Model Extraction, Transformation and Load (ETL)**
   This step manages the process of extracting and transforming local data from engineering tools in spreadsheet format into the local ontologies previously defined in Step 1. In our prototype, the data is cleaned and transformed into these

local ontologies using the combination of OpenRefine[12] tool, CommonCSV[13] library and Jena[14] API.

4. **Data Transformation**

   This step conducts the data transformation process from local ontologies into common data model based on the mapping definition. Here, we utilized SPARQL constructs mentioned in Sect. 12.6.2 to execute this transformation. It has to be noted that in our prototype we do not consider data propagation, since it is a complex process and will be addressed separately. In addition to the engineering data, we also attach metadata information such as timestamp and data source.
5. **Data Validation**

   This process step validates newly transformed data according to validation rules previously defined by domain experts. The validation rules need to be transformed into a set of rule languages. In our prototype, the rules are represented as SPARQL queries. In the future, we are considering using Shapes Constraint Validation Language to replace SPARQL queries for data validation.
6. **Data Store and Analysis**

   The process step deals with storing validated data into a semantic triplestore (i.e., Apache, Jena TDB). All data is stored within in-memory datasets of Apache Jena TDB, separated into the three local (Project Management, Mechanical Engineering, and Electrical Engineering) and one common datasets (Common Concept). In the earlier version of the prototype, we used in-memory storage to ease the development process and to allow rapid prototyping. Later on we switched to transactional TDB storage for robustness and scalability reasons. Analysis queries then can be performed against both integrated data in the common and local datasets.

From the resulted integrated data in Common Concept ontology, we can conduct analysis of the overall production system since we can access both common concept and the local ontologies. Several sample analyses that can be conducted are:

- *Which signals contain different information about the same properties?*
- *Which signals have been changed in the last commit?*
- *How many racks have been added in the overall commit from both local ontologies?*

12.7 Summary

This chapter focused on the use of SWTs for data integration in MDE settings specific for the creation of CPPS. The first part of the chapter (Sect. 12.2) identified a set of typical industrial scenarios where data integration is needed and then synthesized

[12]OpenRefine: http://openrefine.org/

[13]CommonCSV: https://commons.apache.org/proper/commons-csv/

[14]Jena API: https://jena.apache.org/

needs for semantic integration. Engineers and managers from engineering domains can use these scenarios to select and adopt appropriate SWT solutions or alternative solution approaches in their own settings. The rest of the sections introduced SWTs (Sect. 12.3) and exemplified their use in selected approaches (Sect. 12.4) as well as in a concrete use case from the area of Hydro power plant development (Sects. 12.5 and 12.6). This material will support engineers and managers from engineering domains to get a better understanding of the benefits and limitations coming from using SWTs for data integration.

The overall conclusion is that Semantic Web technologies are well suited to address a range of various industry needs frequent in use cases and application examples introduced in (VDI/VDE 2014) and by the Industrie 4.0 WG (2013). Thanks to this suitability, SWTs have been adopted by various groups to solve diverse tasks in engineering settings, in particular related to model consistency management, flexible comparison and model integration. Data integration emerges therefore as one of the key valuable capabilities of SWTs especially supported by ontology matching techniques and Linked Data technologies. In the presented multi-disciplinary engineering use case, data integration enabled performing queries over data originating from both contributing engineering disciplines. This provides a basis for project wide data analysis and the identification of cross-disciplinary issues and anomalies.

Data integration solutions inherently bring up the challenge of how to propagate changes across the data from the various disciplines and how to manage these changes. Therefore, in future work we will investigate topics related to change management and propagation within datasets integrated using hybrid-ontology based approaches. Other topics of interest include easy integration with legacy systems and providing methods to more easily build and extract local and global ontologies.

Acknowledgments This work was supported by the Christian Doppler Forschungsgesellschaft, the Federal Ministry of Economy, Family and Youth, and the National Foundation for Research, Technology and Development in Austria.

References

Arp, R., Smith, B., Spear, A.D.: Building Ontologies with Basic Formal Ontology, 248p. MIT Press, Cambridge, MA (2015). ISBN: 978-0262527811

Baader, F., Calvanese, D., McGuinness, D.L., Nardi, D., Patel-Schneider, P.F. (eds.): The Description Logic Handbook: Theory, Implementation, and Applications. Cambridge University Press, New York, NY (2003)

Berners-Lee, T., Hendler, J., Lassila, O.: The Semantic Web. Sci. Am. 29–37 (2001)

Biffl, S., Lüder, A., Dietmar Winkler, D.: Multi-disciplinary engineering for Industrie 4.0: semantic challenges and needs. In: Biffl, S., Sabou, M. (eds.) Semantic Web for Intelligent Engineering Applications. Springer, Cham (2016)

Calvanese, D., De Giacomo, G., Lenzerini, M.: Ontology of integration and integration of ontologies. In: International Description Logics Workshop (2001)

Ekaputra, F.J., Sabou, M., Serral, E., Biffl, S.: Knowledge change management and analysis during the engineering of cyber physical production systems: A use case of hydro power plants. In: Proceedings of the 12th International Conference on Semantic Systems (Semantics) (2016)

Euzenat, J., Shvaiko, P.: Ontology Matching. Springer, Berlin (2013)

Feldmann, S., Rösch, S., Legat, C., Vogel-Heuser, B.: Keeping requirements and test cases consistent: towards an ontology-based approach. In: 12th IEEE International Conference on Industrial Informatics, INDIN, pp. 726–732 (2014)

Feldmann, S., Kernschmidt, K., Vogel-Heuser, B.: Applications of Semantic Web technologies for the engineering of automated production systems – three use cases. In: Biffl, S., Sabou, M. (eds.) Semantic Web for Intelligent Engineering Applications. Springer, Cham (2016)

Gangemi, A., Guarino, N., Masolo, C., Oltramari, A.: Sweetening WORDNET with DOLCE. AI Mag. **24**(3), 13–24 (2003)

Gray, A.J.G., Groth, P., Loizou, A., Askjaer, S., Brenninkmeijer, C., Burger, K., Chichester, C., Evelo, C.T., Goble, C., Harland, L., Pettifer, S., Thompson, M., Waagmeester, A., Williams, A.J.: Applying linked data approaches to pharmacology: architectural decisions and implementation. Semant. Web J. **5**(2), 101–113 (2014)

Groth, P., Loizou, A., Gray, A.J.G., Goble, C., Harland, L., Pettifer, S.: API-centric linked data integration: The open PHACTS discovery platform case study. Web Semant. Sci. Serv. Agents World Wide Web. **29**, 12–18 (2014)

Gruber, T.R.: A translation approach to portable ontology specifications. Knowl. Acquis. **5**(2), 199–220 (1993)

Grünwald, A., Winkler, D., Sabou, M., Biffl, S.: The semantic model editor: efficient data modeling and integration based on OWL ontologies. In: Proceedings of the 10th International Conference on Semantic Systems, SEM '14, ACM, pp. 116–123 (2014)

Hausenblas, M., Ruth, L., Wood, D., Zaidman, M.: Linked Data, Manning, 336p (2014). ISBN: 978-1617290398

Hepp, M. (2008) GoodRelations: an ontology for describing products and services offers on the web. In: Gangemi, A., Euzenat, J. (eds.) Proceedings of the 16th International Conference on Knowledge Engineering and Knowledge Management (EKAW), Acitrezza, Italy. LNCS, vol. 5268, pp. 332-347. Springer, Berlin

Hyvönen, E.: Publishing and Using Cultural Heritage Linked Data on the Semantic Web, Series: Synthesis Lectures on Semantic Web, Theory and Technology. Morgan and Claypool, San Rafael, CA (2012)

Industrie 4.0 WG: Industrie 4.0 Working Group. Securing the future of German manufacturing industry. Recommendations for implementing the strategic initiative. http://www.acatech.de/fileadmin/user_upload/Baumstruktur_nach_Website/Acatech/root/de/Material_fuer_Sonderseiten/Industrie_4.0/Final_report__Industrie_4.0_accessible.pdf (2013)

Kobilarov, G., Scott, T., Raimond, Y., Oliver, S., Sizemore, C., Smethurst, M., Bizer, C., Lee, R.: Media meets Semantic Web – How the BBC uses DBpedia and linked data to make connections. In: Aroyo, L., Traverso, P., Ciravegna, F., Cimiano, P., Heath, T., Hyvönen, E., Mizoguchi, R., Oren, E., Sabou, M., Simperl, E. (eds.) Proceedings of the 6th European Semantic Web Conference on The Semantic Web: Research and Applications (ESWC), pp. 723–737. Springer, Berlin (2009)

Kovalenko, O., Euzenat, J.: Semantic matching of engineering data structures. In: Biffl, S., Sabou, M. (eds.) Semantic Web for Intelligent Engineering Applications. Springer, Cham (2016)

Kovalenko, O., Debruyne, C., Serral, E., Biffl, S.: Evaluation of technologies for mapping representation in ontologies. In: Meersman, R., Panetto, H., Dillon, T., Eder, J., Bellahsene, Z., Ritter, N., De Leenheer, P., Dou, D. (eds.) On the Move to Meaningful Internet Systems: OTM 2013 Conferences, pp. 564–571. Springer, Berlin (2013)

Legat, C., Lamparter, S., Vogel-Heuser, B.: Knowledge-based technologies for future factory engineering and control. In: Borangiu, T., Thomas, A., Trentesaux, D. (eds.) Service Orientation in Holonic and Multi Agent Manufacturing and Robotics. Studies in Computational Intelligence, vol. 472, pp. 355–374. Springer, Berlin (2013)

Niles, I., Pease, A.: Towards a standard upper ontology. In: Proceedings of the international conference on Formal Ontology in Information Systems (FOIS '01), vol. 2001, pp. 2–9, ACM (2001)

Novák, P., Šindelár, R.: Ontology-based simulation design and integration. In: Biffl, S., Sabou, M. (eds.) Semantic Web for Intelligent Engineering Applications. Springer, Cham (2016)

Noy, N.F.: Semantic integration: a survey of ontology-based approaches. SIGMOD Rec. **33**(4), 65–70 (2004)

Oberle, D.: Ontologies and reasoning in enterprise service ecosystems. In: Informatik Spektrum 37/4 (2014)

Sabou, M.: An introduction to Semantic Web technologies. In: Biffl, S., Sabou, M. (eds.) Semantic Web for Intelligent Engineering Applications. Springer, Cham (2016)

Sabou, M., Biffl, S.: Conclusions. In: Biffl, S., Sabou, M. (eds.) Semantic Web for Intelligent Engineering Applications. Springer, Cham (2016)

Sabou, M., Kovalenko, O., Ekaputra, F.J., Biffl, S.: Semantic Web solutions in engineering. In: Biffl, S., Sabou, M. (eds.) Semantic Web for Intelligent Engineering Applications. Springer, Cham (2016)

Shadbolt, N., Berners-Lee, T., Hall, W.: The semantic web revisited. IEEE Intell. Syst. **21**(3), 96–101 (2006)

Simperl, E.: Reusing ontologies on the Semantic Web: a feasibility study. Data Knowl. Eng. **68**(10), 905–925 (2009)

Steyskal, S., Wimmer, M.: Leveraging Semantic Web technologies for consistency management in multi-viewpoint systems engineering. In: Biffl, S., Sabou, M. (eds.) Semantic Web for Intelligent Engineering Applications. Springer, Cham (2016)

Tudorache, T., Alani, L.: Semantic Web solutions in the automotive industry. In: Biffl, S., Sabou, M. (eds.) Semantic Web for Intelligent Engineering Applications. Springer, Cham (2016)

VDI/VDE: Industrie 4.0 – CPS-basierte Automation, Forschungsbedarf anhand konkreter Fallbeispiele. VDI/VDE Gesellschaft Mess- und Automatisierungstechnik, Statusreport Juli 2014. https://www.vdi.de/technik/fachthemen/mess-und-automatisierungstechnik/artikel/cps-basierte-automation-forschungsbedarf-anhand-konkreter-fallbeispiele/ (2014)

Wache, H., Voegele, T., Visser, U., Stuckenschmidt, H., Schuster, G., Neumann, H., Hubner, S.: Ontology-based integration of information – a survey of existing approaches. In Stuckenschmidt, H. (ed.) Proceedings of IJCAI Workshop: Ontologies and Information, pp. 108–117 (2001)

Willmann, R., Kastner, W.: Product ramp-up for semiconductor manufacturing: automated recommendation of control system setup. In: Biffl, S., Sabou, M. (eds.) Semantic Web for Intelligent Engineering Applications. Springer, Cham (2016)

Chapter 13
Patterns for Self-Adaptation in Cyber-Physical Systems

Angelika Musil, Juergen Musil, Danny Weyns, Tomas Bures, Henry Muccini, and Mohammad Sharaf

Abstract Engineering Cyber-Physical Systems (CPS) is challenging, as these systems have to handle uncertainty and change during operation. A typical approach to deal with uncertainty is enhancing the system with self-adaptation capabilities. However, realizing self-adaptation in CPS, and consequently also in Cyber-Physical Production Systems (CPPS) as a member of the CPS family, is particularly challenging due to the specific characteristics of these systems, including the seamless integration of computational and physical components, the inherent heterogeneity and large-scale of such systems, and their open-endedness.

In this chapter we survey CPS studies that apply the promising design strategy of combining different self-adaptation mechanisms across the technology stack of the system. Based on the survey results, we derive recurring adaptation patterns that structure and consolidate design knowledge. The patterns offer problem-solution pairs to engineers for the design of future CPS and CPPS with self-adaptation capabilities. Finally, the chapter outlines the potential of collective intelligence systems for CPPS and their engineering based on the survey results.

Keywords Collective intelligence systems • Cyber-physical systems • Patterns • Self-adaptation • Software architecture • Systematic study

---

A. Musil (✉) • J. Musil
Institute of Software Technology and Interactive Systems, Technische Universität Wien, Wien, Austria
e-mail: angelika@computer.org; jmusil@computer.org

D. Weyns
Department of Computer Science, KU Leuven, Leuven, Belgium

Department of Computer Science, Linnaeus University, Växjö, Sweden
e-mail: danny.weyns@kuleuven.be

T. Bures
Department of Distributed and Dependable Systems, Charles University Prague, Prague, Czechia
e-mail: bures@d3s.mff.cuni.cz

H. Muccini • M. Sharaf
DISIM Department, University of L'Aquila, L'Aquila, Italy
e-mail: henry.muccini@univaq.it; massharaf@yahoo.com

© Springer International Publishing AG 2017
S. Biffl et al. (eds.), *Multi-Disciplinary Engineering for Cyber-Physical Production Systems*, DOI 10.1007/978-3-319-56345-9_13

13.1 Introduction

Cyber-Physical Production Systems (CPPS) form a distinct sub-category of the more general family of *Cyber-Physical Systems (CPS)*. Though distinctively focused on production, CPPS, as a member of the CPS family, share many common traits with other types of CPS, such as distributed robotics, autonomous vehicular systems, smart grid, and smart spaces. These common traits include the strong coupling of physical environment and the computing system via sensors and actuators, involvement of humans-in-the-loop, the necessity of coping with a multitude of heterogeneous models (e.g., physical, electrical, mechanical, control), the need of real-timeness, and strong requirements on dependability.

The close relation to the environment, humans-in-the-loop and the complex interplay of the heterogeneous models brings a high level of uncertainty as a critical factor to be taken into account and addressed when designing CPS, and consequently also CPPS. Examples of uncertainty include the unpredictability of human actions, unexpected emergent behavior of the environment (typically stemming from unanticipated interactions among constituents of the environment and the CPS due to the fact that a CPS is an inherent part of the environment it observes and controls), unexpected or faulty interplay between CPS components, and incomplete requirements. The presence of uncertainty makes it difficult to design the complete behavior of a very complex CPPS with guaranteed dependability, as parts of the knowledge required for such a design may only become available at run time.

A viable software-based solution to the problem of uncertainty lies in equipping the system with self-adaptation capabilities. Self-adaptation adds introspective capabilities to the system allowing it to be aware of its internal state and structure, reason about itself and its goals, identify potential problems in its ability to dependably achieve its goals, and adapt itself to cope with the identified problems. Self-adaptation was already introduced in the area of enterprise systems by IBM in 2003 (Kephart and Chess, 2003). Similar concepts are nowadays regularly applied also for instance in cloud computing where an application automatically reconfigures to scale with the current load and to avoid virtual machines that perform badly due to resource sharing. For a guided tour through the history of the field of self-adaptation, we refer the interested reader to Weyns (2017).

As outlined in Chap. 1 of this book, the key research question addressing the *modelling of CPPS flexibility and self-adaptation capabilities (RQ C1)* discusses a very relevant topic for CPPS engineers. From the perspective of this research question, this chapter elaborates concretely on effective architectural approaches and best practices to combine self-adaptation mechanisms to handle uncertainty challenges and concerns. Although other chapters of this book also address a CPPS architecture perspective, we focus on the design of self-adaptation capabilities.

In this chapter, we aim at providing insight on how self-adaptation can be used in addressing uncertainty in CPPS. Since there is rather a general lack of knowledge on self-adaptation specifically in CPPS, we take a generalization step and overview self-adaptation related to the larger family of CPS. Since CPS are systems that do not focus on one layer of the technology stack, but their engineering crosses all layers, self-adaptation mechanisms are also relevant to be considered on all layers. This claim is supported by the results of a recent systematic literature review aiming at assessing state-of-the-art approaches to handle self-adaptation in CPS at an architectural level. The study revealed that, remarkably, 36% of the investigated studies combine different adaptation mechanisms across the technology stack to realize adaptation in a CPS (Muccini et al., 2016). Therefore, this chapter follows this promising architecture design strategy for CPS and focuses explicitly on combinations of different types of adaptation mechanisms that may span various layers within a system. We do so by the means of a systematic literature mapping with the goal to identify recurring adaptation patterns used in addressing uncertainty by self-adaptation. We further relate these patterns to the specific field of CPPS to give an insight on how to exploit self-adaptation to CPPS. Finally, we outline the potential of *Collective Intelligence Systems (CIS)* for CPPS and their engineering based on the study results by presenting three emerging research directions.

The remainder of this paper is organized as follows: Sect. 13.2 introduces background information about uncertainty types, self-adaptation approaches, and collective intelligence systems. In Sect. 13.3 the research questions and research methodology we used are presented. Section 13.4 summarizes the method used to conduct the systematic mapping study. The results of the systematic mapping study are presented in Sect. 13.5, followed by a summary of threats to its validity in Sect. 13.6 and a reflection on the results in Sect. 13.7. Section 13.8 describes and discusses the three identified adaptation patterns in CPS. We further explore the potential of collective intelligence systems for CPS and CPPS in Sects. 13.9 and 13.10 summarizes related work. Finally, Sect. 13.11 draws conclusions and outlines future work.

13.2 Background

This section provides a general introduction to uncertainty types in adaptive systems, different adaptation approaches, its purpose and different methods, as well as collective intelligence systems as a promising enhancement to CPS and CPPS architectures.

13.2.1 Uncertainties

When designing CPS the available knowledge is often not adequate to anticipate all the run time conditions the system will encounter (e.g., missing or inaccurate knowledge regarding the availability of resources, concrete operating conditions that the system will face at run time, and the emergence of new requirements while the system is operating). To that end, Garlan (2010) argues that in today's software

Table 13.1 Uncertainty dimensions (Mahdavi-Hezavehi et al., 2016)

Uncertainty dimension	Description	Options
Location	Refers to the locale, where uncertainty manifests itself within the whole system	Environment, model, adaptation functions, goals, managed system, resources
Nature	Specifies whether the uncertainty is due to the imperfection of available knowledge, or is due to the inherent variability of the phenomena being described	Epistemic, variability
Level/Spectrum	Indicates the position of uncertainty along the spectrum between deterministic knowledge and total ignorance	Statistical uncertainty, scenario uncertainty
Emerging time	Refers to time when the existence of uncertainty is acknowledged or uncertainty is appeared during the life cycle of the system	Run time, design time
Sources	Refers to a variety of circumstances affecting the adaptation decision, which eventually deviate system's performance from expected behavior	Variety of options based on the sources of uncertainty (e.g., abstraction, model drift, etc. for model uncertainty; sensing, effecting etc. for adaptation functions)

systems uncertainty should be considered as a first-class concern throughout the whole system life cycle. In the context of adaptive systems, Ramirez et al. (2012) provide a taxonomy for uncertainty that describes common sources of uncertainty and their effect on requirements, design and run time phases of the system. Esfahani and Malek (2013) present an extensive list of sources of uncertainties with examples. Moreover, these authors investigate uncertainty characteristics, i.e., reducibility versus irreducibility, variability versus lack of knowledge, and spectrum of uncertainty. Perez-Palacin and Mirandola (2014) present another taxonomy for uncertainty for adaptive systems based on three dimensions: location, level, and nature of uncertainty. Mahdavi-Hezavehi et al. (2016) present a classification framework for uncertainty in adaptive systems, which is based on a systematic review of the literature. This classification is shown in Table 13.1.

One way to deal with uncertainties is to design systems that adapt themselves during run time, when the lacking knowledge becomes available. Adaptive systems are capable of autonomously modifying their run time behavior to deal with dynamic system context, and changing or new system requirements in order to provide dependable systems. However, realizing adaptation in CPS is particularly challenging due to specifics of these systems include the blurring boundaries between the system and its environment, large scale and inherent complexity, the role of end-users, multi-level uncertainty, open-endedness, among others (Bures et al., 2015).

13.2.2 Adaptation

Adaptive systems are capable of modifying their run time behavior in order to achieve systems objectives. Unpredictable circumstances such as changes in the system's environment, system faults, new requirements, and changes in the priority of requirements are some of the reasons for triggering adaptation action in a self-adaptive system. To deal with these uncertainties, an adaptive system continuously monitors itself, gathers data, and analyzes them to decide if adaption is required. Different paradigms for realizing adaptation have been developed. We summarize three paradigms that appeared in the study presented in this paper: architecture-based adaptation, multi-agent based approaches, and self-organizing based approaches. Examples of other adaptation approaches, out of scope of this chapter, are computational reflection and approaches based on principles from control theory.

13.2.2.1 Architecture-Based Adaptation

Architecture-based adaptation (Oreizy et al., 1998; Garlan et al., 2004; Kramer and Magee, 2007; Weyns et al., 2012) is one well-recognized approach that deals with uncertainties at run time. The essential functions of architecture-based self-adaptation are defined in the MAPE-K (i.e., Monitor, Analyze, Plan, Execute, and Knowledge component) reference model (Kephart and Chess, 2003). By complying with the concept of separation of concerns (i.e., separation of domain-specific concerns from adaptation concerns that deal with uncertainties), the MAPE-K model has shown to be a suitable approach for designing feedback loops and developing self-adaptive systems (Weyns et al., 2013a). One well-known architecture-based self-adaptive framework is Rainbow (Garlan et al., 2004). Rainbow uses an abstract architectural model to monitor software system run time specifications, evaluates the model for constraint violations, and if required, performs global or module-level adaptations. Calinescu et al. (2011) present a quality of service management framework for self-adaptive services-based systems, which augments the system architecture with the MAPE-K loop functionalities. In their framework, the high-level quality of service requirements are translated into probabilistic temporal logic formulae which are used to identify and enforce the optimal system configuration while taking into account the quality dependencies. Moreover, utility theory can be used (Cheng et al., 2006) to dynamically compute trade-offs (i.e., priority of quality attributes over one another) between conflicting interests, in order to select the best adaptation strategy that balances multiple quality requirements in the self-adaptive system.

13.2.2.2 Multi-Agent Based Approaches

Multi-agent systems belong to a class of decentralized systems in which each component (agent) is an autonomous problem solver, typically able to operate successfully in various dynamic and uncertain environments (Wooldridge, 2001). These agents interact to solve problems that are beyond their individual capabilities or knowledge. Multi-agent systems have features that are key to engineering adaptive systems, specifically loose coupling, context sensitivity, and robustness to failures and unexpected events (Weyns, 2010; Weyns and Georgeff, 2010). Agents are self-contained, goal-directed entities. They get their adaptability from goals. When multiple agents are available, a goal can be achieved by selecting among the agents at run time, for example using negotiation (Fatima et al., 2006), rather than requiring a hardwired design. An agent includes a specification of the situation or context in which it is appropriate or expected to achieve its target goal. A calling agent can simply post the goals it wishes to achieve and select only those agents appropriate to the goal and current processing context: the right agent at the right time in the right circumstances. Similarly, an agent's internal processes are typically associated with a context condition describing the situations in which the process can achieve its specified goal. This means that processes "self select" according to the desired goal and prevailing situation. Goal-directed multi-agent systems eliminate most of the complexity needed for handling failures (Minsky and Murata, 2004). Failures and unexpected events cause the original goal to be reposted and tried again, without the need for explicit exception handling. The goal-directed mechanism will automatically try them until success or ultimate failure.

13.2.2.3 Self-Organizing Based Approaches

Self-organization is a dynamic and adaptive process where a system acquires and maintains structure itself, without external control (De Wolf and Holvoet, 2004). The essence of self-organization is an adaptable behavior that autonomously acquires and maintains an increased order. Self-organizing systems exhibit the following essential properties: increase in order (exhibiting useful behavior), autonomy (absence of external control), robustness (adaptability in the presence of perturbations), and dynamicity (dynamics that handle changes). Self-organizing systems may expose emergent behavior at the global level that dynamically arises from the interactions between the parts at the local level. The engineering of self-organizing systems if often inspired by natural phenomena, for example from biology such as ant behavior and swarms (Di Marzo Serugendo et al., 2006). The principle idea is to exploit the robustness and flexibility of these natural systems as a metaphor for engineering computing systems. As an example, field-based coordination relies on virtual computational fields (e.g., distributed data structures), mimicking gravitational and electromagnetic fields, as the basic mechanisms with which to coordinate activities among open and dynamic groups of application components. This enables components to spontaneously interact with each other via the mediation of fields to self-organize their activity patterns in an adaptive way (Mamei et al., 2006; Weyns et al., 2008). In a recent paper, Bures et al. (2013) propose a component-based approach that exploits principles of self-organization. In this approach, autonomic components dynamically form so called ensembles that share data to organize themselves on the fly. The authors present the DEECo component model, a concrete realization of the approach.

13.2.3 Collective Intelligence Systems

In the last decades, a form of user-contribution-driven web platforms has intensively influenced the way of today's knowledge creation and sharing processes. Today, this kind of software systems is very popular and widespread in use in our daily lives. Well-known examples of such CIS include Facebook,[1] Wikipedia,[2] YouTube,[3] and Yelp.[4] CIS are *socio-technical multi-agent systems* that aim to harness the collective intelligence of interacting human actors by providing a web-based environment for sharing, distributing and retrieving topic-specific information in an efficient way (Musil et al., 2015a). A CIS posses a characteristic system model that is illustrated by Fig. 13.1. It consists of three layers: (1) a *proactive actor base*, (2) a *passive CI artifact network*, and (3) a *reactive/adaptive computational analysis, management and dissemination (AMD) system* (Musil et al., 2015b). Between the layers, the CIS realizes a perpetual feedback loop connecting the human actor base and the reactive computational coordination environment and consisting of two essential phases: aggregation and dissemination (Musil et al., 2015b). In the aggregation phase, the individual actor contributes explicitly or implicitly new content to so-called CI artifacts by performing defined local activities. These CI artifacts store the aggregated information in a defined structure and are part of a passive artifact network. Defined rules of coordination in the reactive and adaptive AMD system govern the processing and analysis of the artifact data as well as the extraction of consolidated information. In the following dissemination phase, the AMD system uses both active and passive

[1]http://www.facebook.com/ (last visited 01/15/2017).

[2]http://www.wikipedia.org/ (last visited 01/15/2017).

[3]http://www.youtube.com/ (last visited 01/15/2017).

[4]http://www.yelp.com/ (last visited 01/15/2017).

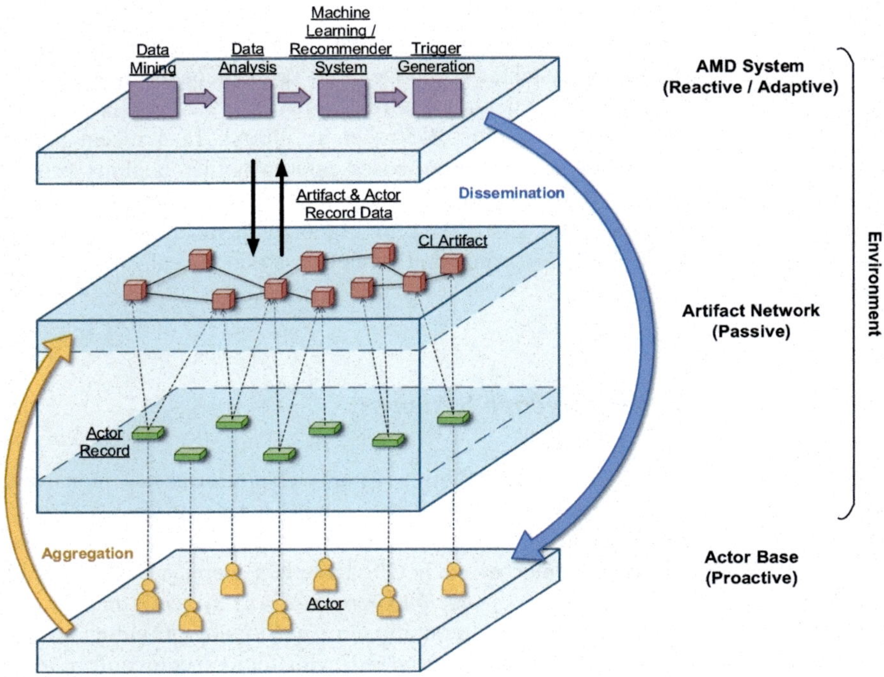

Fig. 13.1 Multi-layer CIS model with three main components and the stigmergic process (Musil et al., 2016a)

dissemination mechanisms to make the actors aware about artifact content changes and overall actor activities in the system environment as well as stimulate subsequent actor interaction. Thus the resulting bottom-up feedback loop constitutes a *stigmergic process* (Heylighen, 2016) enabling indirect, environment-mediated communication and coordination (Musil et al., 2015b). In addition, the stigmergic process enables self-organization which realizes adaptation within the CIS environment.

To support software architects in the design of CIS architectures, Musil et al. (2015c) proposed the *architecture framework for collective intelligence systems (CIS-AF)* as one approach that provides guidance for realizing new CIS solutions. The CIS-AF is developed as a methodology to efficiently describe the core elements of a CIS architecture, which are documented in the *Stigmergic Information System (SIS) architecture pattern* (Musil et al., 2015b), without being limited in its technical implementation.

13.3 Research Questions

This chapter basically focuses on contributions to answer *RQ C1—Modelling of CPPS flexibility and self-adaptation capabilities* specified in Chap. 1 of this book. Concretely, we aim to consolidate existing design knowledge on self-adaptation strategies to address uncertainty in CPPS and to identify novel and promising approaches that need further research. Since there is rather a general lack of knowledge on self-adaptation specifically in CPPS, we broaden the scope of our investigation and provide insight on how self-adaptation capabilities of the more general family of CPS are designed. Thus CPPS engineers can learn from application experiences with CPS for dealing with adaptation challenges and concerns in CPPS.

To address this goal, we identified the following research questions:

RQ1. *How is self-adaption applied in cyber-physical systems in general?*
We aim to analyze how state-of-the-art approaches make use of self-adaptation mechanisms and models to handle uncertainty while architecting CPS. In addition, we focus on self-adaptation applied in CPS in the manufacturing domain.

RQ2. *How can this knowledge be applied and exploited to cyber-physical production systems and their engineering?*
Based on a better understanding of existing developed adaptation strategies to address challenges and concerns of CPS, we seek to closely examine common approaches, considerations and advances to identify recurring patterns, models or tactics. The documentation of such architectural knowledge should support CPPS engineers with the realization and coordination of self-adaptation. In addition, this consolidated design knowledge base can provide a strong foundation for designing self-adaptation capabilities in CPPS engineering that can be further researched and extended by CPPS researchers.

RQ3. *Can principles from collective intelligence systems provide innovation for adaptive CPPS and CPPS engineering?*
CIS are complex adaptive socio-technical systems that apply stigmergic adaptation with humans-in-the-loop. They represent a well-known approach for adaptation used in particular, predominantly social, domains. This research question aims to go beyond the typical application contexts of CIS and to explore CIS capabilities applied in the environment of CPPS. This contributes to a new perspective on CPPS with the focus on social interactions and social dynamics as innovative enhancements.

To answer these research questions we applied an iterative research approach with three steps. In the first step, we reviewed the state-of-the-art in literature using a *systematic mapping study* method and consolidated existing design knowledge on self-adaptation strategies in CPS. The goal of the first step is to answer RQ1. In the second step, we synthesized and analyzed the collected knowledge to derive recurring adaptation *patterns* that can be applied for engineering CPPS.

The goal of the second step is to answer RQ2. In the third step, we explored a new perspective on adaptive CPPS by introducing *collective intelligence system principles* with humans, but also machines-in-the-loop. The goal of the third step is to answer RQ3.

13.4 Systematic Mapping Study Method

In order to get an overview of the current state of primary studies focusing on self-adaptation approaches in CPS on an architectural level and get insights into recurring patterns and models, we performed a systematic mapping study (SMS). To apply this research method in an unbiased, objective and systematic way, we followed the guidelines by Kitchenham and Charters (2007). In contrast to a systematic literature review, a SMS is applied to review a specific software engineering topic area and classifies primary research papers in that specific domain (Kitchenham et al., 2011). Thus the research questions for such a study are generally broader defined and more high level to provide an overview of a certain topic (Kitchenham et al., 2011). In the following we briefly summarize the performed study process and activities.[5] For detailed information about the study and its results, we refer the interested reader to the study protocol (Musil et al., 2016c) and the study website.[6]

The study started with defining an initial study protocol. Since the protocol is a critical element of a systematic study, it was piloted by reviewing a sample of 4 papers. In the following, the study protocol was revised with respect to the pilot results. Once all reviewers agreed on the protocol, the phase of conducting the SMS started by applying the search strategy and selection criteria, data extraction form, data analysis methods, and reporting strategy defined in the protocol.

Figure 13.2 shows the overall systematic mapping study process that we applied with the number of remaining papers after each phase of the study. The study was conducted by six researchers. Two reviewers defined the initial protocol. The retrieving and selecting publications process was performed by two other reviewers. Four reviewers extracted the data from the selected studies. Finally, they synthesized and analyzed the data as well as prepared the final study report. These final steps were crosschecked by the other two reviewers.

---

[5]If the reader is familiar with this research method, this section can be skipped.

[6]Supplementary material of the study is available at: http://qse.ifs.tuwien.ac.at/ci/material/pub/mde-cpps17/

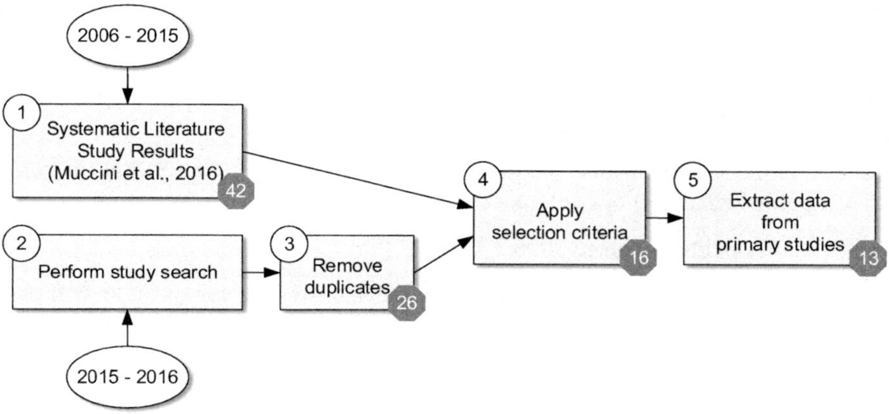

Fig. 13.2 Applied systematic mapping study process

13.4.1 Search and Selection Strategy

The initial set of primary studies under investigation is based on the replication package of the study *"Self-Adaptation for Cyber-Physical Systems: A Systematic Literature Review"* by Muccini et al. (2016), where the search and selection strategy as well as the inclusion/exclusion criteria are defined, that were used for retrieving the studies. The scope of the systematic literature review includes studies from 2006 to mid 2015. Therefore, the SLR protocol is reused to extend the set of primary studies by searching and selecting studies that were published since mid 2015 (end of scope of the SLR) and are relevant for this SMS.

In order to cover as many as possible relevant studies about self-adaptation approaches applied in CPS on an architectural level, we performed searches in four of the largest scientific online databases as sources of primary studies: *IEEE Xplore Digital Library*, *ACM Digital Library*, *SpringerLink*, and *ScienceDirect*. For these searches, we used defined keywords and combinations of them to identify candidate papers, e.g., *software*, *architect*, *cyber-physical*, *control system*. The involved reviewers applied the search strategy to identify potential study candidates. The search results are documented in a spreadsheet where the identified candidate studies are collected and stored. In addition, duplicates are removed. Each paper is indexed by a unique identifier and title.

The identified set of candidate studies is carefully assessed and filtered for their actual relevance to answer RQ1 by two reviewers. Therefore, the study goals and well-defined study selection criteria are used to determine which studies to include or exclude. Hence the inclusion and exclusion criteria defined in the SLR protocol (Muccini et al., 2016) are extended. A study is included if it is compliant to the following inclusion and exclusion criteria:

IC 1 Studies proposing, leveraging, or analyzing an architectural solution, architectural method or technique specific for CPS.

IC 2 (updated) Studies in which multiple types of self-adaptation are explicitly used as an instrument to engineer CPS.

IC 3 Studies subject to peer review (Wohlin et al., 2012) (e.g., journal papers, papers published as part of conference proceedings).

IC 4 Studies published since 2006.

IC 5 (new) Studies in which self-adaptation mechanisms are applied at least at two layers of the technology stack.

IC 6 (new) Studies comprising at least a minimal description of a concrete scenario or use case.

EC 1 Studies that are written in a language other than English, or that are not available in full-text.

EC 2 Secondary studies (e.g., systematic literature reviews, surveys, etc.).

EC 3 (new) Studies of poor quality (e.g., poorly described architecture or use case).

Results of selections and rejections are crosschecked by two other reviewers and any disagreements are discussed and resolved. Finally, the set of studies to be included in the data collection process is finalized.

13.4.2 Data Extraction

For each study remaining after the selection process, we independently investigated and extracted pre-defined data. In addition to including all the data items needed to answer RQ1, the data extraction form provides standard information about the publication. The definition of pre-defined extraction forms with data items allows to survey each study in the same way (objectively) and reduces the room for bias. Table 13.2 gives an overview of the data items that were collected from the primary studies to answer the research question. Each primary study was assigned to and reviewed by at least two reviewers. After discussion of the individual results for each study with the other reviewers, the extracted data were collected and documented in a spreadsheet in a consistent manner.

13.4.3 Data Analysis and Reporting

The process of analyzing and synthesizing the collected data of the SMS includes the application of descriptive statistics and representation and interpretation of the

Table 13.2 Data extraction form

Data item	
(D1) Study title	(D2) Publication year
(D3) Venue	(D4) Country
(D5) Application domain	(D6) Overall architectural style
(D7) Overall system goal	(D8) Type of distribution
(D9) Uncertainties considered	(D10) Adaptation purposes/goals
(D11) Adaptation mechanisms applied	(D12) Location of the adaptation mechanisms in the technology stack
(D13) Inter-adaptation coordination mechanisms	

results with respect to RQ1. Besides standard information about each included paper (study title, publication year, venue, country), data items needed to answer RQ1 were collected and analyzed. Data item (D5) captures the reported application domain of the study to ensure representative and evaluated results. In addition, the application description supports a better understanding of the approach, and maybe allows to draw conclusions about a more beneficial application of particular adaptation mechanisms in one domain. Data item (D9) focuses on different types of addressed uncertainties in the environment, in parts of the system itself, and in requirements/goals for which adaptation is applied. This knowledge supports a better understanding of the focus of current research and shows what uncertainty types are mostly addressed and what areas of uncertainties are not yet addressed at all. Data item (D10) summarizes different purposes and goals of applying adaptation mechanisms in a CPS, while data item (D11) is used to identify and investigate the types of adaptation mechanisms applied in the study. To generalize the technology stack that is commonly used for applying adaptation mechanisms, data item (D12) is used to create a general layer model. Finally, data item (D13) captures the interaction and coordination between different adaptation mechanisms across the layers.

The analysis results and their visual representation are documented in a spreadsheet that is available at the study website.[7]

13.5 Adaptation in Cyber-Physical Systems

After applying inclusion and exclusion criteria to the initial set of 42 primary studies from the SLR by Muccini et al. (2016) as well as to the extended set of 26 studies, data were extracted from a total number of 13 primary studies to answer RQ1. An overview list of all selected primary studies is shown in Table 13.3. After finishing

[7]Supplementary material of the study is available at: http://qse.ifs.tuwien.ac.at/ci/material/pub/mde-cpps17/

Table 13.3 Final list of primary studies retrieved in the systematic mapping study

ID	Title	Reference
1	An Architecture of Cyber Physical System based on Service	Yu et al. (2012)
2	An Architecture Framework for Experimentations with Self-Adaptive Cyber-Physical Systems	Kit et al. (2015)
5	C-MAP: Framework for Multi-agent Planning in Cyber Physical Systems	Mukherjee and Chaudhury (2013)
7	Context-Aware Vehicular Cyber-Physical Systems with Cloud Support: Architecture, Challenges, and Solutions	Wan et al. (2014)
9	Towards Context-aware Smart Mechatronics Networks: Integrating Swarm Intelligence and Ambient Intelligence	Gupta et al. (2014)
13	A multi-agent RFID-enabled distributed control system for a flexible manufacturing shop	Barenji et al. (2014)
19	Multi-Agent Control System for Real-time Adaptive VVO/CVR in Smart Substation	Nasri et al. (2012)
37	Coupling heterogeneous production systems by a multi-agent based cyber-physical production system	Vogel-Heuser et al. (2014)
41	Cloud Robotics: Architecture, Challenges and Applications	Hu et al. (2012)
51	Continuous Collaboration: A Case Study on the Development of an Adaptive Cyber-physical System	Hölzl and Gabor (2015)
62	Cloud-Assisted Context-Aware Vehicular Cyber-Physical System for PHEVs in Smart Grid	Kumar et al. (2015)
63	Cyber-physical-social system in intelligent transportation	Xiong et al. (2015)
67	Cross-layer Virtual/Physical Sensing and Actuation for Resilient Heterogeneous Many-core SoCs	Sarma et al. (2016)

the data collection, the results were checked for consistency and completeness as well as documented in a spreadsheet. This section presents the results of the conducted systematic mapping study.

Figure 13.3 presents the variety of application domains (D5) where proposed self-adaptation approaches were applied to evaluate their efficiency and performance. The results show that transportation (23%) is the dominant application

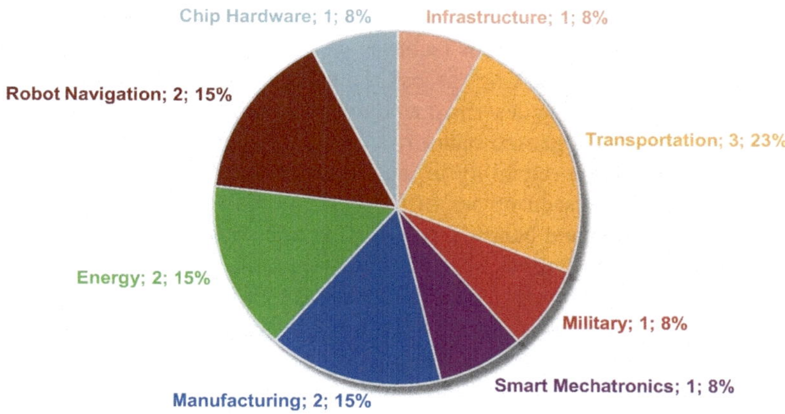

Fig. 13.3 Overview of identified application domains (D5)

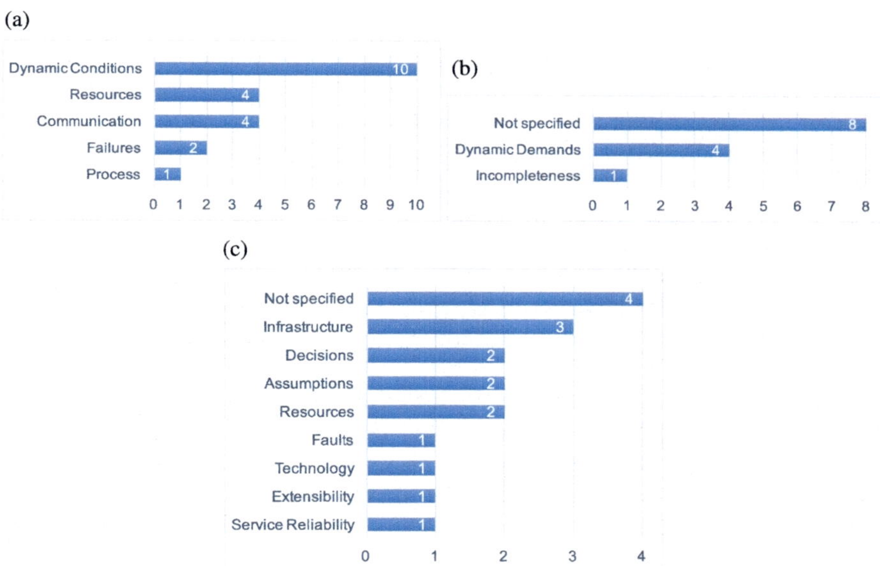

Fig. 13.4 Types of uncertainties addressed by existing approaches for self-adapting CPS (D9). (**a**) Uncertainties in the environment. (**b**) Uncertainties in requirements and goals. (**c**) Uncertainties in parts of the system itself

domain, followed by robot navigation (15%), energy (15%) and manufacturing (15%).

All studies report to enable adaptation in the CPS to address uncertainties (D9) in the environment and to make context-aware decisions. The identified clusters of uncertainty types in the environment are presented in Figure 13.4a. As CPS operate in real-time and thus have to deal with environments that are usually subject to volatile and dynamic conditions (such as weather conditions, road conditions, traffic flow, danger zones, parking spaces), the most dominant uncertainty type that need to be addressed can be described as dynamic conditions (77%). For enabling the exchange of knowledge and data or negotiations between components of the CPS, the communication reliability in the environment is also a relevant issue to consider. Due to unforeseen noises in the environment, CPS need to deal with limited communication and bandwidth as well as latencies (31%). In addition, not predictable resource constraints (such as fuel, ammunition, personnel, ingredients) in the environment (31%) can lead to obstacles for the correct operation of CPS that need action to be taken. Further identified uncertainty types related to the environment are failures (such as production system breakdown, infrastructure failure) reported by 15% of the studies and process (such as unforeseen changes of the requested production process) reported by 8%.

Only five studies mention uncertainties (D9) in requirements and goals that are affected by their proposed adaptation approaches. In particular, they describe dynamic demands by customers or the market (31%) and incompleteness of specified requirements (8%) as uncertainty types. The results are summarized by Fig. 13.4b.

Most of the studies also considered uncertainty in parts of the CPS itself to be addressed by adaptation approaches, but the results are quite diverse as Fig. 13.4c illustrates. The most frequently mentioned uncertainty type is infrastructure (such as broken hardware, aging effects) considered by 23% of the studies. Other uncertainty types are system decisions (15%), assumptions of behavior and knowledge of other components (15%), system resources (15%), internal faults (8%), changing technology (8%), extensibility to integrate unknown features into the system (8%), and service reliability (8%).

In our mapping study we further investigated the purposes and goals of designing self-adaptive CPS (D10). We clustered the results of the identified adaptation purposes and goals as presented in Fig. 13.5. As the dominant adaptation purpose, we identified performance in 77% of the studies. Other stated adaptation purposes and goals are efficiency (46%), flexibility (31%), and reliability (23%).

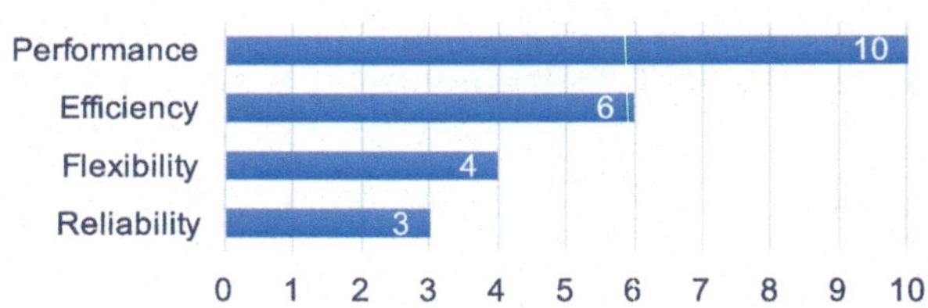

Fig. 13.5 Purposes and goals of adapting CPS (D10)

The analysis of the technology stacks that are used in the studies for applying adaptation mechanisms (D12) revealed a general architecture model of CPS comprising the following 6 different layers:

1. *Physical Layer:*
   This layer represents physical real-world components of the CPS that interact with humans. Examples include vehicles, production systems, robots, road infrastructure, parcels, and smart meter. The physical components monitor the environment, collect information with sensors and take actions to modify the environment with actuators.
2. *Proxy Layer:*
   This layer constitutes the transition from the real-time physical world to the virtual world where the physical components are represented by intelligent mechanisms. Examples include interfaces, software agents, and smart components. These mechanisms communicate the collected context information to the computational system in the upper layers and receive responses based on the sent data.
3. *Communication Layer:*
   The interaction between the proxy layer and the upper layers is enabled by this layer. A variety of technologies are available (wired/wireless, short/long-range) to use for the communication process. Examples include Bluetooth, ZigBee, WLAN, LAN, and special communication protocols.
4. *Service and Middleware Layer:*
   This layer provides context-aware services and middleware to process and analyze the collected data according to defined goals. In some studies these services are located in the cloud. Examples include traffic cloud services, controller intelligent agents, component framework, and optimization unit.
5. *Application Layer:*
   This layer represents the domain-specific application that is in charge of system control and responsible for realization of the system goals. Therefore, it has to acquire all required resources and make justified decisions to achieve the demanded Quality-of-Service. Examples include software agents and control applications.
6. *Social Layer:*
   This layer represents an optional extension of the common CPS architecture. It integrates social systems into a CPS architecture by providing specific mechanisms to humans, organizations, and societies so that they can offer knowledge and feedback. The combination of social and physical sensing information can then achieve more intelligent and improved decisions by a CPS. An example could be a social network.

Figure 13.6 illustrates the identified general multi-layer CPS architecture.

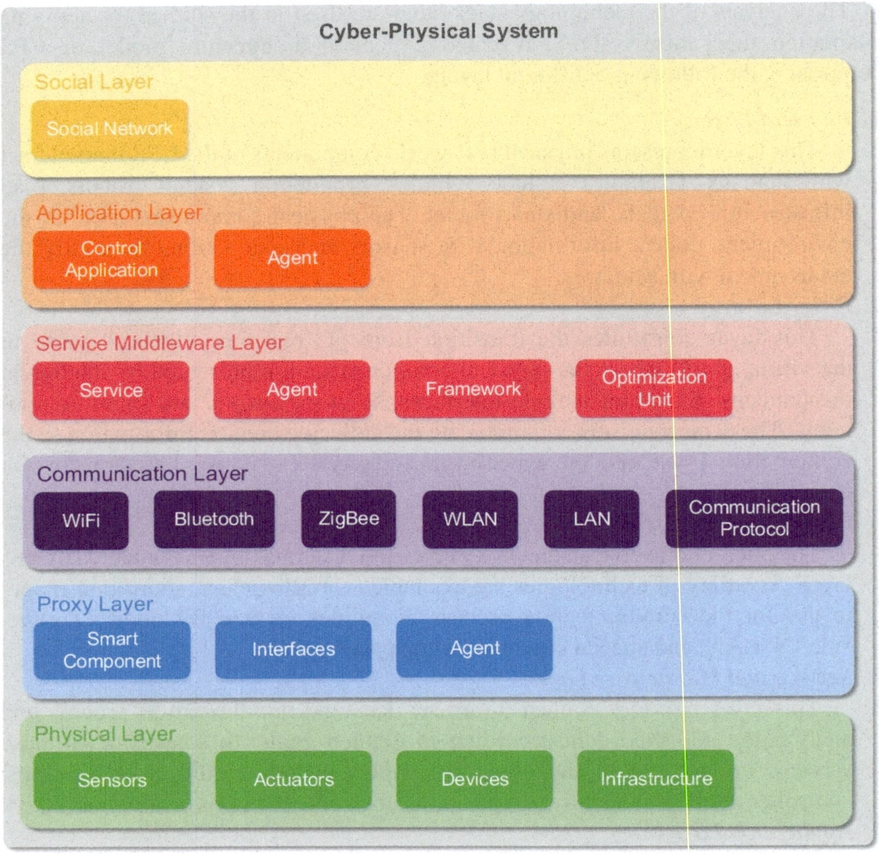

Fig. 13.6 General multi-layer architecture of CPS

We further investigated the types of adaptation mechanisms that are applied in proposed CPS architecture designs (D11). Figure 13.7 shows the frequencies of different self-adaptation mechanisms as well as their locations in the technology stack (D12). Smart elements are mostly applied (77%) and always located at the proxy layer. They represent the physical components capable of self-adaptation. Other types of adaptation mechanisms are broader distributed across the technology stack. Multi-agent systems (69%) are applied at the proxy layer, service middleware layer and application layer, followed by MAPE (54%) at the proxy layer and service middleware layer, autonomous entities (38%) at the proxy layer, service middleware layer and application layer, collaborative entities (23%) at the proxy layer, service middleware layer and application layer, swarm (15%) at the service middleware layer and application layer. Self-organization (at the application layer) and social network (at the social layer) are equally applied (both 8%) as adaptation mechanism.

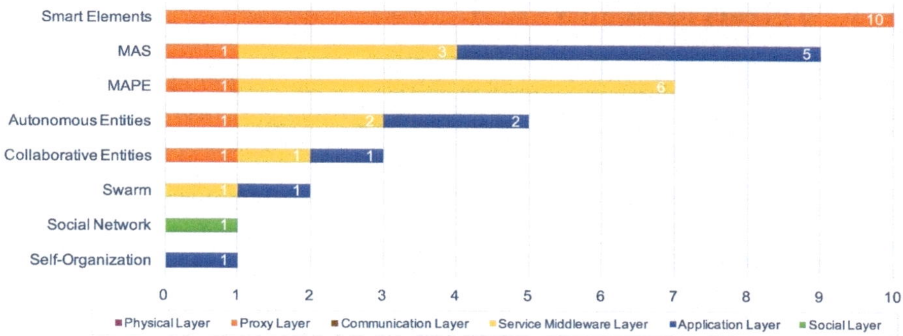

Fig. 13.7 Self-adaptation mechanisms applied at multiple layers in CPS architectures (D11/D12)

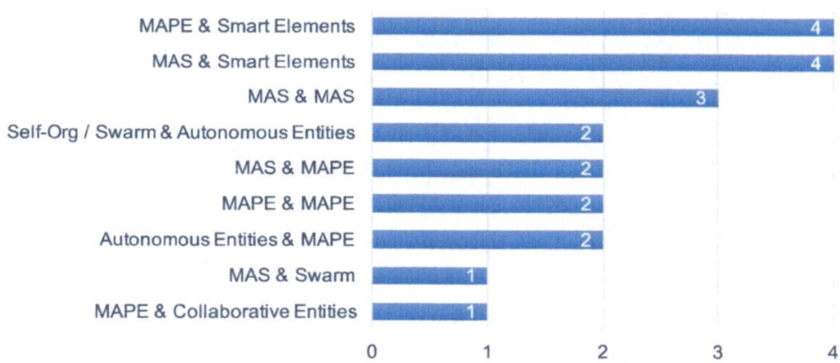

Fig. 13.8 Combinations of adaptation mechanisms (D13)

Each primary study proposed an application of adaptation at the proxy layer and service middleware layer.

In the primary studies we observed the application of combinations of different types of adaptation mechanisms that interact and coordinate across multiple layers of the technology stack (D13). By applying such solution approaches the CPS is capable to deal with different uncertainties and concerns at a time using adaptation. Figure 13.8 presents the observed combinations of adaptation mechanisms. The results show that the majority of primary studies combine MAPE with smart elements or MAS with smart elements (both 31%) in a CPS architecture design. Combinations of multiple multi-agent systems were realized in 23% of the primary studies. 15% of the primary studies equally combined self-organization or swarm with autonomous entities, MAS with MAPE, MAPE with MAPE, and autonomous entities with MAPE. Only 8% of the primary studies combined MAS with swarm and MAPE with collaborative entities.

13.6 Threats to Validity

As with any empirical research, there are threats to the validity of this study that need to be considered. The following potential validity threats were identified and discussed how to mitigate them in order to strengthen the outcomes of the study.

- *Quality of the selected primary studies.* We defined several inclusion/exclusion criteria to ensure sufficient quality of the selected primary studies for the mapping study, but we did not apply a systematic and detailed quality assessment procedure in order to critically evaluate the quality of each paper as it is common in systematic literature reviews. To mitigate this weakness to some extent, we particularly added the inclusion criterion that each primary study must provide a description of a concrete scenario or use case to draw conclusions for real-world applications. In addition, we discussed and excluded some papers with determined poor quality during the mapping study.
- *Adaptation of data items.* Based on our expertise, we defined a set of data items for the data collection of the mapping study. During piloting the data extraction form, we identified some data items that did not always fit for the selected primary studies or were omitted. Based on the results of this initial review, we updated the data extraction form for the remaining studies to ensure a consistent data collection and analysis of the results.
- *Limited CPPS expertise of research team.* The research team consisted of experts in the fields of self-adaptation, software architecture, cyber-physical systems and collective intelligence systems without special experience with cyber-physical production systems. In order to reduce bias, we consolidated other researchers with expertise in the CPPS domain and discussed the outcomes.
- *Generality of the results for CPPS.* Due to a general lack of knowledge on self-adaptation specifically in CPPS and the nature of a mapping study to provide a broad overview of a research topic area, we reviewed state-of-the-art self-adaptation approaches related to the larger family of CPS. However, during the mapping study we focused on self-adaptation applied in CPS in the manufacturing domain and eventually included some primary studies with described use cases in this domain. During the data analysis, we came to the conclusion that these studies are no outliers and informally discussed the results with CPPS experts. Nevertheless we are careful to generalize the results for CPPS and see a need for further research in this direction to enhance the validity of the results.
- *Small number of primary studies.* Since CPS and specifically CPPS is still a quite young research field, the number of primary studies providing insight on how self-adaptation can be used in addressing uncertainty in these systems is small. Thus the data extracted from this mapping study can only be considered as evaluation of the current state-of-the-art. This study should be replicated after a period of time.

13.7 Reflection of the Systematic Mapping Study Results

The goal of the systematic mapping study was to capture, consolidate and document state-of-the-art design strategies and best practices how engineers currently deal with adaptation across the technology stack in CPS. Based on the analysis results of the collected data, we were able to synthesize this knowledge to derive recurring patterns in CPS architectures that can be reused to engineer adaptive CPPS to handle uncertainty. In particular, we studied the interaction and coordination between different applied adaptation mechanisms across the layers. Finally, the results of this investigation supports the proposal of three identified adaptation patterns to answer RQ2, whereby each pattern is applicable for a different purpose. The consolidation and documentation of this design knowledge represents a valuable contribution for CPPS engineers as a useful starting point for the system's design with respect to adaptation. The provided insights into evaluated and effective design strategies, that enable self-adaptation in CPS, are promising to be useful for reuse in CPPS architecture design as well. In evaluations of the proposed approaches with scenarios in different application domains, the solution designs demonstrated their applicability and effectiveness to address particular uncertainties and concerns related to self-adaptation. In addition, they were observed to support the CPS in the realization of the stated adaptation purposes and goals.

However, these results require further investigation with specific focus on the CPPS domain and related application scenarios, but our contributions can serve as a useful basis for future research. The identified patterns are introduced in the following section in more detail.

13.8 Patterns for Self-Adaptation

To derive the patterns, we carefully studied the collected data and analysis results of the conducted systematic mapping study. We created a comprehensive table[8] presenting the concrete designs of self-adaptation mechanisms applied across the technology stack for the application scenario of each investigated study. The comparison across all represented solution designs highlighted areas that follow similar or equal strategies, enabling us to identify three multi-adaptation patterns with different combinations of multiples types of self-adaptation within a system: SYNTHESIZE-UTILIZE, SYNTHESIZE-COMMAND, and COLLECT-ORGANIZE. In particular, a multi-adaptation pattern provides knowledge about (1) the kinds of used adaptation mechanisms, (2) their layer locations, and (3) the cross-layer inter-adaptation interactions between the respective mechanisms. This section describes

[8]Supplementary material of the study is available at: http://qse.ifs.tuwien.ac.at/ci/material/pub/mde-cpps17/

each pattern according to the pattern writing form provided by Meszaros and Doble (1997).

A pattern aims to capture best practices that address certain recurring problems in a specific context for reuse and guidance (Meszaros and Doble, 1997). Thereby it is important to communicate the purpose of the pattern, the concrete context in which to apply it, the problem it addresses, a description of the solution to solve the problem, and the effects and consequences it creates in detail. Such a detailed pattern description efficiently supports the reader to decide about the applicability of the pattern of interest in a specific scenario and eventually also to guide the application. The patterns are structured using the following template based on Meszaros and Doble (1997):

- **Name:** a name to refer to the pattern
- **Context:** a short description of the situation in which to apply the pattern
- **Problem:** a short description that describes a specific recurring problem that the pattern solution aims to solve
- **Solution:** kind of a reusable model that solves the problem
- **Consequences:** set of rationales why the proposed solution is most appropriate for the stated problem (benefits) as well as a set of other effects and limitations to make clear where the pattern is not applicable
- **Known Uses:** a short description of representative use cases that illustrate the application of the solution and the positive effects to address the problem

13.8.1 Synthesize-Utilize Pattern

Context: A distributed application is composed of diverse physical resources and application entities, whereby each of which possesses data that is heterogeneous, spatially distributed and continuously changing.

Problem: A distributed application seeks to improve the utility of its services to the physical resources by dynamically exploiting rich context information.

Solution: SYNTHESIZE-UTILIZE is composed of a MAPE-like adaptation mechanism on the service middleware layer and autonomous entities on the application layer. The pattern is illustrated in Fig. 13.9, which also depicts the concrete workflow steps across the layers. The characteristic workflow between the layers is as follows: (1) The service middleware layer receives data from the physical resources via the proxy layer and requests data from the autonomous entities on the application layer. (2) The service middleware layer uses MAPE-like adaptation for the continuous collection and synthesis of data. (3) The consolidated data is then provided to the autonomous entities on the application layer. (4) The autonomous entities dynamically optimize their services to the physical resources by collaborating with each other and by using the integrated data that is offered from the service layer.

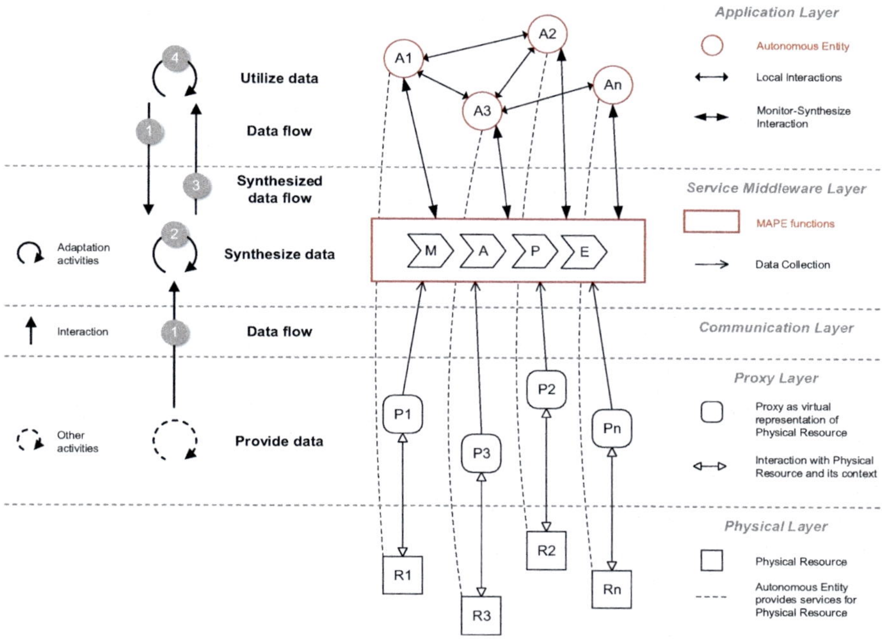

Fig. 13.9 Synthesize-Utilize pattern with adaptation mechanisms (*red*) and characteristic workflow steps across layers

Consequences:

+ Efficient sourcing of heterogeneous data from a diverse set of systems and its consolidation that can be used for situated optimization.
+ Reduced effort for adding and removing data sources.
– Timing aspects.
– Varying quality and granularity of data requires additional data acquisition and integration effort.

Example "Intelligent Transportation System": In this example the physical resources include vehicles and sensors that are positioned along the road infrastructure. On the application layer, services are provided to traffic participants on a global scale (traffic-related information) and individual scale (smart parking, routing/navigation), as well as to traffic management authorities (traffic performance, congestion control). By collecting data from individual vehicles and road sensors, the service middleware layer can monitor traffic-related information. It then continuously integrates the data into traffic models that are created from dynamic simulations and experiments. Based on the evaluated models the traffic information is consolidated and forwarded with respect to the individual traffic services. The consolidated data is used for optimization and if necessary personalized to the individual recipient (e.g., traffic participant). Further the services locally interact (e.g., via Vehicle-2-Vehicle networking, flexible web service orchestration) with each other in order to factor in additional context information.

Identified in Primary Studies: ID-07 Dynamic Parking Service (Wan et al., 2014), ID-63 Intelligent Transportation System (Xiong et al., 2015).

13.8.2 Synthesize-Command Pattern

Context: A distributed application produces its functionality by employing an assembly of heterogeneous, physical resources which are independent and have different capabilities and capacities.

Problem: A distributed application exploits data of individual resource to improve its overall utility by changing the resource configuration that produces the application's functionality.

Solution: SYNTHESIZE-COMMAND is composed of a MAPE-like adaptation mechanism on the service middleware layer and a multi-agent system (MAS) on the application layer which manages the physical resources. The pattern is illustrated in Fig. 13.10, which also depicts the concrete workflow steps across the layers. The

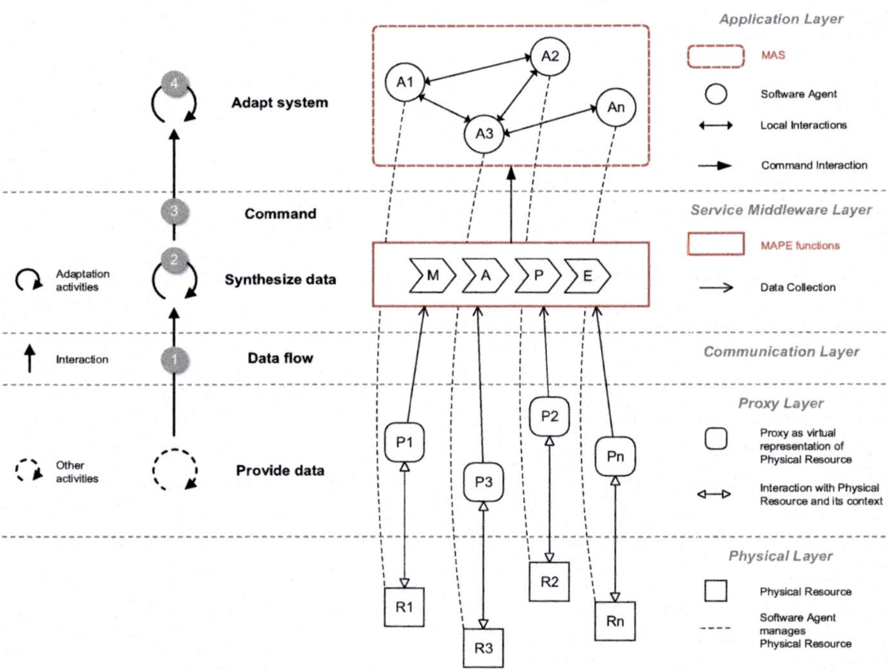

Fig. 13.10 Synthesize-Command pattern with adaptation mechanisms (*red*) and characteristic workflow steps across layers

characteristic workflow between the layers is as follows: (1) The service middleware layer receives data from the physical resources via the proxy layer. (2) The service middleware layer uses MAPE-like adaptation for the continuous collection and synthesis of physical resource data. (3) The consolidated data is then used to derive commands which are sent to the MAS on application layer. (4) The agents in the MAS locally interact and reorganize so that the commands are performed in an efficient way on the physical layer.

Consequences:

+ MAPE-based "plug-in" model allows selection of appropriate adaptation function.
+ Separation of concerns: reconfiguration of request and its execution.
+ Easy extensibility of resources on the physical layer.
– Cross-layer coordination is complex.
– Reduced autonomy of physical resources due to high dependence on central command coordination.

Example "Flexible Manufacturing Shop": In this example the physical resources include various kinds of manufacturing equipment which are grouped into stations. The application layer consists of manufacturing resource agents, whereby the agents are connected to the physical resources with specific agent-machine interfaces. On the service middleware layer, the MAPE-like adaptation mechanism is realized in a station controller which collects data on realizable capabilities from the station and compares it with a product capability list in order. If the product capabilities are realizable on the station, the station controller assigns information on related manufacturing resources and process steps for efficient production to the resource manufacturing agents. The resource manufacturing agents forward the production information to the associated manufacturing equipments and continuously send feedback about the status of the manufacturing process back to the station controller, which is case of constraint conflicts would adapt the process on the respective station.

Identified in Primary Studies: ID-01 Water Resource Management (Yu et al., 2012), ID-13 Flexible Manufacturing Shop (Barenji et al., 2014)

13.8.3 Collect-Organize Pattern

Context: A distributed application provides services to autonomous, cyber-physical entities, whereby each entity generates their own, local models and collects data that is spatially distributed and continuously changing.

Problem: A distributed application seeks to improve the overall utility of its service, which requires the autonomous entities to efficiently share information and coordinate their tasks on a local basis.

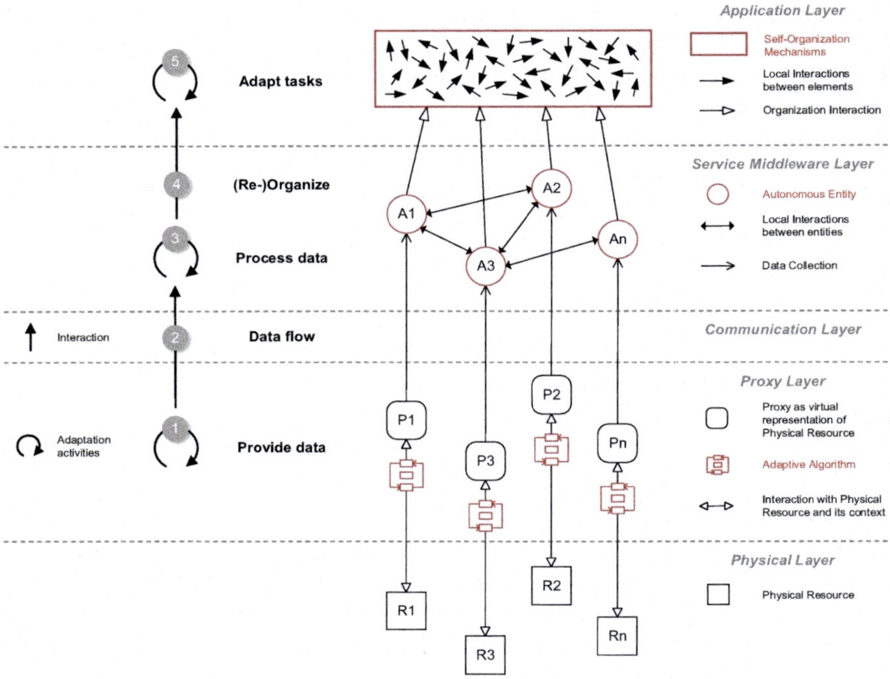

Fig. 13.11 Collect-Organize pattern with adaptation mechanisms (*red*) and characteristic workflow steps across layers

Solution: COLLECT-ORGANIZE is composed of adaptive algorithms on the proxy layer, autonomous entities on the service middleware layer and self-organization mechanisms on the application layer. The pattern is illustrated in Fig. 13.11, which also depicts the concrete workflow steps across the layers. The characteristic workflow between the layers is as follows: (1) On the proxy layer, adaptive algorithms continuously integrate data from physical resources into local models. (2) Autonomous entities on the service middleware layer exchange local information in order to generate and update global models. (3) On the application layer, self-organization mechanisms facilitate the adaptation with regard to situated tasks among the individual entities based on the local and global models.

Consequences:

+ Efficient organization in highly dynamic application scenarios.
+ Cyber-physical entity has autonomy and maintains operational, even if service middleware layer adaptation fails.
– Cross-layer coordination is complex, in particular with regard to emergent behavior induced by self-organization mechanisms.
– Pattern requires smart systems at the bottom of the architecture instead of "dumb" physical systems.

Example "Smart Parking": In this example the physical resources include smart vehicles which are equipped with intelligent sensors that collect data about available parking space. The service provided on the application layer is a smart parking service, which supports the vehicles in the discovery of parking space and with negotiating the allocation of suitable space. On the proxy layer, the data from the sensors is integrated into models using adaptive algorithms (like MAPE-K loops) and real-time CPS control logic. On the service middleware layer, an autonomous entity, representing the vehicle, receives the model information. These entities are realized as autonomous components which are dynamically grouped into ensembles (subset of vehicles in proximity) consisting of a coordinator and multiple members. Within an ensemble the autonomous components continuously exchange their partial models in order to collectively produce a more complete, global model of the situation. The consolidated model is then forwarded to the application layer, where it is used by the smart parking service to derive a list of suitable parking spaces. These spaces are particular for the individual vehicle with respect to the other vehicles, leading to emergent self-organizational interaction, since the ensembles are highly temporal.

Identified in Primary Studies: ID-02 Smart Parking (Kit et al., 2015), ID-51 Rescue Robot Navigation (Hölzl and Gabor, 2015).

13.9 Potential of Collective Intelligence Systems for Cyber-Physical Systems and Cyber-Physical Production Systems

This section presents an overview about emerging directions of how collective intelligence systems are used in CPS and CPPS. Based on the background of CIS in Sect. 13.2.3 and the findings of our systematic mapping study in Sect. 13.5, we describe the three directions of capability augmentation, emergent machine-to-machine interactions, and multi-disciplinary knowledge integration and coordination.

13.9.1 Collective Intelligence Systems for Capability Augmentation

The results of the systematic mapping study reveal a recent trend to add an additional "social" layer in a CPS architecture that involves components in a CPS to address human and social factors. So-called *Cyber-Physical-Social Systems (CPSS)* (Wang, 2010) consider social and human dynamics as an essential part of an effective CPS design and operation. A CPSS is defined as a complex system that is constituted by three parts: a physical system, a social system including human beings, and a cyber system that connects both of them (Xiong et al., 2015). The social system is a human-centered system, like social networking sites and social media platforms, that explicitly involves individuals, organizations, and societies in the processes of a CPS for information exchange and feedback. One application scenario of a CPSS architecture is an intelligent transportation system (Xiong et al., 2015) where the social system aims to effectively aggregate (by so-called *social sensor network*) and disseminate up-to-date travel-related information and resources from multiple sources. Examples of this shared information are emergency events, traffic jams, navigation, road conditions, and car-sharing information. Such useful information can have influences on the decisions and behavior of individuals as well as on transportation authorities who can use this information to improve their services and management. As one successful example of such a CIS which support a transportation system is *Waze*.[9] They concluded that CPSS have a significant value, but there are existing challenges to control and manage them by applying traditional theories and methods which demand the research of novel approaches.

Considering and utilizing the potential of humans-in-the-loop, their interactions and the created social dynamics offer new opportunities for CPPS as well. According to the scenario of a "Cyber-Physical System for the factory of the future" investigated by the German acatech – National Academy of Science and Engineering (2011), industrial production systems should be able to react virtually in real-time to changing customer demands as well as changes in the market and the supply chain. A CIS in the social layer of a CPPS architecture can support this vision by facilitating to effectively incorporate humans into the CPPS processes and by introducing CIS-specific capabilities. Companies as well as customers (local or globally distributed) can interact with each other by using a CIS for sharing opinions, experiences, requirements as well as discussing new ideas and designs for future products. The combination of information and feedback from multiple sources (different sensors in the physical system and different human groups in the CIS) enables a more efficient sensing of the CPPS and thus improves its decision-making processes. For example, based on the collected knowledge a CPPS can react rapidly to customer feedback and optimize the manufacturing of tailored customer products or correct defective production models.

13.9.2 Collective Intelligence Systems as Enabler for Emergent Machine-To-Machine Interactions

CIS-based CPPS architectures highlights the potential of a new kind of social interactions that goes beyond the typical human-to-human interactions. So far CIS approaches consider humans as essential entities in the critical feedback loop to be

[9]http://www.waze.com/ (last visited 01/15/2017).

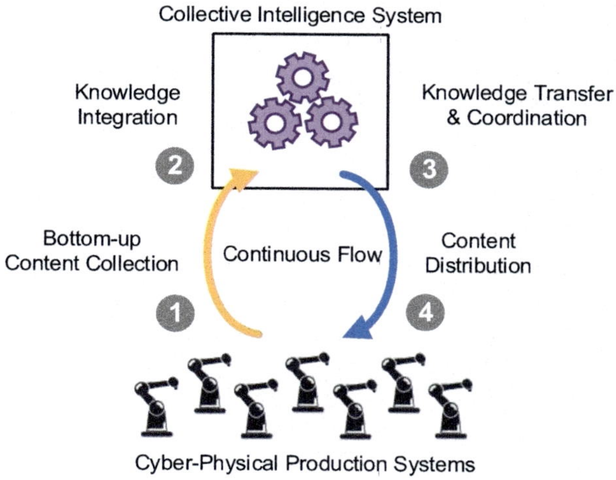

Fig. 13.12 Overview of a CIS as enabler for emergent machine-to-machine interactions with feedback loop of information aggregation (*yellow*) and distribution (*blue*)

successful and effective (e.g., *Yelp*[10] in the domain of business ratings and reviews or *Facebook*[11] for creating a social network of friend relationships). But if humans were regarded as one of many variability points in a CIS architecture, they could be replaced by machines as for instance robots or CPPS to realize machine-to-machine configurations (Musil et al., 2015a). The integration of a CIS with machines that represent its actor base into a manufacturing environment enables the connection and communication of several groups of CPPS or single systems to support the machines to share their information and experiences among each other, which is illustrated by Fig. 13.12. Single globally distributed, adaptive and evolutionary production units that belong to different operators are situated in a specific, local context and thus have only awareness about local available information but have difficulties to access remote information from other machines involved in the production processes. The creation of a global network of cooperating and interacting industrial plants as well as the aggregation and coordination of their collective intelligence by applying CIS mechanisms, that have proven to be effective, offers the possibility of a global access to relevant and critical information and data of individual machines with respect to context, status, defects, diagnoses, experiences, influences, effects, analytics, and learned capabilities. In their work Bauernhansl et al. (2014) recognize on numerous occasions the potential of social media platforms, besides data mining and mobile, as a pivotal enabling technology for smart factories of the future. *Factory Social Media* is expected to play an

[10]http://www.yelp.com/ (last visited 01/15/2017).

[11]http://www.facebook.com (last visited 01/15/2017).

important role in future CPPS process improvement efforts by enabling effective and efficient bottom-up information collection and dissemination capabilities in human-human, human-machine and machine-machine scenarios. But the authors also mention the lack of their current application, although there is a clear need to support the increased computerization of physical systems and address the resulting need to organize and coordinate.

Approaches in this promising direction can be found in the work of Mukherjee and Chaudhury (2013) who elaborated on this topic and proposed a novel multi-agent planning framework. For illustration they used the scenario of a network centric battle space with military CPS. The challenge here is the collaboration and coordination of plans between multiple planner agents. To deal with uncertainty while planning, Mukherjee and Chaudhury (2013) explored (1) bottom-up biologically inspired continuous planning in order to adapt to changing environment and (2) Blackboard-based multi-agent coordination. Similar to this approach, CIS use the nature-inspired coordination mechanism of stigmergy (Heylighen, 2016) which enables bottom-up, environment-mediated coordination and indirect communication of agents via traces in the environment (Musil et al., 2015a). Thus the stigmergic process creates a positive feedback loop that promotes awareness among agents about the activities of others and stimulates further agent activities. The resulting feedback loop provides CIS with emergent, self-organizational capabilities and allows the system to adapt (Musil et al., 2015a).

The concept of CIS for machine-to-machine communication brings along new architectural design and realization challenges to deal with in an agenda for future research. Aspects like what processes can be supported by a CIS of machines-in-the-loop or what skills provided by machines are needed to play the role of humans in a CIS for CPPS need to be investigated.

13.9.3 Collective Intelligence Systems as Coordinators and Knowledge Integrators Across Heterogeneous, Multi-Disciplinary Domains

Similar to the support of CPPS processes during production, also the engineering of CPPS as group processes can be improved with CIS-based approaches. CPPS engineering processes involve humans from multiple engineering disciplines, such as electrical, mechanical and software engineering, who are essential factors in the planning, design, and realization of a well-functioning CPPS. Consequently the following challenges arise for the multi-disciplinary engineering teams in particular: heterogeneous representations as for instance of engineering data, models and terminologies, weak accumulation and integration of dispersed, local engineering know-how that is instrumental for the engineering processes, lack of traceability and awareness of activities and changes as well as required effective communication, coordination and sharing of knowledge and artifacts between teams across the organization (Jazdi et al., 2010; Musil et al., 2016b).

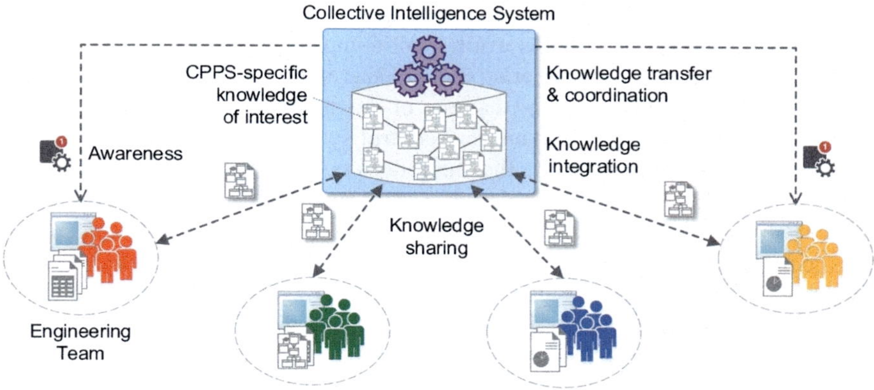

Fig. 13.13 Overview of a multi-disciplinary engineering scenario with a CIS as coordinator and knowledge integrator

In such environments of multi-disciplinary projects in which CPPS are engineered, social interaction and communication between the experts are critical and thus need to be supported by social systems, like a CIS. This kind of a socio-technical system has the potential to enhance current engineering methods and tools to overcome existing challenges and complexities (Musil et al., 2016b). The capabilities of a CIS support engineers to identify important implicit engineering knowledge and to integrate the dispersed, local information into an organization-wide and project-independent knowledge base in order to make it explicitly available (Musil et al., 2016b). The efficient and structured collection and maintenance of project-independent knowledge enabled by a CIS allows the sharing and awareness of engineering know-how without loss of information (Jazdi et al., 2010). So it provides the ability to deal with complexity of a CPPS and to reuse expert knowledge, and thus efficiently and transparently supports engineering, maintenance (Winkler et al., 2016), and modernization activities. Figure 13.13 illustrates a scenario with a CIS as coordinator and knowledge integrator in multi-disciplinary engineering processes.

All three introduced CIS-focused concepts open up different new future research directions to investigate novel theories, methods and technologies, but provide a good starting point for a research agenda.

13.10 Related Work

This section presents an overview of related systematic studies in the areas of agent/multi-agent based, architecture-based, control theory-based, and self-organization-based adaptation. While we could not identify any secondary study directly looking at patterns of combined adaptation mechanisms in the field of CPS, we summarize related research that is worth mentioning.

First we looked at systematic studies surveying agent/multi-agent based approaches. Leitão (2009) presented a state-of-the-art survey on multi-agent system approaches for designing manufacturing control systems that exhibit intelligence, robustness and adaptation. While the study does not deal directly with adaptation patterns for CPS, it reports on holonic- and agent-based architectures based on centralized and decentralized patterns such as agent federation centered in the mediator approach, adaptive control approach, hybrid control architecture, decentralized planning and rough level process planning, and hierarchical structure based on rules. Juziuk et al. (2014) provided an overview of existing design patterns for multi-agent systems collected using a systematic literature review. In their study, the authors identified a total of 206 patterns. One cluster of associated patterns was formed around bio-inspired concepts such as pheromones, ants, and stigmergy. This cluster also included recent patterns related to self-organization and adaptive behavior. In addition, the authors investigated the types of systems for which these design patterns have been applied. They concluded that there is not a dominant type of system, but with regard to industrial applications the main reported application domains are process control and manufacturing, traffic and transportation.

Then we focused on research work in designing architecture-based adaptation. Weyns and Ahmad (2013) conducted a systematic literature review on claims and evidence for architecture-based self-adaptation. The authors discovered that 69% of the studies focus on a single feedback loop, with 37% of the primary studies using distinct components for each of the MAPE functions and 32% using components that mix (some of) the MAPE functions. Only 20% of the studies focus on multiple feedback loops. As commented by the authors, while MAPE serves a reference model, it is not generally considered as a reference architecture. Weyns et al. (2013b) consolidated a number of well-known patterns of decentralized control in self-adaptive systems as well as described them with a simple notation to foster their comprehension. The presented MAPE patterns model different types of interacting MAPE loops with different degrees of decentralization. The set of described patterns include the *coordinated control*, *information sharing*, *master/slave*, and *regional planning*. In addition, the authors discussed drivers that should be considered by designers of self-adaptive systems when choosing one of these MAPE patterns (e.g., optimization, scalability, robustness). Ramirez and Cheng (2010) collected adaptation-oriented design patterns from the literature and open sources that support the development of self-adaptive systems. These patterns aim to facilitate the separate development of the functional and adaptive logic. In their work the authors present 12 adaptation-oriented design patterns for reuse of existing adaptation expertise and cluster them in three groups based on their overall objective in the self-adaptive system: monitoring, decision-making, or reconfiguration. Their patterns are at the level of software design in contrast to our architecture-centric perspective that we adopted in this chapter.

In addition, we had a look at control-based self-adaptation approaches. Patikirikorala et al. (2012) systematically surveyed the design of self-adaptive software systems using control engineering approaches. The authors investigated control methodologies in self-adaptive systems and identified a set of design patterns, that are: feedback control system (with reactive decision making) implemented by 88.2% of the surveyed approaches, feed-forward control system (with proactive control mechanisms) utilized by 10.6% of the analyzed papers, and feedback and feed-forward control systems used by only two over the 161 surveyed papers. A number of adaptive control strategies are also elicited and discussed in their work, being the Fixed and Adaptive the most commonly used (29.7% and 14.9%, respectively).

Research in self-organizing systems has brought forward a number of patterns. De Wolf and Holvoet (2007) consolidated and described a set of patterns to systematically design a self-organising emergent solution such as gradient fields and market-based control. The purpose of the patterns proposed in their work is to help engineers to decide which decentralised coordination mechanisms are promising to solve a certain problem, to provide best practice in using each coordination mechanism and to guide engineers in applying them. The benefits of a self-organising solution, as explained by the authors, are that it is constantly adapting to changes without a central controlling entity and is still robust to failures. In their work, Fernandez-Marquez et al. (2013) provide a catalogue of bio-inspired mechanisms for self-organizing systems in form of modular and reusable design patterns. The authors investigated the inter-relations among self-organising mechanisms for engineering self-systems in order to understand how they work and to facilitate their adaptation or extension to tackle new problems. By analyzing these mechanisms and their behaviors in detail, they identified different levels: lower-level patterns include basic mechanisms (*repulsion, evaporation, aggregation, spreading*) that can be used individually and other more complex mechanisms composed of basic ones (*digital pheromone, gradients, gossip*). Higher-level patterns show different ways to exploit the basic and composed mechanisms (*flocking, foraging, quorum sensing, chemotaxis, morphogenesis*). The presented patterns are best exploited during the design phase.

All these systematic studies have in common that they did not focus on combinations of different types of adaptation mechanisms and cross-layer inter-adaptation interactions as we did in the mapping study and the adaptation patterns described in this chapter.

13.11 Conclusion and Future Work

In this chapter we reported the results of a systematic survey of CPS studies that combine different self-adaptation mechanisms across the technology stack of the system. The results show that the majority of primary studies combine either MAPE with smart elements or MAS with smart elements in CPS architecture design. From the designs of the primary studies, we derived three patterns: SYNTHESIZE-UTILIZE, SYNTHESIZE-COMMAND, and COLLECT-ORGANIZE. These patterns offer problem-solution pairs to engineers for the design of future CPS and CPPS with self-adaptation capabilities, whereby the SYNTHESIZE-COMMAND pattern seems to be particularly relevant for the design of CPPS. Based on the survey results and the background of CIS, we described three emerging directions of how CIS are used in CPS and CPPS: capability augmentation, emergent machine-to-machine interactions, and multi-disciplinary knowledge integration and coordination. We hope that the research results presented in this chapter can contribute to push forward the important field of CPPS in general and the application of self-adaptation to it in particular.

Acknowledgements This work was supported by the Christian Doppler Forschungsgesellschaft, the Federal Ministry of Science, Research and Economy, the National Foundation for Research, Technology and Development in Austria, and Technische Universität Wien research funds.

References

Acatech – National Academy of Science and Engineering: Cyber-Physical Systems: driving force for innovation in mobility, health, energy and production. Tech. rep., http://www.acatech.de/fileadmin/user_upload/Baumstruktur_nach_Website/Acatech/root/de/Publikationen/Stellungnahmen/acatech_POSITION_CPS_Englisch_WEB.pdf (2011)

Barenji, R.V., Barenji, A.V., Hashemipour, M.: A multi-agent RFID-enabled distributed control system for a flexible manufacturing shop. Int. J. Adv. Manuf. Technol. **71**(9), 1773–1791 (2014)

Bauernhansl, T., ten Hompel, M., Vogel-Heuser, B. (eds.): Industrie 4.0 in Produktion, Automatisierung und Logistik. Springer, New York (2014)

Bures, T., Gerostathopoulos, I., Hnetynka, P., Keznikl, J., Kit, M., Plasil, F.: DEECO: an ensemble-based component system. In: Proceedings of the 16th International ACM Sigsoft Symposium on Component-based Software Engineering (CBSE '13), pp. 81–90. Association for Computing Machinery, New York (2013)

Bures, T., Weyns, D., Berger, C., Biffl, S., Daun, M., Gabor, T., Garlan, D., Gerostathopoulos, I., Julien, C., Krikava, F., Mordinyi, R., Pronios, N.: Software engineering for smart Cyber-Physical Systems – towards a research agenda. ACM SIGSOFT Softw. Eng. Notes **40**(6), 28–32 (2015)

Calinescu, R., Grunske, L., Kwiatkowska, M., Mirandola, R., Tamburrelli, G.: Dynamic QoS management and optimization in service-based systems. IEEE Trans. Softw. Eng. **37**(3), 387–409 (2011)

Cheng, S.W., Garlan, D., Schmerl, B.: Architecture-based Self-adaptation in the presence of multiple objectives. In: Proceedings of the International Workshop on Self-adaptation and Self-managing Systems (SEAMS '06), pp. 2–8. Association for Computing Machinery, New York (2006)

De Wolf, T., Holvoet, T.: Emergence and Self- Organisation: a statement of similarities and differences. In: Proceedings of the 2nd International Workshop on Engineering Self-Organising Applications, pp. 96–110 (2004)

De Wolf, T., Holvoet, T.: Design patterns for decentralised coordination in Self-organising emergent systems. In: Brueckner, S.A., Hassas, S., Jelasity, M., Yamins, D. (eds.) Engineering Self-Organising Systems – 4th International Workshop on Engineering Self-Organising Applications (ESOA '06). Lecture Notes in Computer Science, vol. 4335, pp. 28–49. Springer, Berlin/Heidelberg (2007)

Di Marzo Serugendo, G., Gleizes, M.P., Karageorgos, A.: Self-organisation and emergence in MAS: an overview. Informatica **30**(1), 45–54 (2006)

Esfahani, N., Malek, S.: Uncertainty in Self-adaptive software systems. In: De Lemos, R., Giese, H., Müller, H.A., Shaw, M. (eds.) Software Engineering for Self-Adaptive Systems II. Lecture Notes in Computer Science, vol. 7475, pp. 214–238. Springer, Berlin/Heidelberg (2013)

Fatima, S.S., Wooldridge, M., Jennings, N.R.: Multi-issue negotiation with deadlines. J. Artif. Intell. Res. **27**(1), 381–417 (2006)

Fernandez-Marquez, J.L., Di Marzo Serugendo, G., Montagna, S., Viroli, M., Arcos, J.L.: Description and composition of bio-inspired design patterns: a complete overview. Nat. Comput. **12**(1), 43–67 (2013)

Garlan, D.: Software engineering in an uncertain world. In: Proceedings of the FSE/SDP Workshop on Future of Software Engineering Research (FoSER '10), pp. 125–128. Association for Computing Machinery, New York (2010)

Garlan, D., Cheng, S.W., Huang, A.C., Schmerl, B., Steenkiste, P.: Rainbow: architecture-based self-adaptation with reusable infrastructure. Computer **37**(10), 46–54 (2004)

Gupta, A., Pandey, O.J., Shukla, M., Dadhich, A., Ingle, A., Gawande, P.: Towards context-aware smart mechatronics networks: integrating swarm intelligence and ambient intelligence. In: Proceedings of the International Conference on Issues and Challenges in Intelligent Computing Techniques (ICICT '14), pp. 64–69. IEEE, Piscataway, NJ (2014)

Heylighen, F.: Stigmergy as a universal coordination mechanism I: definition and components. Cogn. Syst. Res. **38**, 4–13 (2016)

Hölzl, M., Gabor, T.: Continuous collaboration: a case study on the development of an adaptive Cyber-Physical System. In: Proceedings of the IEEE/ACM 1st International Workshop on Software Engineering for Smart Cyber-Physical Systems (SEsCPS '15), pp. 19–25. IEEE, Piscataway, NJ (2015)

Hu, G., Tay, W., Wen, Y.: Cloud robotics: architecture, challenges and applications. IEEE Netw. **26**(3), 21–28 (2012)

Jazdi, N., Maga, C., Göhner, P., Ehben, T., Tetzner, T., Löwen, U.: Improved systematisation in plant engineering and industrial solutions business – increased efficiency through domain engineering. at-Automatisierungstechnik **58**(9), 524–532 (2010)

Juziuk, J., Weyns, D., Holvoet, T.: Design patterns for multi-agent systems: a systematic literature review. In: Shehory, O., Sturm, A. (eds.) Agent-Oriented Software Engineering, pp. 79–99. Springer, Berlin/Heidelberg (2014)

Kephart, J.O., Chess, D.M.: The vision of autonomic computing. Computer **36**(1), 41–50 (2003)

Kit, M., Gerostathopoulos, I., Bures, T., Hnetynka, P., Plasil, F.: An architecture framework for experimentations with self-adaptive Cyber-Physical Systems. In: Proceedings of the IEEE/ACM 10th International Symposium on Software Engineering for Adaptive and Self-Managing Systems (SEAMS '15), pp. 93–96. IEEE, Piscataway, NJ (2015)

Kitchenham, B.A., Charters, S.: Guidelines for performing systematic literature reviews in software engineering. Tech. Rep. EBSE-2007-01, Software Engineering Group, School of Computer Science and Mathematics, Keele University, UK, and Department of Computer Science, University of Durham, UK. http://www.cs.auckland.ac.nz/~norsaremah/2007GuidelinesforperformingSLRinSEv2.3.pdf (2007)

Kitchenham, B.A., Budgen, D., Pearl Brereton, O.: Using mapping studies as the basis for further research – A participant-observer case study. Inf. Softw. Technol. **53**(6), 638–651 (2011)

Kramer, J., Magee, J.: Self-managed systems: an architectural challenge. In: Future of Software Engineering (FOSE '07), pp. 259–268. IEEE Computer Society, Los Alamitos, CA (2007)

Kumar, N., Singh, M., Zeadally, S., Rodrigues, J.J.P.C., Rho, S.: Cloud-assisted context-aware vehicular Cyber-Physical System for PHEVs in smart grid. IEEE Syst. J. **PP**(99), 1–12 (2015)

Leitão, P.: Agent-based distributed manufacturing control: a state-of-the-art survey. Eng. Appl. Artif. Intel. **22**(7), 979–991 (2009)

Mahdavi-Hezavehi, S., Avgeriou, P., Weyns, D.: A classification framework of uncertainty in architecture-based Self-adaptive systems with multiple quality requirements. In: Mistrik, I., Ali, N., Kazman, R., Grundy, J., Schmerl, B. (eds.) Managing Trade-offs in Adaptable Software Architectures, pp. 45–78. Morgan Kaufmann, Cambridge (2016)

Mamei, M., Menezes, R., Tolksdorf, R., Zambonelli, F.: Case studies for Self-organization in computer science. J. Syst. Archit. **52**(8–9), 443–460 (2006)

Meszaros, G., Doble, J.: A pattern language for pattern writing. In: Martin, R.C., Riehle, D., Buschmann, F. (eds.) Pattern Languages of Program Design 3, pp. 529–574. Addison-Wesley Longman Publishing Co., Inc., Reading, MA (1997)

Minsky, N.H., Murata, T.: On manageability and robustness of open multi-agent systems. In: Lucena, C., Garcia, A., Romanovsky, A., Castro, J., Alencar, P.S.C. (eds.) Software Engineering for Multi-Agent Systems II. Lecture Notes in Computer Science, vol. 2940, pp. 189–206. Springer, Berlin/Heidelberg (2004)

Muccini, H., Sharaf, M., Weyns, D.: Self-adaptation for Cyber-Physical Systems: a systematic literature review. In: Proceedings of the 11th International Symposium on Software Engineering for Adaptive and Self-Managing Systems (SEAMS '16), pp. 75–81. Association for Computing Machinery, New York (2016)

Mukherjee, S., Chaudhury, S.: C-MAP: framework for multi-agent planning in Cyber physical systems. In: proceedings of the 5th International Conference on Pattern Recognition and Machine Intelligence (PReMI '13). Lecture Notes in Computer Science, vol. 8251, pp. 237–242. Springer, Berlin/Heidelberg (2013)

Musil, J., Musil, A., Biffl, S.: Introduction and challenges of environment architectures for collective intelligence systems. In: Weyns, D., Michel, F. (eds.) Agent Environments for Multi-Agent Systems IV. Lecture Notes in Computer Science, vol. 9068, pp. 76–94. Springer International Publishing, Cham (2015a)

Musil, J., Musil, A., Biffl, S.: SIS: an architecture pattern for collective intelligence systems. In: Proceedings of the 20th European Conference on Pattern Languages of Programs (EuroPLoP '15), pp. 20:1–20:12. Association for Computing Machinery, New York (2015b)

Musil, J., Musil, A., Weyns, D., Biffl, S.: An architecture framework for collective intelligence systems. In: Proceedings of the 12th Working IEEE/IFIP Conference on Software Architecture (WICSA '15), pp. 21–30. IEEE, Piscataway, NJ (2015c)

Musil, A., Musil, J., Biffl, S.: Major variants of the SIS architecture pattern for collective intelligence systems. In: Proceedings of the 21st European Conference on Pattern Languages of Programs (EuroPLoP '16), pp. 30:1–30:11. Association for Computing Machinery, New York (2016a)

Musil, A., Musil, J., Biffl, S.: Towards collective intelligence system architectures for supporting multi-disciplinary engineering of Cyber-physical production systems. In: Proceedings of the 1st International Workshop on Cyber-Physical Production Systems (CPPS '16), pp 1–4. IEEE, Piscataway, NJ (2016b)

Musil, A., Musil, J., Weyns, D., Bures, T., Muccini, H., Sharaf, M.: Protocol for: patterns for Self-adaptation in Cyber-Physical Systems – a systematic mapping study. Tech. rep., IFS-CDL 16-02, Vienna University of Technology. http://qse.ifs.tuwien.ac.at/publication/IFS-CDL-16-02.pdf (2016c)

Nasri, M., Farhangi, H., Palizban, A., Moallem, M.: Multi-agent control system for real-time adaptive VVO/CVR in smart substation. In: Proceedings of the IEEE Electrical Power and Energy Conference (EPEC '12), pp. 1–7. IEEE, Piscataway, NJ (2012)

Oreizy, P., Medvidovic, N., Taylor, R.N.: Architecture-based runtime software evolution. In: Proceedings of the 20th International Conference on Software Engineering (ICSE '98), pp. 177–186, IEEE Computer Society, Washington, DC (1998)

Patikirikorala, T., Colman, A., Han, J., Wang, L.: A systematic survey on the design of Self-adaptive software systems using control engineering approaches. In: Proceedings of the 7th

International Symposium on Software Engineering for Adaptive and Self-Managing Systems (SEAMS '12), pp. 33–42. IEEE, Piscataway, NJ (2012)
Perez-Palacin, D., Mirandola, R.: Uncertainties in the modeling of Self-adaptive systems: a taxonomy and an example of availability evaluation. In: Proceedings of the 5th ACM/SPEC International Conference on Performance Engineering (ICPE '14), pp. 3–14. Association for Computing Machinery, New York (2014)
Ramirez, A.J., Cheng, B.H.C.: Design patterns for developing dynamically adaptive systems. In: Proceedings of the ICSE Workshop on Software Engineering for Adaptive and Self-Managing Systems (SEAMS '10), pp. 49–58. Association for Computing Machinery, New York (2010)
Ramirez, A.J., Jensen, A.C., Cheng, B.H.C.: A taxonomy of uncertainty for dynamically adaptive systems. In: Proceedings of the 7th International Symposium on Software Engineering for Adaptive and Self-Managing Systems (SEAMS '12), pp. 99–108. IEEE, Piscataway, NJ (2012)
Sarma, S., Muck, T., Shoushtari, M., BanaiyanMofrad, A., Dutt, N.: Cross-layer virtual/physical sensing and actuation for resilient heterogeneous many-core SaCs. In: Proceedings of the 21st Asia and South Pacific Design Automation Conference (ASP-DAC '16), pp. 395–402. IEEE, Piscataway, NJ (2016)
Vogel-Heuser, B., Diedrich, C., Pantförder, D., Göhner, P.: Coupling heterogeneous production systems by a multi-agent based cyber-physical production system. In: Proceedings of the 12th IEEE International Conference on Industrial Informatics (INDIN '14), pp. 713–719. IEEE, Piscataway, NJ (2014)
Wan, J., Zhang, D., Zhao, S., Yang, L.T., Lloret, J.: Context-aware vehicular Cyber-Physical Systems with cloud support: architecture, challenges, and solutions. IEEE Commun. Mag. **52**(8), 106–113 (2014)
Wang, F.Y.: The emergence of intelligent enterprises: from CPS to CPSS. IEEE Intell. Syst. **25**(4), 85–88 (2010)
Weyns, D.: Architecture-based design of multi-agent systems. Springer, Berlin/Heidelberg (2010)
Weyns, D.: Software engineering of Self-adaptive systems: an organised tour and future challenges. In: Taylor, R., Kang, K.C., Cha, S. (eds.) Handbook of Software Engineering, Springer (2017, to appear)
Weyns, D., Ahmad, T.: Claims and evidence for architecture-based Self-adaptation: a systematic literature review. In: Proceedings of the 7th European Conference on Software Architecture (ECSA '13), pp. 249–265. Springer, New York (2013)
Weyns, D., Georgeff, M.: Self-adaptation using multiagent systems. IEEE Softw. **27**(1), 86–91 (2010)
Weyns, D., Boucké, N., Holvoet, T.: A field-based versus a protocol-based approach for adaptive task assignment. Auton. Agent. Multi-Agent Syst. **17**(2), 288–319 (2008)
Weyns, D., Malek, S., Andersson, J.: FORMS: unifying reference model for formal specification of distributed Self-adaptive systems. ACM Trans. Auton. Adap. Syst. **7**(1), 8:1–8:61 (2012)
Weyns, D., Iftikhar, M.U., Söderlund, J.: Do external feedback loops improve the design of Self-adaptive systems? A controlled experiment. In: Proceedings of the 8th International Symposium on Software Engineering for Adaptive and Self-Managing Systems (SEAMS '13), pp. 3–12. IEEE, Piscataway, NJ (2013a)
Weyns, D., Schmerl, B., Grassi, V., Malek, S., Mirandola, R., Prehofer, C., Wuttke, J., Andersson, J., Giese, H., Göschka, K.M.: On patterns for decentralized control in Self-adaptive systems. In: de Lemos, R., Giese, H., Müller, H.A., Shaw, M. (eds.) Software Engineering for Self-Adaptive Systems II. Lecture Notes in Computer Science, vol. 7475, pp. 76–107, Springer, Berlin/Heidelberg (2013b)
Winkler, D., Musil, J., Musil, A., Biffl, S.: Collective intelligence-based quality assurance: combining inspection and risk assessment to support process improvement in multi-disciplinary engineering. In: Kreiner, C., O'Connor, R.V., Poth, A., Messnarz, R. (eds.) Systems, Software and Services Process Improvement: Proceedings of the 23rd European System, Software and Service Process Improvement and Innovation Conference (EuroSPI '16). Communications in Computer and Information Science, vol. 633, pp. 163–175. Springer International Publishing, Cham (2016)

Wohlin, C., Runeson, P., Höst, M., Ohlsson, M.C., Regnell, B., Wesslén, A.: Experimentation in Software Engineering. Springer, Berlin/Heidelberg (2012)

Wooldridge, M.: Multi-Agent Systems: An Introduction. Wiley, Harlow (2001)

Xiong, G., Zhu, F., Liu, X., Dong, X., Huang, W., Chen, S., Zhao, K.: Cyber-Physical-Social system in intelligent transportation. IEEE/CAA J. Automat. Sin. **2**(3), 320–333 (2015)

Yu, C., Jing, S., Li, X.: An architecture of Cyber Physical System based on service. In: Proceedings of the International Conference on Computer Science and Service System (CSSS '12), pp. 1409–1412. IEEE, Piscataway, NJ (2012)

Chapter 14
Service-Oriented Architectures for Interoperability in Industrial Enterprises

Ahmed Ismail and Wolfgang Kastner

Abstract This chapter focuses on the technological aspects involved in developing a service-oriented solution for interoperability in the context of cyber-physical production systems (CPPS). It addresses the typical state of industrial enterprises and the core technologies currently available for the development of a service-oriented (SO) solution for agile environments. The chapter therefore discusses features of the service-oriented paradigm as well as aspects related to enterprise and network architectures, constraints, and technologies to discern the current challenges facing modern enterprises. The chapter also explores the service-oriented reference architectures of recent EU projects to highlight their main characteristics. Finally, their respective realizations are decomposed to discern the connectivity strategies and standards employed by each to achieve an interoperability-focused technology stack for the operation of agile and flexible industrial plants.

Keywords Horizontal integration • Interoperability • Service-oriented architectures • Technology stack • Vertical integration

14.1 Introduction

The increasing rapidity of change in the environmental factors of modern enterprises requires that they be able to adapt to external and internal stimuli over increasingly shorter timescales (Corrêa, 2001). Industrial enterprises must be agile to withstand and thrive in such dynamic environments. The defining characteristics of their systems should include having "easy access to integrated data whether it is customer driven, supplier driven, or product and process driven", "modular production facilities that can be organized into ever-changing manufacturing nodes", and "data that is rapidly changed into information [for the expansion of] knowledge", amongst other things (Choudri, 2001). Together, such features may fuel a successful agile enterprise.

A. Ismail (✉) • W. Kastner
Institute of Computer Aided Automation, Technische Universität Wien, Wien, Austria
e-mail: aismail@auto.tuwien.ac.at; k@auto.tuwien.ac.at

© Springer International Publishing AG 2017

S. Biffl et al. (eds.), *Multi-Disciplinary Engineering for Cyber-Physical Production Systems*, DOI 10.1007/978-3-319-56345-9_14

From a technical perspective, the pursuit of such characteristics can be supplemented using a number of techniques from the domain of service-oriented architectures (SOA). This is a field that is concerned with the creation of modular IT and productions systems that enhance an enterprise's capabilities for information exchange, technological independence, and component reuse. The result would effectively be an industrial environment of operational flexibility and responsiveness (Valipour et al., 2009).

This chapter focuses on service-oriented architectures and how they may be applied in current enterprises to achieve flexibility, agility and interoperability, all within the context of research question ***RQ I1*** of Chap. 1. As such, it will provide an understanding of the technical state of present industrial enterprises and detail the characteristics of the SO approach in order to highlight the competitive advantage possible through service orientation. A sizeable part of this chapter is dedicated to presenting preliminary SO reference architectures delivered by major European Union research projects. Respective realizations of these architectures will be discussed to underline their choices in technologies and their delivered technical innovations.

14.2 Technical Features of the Industrial Enterprise

Typically, an industrial enterprise undergoes functional segmentation creating layers that distinguish between the components of the process, control, operations, business, and Internet-based systems present in or interacting with the enterprise. Normally, a demilitarized zone (DMZ) is also included to manage access from the uppermost business and enterprise layers to the lower process-focused layers' network and data. This kind of segmentation is done to increase the manageability and security of the enterprise. However, further impositions on the enterprise exist due to the physical and logical constraints of the enterprise's assets. Physically, devices may be connected using legacy serial interfaces (e.g. EIA-232/422/485), fieldbus systems, and wired and wireless Ethernet and IP-based technologies. Although devices with both interface types may be used to physically bridge the two networks together, the protocols may have different demands in real-time (RT), bandwidth, latency, and other communication-related requirements. At the messaging layer, these protocols may also differ in their message formats and exchange mechanisms, as well as in other features. The process of bridging together these various systems requires special devices and techniques, designed and implemented carefully to safely allow them to share data (Ismail and Kastner, 2016; Knapp, 2014).

By convention, there are two approaches for managing this heterogeneity and technical complexity, namely tunneling and translation. Tunneling involves the use of routers for the encapsulation of one protocol's data inside the payload of another and treating the channel as a transparent communication medium. Translation, on the other hand, uses gateways as intermediaries to carry out data mappings on behalf of communicating devices to allow them to exchange information using their native protocols. Each of these methods has its own drawbacks in relation to capabilities and implementation complexity, and is often considered to be costly in engineering efforts (Sauter et al., 2011).

These costs may be compounded by the technological choices and implementations of enterprise infrastructures, which commonly use monolithic applications to achieve their functional goals. The resulting system involves many implicit and explicit dependencies (i.e., to a technology stack) that introduce stiff resistance to change. In such a case, the enterprise cannot be described as agile or flexible. In fact, it is properties such as these and their implications that have become major arguments used by the proponents of service oriented architectures to effect infrastructural changes in enterprises (Krafzig et al., 2005).

Take for example the concept of a service-oriented CPPS. CPPS are physical and computational resources that are tightly bound in coordinated and controlled relationships and embedded in a socio-technical context. The functionalities originally addressed by monolithic applications may be decomposed and distributed across the system's member devices. That is, by applying the service-oriented design pattern, a large business problem can be fragmented into smaller problems that may then be solved using a number of small and related units of logic, termed services, rather than through a single monolithic application. Distributing these services across the system would create networks of smart devices that are inherently resilient due to their lack of dependence on any central component. Furthermore, as services are typically designed with standardized interfaces, system-wide interoperability is guaranteed. The internals of these services, such as how they are implemented, or what technology they use is hidden behind the service interface therefore also affording the system technological independence and flexibility. Services are also typically designed with functional agnosticism to allow for their reuse, reducing future application development efforts. Together, the concepts that define the SO approach, all of which are summarized from Valipour et al. (2009), Erl (2009), and Erl et al. (2014) and presented in Table 14.1, afford an enterprise the agility it pursues and minimizes the need for integration (Krämer, 2014; Rubio, 2011; Erl, 2009; Valipour et al., 2009).

For reasons such as those mentioned above, several research projects have pursued and outlined reference architectures for highly interoperable industrial environments based on SOAs. These efforts have been extensively documented to facilitate future system implementations. In the coming section, we highlight a number of these projects, summarizing and evaluating their reference architectures to give a concise understanding of their details.

Table 14.1 Features of SOAs (Valipour et al., 2009; Erl, 2009; Erl et al., 2014)

Characteristic	Sub-characteristic	Explanation
Discoverability		Services are supported using metadata that allows them to be discovered and interpreted
Modularity	Modular decomposability	The equivalent concept of functional decomposition as applied to modules
	Modular composability	The ability to create a software service or system by freely combining reusable services
	Modular understandability	The function of a service should be comprehensible without requiring knowledge on any other services
	Modular continuity	A service interface should conceal service implementation details to allow changes to the service to occur without them requiring changes in other services
	Modular protection	Perimeterisation of modules to prevent the cascading of faults unto other services
Interoperability		Ensured ability for different modules to communicate with each other
Loose coupling		Appropriately defined service contracts that increase the independence of services from their implementations and from each other
Location transparency		The decoupling of a service from a specific location allowing dynamic service lookups and runtime binding that enhance the system's flexibility, availability and performance
Composability	Application	The piecing together of services and their orchestration using application logic to achieve specifically set goals
	Service federation	The aggregation of services under a single service representation
	Service contracts	The explicit definition of a service's features and parameters as contractual terms and conditions in a granular form accessible by service requesters. This may include the definition of supported data types, data models, policies and other features that declare a service's interaction requirements
	Service orchestration	Service execution as part of an application should be sub-transactional and not permitted to perform data commits. This allows the system to rollback to the pre-transactional state in case of service failure

14.3 Service-Oriented Architectures and the Industrial Enterprise

In this section, we look at a total of five service-oriented reference architectures that resulted from collaborations between research, vendor, and user organizations These are the Internet of Things at Work (IoT@Work), Embedded systems Service-based Control for Open manufacturing and Process automation (eScop), Production Logistics and Sustainability Cockpit (PLANTCockpit), ArchitecturE for Service-Oriented Process—Monitoring and Control (IMC-AESOP), and Arrowhead Framework projects; the inferred or explicitly stated purpose of each of these projects is summarized in Table 14.2. This list of projects is not comprehensive as there exists a larger number of SOA-driven projects both old, such as SOCRADES and SODA, and new, such as ProSEco and CREMA, that are not covered in this chapter. Nor is it the point of this chapter to give such a viewpoint, but the purpose here is to bring attention to the main features of a sample of SO reference architectures to underline their strategies for the modernization of industrial enterprises. As such, our analysis is limited to projects that started and completed within the period of 2010–2016. Arrowhead does not meet this timeline, as it is scheduled for completion in 2017, however, an exception is made in this case as, with a 77-member consortium and 69

Table 14.2 A summary of project durations and objectives

Name	Duration	Objective
IoT@Work	Jun 2010–Jun 2013	Use IoT technologies to decouple application/control programming from the network, enable communication-centric plug & work capabilities, and enhance the system security (Rotondi et al., 2013)
PLANTCockpit	Sept 2010–Dec 2013	Creating a SO and centralized plant-wide human-machine interface (HMI) (PLANTCockpit Consortium, 2011a)
IMC-AESOP	Sept 2010–Dec 2013	Using SO approach for supervisory control and data acquisition/distributed control system (SCADA/DCS) in large-scale process control systems (Colombo et al., 2014a)
eScop	Mar 2013–Feb 2016	System integration using ontology based knowledge-management, embedded devices, and SOA (eScop Consortium, 2013)
Arrowhead framework	Mar 2013–Feb 2017	Providing a SO technical framework for cooperative automation in technologically heterogeneous systems (Blomstedt et al., 2014b)

million Euros in funding, it is one of the largest EU projects in the industrial domain (Nagorny et al., 2014).

14.3.1 IoT@Work

The IoT@Work project represents its reference architecture using layers and planes. In terms of the former, three layers are used; the physical, abstraction, and composite service layers. The first of these, the physical layer, is the physical world and is therefore composed of physical devices. The second layer is an abstraction of the physical devices as resources and services. In the context of the IoT@Work architecture, a resource is an object representing a specific physical or virtual element, while a service gives access to a resource by specifying the type, identifier and interface. Effectively, a single device may be represented using one or more resources and services. The third and final layer is that of composite services. These group together the elements of the second layer to hide their complexity and deal with context, contention over resources, and access rights. It is this third layer atop which applications such as event notification, complex event processing (CEP), network access control (NAC), and controller I/O applications run (Rotondi et al., 2013).

To address the functional aspects of these three layers, the IoT@Work project defines a set of core services, listed and defined in Table 14.3, and organizes the large number of functional components they are composed of into three planes; the communication, security, and management planes. The communication plane is concerned with the orchestration of network resources and communication to resolve access contention issues and provide support for Quality of Service (QoS) guarantees. The security plane, as the name implies, manages and integrates security into the overall system. Lastly, the management plane, attends to device, service, and configuration management with a focus on their inter-relations (Rotondi et al., 2013).

14.3.2 PLANTCockpit

The PLANTCockpit system architecture is composed of an internal and external system. The external system refers to the data sources connected to the PLANT-Cockpit using proprietary or open interfaces, such as an Enterprise Resource Planning (ERP) system, OPC server, or sensors & actuators. As for the internal system, this is made up of five layers; the system connector, function engine, persistence, visualization engine, and presentation engine layers (PLANTCockpit Consortium, 2011a).

The first of these, the system connector layer, is primarily concerned with interfacing with external data sources. It provides the configurable *adapter* modules

Table 14.3 IoT@Work core services (Rotondi et al., 2013)

Core service	Explanation
Event notification service (ENS)	A common functional component that collects and distributes events
ENS access request broker (ARB)	A broker between ENS clients attempting to access namespaces and the ENS AS
ENS authorisation service (AS)	Decision point for access requests sent to the ENS ARB
Policy decision point (PDP)	Evaluates the status of capability tokens and policies to approve or refuse access requests
Revocation service	Manages capability revocation requests and capability revocation life cycles
ENS namespace management service	A service for the management of hierarchical structures used for the organization of event publishing
Slice management system	A three part service consisting of a communication service interface (CSI), slice enforcement point (SEP) and slice manager. Used for the creation of a 'slice'[a]
Embedded application configuration service	Provides devices with the configurations required by their applications
Directory service (DS)	Stores device information in an ontology-based DS data model
Orchestrated management	Orchestrated management authoring support: a lightweight algorithm and API
	Orchestrated management scheduling service: algorithms to produce management plans and schedule operations
	Management services: a wrapper around existing operations in the three planes so that they may be used and executed in Orchestrated Management scenarios
	Context services: capture constraint values to provide context. May be a parameter of the manufacturing execution system (MES) or ERP system
Complex event monitoring service	Responsible for the verification of rule compliance to allow the system to meet safety and security goals

[a]A slice is a virtual network with QoS guarantees and policies

required to allow the PLANTCockpit to access and communicate with these sources. Due to their configurability, an *adapter manager* is included in the architecture to oversee the entire life cycle of adapters. As for the external data structures acquired through the adapters, these are transformed to an internal data structure using a *mapper* module. Finally, the layer uses two generic components, the *subscriber* and *publisher*, to query the external systems via the adapters and push the data retrieved by way of the adapter and mapper components to the function engine layer, respectively (PLANTCockpit Consortium, 2011a).

The function engine layer, receiving these data, provides a platform atop where analytics and functions may be executed. It is based on the concept of *function blocks*, which, inspired by the object oriented paradigm and the IEC 61499 standard, are reconfigurable and encapsulated blocks of program code with clearly defined interfaces to allow for reuse and composability. These blocks' life cycles are managed using a *function manager*, while a *pub/sub broker* (publisher/subscriber) provides them with secure, reliable, and event-driven mechanisms through which they may communicate with each other (PLANTCockpit Consortium, 2011a).

Any data relevant to the function engine or any other layer's workings are managed and stored using the persistence layer. This subsystem is composed of three components; the *data persistence manager*, *configuration repository*, and *data repository*. The manager administers the storage, archiving, retrieval, and deletion of data. The configuration repository maintains all of the data needed to configure the internal components of the PLANTCockpit system at design and run time. Finally, the data repository stores all of the data required by analysis processes in the PLANTCockpit system. It includes a cache that can temporarily store data to improve the system performance, and a more permanent store that archives historical data (PLANTCockpit Consortium, 2011a).

Finally, any data to be presented via the HMI interface is prepared for visualization using the visualization engine layer. It consists of a service engine which is an aggregation of a runtime and design engine. The former contains the configurable user interface (UI), while the latter configures the interface using a composition of *building blocks* (visualization elements) and their associated data points. A *building block browser* and *function block browser* is used to make all possible building blocks and all available data points provided by the data persistence manager accessible by the design engine for the UI's configuration, respectively. Finally, a *data provider* component subscribes to the Pub/Sub Broker for data and events that it delivers to the runtime engines. The presentation engine layer, which is composed solely of a *presentation runtime engine*, then graphically presents the configured building blocks (PLANTCockpit Consortium, 2011a).

14.3.3 IMC-AESOP

As opposed to the PLANTCockpit framework, the IMC-AESOP architecture attempted to provide a generic architecture to support multiple applications, with the HMI only being one of these applications. As such, the framework is in fact a behemoth of services, service groups, and interactions presented using both natural language and semi-formal descriptions based on the Fundamental Modeling Concepts (FMC) graphical notation. The abridged description of the framework's components are shown in Table 14.4 (Colombo et al., 2014b).

Based on the architectural overview given in Colombo et al. (2014b), the framework differentiates between four system components. The first of these is the user roles, which designates users *business*, *operations*, *engineering*, *maintenance*, or *training* roles. These roles interact with or impact the architecture directly or indirectly as they take part in plant processes. The second system component is that of the service groups themselves. These act as the glue binding together the user roles, the external systems (the third component), and the plant data itself (the fourth) (Colombo et al., 2014b).

14.3.4 eScop

The eScop project is composed of five layers, a *physical* (PHL), *representation* (RPL), *orchestration* (ORL), *visualization* (VIS), and *interface* (INT) layer. The PHL is concerned with the physical equipment in the eScop system and therefore provides device and service descriptions. The information provided by the physical layer is consumed by the RPL, which is responsible for knowledge representation. Syntactic and semantic service descriptions based on the service implementations are mapped and stored in the RPL. The ORL, which coordinates and executes service compositions, in addition to requiring service descriptions from the RPL, may also need input from the physical layer for it to be able to successfully orchestrate the execution of services. The VIS, configured by the RPL, then provides an interface for the user to interact with the system data accessible via the PHL. Finally, the INT acts as the entry point for external systems and services in the eScop architecture and is functionally concerned with the provision of technology adapters and access control measures (Iarovyi et al., 2015b).

As for the features of the reference architecture, the system's services define concrete technologies for implementation. To exemplify, the various components of each of the layers are as follows.

Starting from the bottom up, the PHL consists of an *I/O module*, *runtime core*, and *Web Services (WS) toolkit*. These support connections to the physical devices, the definition of applications on controllers, and provide the devices with web services and notification mechanisms, respectively (Iarovyi et al., 2015b).

The RPL achieves its goal of knowledge representation using an *ontology service*, a set of functions, and the ontology itself. The ontology service is in fact made up of four modules: *device registration*, *visualization provider*, *ontology manager*, and *service handler* modules. These handle device registration and de-registration in the ontology, assist with visualization, provide an interface for the

Table 14.4 IMC-AESOP service groups and services (Colombo et al., 2014b)

Service group	Component services	Explanation
Alarms	Alarm configuration Alarm & event processing	Defines, maps, filters, and aggregates alarms based on the principles of CEP. Supports simple alarms. May be time and/or event/alarm-triggered
Configuration & Deployment	System configuration service Configuration service Configuration repository	Responsible for the configuration, deployment, and enforcement of configurations on the various plant elements. The model repository service provides it with a structure for saving hierarchical configuration structures of nodes. The service may also manage the versioning of services and the instantiation of plant meta-models
Control	Control execution engine	Typically a distributed service, it can execute process models and configurations and support the online reconfiguration of processes and hot-standby redundancy
Data management	Event broker Data consistency Actuator output Historian Sensory data acquisition	Responsible for "data retrieval, consistency checking, storage, searching", basic eventing, and actuator output control. The service connects data producers with higher level services and provides methods for mapping data to the appropriate data models and ontologies
Data processing	Filtering Calculation engine CEP service	Provides services for basic filtering, CEP and calculations. The CEP engine allows for low-latency analysis of event data. It provides a management interface that allows for the creation, update or removal of rules used for processing events. The calculation engine supports users in executing numerical or logical operations over process data
Discovery	Discovery service Service registry	Supports the discovery of system components by type and location. Uses a registry to support the discovery of services implemented using technologies without inherent discovery capabilities, and to allow for discovery by remote entities where multicast and broadcast based discovery would be of limited usefulness
HMI	Graphics presentation	Provides a generic web interface where graphical tools may be embedded
Integration	Composition service Service mediator Gateway Business process management and execution service Model mapping service	Responsible for ensuring interoperability between heterogeneous components using translation and encapsulation. Also serves as a platform for the execution of business processes as service compositions and their presentation as higher level services

Lifecycle management	Code repository Lifecycle management service	Covers "aspects such as maintenance policies, versioning, service management and also concepts around staging (e.g. test, validation, simulation, production)". It also contains a code repository service which maintains the source code of services to allow for service maintenance, deployment, upgrade, and other functions
Migration	Infrastructure migration solver Migration execution service	Responsible for the migration of legacy systems to the SOA-based approach by identifying system dependencies, offering migration strategies, and executing them
Mobility support	Mobile service management	Concerned with the management of mobile assets and so is responsible for asset tracking, address mapping, data synchronization, and similar supporting functions
Model	Model management service Model repository service	Consists of generic services for the management of models, a repository to structure these models, and an interface to the repository
Process monitoring	Monitoring service	Provides HMIs with an entry point into the system. Responsible for gathering information from the physical process using other system components and semantically enriching it, and for handling alarms and events
Security	Security management service Policy management service	Enforces and executes security measures for confidentiality, authentication, and other features. Also defines and manages security rules/policies for identity or role-based access control and for the definition of identity federation
Simulation	Constraint evaluation Simulation scenario manager Simulation execution Process simulation service	Is connected to almost every other service group as it simulates systems and their processes, evaluating constraints and simulating execution. It consumes other systems' exposed simulation endpoints to imitate characteristic system performance and behavior features
System diagnostics	Asset diagnostics management Asset monitor	Concerned with monitoring the status and health of plant assets and mainly used for maintenance and planning
Topology	Naming service Network management service Location service	Allows for reporting and management on the system's physical and logical features. It includes domain name service (DNS) functionality, device discovery and management, and asset location services

configuration or editing of the model, or manage the RPL's connections to the various components of other layers. The governance of access to the ontology in the triplestore, on the other hand, is done by the RPL's *SPARQL query factory* and *ontology connector* internal functions, and it is suggested that, once secured, SPARQL-over-HTTP may be used for these factors. As for the ontology itself, this in fact is stipulated as being the eScop Manufacturing Systems Ontology (MSO), which is a proprietary component created by one of the designing members of the architecture (Iarovyi et al., 2015b; eScop Consortium, n.d.(a)).

The ORL coordinates the various components in the architecture using a *service composer* and an *orchestration engine*. The former maps process definitions to configurations applicable to the system, and the latter executes them (eScop Consortium, n.d.(a)).

The VIS aims to allow for flexible and generic graphical interfaces. For this it needs a *dynamic composition* module, a *symbol library*, and *visualization agent(s)*. Together, the VIS is able to map descriptions from the RPL to visualization elements from the symbol library which are then transformed by the agent(s) into web pages that can be displayed using a web browser (Iarovyi et al., 2015b; eScop Consortium, n.d.(a)).

14.3.5 Arrowhead Framework

Finally, the Arrowhead project divides its framework into three parts that are *design guidelines*, *documentation guidelines*, and a *software framework*. The first provides a description of design patterns for making legacy or newly created application systems compliant with the Arrowhead Framework. The documentation guidelines provide templates for the description of services, systems and system-of-systems. As for the software framework, this is the main concern of this section and is described in more detail below (Blomstedt et al., 2014b).

The software architecture defines a grouping of core services. These services are meant to support communication exchanges between domain-specific application services. These core services and systems are effectively divided into three groups: *Information Infrastructure* (II), *Systems Management* (SM), and *Information Assurance* (IA). II provides service descriptions and information on how to connect to services and systems. SM core services are concerned with orchestration and system-of-systems composition. Finally, those of IA address security and safety factors in information exchange processes. The categorization of core services and systems identified by the framework under these three groups is shown in Fig. 14.1 (Blomstedt et al., 2014a,b).

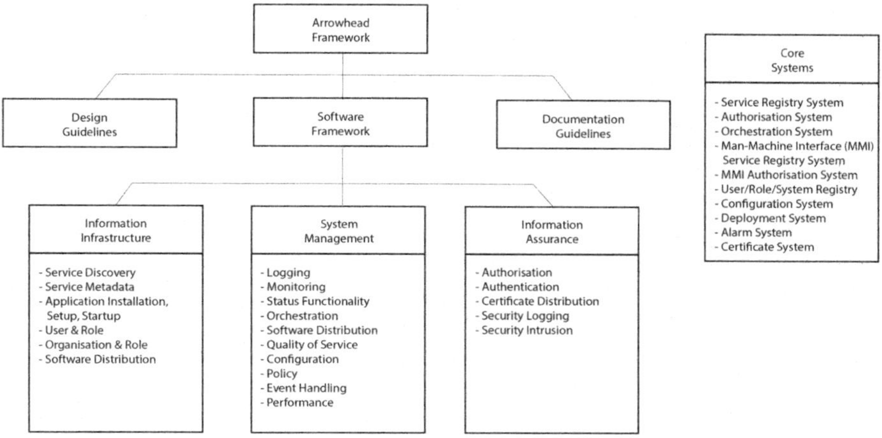

Fig. 14.1 The Arrowhead framework (Blomstedt et al., 2014b; Varga et al., 2014)

14.4 Realizations of the Reference Architectures

So far, the reference architectures have been described in an abstract manner. This portion of the chapter inspects the technology stacks implemented by each of the architectures. For comparability, we segregate these technologies into categories that address the various functional aspects addressed by all of the architectures. We provide brief descriptions on both the mature and novel technologies implemented for each category to give a succinct overview of the technical properties of each project in the pursuit of achieving interoperability.

14.4.1 Service Discovery

An essential aspect present in any service-oriented architecture is that of service discovery. Due to the close association of the device profile for web services (DPWS) with SOAs, the use of the web services dynamic discovery (WS-Discovery) specification is common. Four out of the five architectures, excepting eScop, either directly implemented or discussed methods to allow for the use of the WS-Discovery protocol.

The WS-Discovery protocol is based on the use of multicast messages (typically SOAP-over-UDP) to announce or probe for services using specially crafted eXtensible Markup Language (XML) documents. Announcements operate using multicast *Hello* and best-effort *Bye* messages. Likewise, *Probe* and *Resolve* messages are also multicast; the former is used to locate services based on service types and/or scopes, while the latter searches for a specific service by name. The use of a *Discovery Proxy* is encouraged to allow for the active suppression of multicast traffic in the network.

The specification also endorses the caching of multicast service advertisements to incur further savings. Finally, with respect to securing the discovery process, the specification does not require, but recommends the use of unique XML signatures and a number of other properties to mitigate against a variety of attacks (Bullen et al., 2009).

The IMC-AESOP is one of the architectures implementing the WS-Discovery protocol directly, for example, to allow Service Bus instances to discover each other (Nappey et al., 2013). However, one of the main contributions of IMC-AESOP hinges on its bridging of DPWS with the industry-focused OPC Unified Architecture (OPC UA) standard. The architecture therefore presents concepts for supplementing OPC UA's discovery mechanisms using WS-Discovery. To elaborate, the OPC UA discovery protocol requires the use of a discovery server. The address of the server must be known beforehand by participating OPC UA clients and servers. The IMC-AESOP approach presents two methods for auto-discovery in OPC UA systems using WS-Discovery. The first involves the use of WS-Discovery to allow OPC UA clients and servers to automatically find the OPC UA Discovery Server, while the second approach involves replacing the OPC UA Discovery Server with the WS-Discovery protocol to allow OPC UA clients to find OPC UA servers directly (Bony et al., 2011).

A second core technology in IMC-AESOP is the *Constrained Application Protocol* (CoAP). Identified as a suitable protocol for device-level integration of constrained devices, such as those belonging to wireless sensor networks (WSN), the IMC-AESOP approach discusses a reliance on the discovery mechanisms of CoAP, CoAP multicast, and the Constrained RESTful Environments (CoRE) Resource Directory (RD), for the location of services and resources hosted on resource-limited clients (Eliasson et al., 2013; Kyusakov et al., 2014).

As for PLANTCockpit, as the architecture is dependent on the concepts of adapters to interface with the various systems and function blocks, the discovery mechanism employed is dependent on the system being interfaced. For example, the system implements a DPWS adapter to allow for the discovery of DPWS devices. The adapters themselves, however, are implemented as function blocks based on the concepts of IEC 61499 function blocks. The identification of FBs depends on FB Service Interfaces, and these are implemented using the OSGi framework. As such, although not explicitly stated, it may be the case that the implementation depends on OSGi's service registry to register, get, or listen for services (Iarovyi et al., 2013; Dennert et al., 2013).

Although the IoT@Work approach explored and compared the WS-Discovery and OPC-UA's discovery mechanisms, it did so within the context of auto-configuration. Instead, the IoT@Work system uses a configurable RESTful Directory Service (DS) as a form of service registry. Devices interact with the DS using a RESTful API to retrieve, submit or delete device and service information using HTTP GET, PUT, POST and DELETE requests. The RESTful nature of the service allows every service to be modeled as a URL-accessible resource. The system also supports the use of QR codes and NFC tags for the identification of devices (Rotondi et al., 2013).

Similar to IoT@Work, the eScop project generates its own discovery mechanisms. Discovery here is based on the multicasting of Hello, Bye and Probe messages, and in this respect, it is similar to the WS-Discovery specification (Iarovyi et al., 2016). Inspection of the source code,[1] however, shows that the protocol does not, in fact, follow the WS-Discovery specification. This is as a number of critical differences exist, such as the use of JavaScript Object Notation (JSON) encoding for messages, and of multicast IPs and ports different than those stipulated for use by WS-Discovery.

Finally, the Arrowhead project defines three approaches for service discovery. The first is a service registry functionality based on the Domain Name System (DNS) and DNS Service Discovery (DNS-SD). Effectively, DNS is a hierarchical database mechanism that can store any kind of data and DNS-SD is a method for specifying how DNS resource records may be named, structured, and browsed. These records may be accessed using unicast DNS requests or multicast DNS (mDNS). The Arrowhead framework applies mDNS for constrained devices, such as those belonging to WSNs. The second and third approaches for discovery in the Arrowhead projects are based on the use of XML-over-HTTP and JSON-over-HTTP for RESTful web services. The former uses a DNS protocol specific to Arrowhead to allow for service discovery and the retrieval of service and data descriptions. The JSON-over-HTTP approach is marked to be done and an implementation still remains to be published. The framework does however discuss the prospect of implementing a translation service for integration between the XML/JSON and DNS-SD registry systems (Delsing, 2015; Cheshire and Krochmal, 2013; Arrowhead Consortium, 2014a; Blomstedt and Olofsson, 2016b; Blomstedt, 2016a,b; Blomstedt and Olofsson, 2016c).

14.4.2 Service Description

For this subsection, we focus specifically on the aspect of service contracts as defined in Table 14.1. The goal of service contracts is to define a minimal level of interoperability and thereby reduce the need for integration. It may do so by making available definitions of the service's functionality, data model, data transfer mechanisms and encodings, and policies for security and quality of service, amongst other things. Naturally, these contracts themselves need to be interpretable by all services available in the registry or operating environment. Similar to what was the case for service discovery, a web services technology, the Web Services Description Language (WSDL), is employed by a number of projects (Erl, 2008).

The WSDL language is a machine-readable XML-based language for the definition of interfaces. It is capable of describing all of a service's operations, the data required and output by each operation, and their respective data types.

[1] http://www.escop-project.eu/tools/.

It may also provide addressing and networking information to support inter-service connectivity. Both of the IMC-AESOP and PLANTCockpit projects use WSDL files for service descriptions. However, for describing sensor services, the IMC-AESOP project uses the JSON, XML and EXI-compatible Sensor Markup Language (SenML) (PLANTCockpit Consortium, 2012c; Colombo et al., 2014b).

The eScop project also employs a WS-based approach and develops a WS-enabled Remote Terminal Unit (RTU) titled the eScopRTU. Within the eScopRTU, all services are IEC 61131 Structured Text Language (STL) functions that are used to execute operations on resources. They are RESTful, hypermedia-driven, accessible via a REST API, and the API documentation, read "service descriptions", are created using Swagger. Swagger is a specification for the definition of language-agnostic, human and machine-readable representations of RESTful APIs. The specification requires that the API be described using either JSON or YAML[2]; the resulting files may then be processed by tools that can generate clients in a variety of languages. The Swagger ecosystem also includes tools to display and test the API. Swagger has since been renamed the OpenAPI Specification (OAS) (Faist and Štětina, 2015; Ratovsky et al., 2014).

IoT@Work, as previously mentioned, explored the prospects of auto-configuration using WS-Discovery and OPC UA. Part of the procedure outlined involves the acquisition of service descriptions. With WS-Discovery, this was achieved by having metadata on a DPWS-enabled device retrieved by its controller using the WS-Transfer protocol. The metadata that may be included is defined as part of the WS-MetadataExchange specification. This metadata would allow the service to share WSDL definitions, XML schema, policy expressions and so on. For the case of OPC UA, the GetEndpoints Service, which is part of the OPC UA standard, retrieves the information required to allow for secure communication between clients and servers. The information mainly consists of addressing and security policies and definitions (Dürkop et al., 2012; Ballinger et al., 2008; Mahnke et al., 2009).

Service description in Arrowhead is dependent on the method of service discovery implemented. As previously mentioned, the DNS system uses DNS-SD guidelines to organise the resource records. In such a case, the specification already allots a structure for the definition of addressing, service name, and other connection-related information. For any additional requirements, the DNS TXT record is capable of accommodating such information. The XML-based system uses XML schema for the description of data, and the Web Application Description Language (WADL), which is WSDL's counterpart for RESTful services for the description of service interfaces. The JSON-based system, as previously mentioned, is yet to be defined (Blomstedt and Olofsson, 2016b; Blomstedt, 2016a).

[2]http://www.yaml.org/.

14.4.3 Data Representation and Access

The aspect of data representation and access has been covered somewhat partially while discussing realizations of service descriptions. Our focus here is a more in-depth description of the information or data model and the semantics used by the architectures' respective implementations. The main purpose of these models and semantics is to homogenize the representation of information or data in the entire system ensuring accessibility by all participants and avoiding the need for any data transformation (Erl, 2008).

Of the five projects only two, IMC-AESOP and Arrowhead, rely primarily on mature standardized models for the representation of data. In IMC-AESOP, the system applies the OPC UA information model to link up information in the majority of the enterprise, save the lowest of layers, which instead uses a custom data model based on the Sensor Markup Language (SenML). The former, OPC UA, contains a flexible address space that can be used to create information models that capture objects, their attributes, and relationships. These objects are known as nodes in the OPC UA address space and can be used to represent physical or virtual components. The resulting information models create full-mesh networks of these nodes, with associated properties and relations, and are exposed to applications through OPC UA servers (Colombo et al., 2014b; Mahnke et al., 2009; Henßen and Schleipen, 2014).

As for SenML, the associated specification defines a data model suitable for highly constrained devices, such as sensors and actuators. It does so by having a minimalist approach where the goal is to maximize the amount of information not included in a message while still allowing for self-describing data that includes measurements and meta-data. The result is a single array data model that contains a series of data records. The records can contain the device's unique identifier, a time stamp, the measurement value and unit, amongst other details, for example, to allow for the description of the measurements and device, in addition to the measurement values themselves. The IMC-AESOP project, however, claims that the level of granularity of information carried by SenML's data model are insufficient for its needs and therefore create a custom data model that is primarily and heavily based on SenML to achieve this granularity (Jennings et al., 2016; Colombo et al., 2014b).

Similar to IMC-AESOP, Arrowhead also identifies OPC UA and SenML as suitable candidates for the implementation of data structures and semantics. However, it also highlights the Home Performance XML standard and the CoRE Link Format as appropriate for its needs. The former is a set of data standards that define a number of XML schema and associated data elements to allow for the description of customers, contractors, buildings (and their components and systems), and energy performance factors such as conservation, consumption, and savings, both as actual readings and as estimates. The goal of these standards is therefore, as the name implies, standardization in the collection and transfer of information in the domain of home performance. The latter, the CoRE Link Format, is a realization for exposing the

URIs of resources on constrained devices and networks. It does so by extending the HTTP Link Header format to include the URI descriptions, such as resource relations and attributes, as a message payload, and specifying an entry point URI as a default request path for the retrieval of these URIs. We do note, however, the existence of divergences from the semantic and modeling technologies listed in the Arrowhead guidelines as the energy production demonstrator pilot, for example, adopts the domain-specific language, Thing Markup Language (ThingML), instead of the ones listed above for its semantic needs (Varga et al., 2014; Building Performance Institute, 2013a,b; Shelby, 2012; Arrowhead Consortium, 2014b).

Opposed to IMC-AESOP and Arrowhead are the ontology-driven eScop and IoT@Work models. eScop, as previously mentioned, develops a proprietary Manufacturing System Ontology (MSO) based on OWL to describe the system components, their attributes and relationships. The MSO is in fact the evolved form of the Politecnico di Milano Production Systems Ontology (P-PSO), which is a general taxonomy for discrete manufacturing systems. The MSO extends the P-PSO to include logistics and process production from the perspective of control. The MSO also incorporates concepts to allow for the visualization of the respective systems and their data. The information is stored in an RDF triple-store database that supports SPARQL-over-HTTP to allow for web-based interactions with the ontology (Fumagalli et al., 2014; Negri et al., 2015; eScop Consortium, n.d.(c)).

The IoT@Work project also follows an ontology based approach by storing information on devices in a DS-specific data model (ontology) that uses an RDF-triple-store. In addition to RDF, the model is also inspired by the uCode Relation Model that models device profile attributes as subject-predicate-object triples. Effectively, the resulting DS model is a connected directed graph where the vertices are physical or virtual entities or primitive elements and the edges in the graph represent the relationships between the various entities and elements. The DS is also capable of validating and handling requests for information on devices acquired through an exposed RESTful interface. This information may be retrieved from the database or by collecting this information directly from connected devices. The IoT@Work-compliant devices, unlike the DS, use the OPC UA address space and information model. The project also supports the retrieval of information from SNMP compliant devices. Mappings between the respective device and DS models are therefore a necessity (Rotondi et al., 2013; Imtiaz et al., 2013).

The PLANTCockpit project presents its own metamodel for a database schema and an XML schema for the storage of visualization engine configurations. The database schema consists of four customizable data types and runtime data types to allow for the persistence of analytics-relevant data. The XML schema contains a number of elements to allow for the rendering of SVG components in HTML5 pages, and their linking to data points to create a configurable and compound HMI made up of different graphical elements (PLANTCockpit Consortium, 2012b).

14.4.4 Information and Message Encoding

For message encoding, all of the projects employ XML and extend support for one or more other specifications. XML is in fact a platform-independent data structuring format that defines rules for the textual encoding of human and machine-readable data. It allows for user-defined tags and different data types and processing methods. By allowing for the definition of syntax rules and standardized contracts through the use of document type definition (DTD) or XML schema definition (XSD) descriptions, the validation and verification of encoded data structures are possible. Several specifications have since been defined for the binary-encoding of XML documents to address the overhead and performance issues associated with XML. Of the possible choices, the Efficient XML Interchange (EXI) specification is employed by the IMC-AESOP and Arrowhead projects for the compact exchange of information (PLANTCockpit Consortium, 2012d; Hill, 2015; Varga et al., 2014; Colombo et al., 2014b; Rotondi et al., 2013; Negri et al., 2016).

Other than XML, JSON is also widely employed, being used by all but PLANTCockpit. Stipulations in the PLANTCockpit approach do however allow for the inclusion of JSON. This is the case, for example, with the configuration connector module which is required to be format-agnostic in handling configuration data. As for the JSON specification itself, JSON, like XML, is a data structuring specification that defines rules for the formatting of exchangeable and human and machine-readable data. Tools for the parsing and generation of JSON exist for a large number of programming languages therefore making it a popular alternative to XML. It follows a minimalist encoding approach, using a small number of characters to denote the structure and value of data. Similar to XML, JSON allows for the definition of JSON-based schema for the validation of resulting encodings. Binary encodings, such as the Concise Binary Object Representation (CBOR) specification, exist for JSON. However, aside from the mention of CBOR support as a long-term goal for Arrowhead's historian, it does not appear as though any of the projects include a binary encoding for JSON in their respective stacks (Colombo et al., 2014b; Rotondi et al., 2010; Varga et al., 2014; PLANTCockpit Consortium, 2011b; eScop Consortium, n.d.(b); Galiegue et al., 2013a,b; Eliasson, 2015).

Aside from XML, EXI, and JSON, three other formats used are OPC UA Binary, HTML, and XML-binary Optimized Packaging and Message Transmission Optimization Mechanism (XOP/MTOM). OPC UA Binary, as the name implies, is the binary protocol for OPC UA. Like other binary representations, it is the performance and overhead-sensitive data format for OPC UA. It is used in both IoT@Work and IMC-AESOP and is considered to be part of the PLANTCockpit OPC UA adapter. HTML, on the other hand, is used for the structuring and presentation of multimedia web content using human and machine-readable semantic descriptions. It is explicitly stated as part of the visualization layers of IoT@Work and PLANTCockpit. Finally, XOP/MTOM, is used by IMC-AESOP for the transmission and reception of binary data in SOAP messages (Rotondi et al.,

2013; PLANTCockpit Consortium, 2012a,b; Colombo et al., 2014b; Imtiaz and Jasperneite, 2013; PLANTCockpit Consortium, 2012c).

Further specifications include security relevant ones, such as the XML-based Security Assertion Markup Language (SAML) and eXtensible Access Control Markup Language (XACML), which are employed in IoT@Work, SOAP for the structuring of messages, and the previously mentioned YAML for the description of RESTful APIs. The first of these, SAML, is a standard for the communication of data for authentication and authorization, while XACML handles the definition of access policies. How these are applied as part of IoT@Work will be discussed later on in this chapter. SOAP defines a platform independent XML-based framework for message structuring, encoding, and processing and for the representation of remote procedure calls and responses. The SOAP message structure is made up of a SOAP envelope, SOAP body, and, optionally a SOAP header. The first, the SOAP envelope, is used to represent the message itself, while the SOAP header can be used to add features and their associated attributes to a message. Lastly, the SOAP body contains the message contents to be conveyed to the other communicating parties. The platform-agnostic nature of the SOAP protocol is one of the driving factors behind its popularity in the web services community. As for YAML, this was discussed earlier as the language of choice for the configuration of Swagger files under the eScop project. YAML, like its counterparts, is a serialization language aimed at minimizing the number of characters required to indicate the structure and value of data while maximizing human readability in the resulting data interchange format. It is built around a typing system, an aliasing mechanism and primitives such as mappings, scalars and sequences. According to the YAML specification, compared to JSON, it is more difficult to generate and parse but more legible to humans than JSON. The specification also states that there is no direct correlation between XML and YAML (Rotondi et al., 2013; OASIS Security Services (SAML) TC, n.d.; Iarovyi et al., 2015a; PLANTCockpit Consortium, 2012d; Ben-Kiki et al., 2009; Box et al., 2000).

14.4.5 Message Exchange

For message exchange, we note a preference for web-based solutions, such as HTTP and CoAP, and message-oriented middleware (MOM), such as JMS, AMQP, XMPP, and MQTT.

Starting with HTTP, this application-level, stateless, and generic protocol is used by all projects for the transfer of hypermedia across networks. The protocol typically runs over TCP, employs MIME-like messages for communication, and uses URIs to provide access to resources. For its secure equivalent, HTTPS, HTTP is transported over a TLS tunnel. In several instances, such as IMC-AESOP and PLANTCockpit, implementations used HTTP/HTTPS for the conveyance of SOAP messages. CoAP, on the other hand, is a low overhead URI-based web protocol for machine-to-machine (M2M) communication over UDP with support for unicasting, multicasting, proxying, caching, stateless HTTP mapping, and binding to DTLS. Due to properties such as these, a number of projects, namely IMC-AESOP and Arrowhead, have favored the use of the CoAP protocol for the access of constrained devices and network implementations in their realizations (Fielding and Reschke, 2014; Rescorla, 2000; Varga et al., 2014; Rotondi et al., 2010; Colombo et al., 2014b; PLANTCockpit Consortium, 2012c; Severa et al., n.d.; Shelby et al., 2014; Eliasson, 2015; Derhamy, 2015).

As for the MOMs employed, PLANTCockpit uses JMS, IoT@Work employs AMQP, and certain Arrowhead demonstrators implement XMPP or MQTT. MOMs are a paradigm for asynchronous, loosely coupled and reliable communication in distributed systems. These properties, as well as others such as high scalability and availability, are enabled through the use of an intermediate layer, the middleware, that handles the messaging process on behalf of the communicating parties. The first MOM to be discussed is JMS, which is a vendor-agnostic standard that defines a Java API and semantics for the description of the interface and the messaging system behavior It therefore allows applications to communicate with heterogeneous enterprise messaging systems, and simplifies their development process. However, the implementation of the messaging service itself is not defined by the standard. As such, only a very general structure for the JMS message is defined by the standard necessitating the inclusion of integration techniques if multiple implementations of MOM exist within the same system (Mahmoud, 2004; Richards et al., 2009; PLANTCockpit Consortium, 2012a,d; Imtiaz et al., 2013; Skou et al., 2014; Eliasson, 2015; Derhamy, 2015).

In contrast to JMS, AMQP defines an open-standard messaging protocol that includes the networking protocol and message structure and remains agnostic towards the client API and message broker employed. Its protocol provides flow control features, message delivery guarantees, and highly flexible routing mechanisms for communicating parties. It appears that IoT@Work based its decision on using AMQP for its ENS implementation on the fact that it addresses, as a standard, both aspects of high-level modeling and wire-level communication concepts (OASIS, 2012; Luzuriaga et al., 2015; Richards, 2011; Imtiaz et al., 2013).

In terms of Arrowhead's choices, the XMPP and MQTT protocols are either highlighted for use or directly implemented in a number of its services and demonstrator pilots. The former, XMPP is a widely-implemented XML and TCP/IP based protocol for near-real-time communication. The protocol addresses aspects related to connection establishment and teardown, security, discovery, reliability, messaging, and inter-entity interactions. MQTT, on the other hand, is a lightweight protocol for constrained and unreliable systems that is more efficient than HTTPS, but is not extensible, and does not natively include connection security, transactions, discovery, or message fragmentation features. Multiple instances in the Arrowhead documents list the desire for services to support both MQTT and XMPP, or, in the case of the mediator service, to support translation between the two protocols (Le Pape et al., 2014; Skou et al., 2014; Eliasson, 2015; Derhamy, 2015; Gazis et al., 2015; Saint-Andre, 2011).

Other protocols noted include DPWS and OPC UA. DPWS uses SOAP-over-HTTP and SOAP-over-UDP bindings, yet certain member services, such as WS-Discovery and WS-Transfer, are transport independent. With SOAP-over-HTTP the SOAP message is placed inside the HTTP payload field for request/response messaging allowing for features from both specifications. Similarly, the SOAP-over-UDP binding allows for the inheritance of UDP's messaging, encoding, security and other mechanisms.

Regarding OPC UA, four combinations of encoding, security, and transport protocols are possible:

- UA Binary + UA-SecureConversation + UA-TCP
- UA Binary + HTTPS
- UA XML + SOAP + HTTPS
- UA XML + WS-SecureConversation + SOAP + HTTP

Of these four, the first of these compositions, referred to as native UA Binary, is mandatory for implementation. Both of IMC-AESOP and IoT@Work use the native UA Binary profile for the transport layer of their OPC UA implementations. IMC-AESOP, however, also includes a communication interface in its DPWS stack that implements the third profile made up of UA XML, SOAP and HTTPS, labeling it as OPC UA over WS. Arrowhead, on the other hand, presents a proof of concept of a condition monitoring system that employs the OPC UA SDK from Unified Automation, and discusses its use for the integration of legacy components in its framework. Finally, although OPC UA is highlighted as a development technology for PLANTCockpit, the details of the implementation is unclear from the public deliverables (Jeyaraman et al., 2008; Zurawski, 2005; Moreau et al., 2007; Colombo et al., 2014b; Mahnke et al., 2009; Bony et al., 2011; Varga et al., 2014; Le Pape et al., 2014; Hästbacka et al., 2014; PLANTCockpit Consortium, 2012c).

14.4.6 Networking, Data Link and Media

As may have been partially visible from the previous parts of this chapter, a wide and heterogeneous variety of media and their associated specifications are used, required, or discussed for possible implementations. Effectively, however, only three projects, Arrowhead, IMC-AESOP, and IoT@Work, explicitly state the technologies implemented at the networking, data link and media layers.

The Arrowhead framework's stack is declared as being made up of IPv4/IPv6, 6LoWPAN, 802.11p, 802.15.4, NFC, UWB, and NTP. However, we also note that, other than the aforementioned protocols, several more are used in one of the pilot demonstrations. Specifically, the GSM (2G), GPRS (2.5G), UMTS (3G), WiFi and RS-485 CAN standards are highlighted by the Arrowhead project as possible solutions for communication in electrical vehicle charging infrastructure. The wireless specifications are to allow users to communicate with charging stations using devices separate from the vehicle, such as mobile devices. So far, the pilot demonstration has limited its implementation to UMTS and WiFi. As for the communication channel between the vehicle and the charging station, the Arrowhead project uses CAN, basing it on the CHAdeMO standard (Varga et al., 2014; Bellavista and Ornato, 2014; Arrowhead Consortium, 2014c; Bocchio and Ornato, 2014; Kleyko, 2016).

The IMC-AESOP project declares a stack somewhat similar to Arrowhead, using IPv4, IPv6, TCP, UDP, 6LoWPAN, IEEE 802.15.4, IEEE 802.11 and NTP. However, it uses RS-485 Modbus instead of CAN, and also includes Profibus, UA Native, and IEEE 1588 PTP. The majority of these protocols, namely IPv4, IPv6, TCP, UDP, 6LoWPAN are part of the DPWSCore stack in IMC-AESOP. The component responsible for bridging the DPWS and OPC UA stacks implements the UA Native protocol. A pilot demonstrating the migration of a plant's lubrication system to the IMC-AESOP approach used the Modbus protocol to connect to the distributed control system and a specific stack consisting of XML/EXI, CoAP, NTP, UDP, IP, 6LoWPAN, IEEE 802.15.4. A second pilot, for building system of systems with SOA technology highlights the integration of smart home systems with communication infrastructure using SenML, EXI, CoAP, IPv6, IEEE 802.11, IEEE 802.15.4 and cellphone communication technologies (agnostic) (Colombo et al., 2014b).

The IoT@Work approach necessitates the use of IPv6 and designates IPv4 as optional. For its DS, as previously discussed, both NFC and QR codes are supported. For timing, both NTP or IEEE 1588 PTP are possible. To demonstrate the auto-configuration system, it uses the RT Ethernet standard Profinet. Its network slices technology is mapped using Ethernet VLAN. Where IoT@Work differs from other approaches is its focus on discussing facilitating protocols such as SNMP, DHCP, DNS, LLDP, and STP to support the network-level autoconfiguration of devices (Rotondi et al., 2013; Houyou et al., 2011; Imtiaz et al., 2013).

14.4.7 Security

The IoT@Work approach to security is founded on capability-based access control. This mechanism centers around the use of transmissible tokens that reference an element and its access rights. A process in possession of a valid token may therefore interact with the referenced element within the constraints of its access rights. These capabilities may be forged or revoked in the form of XML documents following specific schema with elements borrowed from the SAML, XACML, Digital Signature and XML encryption schema. IoT@Work also requires that devices have secure identifiers based on the IEEE 802.1AR specification, and that authentication for NAC be carried out based on IEEE 802.1X. With these mechanisms in hand, the system designs and employs multiple components in a SO fashion to perform NAC and to secure the system's event namespace (Gusmeroli et al., 2013; Rotondi et al., 2013; Fischer and Gesner, 2012).

The PLANTCockpit project explored the possibility of using single-sign on solutions for a security service. Ultimately, the PLANTCockpit approach employs an LDAP service for access control and the management of user rights. Interactions with the LDAP service are through a Java client implemented using the JLDAP library. User access rights govern the components and data sources that a user is permitted to interact with. As for the security aspects related to PLANTCockpit's use of JMS, notes in the deliverables claim that PLANTCockpit permits the encryption of the JMS message body, with the encryption to be managed using a central component and that JMS security is addressed using 'interceptors'. Unfortunately, the security aspects of PLANTCockpit are addressed in a non-public deliverable (D3.2) and, as such, details on the interceptors and other system mechanisms for security are not available for us to elaborate further (PLANTCockpit Consortium, 2011b, 2014, 2012d).

As far as the published materials go, Arrowhead instills security measures for its Service Discovery and Authorization Control, for mediation in legacy systems, and for communication in general. The Service Discovery service is secured by using DNS TSIG keys for DNS updates and DNS Security Extensions (DNSSEC) for queries. Authorization Control is dependent on the use of X.509 certificates, and provides TLS security for the establishment of secure communication links. In the case of Arrowhead's Virtual Market of Energy pilot, the Authorization module uses Public Key Infrastructure (PKI) and X.509 certificates over REST for authentication and XMPP-over-TLS for encrypted communications. Generally, the consensus throughout the Arrowhead framework is to secure information exchange using TLS. In the case of UDP communication, DTLS is highlighted as the applicable counterpart. The project has also expressed a desire to develop a secure NFC interface for industrial applications. Based on a publication, this aspect may have been addressed through the 'ESTADO' system for smart maintenance. This system uses a 'CUT-IN' module made up of a secure and a non-secure but more powerful controller. The module is designed to be an add-on that would provide security features to 'non-smart' and legacy devices. This would include abilities for the secure storage and execution of data, integrity checks, encrypted memory and encrypted in-CPU calculations. Finally, we note the use of the Message Passing Interface (MPI) with Secure Shell (SSH) for protected communications in the implementation of a distributed framework for 3D swarming systems such as aerial vehicles and WSNs (Blomstedt and Olofsson, 2016a,b; Varga et al., 2014; Chrysoulas and Jansson, 2016; Krimm et al., 2014; Lesjak et al., 2014; Sadrollah et al., 2014).

In the case of IMC-AESOP, by self-admission, "cyber-security was not at the heart of [this] project" (Colombo et al., 2014b). As such, IMC-AESOP states that WS-Security and WS-ReliableMessaging were not implemented as part of its DPWS communication stack, and neither was IPsec included in the IPv6-based stack for WSANs, claiming them all as planned additions. Furthermore, the service bus in IMC-AESOP only implemented HTTP basic authentication and Role-Based Access Control (RBAC) for service calls with such rights only being given to administrative users. In accordance with RFC 7617, unless communication takes place within a secure system (i.e., over TLS), basic authentication is not to be considered secure as credentials are transferred in clear-text. It is unclear if IMC-AESOP implements basic authentication over a secure channel (Colombo et al., 2014b; Reschke, 2015).

Similarly, other than security testing and a high level discussion of defining and applying a security model as part of an SOA ecosystem, eScop did not address the aspect of security. In the case of the former, security testing of the RPL is defined as a method for verifying the robustness of the ontology by employing "ad-hoc offensive queries". For the VIS layer, testing is to be applied to access control measures. As for the testing of the integration of the PHL, RPL, VIS and ORL, this entailed ensuring that data in the system is secured and that the functionality of the system cannot be misused (eScop Consortium, 2014; Simone et al., n.d.).

14.5 Discussion

The concept of SOAs has been in existence for well over a decade and is generally considered to be a stable and tried architectural design pattern. The evolution of SOAs and alternatives to the SO approach within the context of industrial systems are partially addressed in Chaps. 8 and 13. This chapter, on the other hand, inspects the architectural designs and technological choices of five preliminary SO RAs. Based on this examination, a number of features and limitations common to all five projects are apparent. One such observation is the lack of homogeneity in approach by all five projects in defining their respective architectures. Differences in the levels of abstraction, the aspects chosen for specification, and the selected forms of representation are some noted dissimilarities. For example, IMC-AESOP pursues a high level of abstraction, and therefore limits its reference architecture to a definition of service groups, member services, dependencies, and possibilities for inter-service interactions. It does not specify any protocols and standards as part of the reference architecture and instead addresses these aspects through prototypical demonstrators and pilot projects. In direct contrast are the IoT@Work and Arrowhead projects which provide concrete and detailed architectures with specific protocols, standards and specifications selected for various elements of their architectures. In terms of representation, it may be observed that only IMC-AESOP, eScop and Arrowhead declare their use of semi-formal notation in their respective documents, while IoT@Work and PLANTCockpit lack such explicit statements. Note that a related study on software reference architectures concludes that informal specification is found to introduce a lack of clarity and precision in architectures by leaving room for interpretation (Angelov et al., 2012).

Further issues include missing real-time communication features critical to industrial applications. IMC-AESOP, IoT@Work, and Arrowhead all include protocols to interface with RT systems in their demonstrators or pilot projects. They also have a number of publications that take steps towards addressing real-time aspects in processing, communication, and applications such as auto-configuration (Lindgren et al., 2013; Jammes, 2011; Pietrzak et al., 2011; Dürkop et al., 2012; Durkop et al.,

2013; Garibay-Martínez et al., 2013, 2014a,b). Yet, the services themselves, and the majority of the selected technologies, remain largely without RT-capabilities. This introduces a concern as to their actual ability to operate in safety-critical CPPS environments.

Finally, in the case of interoperability, it appears that integration is still a necessity. This is since a number of the reviewed projects, in multiple instances, have chosen to select several technologies to address certain critical features in their architectures in the pursuit of flexibility. This seems to be the case, for example, with Arrowhead's approach for discovery and IoT@Work's requirement for mappings for its novel internal data model. Determining the overall impact of the continued necessity for integration in SOAs, as well as the effect of the missing features, the selected levels of abstraction and representation is a matter than can be addressed through follow-up surveys. Such studies may aim to draw conclusions on the success of the architectures' choices by determining adoption rates and stakeholder criticisms. Unfortunately, while such information would be of great benefit to both researchers and practitioners alike, it is largely absent from literature.

In sum, it is undeniable that interoperability in the industrial domain has been marred in complexity since its conception. It therefore appears as though integration is here to stay for the foreseeable future. Yet, the SO approach, as presented in this chapter, may be used to strengthen and structure the overall environment through clearly defined and documented systems, components, boundaries and interfaces, protocols, guidelines and policies. It is therefore encouraged that SO solutions continue development in the industrial domain. This is both to address the shortcomings of existing projects and to further the pursuit of agility in enterprises in the face of turbulent technical, economic, and legal environments.

Acknowledgements This paper is supported by Technische Universität Wien research funds as part of the Doctoral College Cyber-Physical Production Systems.

References

Angelov, S., Grefen, P., et al.: A framework for analysis and design of software reference architectures. Inf. Softw. Technol. **54**(4), 417–431 (2012)

Arrowhead Consortium: Arrowhead demo T4.2. Technical report (2014a)

Arrowhead Consortium: WP 4—T4.1 Demonstrator (1st Generation): The Safe Home. Technical report (2014b)

Arrowhead Consortium: WP3.1.2 PO 002 - Requirements definition - Communication and services. Technical report, Arrowhead Consortium (2014c)

Ballinger, K., Bissett, B., et al.: Web Services Metadata Exchange 1.1 (WS- MetadataExchange). (2008)

Bellavista, P., Ornato, M.: Arrowhead D3.3 Appendix 3.3b PO 3.3-002 and PO 3.3-004 of Task 3.3 Details about First Generation Pilot Results. Technical report (2014)

Ben-Kiki, O., Evans, C., et al.: YAML Ain't Markup Language (YAML) (tm) Version 1.2., YAML.org. (2009)

Blomstedt, F.: Service Discovery REST_WS-XML-SPSDTR Version 1.1. Technical report, The Arrowhead Consortium (2016a)

Blomstedt, F.: ServiceRegistryBridge Version 1.1. Technical report, The Arrowhead Consortium (2016b)
Blomstedt, F., Olofsson, P.: Orchestration System Version 1.1. Technical report, The Arrowhead Consortium (2016a)
Blomstedt, F., Olofsson, P.: Service Discovery DNS-SD-TSIG-SPDNS Version 1.3. Technical report, The Arrowhead Consortium (2016b)
Blomstedt, F., Olofsson, P.: Service Discovery REST_WS-JSON-SPSDTR Version 1.1. Technical report, The Arrowhead Consortium (2016c)
Blomstedt, F., Contini, L., et al.: Deliverable D8.3. Technical report, Arrowhead Consortium (2014a)
Blomstedt, F., Ferreira, L., et al.: The arrowhead approach for SOA application development and documentation. In: 40th Annual Conference of the IEEE Industrial Electronics Society (IECON), pp. 2631–2637 (2014b)
Bocchio, S., Ornato, M.: Arrowhead D3.3 Appendix 3.1.1.c PO 002 of Task 3.1.1 Task and concepts. Technical report (2014)
Bony, B., Harnischfeger, M., et al.: Convergence of OPC UA and DPWS with a cross-domain data model. In: 9th IEEE International Conference on Industrial Informatics (INDIN) (2011)
Box, D., Ehnebuske, D., et al.: Simple Object Access Protocol (SOAP) 1.1. W3C Note. World Wide Web Consortium (2000)
Building Performance Institute, Inc.: BPI-2100-S-2013 Standard for Home Performance-Related Data Transfer. (2013a)
Building Performance Institute, Inc.: BPI-2200-S-2013 Standard for Home Performance-Related Data Collection. (2013b)
Bullen, G., Carter, S., et al.: Web Services Dynamic Discovery (WS-Discovery) Version 1.1. Organization for the Advancement of Structured Information Standards (2009)
Cheshire, S., Krochmal, M.: RFC 6762: Multicast DNS. Technical report (2013)
Choudri, A.: The agile enterprise. In: ReVelle, J. (ed.) Manufacturing Handbook of Best Practices: An Innovation, Productivity and Quality Focus, pp. 3–23. CRC Press, Boca Raton (2001)
Chrysoulas, C., Jansson, O.: Arrowhead System Description (SysD) - Translation System Version 0.4. Technical report, The Arrowhead Consortium (2016)
Colombo, A., Bangemann, T., Karnouskos, S.: IMC-AESOP outcomes: Paving the way to collaborative manufacturing systems. In: 12th International Conference on Industrial Informatics (INDIN) (2014a)
Colombo, A., Bangemann, T., Karnouskos, S., et al., (eds.): Industrial Cloud-Based Cyber-Physical Systems. Springer, Cham (2014b). ISBN:978-3-319-05623-4 978-3-319-05624-1
Corrêa, H.: Agile manufacturing as the 21st century strategy for improving manufacturing competitiveness. In: Gunasekaran, A. (ed.) Agile Manufacturing: The 21st Century Competitive Strategy, pp. 3–23. Elsevier Science Ltd, Oxford (2001)
Delsing, J.: Building automation systems from Internet of Things. In: Presented as an 20th IEEE International Conference on Emerging Technologies and Factory Automation Keynote Presentation, Luxembourg (2015)
Dennert, A., Montemayor, J., et al.: Advanced concepts for flexible data integration in heterogeneous production environment. In: IFAC Proceedings Volumes 46(7), 348–353 (2013)
Derhamy, H.: Arrowhead transparency protocol translation. In: Presented at the Arrowhead Budapest Meeting (2015)
Dürkop, L., Imtiaz, J., et al.: Service-oriented architecture for the autocon- figuration of real-time Ethernet systems. In: 3rd Annual Colloquium Communication in Automation (KommA) (2012)
Durkop, L., Imtiaz, J., et al.: Using OPC-UA for the autoconfiguration of real-time Ethernet systems. In: 11th International Conference on Industrial Informatics (INDIN) (2013)
Eliasson, J.: Arrowhead historian system. In: Presented at the Arrowhead Budapest Meeting (2015)
Eliasson, J., Delsing, J., et al.: A SOA-based framework for integration of intelligent rock bolts with Internet of things. In: IEEE International Conference on Industrial Technology (ICIT) (2013)
Erl, T.: SOA: Principles of Service Design. Prentice Hall, Upper Saddle River, NJ (2008)

Erl, T.: SOA Design Patterns, 1st edn. Prentice Hall Service-Oriented Computing Series from Thomas Erl. Prentice Hall, Upper Saddle River, NJ (2009)

Erl, T., Gee, C., et al.: Next Generation SOA: A Concise Introduction to Service Technology and Service-Orientation. Pearson Education, Upper Saddle River (2014)

eScop Consortium: 1st Annual Report. Technical report. http://www.escop-project.eu/wp-content/uploads/2013/05/D14_eScop_Annual_Report1_publishable-summary.pdf (2013). Visited on 08 Dec 2016

eScop Consortium: D2.4 General specification and design of eScop reference architecture. Technical report (2014)

eScop Consortium: eScop Architecture. Technical report (n.d.[a])

eScop Consortium: Orchestration Layer Training Material. Technical report (n.d.[b])

eScop Consortium: Semantic Workbench. Technical report (n.d.[c])

Faist, J., Štětina, M.: Webservice-ready configurable devices for intelligent manufacturing systems. In: IFIP International Conference on Advances in Production Management Systems (APMS). Springer, Heidelberg (2015)

Fielding, R., Reschke, J.: RFC 7235: Hypertext Transfer Protocol (HTTP/1.1): Authentication. Technical report (2014)

Fischer, K., Gesner, J.: Security architecture elements for IoT enabled automation networks. In: IEEE 17th Conference on Emerging Technologies Factory Automation (ETFA) (2012)

Fumagalli, L., Pala, S., et al.: Ontology-based modeling of manufacturing and logistics systems for a new MES architecture. In: IFIP International Conference on Advances in Production Management Systems (APMS), pp. 192–200. Springer, Heidelberg (2014)

Galiegue, F., Zyp, K., et al.: JSON Schema: core definitions and terminology. Technical report, Internet Engineering Task Force (2013a)

Galiegue, F., Zyp, K., et al.: JSON Schema: interactive and non interactive validation. Technical report, Internet Engineering Task Force (2013b)

Garibay-Martínez, R., Nelissen, G., et al.: Task partitioning and priority assignment for hard real-time distributed systems. In: 2nd International Workshop on Real-Time and Distributed Computing in Emerging Applications (2013)

Garibay-Martínez, R., Ferreira, L., et al.: Towards holistic analysis for fork-join parallel/distributed real-time tasks. In: 26th Euromicro Conference on Real-Time Systems (2014a)

Garibay-Martínez, R., Nelissen, G., et al.: On the scheduling of fork-join parallel/distributed real-time tasks. In: 9th IEEE International Symposium on Industrial Embedded Systems (SIES) (2014b)

Gazis, V., Görtz, M., et al.: A survey of technologies for the internet of things. In: 2015 International Wireless Communications and Mobile Computing Conference (IWCMC) (2015)

Gusmeroli, S., Piccione, S., et al.: A capability-based security approach to manage access control in the Internet of Things. Math. Comput. Model. **58**(5-6), 1189–1205 (2013)

Hästbacka, D., Barna, L., et al.: Device status information service architecture for condition monitoring using OPC UA. In: 19th International Conference on Emerging Technology and Factory Automation (ETFA) (2014)

Henßen, R., Schleipen, M.: Interoperability between OPC UA and AutomationML. In: Procedia CIRP, vol. 25, pp. 297–304 (2014)

Hill, B.: Evaluation of efficient XML interchange (EXI) for large datasets and as an alternative to binary JSON encodings. Technical report, DTIC Document (2015)

Houyou, A., Huth, H., et al.: D2.2- Bootstrapping Architecture. Technical report, IoT@Work Consortium (2011)

Iarovyi, S., Garcia, J., et al.: An approach for OSGi and DPWS interoperability: Bridging enterprise application with shop-floor. In: 11th International Conference on Industrial Informatics (INDIN) (2013)

Iarovyi, S., Ramis, B., et al.: Representation of manufacturing equipment and services for OKD-MES: from service descriptions to ontology. In: 13th International Conference on Industrial Informatics (INDIN) (2015a)

Iarovyi, S., Xu, X., et al.: Architecture for open, knowledge-driven manufacturing execution system. In: IFIP International Conference on Advances in Production Management Systems (APMS), vol. 460. Springer, Cham (2015b)

Iarovyi, S., Mohammed, W., et al.: Cyber-physical systems for open-knowledge-driven manufacturing execution systems. Proc. IEEE **104**(5), 1142–1154 (2016)

Imtiaz, J., Jasperneite, J.: Scalability of OPC-UA down to the chip level enables "Internet of Things". In: 11th International Conference on Industrial Informatics (INDIN), pp. 500–505 (2013)

Imtiaz, J., Dürkop, L., et al.: D. 2.5 - Integrated Secure Plug&Work Frame-work. Technical report, IoT@Work Consortium (2013)

Ismail, A., Kastner, W.: Vertical integration in industrial enterprises and distributed middleware. Int. J. Internet Protoc. Technol. **9**(2/3), 79–89 (2016)

Jammes, F.: Real time device level service-oriented architectures. In: IEEE 20th International Symposium on Industrial Electronics (ISIE) (2011)

Jennings, C., Shelby, Z., et al.: Media Types for Sensor Markup Language (SenML). Technical report (2016)

Jeyaraman, R., Modi, V., et al.: Understanding Devices Profile for Web Services, Web Services Discovery, and SOAP-over-UDP. Technical report, Microsoft Corporation (2008)

Kleyko, D.: System-of-Systems Design (SoSDD) LTU Core Framework Description. Technical report (2016)

Knapp, E.: Industrial Network Security: Securing Critical Infrastructure Networks for Smart Grid, Scada, and Other Industrial Control Systems, 2nd edn. Elsevier, Waltham, MA (2014)

Krafzig, D., Banke, K., et al.: Enterprise SOA: Service-Oriented Architecture Best Practices. The Coad Series. Prentice Hall Professional Technical Reference, Indianapolis, IN (2005)

Krämer, B.: Evolution of cyber-physical systems: a brief review. In: Applied Cyber-Physical Systems. Springer, New York (2014)

Krimm, J., Olofsson, P., et al.: Deliverable D8.2 of Work Package 8: Common Service Framework - Generation 1 Version 1.1. Technical report, The Arrow-head Consortium (2014)

Kyusakov, R., Pereira, P., et al.: EXIP: a framework for embedded Web development. ACM Trans.Web (TWEB) **8**(4), 23 (2014)

Le Pape, C., Desdouits, C., et al.: Deliverable D1.3 of WP1. Technical report, Arrowhead Consortium (2014)

Lesjak, C., Ruprechter, T., et al.: ESTADO-Enabling smart services for industrial equipment through a secured, transparent and ad-hoc data transmission online. In: IEEE 9th International Conference for Internet Technology and Secured Transactions (ICITST) (2014)

Lindgren, P., Pietrzak, P., et al.: Real-time complex event processing using concurrent reactive objects. In: IEEE International Conference on Industrial Technology (ICIT) (2013)

Luzuriaga, J., Perez, M., et al.: A comparative evaluation of AMQP and MQTT protocols over unstable and mobile networks. In: 12th Annual Conference on Consumer Communications and Networking Conference (CCNC) (2015)

Mahmoud, Q. (ed.): Middleware for Communications. Wiley, Chichester/Hoboken, NJ (2004)

Mahnke, W., Leitner, S., et al.: OPC Unified Architecture. Springer, Berlin/Heidelberg (2009)

Moreau, J., Nielsen, H., et al.: SOAP Version 1.2 Part 2: Adjuncts, 2nd edn. W3C Recommendation. World Wide Web Consortium (2007)

Nagorny, K., Colombo, A., et al.: A survey of service-based systems-of-systems manufacturing systems related to product life-cycle support and energy efficiency. In: 12th IEEE International Conference on Industrial Informatics (INDIN) (2014)

Nappey P., El Kaed, C., et al.: Migration of a legacy plant lubrication system to SOA. In: 39th Annual Conference of the IEEE Industrial Electronics Society (IECON) (2013)

Negri, E., Fumagalli, L., Macchi, M., et al.: Ontology for service- based control of production systems. In: Advances in Production Management Systems: Innovative Production Management Towards Sustainable Growth, vol. 460, pp. 484–492. Springer, Cham (2015)

Negri, E., Fumagalli, L., Garetti, M., et al.: Requirements and languages for the semantic representation of manufacturing systems. Comput. Ind. **81**, 55–66 (2016)

OASIS: Advanced Message Queuing Protocol (AMQP) Version 1.0. In: Godfrey, R., Ingham, D., et al. (ed.) (2012)

OASIS Security Services (SAML) TC: Technical report (n.d.)

Pietrzak, P., Kyusakov, R., et al.: Roadmap for SOA event processing and service execution in real-time using Timber. In: IEEE 20th International Symposium on Industrial Electronics (ISIE) (2011)

PLANTCockpit Consortium: D3.1 Initial Architectural Components of PLANTCockpit. Technical report (2011a)

PLANTCockpit Consortium: Technical Report Persistency and Synchronization Model. Technical report (2011b)

PLANTCockpit Consortium: PLANTCockpit White Paper. Technical report, http://plantcockpit.eu/fileadmin/PLANTCOCKPIT/user_upload/PLANTCockpit_D3.3_Public.pdf (2012a). Visited on 03 Mar 2015

PLANTCockpit Consortium: Data and Process Model, First Draft. Technical report (2012b)

PLANTCockpit Consortium: External Interface Specification, First Draft. Technical report (2012c)

PLANTCockpit Consortium: Generic Alarms and Events Data Format. Technical report (2012d)

PLANTCockpit Consortium: Project Final Report. Technical report (2014)

Ratovsky R., Gardiner, M., et al.: OpenAPI Specification Version 2.0. Technical report. Open API Initiative (2014)

Reschke, J.: RFC 7617: The 'Basic' HTTP Authentication Scheme. Technical report (2015)

Rescorla, E.: RFC 2818: HTTP over TLS. Technical report (2000)

Richards, M.: Understanding the differences between AMQP & JMS. In: Proceedings of the No Fluff Just Stuff ^{TM} Java Conference Series (2011)

Richards, M., Monson-Haefel, R., et al.: Java Message Service, 2nd edn. O'Reilly, Sebastopol, CA (2009)

Rotondi, D., Galipò, A., et al.: D1.1 State of the Art and Functional Requirements in Manufacturing and Automation. Technical report. IoT@Work Consortium (2010)

Rotondi, D., Piccione, S., et al.: D1.3 - Final Framework Architecture Specification. Technical report, IoT@Work Consortium (2013)

Rubio, J.: Service oriented architecture for embedded (avionics) applications. Ph.D. thesis, Technical University of Catalonia (2011)

Sadrollah, G., Barca, J., et al.: A distributed framework for supporting 3D swarming applications. In: International Conference on Computer and Information Sciences (ICCOINS) (2014)

Saint-Andre, P.: RFC 6120: Extensible Messaging and Presence Protocol (XMPP): Core. Technical report (2011)

Sauter T., Soucek, S., et al.: Vertical integration. In: Wilamowski, B., Irwin, J. (eds.) Industrial Communication Systems, 2nd edn., pp. 1–12. CRC Press, London (2011)

Severa, O., Faist, J., et al.: eScopRTU with Service Manager. Technical report, eScop Consortium (n.d.)

Shelby, Z.: RFC 6690: Constrained RESTful Environments (CoRE) Link Format. Technical report (2012)

Shelby Z., Hartke, K., et al.: RFC 7252: The Constrained Application Protocol (CoAP). Technical report (2014)

Simone, P., Obluska, I., et al.: D6.1 Test strategy. Technical report (n.d.)

Skou, A., Lino Ferreira, L., et al.: Deliverable D5.3 of WP 5. Technical report, Arrowhead Consortium (2014)

Valipour M., AmirZafari, B., et al.: A brief survey of software architecture concepts and service oriented architecture. In: 2nd International Conference on Computer Science and Information Technology (ICCSIT) (2009)

Varga, P., Martínez de Soria, I., et al.: Deliverable D7.3 of WP 7. Technical report, Arrowhead Consortium (2014)

Zurawski, R. (ed.): The Industrial Communication Technology Handbook. Industrial Information Technology Series. Taylor & Francis/CRC Press, Boca Raton (2005)

Chapter 15
A Deterministic Product Ramp-up Process: How to Integrate a Multi-Disciplinary Knowledge Base

Roland Willmann and Wolfgang Kastner

Abstract Ramping up new products to volume production is a challenge for most manufacturing companies. The deviation between plan and reality of costs and duration of ramp-up projects is still significant, and the achievable quality of new products at the start-up of volume production is difficult to predict. Consequently, new products arrive too late at the customer, causing dissatisfaction or even loss of customers, additional operational costs or unplanned enhancement of ramp-up budget. The vertical knowledge exchange between product engineering and process engineering, as well as horizontally along the production process and the supply chain has turned out to be the major reason for deviations. This chapter describes how information from product engineering and process engineering has to be structured for automated recommendation of information reuse during the planning of ramp-up projects. It discusses the involvement of a multi-disciplinary knowledge base in a production environment but also organizational measures to be taken into account in order to address this challenge. Needs for standardization across enterprises is addressed as well. Through thus achievable improvement of planning quality, based on reused production knowledge, ramp-up projects can improve towards deterministic ramp-up processes. The article is of interest for industrial engineers, quality managers and ICT-managers in the industrial field.

Keywords Knowledge management • Product ramp-up • Ontology • Ontology mapping and matchmaking • Multi-disciplinary engineering

R. Willmann (✉)
Carinthia University of Applied Sciences, Europastraße 4, 9524, Villach, Austria
e-mail: r.willmann@fh-kaernten.at

W. Kastner
Technische Universität Wien, Favoritenstraße 9-11, 1040, Wien, Austria
e-mail: k@auto.tuwien.ac.at

© Springer International Publishing AG 2017
S. Biffl et al. (eds.), *Multi-Disciplinary Engineering for Cyber-Physical Production Systems*, DOI 10.1007/978-3-319-56345-9_15

15.1 Introduction

The trend towards individualized products and shorter product life cycles, which is expected to be accelerated by *Industrie 4.0*, enforces companies to improve the performance of product ramp-up projects. The manufacturing industries (Kurttila et al. 2010; Fransoo and de Kok 2007) have to deal with shortening of product life cycles and an increasing number of new products in their production lines. In Chapter 4 this topic is also addressed as an essential part of the product lifecycle. As a consequence, product ramp-up projects have to be better predictable with respect to their costs and durations but also concerning the achieved initial yield. Moreover, costs and duration of ramp-up projects have to be minimized, and the initially achieved yield has to be maximized.

The time span within which a new product is ramped up after the end of product development to stable volume production is called the ramp-up phase. The ramp-up phase usually begins when the development phase of the new product ends (Fig. 15.1). From the development phase there is obtained a so called *bill of material*—abbreviated as *BOM*—of the new product. This is the hierarchical structure of all subparts or subproducts which composes the final product, including the detailed characteristics of all ingredients of the subproducts (e.g., size, weight, uniformity of surface).

Together with the BOM, the development phase provides expectations for each step of the production process which need to be achieved in order to transform subproducts to the next higher composition level. Accordingly, those process expectations are close related to the BOM but do not comprise specific details with respect to the setup of individual devices of a production system. In conjunction

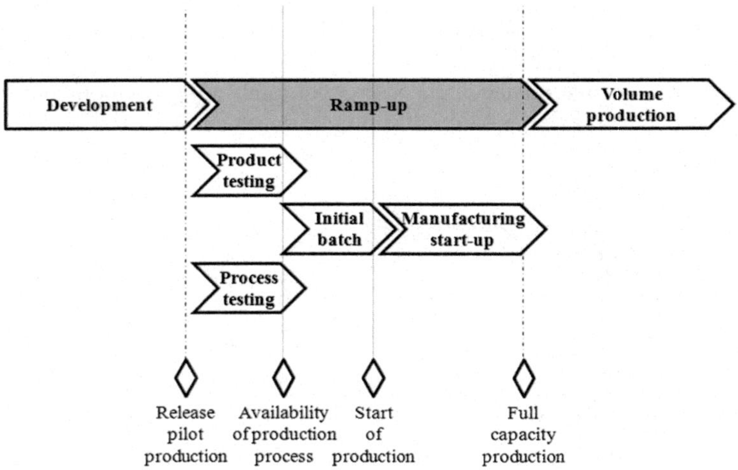

Fig. 15.1 Location of the ramp-up phase of a new product based on (Slamanig and Winkler 2012, p. 484).

with the BOM the expectations of all process steps result in a structure which can be called *device-independent process plan* throughout this chapter.

The *device-independent process plan* has to be enriched with specific handling instructions and information about the setup of devices within the context of each production system. This is the task of the product ramp-up team. It is time-consuming, and its efficiency and effectiveness strongly depends on the interaction between the multi-disciplinary team members as well as the quality of information provided to the team members. However, due to the shortening of product lifecycles and increasing variations of products (individualization of products) it is crucial to improve the predictability of this phase of the product lifecycle.

In accordance to Terwiesch et al. (International product transfer and production ramp-up: a case study from the data storage industry, 2001, p. 435), time-to-volume attracted reasonable more attention than time-to-market already to the beginning of the century. "The fundamental difference between time-to-market and time-to-volume is that the former ends with the beginning of commercial production whereas the latter explicitly includes the period of production ramp-up. Production ramp-up is the period during which a manufacturing process makes the transition from zero to full-scale production at target level of cost and quality."

Delayed completion of product ramp-up (Abele et al. 2008, pp. 96–98) may have significant impact to the success of the product on the market. Customers may migrate to competitive products, thus reducing the planned sales quantity. As a consequence, there is not only a reduction of the planned turnover and profit. The planned production capacity is too large for the now lower sales. Aside of these major reasons for costs and losses due to non-deterministic product ramp-up, the costs for ramping up new products are typically operational expenses of the manufacturing companies and have therefore an immediate impact on their profitability and liquidity. Non-deterministic product ramp-up also causes non-deterministic need for additional budget which is difficult to gain.

Abele et al. are discussing product ramp-up in conjunction with the construction of a new facility. However, the principle statements are also correct for the concrete situation of this chapter, were the production of a new product shall be transferred from one existing production system to another one. Utilization of production resources (staff and equipment) without adding value, additional setup times of actually productive equipment (increased idle times and thus reduced added value), consumption of energy or scraped material (reduced added value, useless emissions to the environment), enhanced risk of negative impact on the quality of simultaneously ongoing volume production of forerunner products (customer dissatisfaction) are additional cost drivers. Personnel costs of process experts for knowhow transfer at far distant locations are another one.

Therefore, the questions to be answered in the following chapters are all related to measures in order to achieve more deterministic product ramp-up projects by reusing existing production knowledge.

Q1: How is a production system organized in order to achieve maximal reuse of production knowledge?
Q2: How is a multi-disciplinary knowledgebase for product ramp-up support integrated with the production system?
Q3: Which scenarios have to be considered in order to match information about a new product with existing information?
Q4: Which information domains needs to be standardized across enterprises?

Q2 contributes to RQ I3 of Chap. 1 and discusses the need to integrate knowledge management for product ramp-up with existing manufacturing execution systems (MES) as well as the multi-disciplinary engineering team being involved. Q3 is related to the research question RQ M1 of Chap. 1 and contributes particularly to the reuse of existing production knowledge of a CPPS for mastering ramp-up scenarios of new products. Q4 addresses RQ I1 of Chap. 1 with respect to the exchange of product-related knowledge as well as a subset of process-related knowledge which is discussed in the sequel in more detail.

For answering those questions, there shall be first a discussion about the relevance of this problem case dependent on the individual strategies of enterprises (Sect. 15.2). The applied terminology with respect to the structure of the production process is introduced in Sect. 15.3. The central activities of a ramp-up team are discussed in Sect. 15.4 including additionally introduced terminology. Section 15.5 answers Q1 by discussing methods to make production systems more agile and how those measures lead to reusable production knowledge during a product ramp-up. Section 15.6 provides an answer to Q2 by the description of an IT-architecture which utilizes *Semantic Web* specifications for information representation and for information exchange. An information model and a categorization of scenarios which may be faced during a new product's ramp-up are introduced in Sect. 15.7 thus answering Q3. It turns out that there is a need for standardization of information models across enterprises in order to maximize the benefit from a multi-disciplinary knowledge base for the purpose of new products' ramp-up support. In Sect. 15.8 those needs are summarized.

15.2 Strategy-Dependent Relevance

Depending on the complexity of the product and the production process accordingly, the ramp-up team has to master a challenging but also error prone task, which is sometimes also based on trial and error. The complexity of the product is driven by the number of components (subproducts) which are used to compose the new product or by high quality demands, while the complexity of the production process

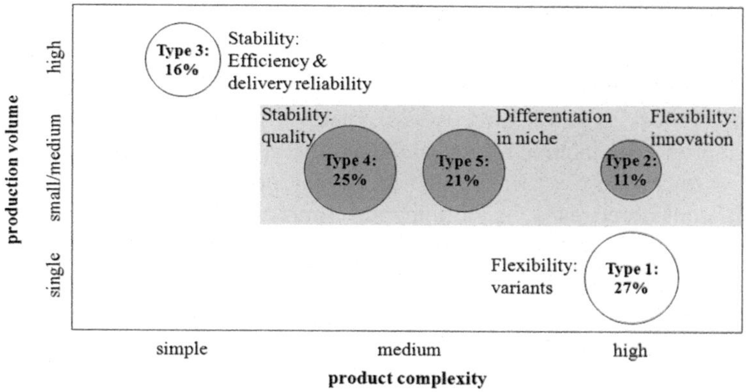

Fig. 15.2 Types of German enterprises according to production volume and product complexity—source (Kinkel and Maloca 2010, p. 3)

depends on the number of single process steps which have to be performed with continuously high precision.

Kinkel and Maloca (2010) published the results of a survey amongst German enterprises in the manufacturing industry. In Fig. 15.2 one result of this report is visualized and the enterprises with the highest relevance for a deterministic product ramp-up process are grayed. According to this assumption, almost 60% of the manufacturing enterprises in Germany are challenged by small to medium production volume with respect to each produced product (10,000 workpieces per product or product variant per year (Schmidt-Dilcher and Minssen 1996, p. 95)) and by medium to high product complexity. Consequently, a deterministic product ramp-up is of particular relevance for these enterprises as there is a significant overhead

- due to a large number of new product ramp-ups (as a consequence of the low production volume per product and the product complexity on a certain level) and
- because of the given potential to categorize production knowledge for reuse.

Typical examples of such enterprises are in the domain of car manufacturing or semiconductor manufacturing (Type 2 or Type 4 depending on strategic priority), or confectioneries (Type 5).

More deterministic product ramp-up may be also of some relevance for enterprises of Type 1 (e.g., plant engineering) or Type 3 (e.g., food and beverage). However, in case of Type 1 it depends strongly on the specific nature of produced products, with respect to the potential for categorization and reuse of production knowledge. For Type 3, the number of ramp-ups of new products may be low and the resulting benefits may have no economical relevance.

15.3 Structure of a Production Process

Process operations (SEMATECH, Inc. 1998, p. 222) represent the atomic part of a production process which is directly associated with a set of particular machines as well as appropriate machine recipes and handling instructions for human operators. A *process operation* represents either a material processing operation (Treating), which is considered as value-adding performance, or a material measurement operation (Inline Metrology), which is considered as support performance. An arbitrary sequence of material processing operation followed by one or more metrology steps makes up a *process segment* (ISA 2001, p. 48) (Fig. 15.3).

In the production system, subproducts are created or treated by manufacturing resources by executing the sequence of *process operations* within a *process segments* one after the other. Therefore, the sequence of *process operations* within a *process segment* depends on the utilized manufacturing resources. Moreover, there are *device-specific setups* on the level of the *process segment* (*process segment setup*) and on the level of each *process operations* (*process operation setup*). *Process segments* and *process operations* as well as their setups are thus concepts of the *device-dependent information* of a production system which cannot be transferred easily between heterogeneously equipped production systems.

The creation of subproducts can be also outsourced to suppliers. In this case, a *process segment* of the production system which is named *goods receipt* measures the incoming subproduct (a.k.a. consumable material) of the suppliers. Such *process segments* comprise only of one or more *process operations* for material measurement (metrology operations) to ensure that the received parts are delivered to the production with the required quality. Therefore, the existence of a metrology operation is in common for all *process segments*, independently whether the *process segment* is treating material as part of the production process or whether the *process segment* does only measure received parts of suppliers during *goods receipt*.

In combination with a particular setup, each *process segment* is qualified for a specific *process capability*. A *process capability* (ISA 2001, p. 94) (SEMATECH,

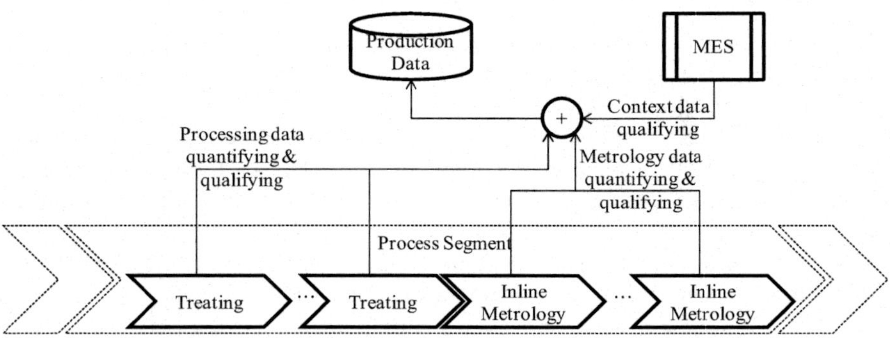

Fig. 15.3 Structure of process segments as the source of quality data (Willmann 2016, p. 16)

Inc. 1998, p. 226) can be treated as an agreement about the result of a *process segment* if a particular setup is applied. A subproduct requires specific *process capabilities* for its creation or its composition from other subproducts. Through the introduction of *process capabilities*, this requirement is completely independent from any production resource of the physical production system as it is probably described through AutomationML (see Chap. 10). Therefore, *process capabilities* are an essential design concept of production systems in order to separate the *device-independent* specifications of subproducts or products from the *device-specific* setup of production resources.

A process plan is a sequence of *process segments* (see process flow context in (SEMATECH, Inc. 1998, p. 221)) which needs to be executed in order to create the new product. The hierarchical composition of a product from subproducts and the specific *process capabilities* which are needed for each composition step, result in a directed path of *process capabilities*. Consequently, this directed path of *process capabilities* also causes a directed path of *process segments* and thus a directed path of *process operations*. Therefore, a process plan comprises all information which is needed for the creation of a product by utilizing available production resources. This set of *process operations* is also known as the *Bill of Operations* (*BOO*). MESs use such process plans and the detailed information which is provided by *process segments* and *process operations* in order to control the logistic flow of material within the physical production system. Therefore, the outcome of the ramp-up project is an essential in order to setup the MES for producing the new product.

15.4 Qualification of a Production Process

Each product has to fulfill a set of functional requirements which satisfy certain customer needs. The satisfaction of functional requirements is measured in every production system with the context of the *process segment* named *quality assurance*. Commonly, this is one of the last segments in the production process. Ramping up a new product at a target production system requires information about the functional requirements of this new product.

During execution of the *quality assurance*, samples are taken and, from these samples, predefined *characteristics* are measured which are specified by functional requirements. It is verified, whether the measured *characteristics* are located within the product-specific specification limits. If this is the case, the group of workpieces, from which the sample was taken, has passed the test. Otherwise, the entire group is either discarded or, possibly, cost-intensive detailed gauging or reworking must be performed.

Similarly, in advanced production processes, samples of the semifinished workpieces are removed and tested almost after every single material processing operation resulting in the structure of *process segments* as discussed in Sect. 15.4.

During these inline metrology steps it is not possible to test the function of the final product. However, it is possible to test *characteristics* of the design specifications of the respective subproducts. The ratio of measured parts depends on the sampling rate, which can be calculated with statistical means as they are for instance described by (Breyfogle III 2003) or (Dietrich and Schulze 2003).

For each *process segment*, the first pass yield (FPY) is calculated as the portion of defect-free workpieces Eq. (15.1) of a specific subproduct which are passing a particular *process segment* (Wappis and Jung 2010, pp. 179–180) at the first pass (without rework). The FPY is calculated from the results of each inline metrology operation and the *quality assurance process segment*.

$$FPY = \left(1 - \frac{Count\ of\ defective\ parts}{Count\ off\ all\ parts}\right).100\% \tag{15.1}$$

With respect to the overall production process (Wappis and Jung 2010, pp. 179–180), the so called rolled throughput yield (RTY) represents the portion of all produced workpieces Eq. (15.2) which are passing the overall production process at the first pass, which means that there is no potential rework required in order to correct defects. The RTY is calculated for each product which is produced in the production system. The $\mathrm{FPY_i}$ are considered to be independent from each other accordingly.

$$RTY = \mathrm{FPY}_1.\ldots.FPY_n \tag{15.2}$$

High quality, being close to the optimum of production costs, requires a high RTY and thus low costs with respect to expensive rework loops (aside of other cost drivers). As RTY strongly depends on the performance of each FPY, the same requirement is also valid for each *process segment*. Keeping the initially achieved yield as performance metric of a ramp-up project in mind, RTY and the $\mathrm{FPY_i}$ are becoming obviously critical performance indicators of any ramp-up project.

At the beginning of a product ramp-up project, it is therefore important to determine the potential of reusable and thus already mastered subproducts and process capabilities with respect to their FPY. However, it has also to be validated whether new or modified *process segments* respectively new or modified subproducts of suppliers meet the expected $\mathrm{FPY_i}$.

This validation is called the "qualification of a subproduct" if the subproduct is provided by a supplier along the supply chain (a.k.a. consumable material or consumable), and it is called "process qualification" in case of a *process segment* which is performed within the domain of the production system. Consumable material is qualified if it satisfies repeatedly the expected specification. *Process segments* are qualified for a particular *process capability* if the specification of this *process capability* is repeatedly satisfied. In case of consumable material, this specification is usually equivalent to the functional specification or the design specification in accordance to agreements with the suppliers. In case of *process segments*, as underlined before, this specification is represented by the *process capability*. A repeatedly achieved specification requires that each gauged *characteristic* of a workpiece which is treated by a *process segment* is within the specified ranges of a *process capability* for which the *process segment* is qualified.

Process capabilities represent *device-independent process-related information* and are used as a contract between products and *process segments*. In order to create or treat some product or subproduct, an appropriate *process capability* is required. On the other hand, *process segments* of a production system are qualified in order to satisfy the specifications of a *process capability*. Consequently, *process capabilities* do not refer anyhow to specific equipment and are therefore *device-independent*. But they are referenced from both sides—from the *product-related information* and from the *device-specific process-related information*. In a product ramp-up scenario, *process capabilities* are also transferred from the original production system.

Products are formally specified by *characteristics* which make up their distinction. Such a *set of specifications* holds *specification ranges* or *specification targets*. Both concepts are generalized as *specifications*. A *specification* connects a specific *characteristic* with a set of values. The set of values comprises a small discrete number of values (*specification target*) or a large or even infinite number of continuous values within a given range (*specification range*). Each *product*, *product category* or *process capability* is specified by more than one *characteristic*, each with its particular *specification range* or *specification target*.

A *specification range* is delimited by specification limits (Fig. 15.4)—commonly an upper specification limit (USL) and a lower specification limit (LSL). In exceptional cases also onesided specification limits are possible if the opposite side is delimited due to physical restrictions for instance.

While the qualification of consumable material is ensured within the production process of the supplier, the qualification of *process segments* has to be ensured within the domain of the own production system. Every workpiece which is produced with at least one *characteristic* outside of the *specification range* is counted as defective part and must be either scrapped or reworked until all *characteristics* are within their *specification ranges*.

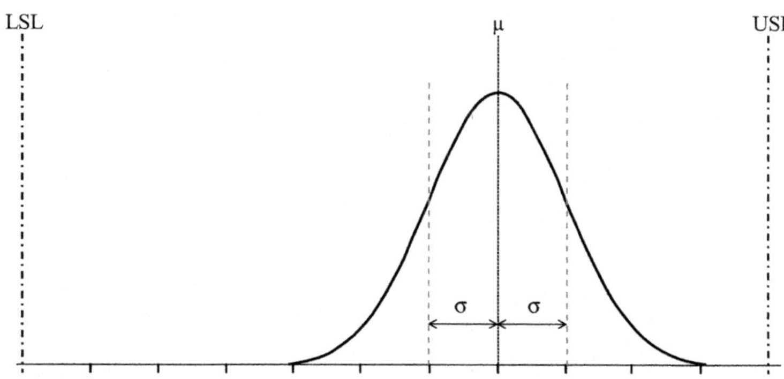

Fig. 15.4 Exemplary normal distribution of measurements between specification limits (*LSL* lower specification limit, *USL* upper specification limit) (Willmann 2016, p. 20)

The critical process capability index (c_{pk}) is commonly used in the industry in order to quantify the capability of a *process segment* to avoid defective parts. Assuming normal distribution of the gauged values of a *characteristic*, the c_{pk} represents a multiple of three standard deviations (σ) from the shortest distance between the process average (μ) and the specification limits (LSL or USL) Eq. (15.3).

$$c_{pk} = \frac{\min(\mu - LSL; USL - \mu)}{3\sigma} \tag{15.3}$$

Based on the c_{pk}, it is immediately possible to derive the probability of measurements outside the *specification range* and thus the number of defective parts due to this particular *characteristic*. In high-quality production, the c_{pk} of each *characteristic* of a *process segment* (and consequently of each acquired consumable material) is 2.0. A c_{pk} of 2.0 is equivalent to 0.002 defective parts per million (ppm). Consequently, the FPY of a *process segment* is in close relationship with the achieved c_{pk}. With respect to the ramp-up of a new product the c_{pk} is therefore the most critical performance indicator which is associated with the achieved initial yield of the volume production. Being able to determine reusable subproducts or *process capabilities* thus helps to save qualification efforts for the production of introduced subproducts.

There is not always a normal distribution of measurements. For this reason the ISO 21747 standard defines time dependent distribution models and standardizes the calculation of the c_{pk} for each type of distribution model (ISO—International Standards Organization 2007).

Also qualifying specifications are possible. For instance, a *specification* with a *characteristic* "surface color" is linked with acceptance criterion "red". However, it could be also more than one acceptance criterion, like the criteria "red" and "dark orange" in order to specify a set of acceptable colors. Such acceptance criteria are targets which must be achieved in order to avoid defective parts. Therefore, this type of *specification* is called *specification target* in the context of the following chapters.

Due to the previous insight, ramping up of a new product in a production system is an interdisciplinary task which involves experts from several domains (ramp-up team). Knowledge about currently produced products and the capabilities of the underlying production processes of a production line have to be combined with the specific requirements for the new product. In Chap. 1 those competences are summarized as *Product Engineer* and *Production System Engineer* (Fig. 1.4). In more complex production systems (e.g. semiconductor, flat panel, photovoltaic, printed circuit board manufacturing) there are dedicated *Process Engineers* or *Production Technology Experts* in addition. During the ramp-up phase, product-specific setup has to be specified for each process operation and process segment, where no immediate reuse is possible. This setup comprises handling instructions, the setup of devices but also the setup of data collection and process control models.

There are a clearly planned budget and a predefined duration which must not be exceeded by the ramp-up team while executing a ramp-up project. After completion of the ramp-up phase, the resulting instructions must enable the available production resources as well as suppliers and the control software to produce instances of the new product (workpieces) with repeatable quality on a certain level. This quality level to be achieved (initial yield) is predefined as well and limits the count of workpieces which are allowed to be scrapped or reworked because of missed quality criteria after the ramp-up project is finished.

Slamanig and Winkler (2012, p. 488) reported that the majority of companies still struggle to perform the ramp-up of new products within the planned costs or budget or to achieve the planned yield after ramp-up. "In the past, almost two-thirds of the companies were unable to meet their time-related targets, nearly 60% of the companies failed to achieve their cost-related goals, and 47% of the companies stated that they could not attain their objectives in process quality". Altogether, the results revealed that the companies within the industries being investigated lack considerable knowledge and expertise in managing their product change projects in their supply chain networks.

A significant proportion of the problem is caused by poor planning and poor information exchange. With increasing complexity of the product and the production process it is assumed that the problem is also valid within a production system and not only across the supply chain. Such complex products are, for instance, integrated circuits (ICs) and their production processes, which are said to be the most complex ones one can imagine in today's manufacturing industry.

For this reason, the ramp-up of a new product is still an individual project (ramp-up project) instead of a routine process, although some companies have developed technical concepts and business models in order to lower the risk of product ramp-up. Such measures include aspired quality gates during the ramp-up project and a modular product design in order to maximize the reuse of existing production knowledge.

15.5 Product Ramp-up and the Agility of Production Systems

The need for *device-independent process plans* is caused by the differences between the equipment of the volume production line (the target production system) where workpieces of the new product shall be produced and the equipment of the pilot line (original production system) where the first workpieces were created, for development and evaluation purpose. The differences between both production systems are caused by equipment which was acquired from different suppliers and at different times. Equipment variations are therefore caused by the variation of equipment structure across vendors as well as by the age and thus the different stages of technical progress of equipment.

In order to face this challenge, some companies are following the strategy of *copy exactly*, were every production system is an exact copy of a production system template which is following enterprise-wide design rules. Using this approach, the complexity of a product ramp-up project is reduced significantly because the information of the original production system simply needs to be transferred to the target production system without modifications. No additional assumptions need to be performed because equipment and control software in both production systems use exactly the same configuration.

However, Terwiesch and Xu (2003) highlight that although the *copy exactly* approach sounds attractive it is coming with a price. *Copy exactly* requires identical production equipment for every production system thus reducing complexity of change in case of transfer of products between production systems. As a consequence, for complex production processes, either significant investment is needed for leading edge production equipment in all production systems simultaneously or increasingly out-dated equipment has consequently to be used. The latter must be then used for production of new leading-edge products as well. Probably not all companies want or can deal with such restriction. Therefore, it is also admitted by Terwiesch and Xu (2003, p. 4) that for instance most semiconductor manufacturers still favor a much more aggressive process change during product ramp-up and do not follow the *copy exactly* approach. Therefore, the discussed approach is beneficial for a broad range of ecosystems of industrial manufacturing.

The ability to adapt a production system due to changes of its environment, like the need to produce a new product, strongly depends on its physical structure and organization. In the following sections, the most essential measures are discussed and how reuse of knowledge can take advantage from the result of such measures. Accordingly, Q1, as introduced in Sect. 15.1, is answered. Willmann (2016, pp. 148–155), discusses already some premises about the support of knowledge reuse during product ramp-up. In this chapter, those ideas are extended by yet published concepts about the agility of production systems.

It is assumed that cyber-physical production systems need to be agile in order to maximize the reuse of existing production knowledge. What does it mean—to be agile? A flexible production system (Nyhuis et al., pp. 24–26) is able to react and adjust itself within planned specification corridors. An agile production system, in the contrary, is rather solution-neutral and does not contain explicit specification limits. The scope for possible changes, however, is premeditated.

A production system has to interact with a turbulent environment (Fig. 15.5). Through adjustment (Cisek et al. 2002) of one or more of a limited set of receptors, it is possible to adapt a production system according to changes in the environment. The need to produce a new product is understood as a change of this environment. The receptors are the *product*, as well as *costs*, *time*, *quantity*, *quality* and *system elements*. The *system elements* comprise the organizational structure, the production resources and processes and therefore all elements which also make up the production system.

The ramp-up of a new product requires by all means interaction through the receptor *product*, by passing the new product's specification—the BOM—to the production system. It requires rather likely adaptations through the receptor *system elements*—particularly the specifications of appropriate handling instructions for

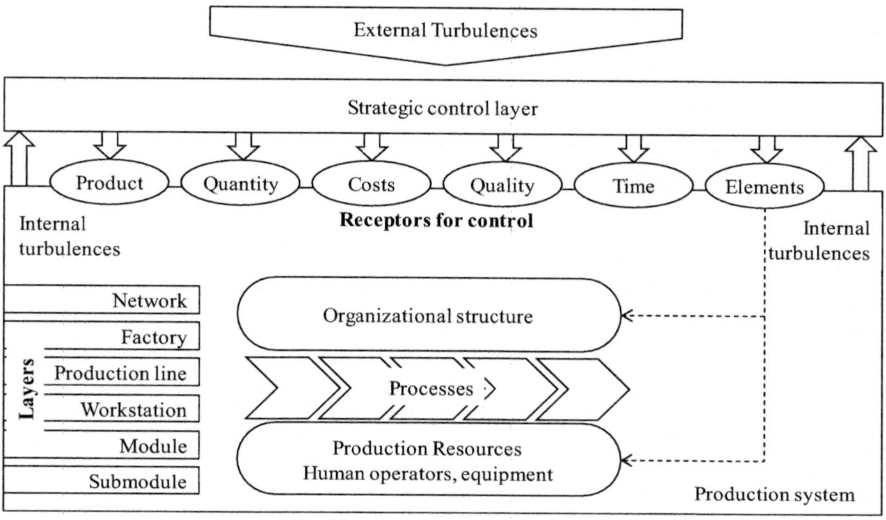

Fig. 15.5 Production system interacting with a turbulent environment—source (Nyhuis et al. n.d)

human operators as well as for machine recipes. Even more complex, the product may require additional qualification of operators, modifications of the equipment setup, adapted sequences of process steps or rearrangement of the organizational structure.

With respect to new products' ramp-up the receptors *costs*, *time*, *quantity* and *quality* are applied on a more strategic level. They are not used due to a single ramp-up event but strategically. Questions to be answered strategically are as follows: What is the targeted time to volume by utilizing the available capabilities of the production system? What is the targeted minimum quantity of individuals of a product for profitable production? Which quality level shall be achieved by consideration of the production system's capabilities? What are the targeted costs to provide the new product with a competitive price?

Answering those questions is close related to the strategic orientation of the respective company and is already discussed in Sect. 15.2. In particular, companies of the types 2, 4 and 5 must optimize the setup as well the monitoring and control of their production systems through those four receptors with respect to the ramp-up of new products.

The strategic design of an agile production system requires the consideration of enablers for agility (conversion enablers). Nyhuis et al. (pp. 26–28) are particularly highlighting *universality*, *mobility*, *scalability*, *modularity* and *compatibility*. With respect to faster ramp-up of new products some of those conversion enablers are discussed in the sequel in conjunction with respective strategies of manufacturers. *The most effective reuse of production knowledge presumes that some of those strategies are implemented first within the structure of the production system.*

By maximizing the potential of reusable subproducts across a large variety of products, the time to volume is reduced thus reducing overhead costs and decreasing the quantity where individuals of a new product can be still produced with profit. At the same time, reinvention of the wheel is avoided by producing new products with a shared set of already qualified *process segments* thus achieving the expected quality and therefore yield soon. Rethinking the design of products and product families with respect to reusability of subproducts leads to the concept of product platforms and addresses the conversion enablers *universality*, *modularity*, *scalability* and *compatibility*.

Matching *product-related information* from the original production system with *product-related information* of the target production system is managed easier if a common product platform is used across both production systems. It is obvious that the chance of matching subproducts increases if new products are designed by reusing subproducts of a common product platform which are already used by forerunners. And with the number of matching subproducts also the number of reusable *process segments* increases in a ramp-up scenario. This is actually the intention of a product platform. *Universality*, *modularity*, *scalability* and *compatibility* shall be first discussed on the level of products and in the sequel on the level of *process segments*.

Ong et al. (2008) describe the design of products for reuse, but also the structure of production basic data as standardized by (ISA 2001) considers agile production systems to certain extend. Closs (2007) introduces *design for supply chain management* and thus the need for comprehensive consideration of the principles of product families respectively product platforms, as well as the principles *modularity* and *universality* (the also introduced principle of *postponement* is not considered in the sequel). Those principles are overlapping with the previously introduced conversion enablers on the product level. Product families, in addition, introduce standardization of subproducts respectively the consideration of standardized parts.

Considering the principle of *universality* during product design implies, for instance, that a power supply is not designed to meet the needs of a product but the product is designed to work with an existing power supply. The power supply is a subproduct with can be used *universally* for a large variety of products. The same is also valid for much less complex subproducts than power supplies.

The principle of *modularity* delegates functionality which is in common for a variety of products to dedicated modules, like, again, a power supply, a display or a cooling system. On the level of single modules it is also easier to consider *scalability* of products as part of the design, like the resolution of a variety of displays.

Finally, the previous principles also include the consideration of the conversion factor *compatibility* where ever it is useful to assemble a product or a subproduct from standardized parts.

Following the meaning of those conversion enablers (respectively the principles of *design for supply chain management*) during the product design has an immediate impact on the organization of the supply chain, the structure of the production process and the physical structure of the production system. The use of standardized parts results in lower vertical integration of the production process, thus less complexity of the own production process and reusable knowledge on the level of standardized products. *Compatibility* on the product level implicitly supports *compatibility* on the process level.

Manufacturing of dedicated modules (Weber 2004) within one production system may lead to a layout of production cells, where dedicated modules are produced by a team of operators by using a set of locally organized machines. It is also possible to establish common *process segments* for varieties of modules which are parameterized through particularly design *characteristics* (e.g., the resolution of the display which shall be mounted to a frame). *Modularity* on the product level therefore implicitly supports the *modularity* on the process level as well. *Scalability* on the product level can be supported by *modularity* on the product level, therefore reduced complexity on the process level and therefore the opportunity to setup common parameterized *process segments*.

Universality of the process level is supported by the use of *universally* usable equipment, which is commonly known as computerized numerical controlled (CNC) equipment. Nowadays, such equipment is not limited to molding machines but it includes a variety of deposition machines, cutters, welding machines, or 3D-printers. All CNC equipment in common is able to provide a particular *production technique* with a huge variety of results (*universality*) based on the loaded instruction set (a.k.a. *machine recipe*). And there is the most universal resource, the human operator, which is able to perform almost every work based on appropriate training and, again, a set of instructions (a.k.a. *handling instructions*).

Mobility on the level of equipment allows rearrangement of production cells in accordance with changing needs due to required *process segments*. And because the need for *process segments* is driven by the need of currently produced products or subproducts, the layout of the production system is driven by them.

In order to reduce complexity during a product ramp-up, IC-manufacturers, for instance, have introduced the concepts of process technologies and silicon intellectual property (silicon IP). Silicon IP (Nenni and McLellan 2013, p. 19) comprises off-the-shelf functions like A/D converters, memory and processors which can be randomly combined during the design of a new IC-product. Silicon IP consists of a particular layout of electronic elements which is mandatory in order to implement some function of an IC-product and helps to prepare the physical setup of production equipment (e.g., reticles) accordingly. Therefore, silicon IP addresses the conversion enablers *modularity*, *compatibility* and *scalability* with respect to IC-products.

Moreover, silicon IP comprises an appropriate stack of material layers which is needed to achieve the expected electrical behavior of such a function. If a function is needed, the appropriate silicon IP is reused to realize this function within an IC-product. The stack of material layers lead to process technologies.

Also process technologies (e.g., CMOS) are commonly used by IC-manufacturers. A process technology is defined by a specific stack of material layers which is built by a particular sequence of *process segments* on a circular ultrathin disk of monocrystalline silicon (a.k.a. silicon wafer). Also the size range of features of each layer is specific for process technologies. Each process technology is therefore linked to a dedicated set of equivalent process plans which can be performed to build the requested stack of layers. Therefore, process technologies represent a link between the process plan and the design of specific IC-products.

To some extent, process technologies can be seen as reusable templates of production processes which can be used to build individual IC-products. Variations of individual products are achieved by modification of the layout of single layers, the thicknesses of layers, controlled impurities of the material on each layer or other means of parameterization. The structure of process plans and the way how *process segments* are utilized for different products is again an example of *modularity*, *compatibility* and *scalability*.

However, even the conversion enabler *universality* is realized in this case. The *universality* of equipment and personnel is achieved, because every layer of an IC-product is created by repetition of the same limited set of *process segments* while the setup of the involved equipment and handling instruction of human operators is adjusted in an appropriate way. Accordingly, the production system of an IC-manufacturer is equipped with rather universally usable machines from the perspective of IC-production.

Because of this branch-specific method of process technologies and modularization of products by the use of silicon IP, the semiconductor industry provides a good pattern of an agile production system by using templates and unified underlying production processes. Some other industries, like the production of printed circuit boards (PCBs) apply this method as well (Macleod 2002).

Also the automotive manufacturing industry uses the concept of product templates. In this industry, product templates are called platforms. An outstanding example of a platform approach is Toyota's policy concerning its car models. According to Hüttenrauch and Baum (2008), Toyota is currently launching new generations of its successful car models that utilize more than 70% of their forerunners' components, and the platforms of these cars have remained largely constant through successive car generations.

The concept of process technologies or product platforms can be considered similar. For instance, Ong et al. (2008, pp. 81–112) are summarizing possible design principles for the design of product platforms independently of a specific branch of industrial manufacturing. The approach of product platforms has therefore general validity in the manufacturing industry for reduction of the complexity of management of product variants and therefore the complexity of product ramp-up projects.

In an agile production system it is therefore necessary to consider *universality*, *modularity*, *scalability* and *compatibility* on the product level as part of the design phase. This first step leads consequently to the consideration of conversion enablers on the process level. To some extend the conversion enablers on the process level are supported by conversion enables on the equipment level (e.g., *universality*, *mobility*).

Answering the question about an appropriate information model for automated recommendation of reusable existing subproducts or process segments, implicitly leads to afore mentioned best practices of agile production systems. The next chapters discuss the architecture of a multi-disciplinary knowledge base and in the sequel the structure of the information model and how it fits to possible scenarios which may be faced during a new product's ramp-up.

15.6 Invoking an Effective Multi-disciplinary Knowledge Base

The brief concept of a multi-disciplinary knowledge base which is in common across production systems in order to improve new product ramp-up is described by Willmann et al. (2013). This concept is enhanced by the general description of an automated ramp-up process which is performed on this knowledge base (Willmann et al. 2014). A detailed ontology model including a detailed automated process for generating recommendations for knowledge reuse in a product ramp-up scenario is provided by Willmann (2016). The architecture of a multi-disciplinary knowledge base which provides those features to connected production systems is also described in this work (Willmann 2016, pp. 96–97) and cited in the following sections in order to answer Q2 as specified in Sect. 15.1.

The overall architecture of a K-RAMP knowledge base is spread across multiple production systems. Individual *knowledge stores*, which reside at each production system, are connected through *intranet or internet communication technologies*. This architecture is shown in Fig. 15.6.

The architectural center of each *K-RAMP knowledge store* is a *Semantic Web database* which supports the widely spread specifications of the *Semantic Web* as there are the Resource Description Framework (RDF) (W3C—World Wide Web Consortium 2014), the RDF-Schema (RDFS) (Brickley et al. 2014), Web Ontology Language (OWL) (World Wide Web Consortium (W3C) 2012), the Semantic Web Reasoning Language (SWRL) (Horroks et al. 2004) and the query language of Semantic Web (SPARQL) (Harris et al. 2013). A detailed description of those specifications is provided in Sect. 13.3.

The *Semantic Web* specification (Allemang and Hendler 2011, p. 20) introduces methods for the creation of distributed information models. Assuming the interaction between a source production system and a target production system, this capability has to be considered as essential. In Chapter 13 the industry needs for *Semantic Web Technologies* in the industry are discussed in detail. The discussed Scenario 3 is close related to the ramp-up scenario which is the central topic of this chapter. Moreover, RDF (W3C—The world wide web consortium 2014), which is actually the foundation of all *Semantic Web* specifications being relevant for K-RAMP, provides means for expressing information and exchange of information without loss of meaning. RDFS provides means of classification of information and OWL introduces means of generalization and certain kinds of reasoning.

Why are *Semantic Web* specifications applied for information storage and exchange in conjunction with K-RAMP? From the perspective of broad applica-

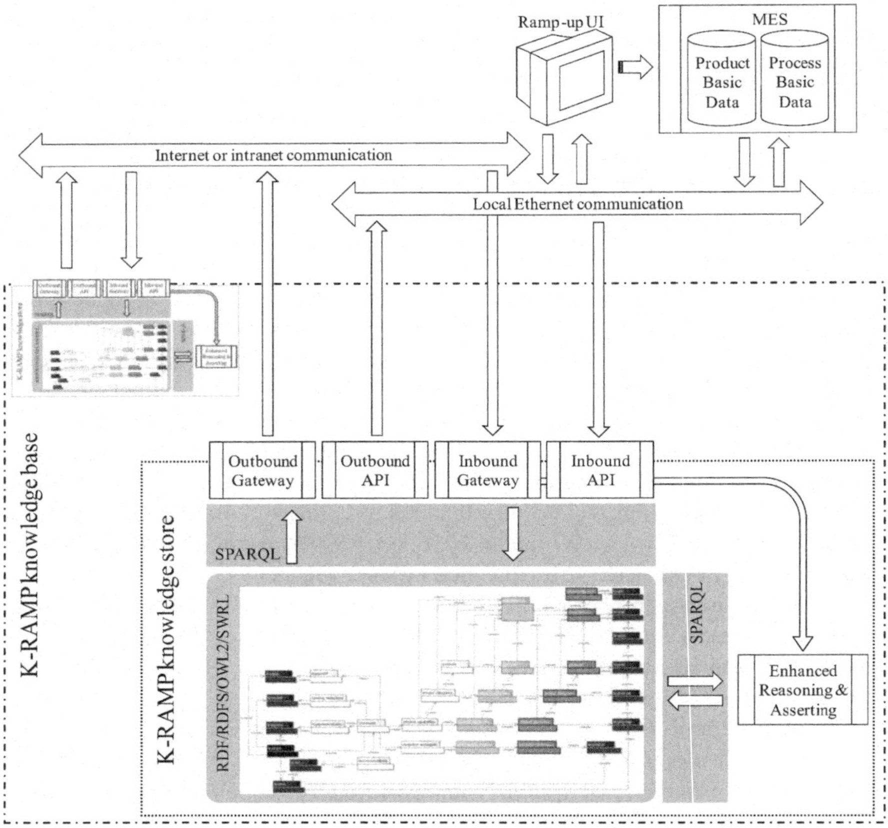

Fig. 15.6 Architectural overview of the K-RAMP knowledge base at a single production system (Willmann 2016, p. 97)

bility in the future, Germany's strategic initiative *Industrie 4.0* (Forschungsunion, National Academy of Science and Engineering 2013, p. 40) indicates the need for a common approach concerning how to see things in production engineering, mechanical engineering, process engineering, automation engineering, as well as IT and the internet. The concept of K-RAMP and the challenges to be discussed in this chapter address such a common approach for the domain of product engineering and process engineering across dislocated production systems. Therefore, on the IT-level a common technology for information exchange and information storage has to be considered as well.

The *K-RAMP knowledge store* comprises this *Semantic Web database* and attached hybrid components which are needed in order to complement the *Semantic Web database* with *enhanced reasoning and asserting* of new information from existing information. These hybrid components are interacting with the *Semantic Web database* by utilizing *SPARQL*.

For the purpose of information exchange the *K-RAMP knowledge stores* across all production systems are connected via the *internet or intranet* through pairs of one *inbound gateway* (incoming information) and *one outbound gateway* (outgoing information). These gateways use *SPARQL* for the interaction with the *Semantic Web database* of the local *K-RAMP knowledge store*.

In order to integrate the local *K-RAMP knowledge stores* with the local production-IT, namely the local *MES*, two complementary application programming interfaces (APIs) are applied for reading respectively for writing of information (*incoming API*, *outgoing API*). Again *SPARQL* is used for the interaction with the *Semantic Web database*. The *MES* or equivalent software is the primary source and target of information at a particular production system. First, it uploads *product-related information* which is converted to the information models being primarily derived from the K-RAMP information model (a.k.a. ontologies). Secondly, it uploads *Process-related information* which is also converted to the information models being derived from the K-RAMP information models.

The ramp-up team interacts through a user interface (*Ramp-up UI*) with the local *K-RAMP knowledge store*. By means of the *Ramp-Up UI* the ramp-up team is able to gather recommendations. After the recommended ramp-up activities are performed the initial updates on *product-related information* and *process-related information* are performed in the local *MES* by the ramp-up team. Again from the local *MES* the updated information is fed back to the *K-RAMP knowledge store*. The local *MES* thus remains the master of *product-related information* and *process-related information*.

Due to the standardization of the applied *Semantic Web* specifications and the exclusive use of *SPARQL* for information exchange, each production system may use its own off-the-shelf software product with respect to the *Semantic Web database*.

15.7 Information Model and Matchmaking Scenarios

In the following sections Q3 is answered by introducing the general concept of K-RAMP's information model first and by discussing the scenarios which may be considered during a product ramp-up next. The following sections of this chapter are based on (Willmann 2016).

At the target production system the *product-related information* and the *device-independent process-related information* has to be interwoven with the respective information about the existing production. Beside *product-related information* about forerunners and *device-independent process-related information*, the target production system also provides *device-dependent process-related information* (Fig. 15.7). This portion of information comprises the sequential order of *process operations* embedded in *process segments* as well as handling instructions or equipment recipes. In a presumed heterogeneous equipment environment between the original production system and the target production system, this portion of

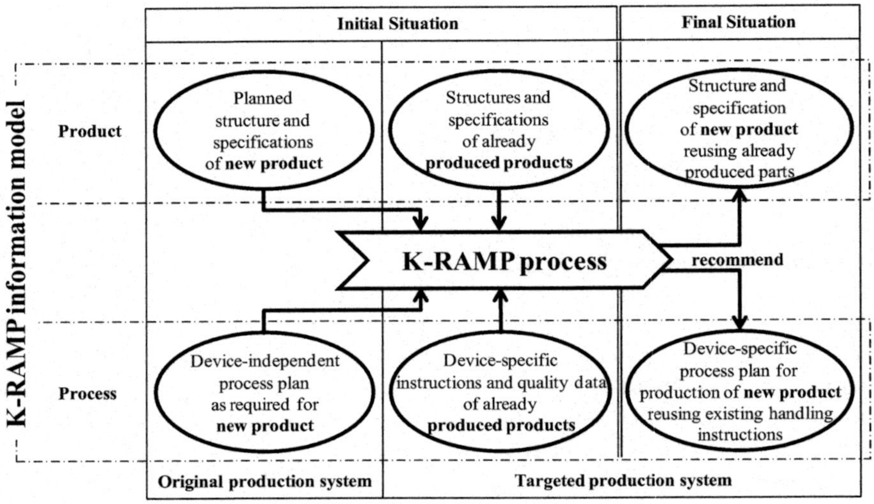

Fig. 15.7 Conceptual overview of the K-RAMP approach (Willmann 2016, p. 34)

information cannot be exchanged easily. This portion of information must be reused through matchmaking of common or similar *product-related information* or *device-independent process-related information*.

The matchmaking process of K-RAMP utilizes two different sources of information—the original production system and the target production system. *Product-related design information* is provided by both sources. *Process-related knowledge* is provided comprehensively (*device-independent* and *device-dependent information*) by the target production system only. The original production system only provides the *device-independent portion* of *process-related information*.

There is another aspect of *device-independent process-related information* beside *process capabilities*. It is possible to categorize *process segments* according to basic *production technique*. Industrial branches may use the basic *production techniques* (Koether and Rau, 2008, p. 15) (Deutsches Institut für Normung e. V. 2003), namely master forming, forming, separating, merging, coating, altering or more specialized derivations of them, like browning, cooking, baking, boiling or heating in restaurant kitchens, or mounting windshield, undercoat painting or thixocasting in automotive manufacturing, or chemical vapor deposition, dry etch or lithography in semiconductor manufacturing. These *production* techniques can be formalized in taxonomy models without consideration of specific equipment. *K-RAMP presumes that such taxonomy of production techniques is applied for categorizing of process segments although the process segments remain individual information of each production system.*

In order to understand the matchmaking between *products*, *product categories* or *process capabilities*, it is first necessary to discuss the common information model which allows their specification. E*ach product, product category or process capability must comprise a set of specifications which describes it uniquely.*

One *specification* may enclose another *specification* or two *specifications* may overlap each other. In case of *specification targets*, these relations are specified easily by the help of set operations. Each *specification target* represents a set of target values. Given two *specifications targets* st_1 and st_2, st_1 encloses st_2 if the set of target values of st_1 is a superset of the set of target values of st_2. Further, st_1 and st_2 are overlapping if the set of target values of st_1 and st_2 have an intersection which is not empty. In case of *specification ranges* the definition of these two relations is similar. However, the set of possible values is defined by LSL and USL. Given two *specification ranges* sr_1 and sr_2, sr_1 encloses sr_2 if its specification range encloses the specification range of sr_2. And both *specification ranges* are overlapping if their two specification ranges are overlapping as well.

Specifications are comparable with each other if both are referring to the identical *characteristic*. For this reason *it is necessary and also a premise of K-RAMP that characteristics are commonly managed across production systems.*

Specification sets—which are represented by *products, product categories* or *process capabilities* in specific—are set of *specification ranges* or *specification targets*. There are also relations between pairs of *specification sets* which are based on afore introduced relations between pairs of *specifications*. This has to be ensured by the information management of each production system.

For a pair of *specifications* from two *specification sets*, one *specification* either *encloses* the other, both specifications *overlap* each other or the pair is not associated at all. Consequently, the two *specification sets* are *enclosing* each other—if all *specifications* of the left *specification* set enclose the respective *specifications* of the right hand *specification set*. In such a case the cardinality of the left hand *specification set* is less than or equal to the cardinality of the right hand *specification set* because the used set of *characteristics* of the left hand *specification sets* is a subset of the set of *characteristics* of the right hand *specification set*.

The *specifications* of both *specification sets* may overlap each other, resulting in a relation "*overlaps*" between the two *specification sets*. This is the case, if there is at least one *specification* of the right hand *specification set* only overlapped by its counterpart of the left hand *specification set*.

Some of the *specifications* of both *specification sets* are not related with each other if no counterpart uses the same *characteristic*. In this case the *specification sets* can be only partially overlapped—except for the last case where not a single *specification* of the one *specification sets* is related with a *specification* of the other *specification set*. In this case there is no relation established between the two *specification sets*. If two *specification sets* are *partially overlapped*, the relation between their cardinalities is not specified because it is not determined how many *specifications* of the one *specification set* and the other *specification set* are not related with each other.

The previously introduced relations between *specification sets* are validated based on the quality of the relation.

1. A ratio of *specifications* which are enclosing *specifications* of the other *specification set* compared to all relations between *specifications* of both *specification sets*. For instance, for two *specification sets* set1 and set2 there is the relationship overlaps (set1, set2)—set1 overlaps set2. Overlaps may also imply a mixture of encloses and overlaps associations between *specifications* as mentioned above. In this example set1 may have 3 *specifications* which enclose *specifications* of set2. Furthermore, set2 has 1 *specification* which encloses a *specification* of set1. From the perspective of set1 this *specification* overlaps in the other direction. Additionally, 2 further *specifications* are overlapping from the perspective of both *specification sets*, and 1 more *specification* in set2 is not related with any *specification* in set1 because there is no common *characteristic* shared.

All in all, this results in $|\text{set1}| = 6$ and $|\text{set2}| = 7$. The relation set1 overlaps set2 is validated with a *ratio of enclosing specifications* of 0.5 (3 enclosing of 6 in total) while in the opposite direction there is a ratio of 0.143 (1 enclosing of 7 in total) for set2 partially set1 because from the perspective of set2 there is one specification which is not associated with any specification of set1.

2. A ratio of *characteristics* which are shared between *specifications* of both *specification sets* compared to the cardinality of the respective *specification sets*.

With respect to the previous example, this *ratio of common characteristics* is 1.0 from the perspective of set1 because $|\text{set1}| = 6$ and there are 3 encloses associations and 3 disjoint overlaps associations encountered. From the perspective of set2, the ratio is 0.43 because $|\text{set2}| = 7$ and there is 1 encloses and 2 overlaps association.

It is recalled at this position that these relations between *specification sets* are specialized as relations between *products*, *product categories* and *process capabilities*. *K-RAMP therefore presumes that beside of products and product categories also process capabilities are managed on top of process segments within production systems.* All three concepts are *specializations* of *specification sets* as mentioned above. Interweaving information of the original production system and the target production system is starting with the assertion of all possible relations between

- pairs of *product categories*—for determining specialization hierarchy,
- *product categories* and *products*—for categorization of *products*,
- pairs of *process capabilities*—for determining appropriate existing *process capabilities*, and
- *product categories* or *products* and *process capabilities*—also for determining appropriate existing *process capabilities*.

A *product category* which encloses another *product category* is a generalization of the enclosed *product category*. *Products* and *product categories* require *process capabilities* in order to produce them. These *process capabilities* must be enclosed by *products* respectively by *product categories* (e.g., p1.1.1 encloses pc2 in Fig. 15.8 because the *specifications* of *characteristics* A, B and C of p1.1.1 enclose the *specifications* of the same *characteristics* of pc2). This is probably not always the

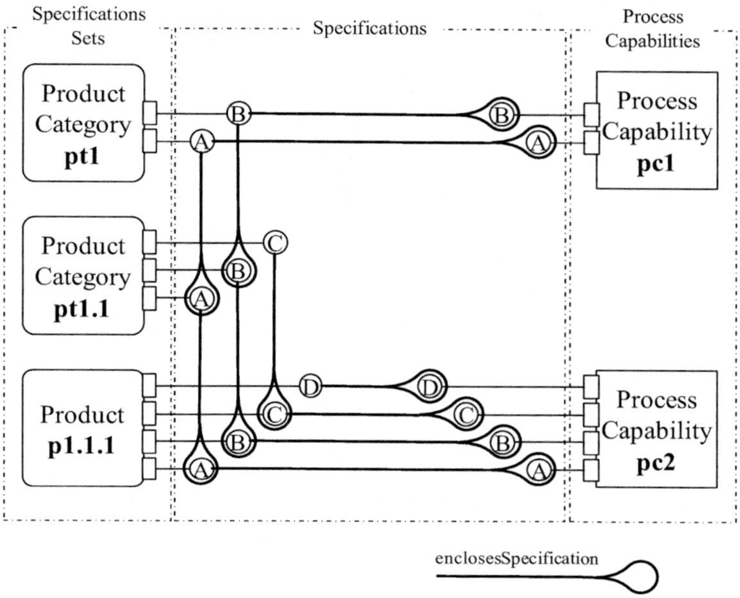

Fig. 15.8 Overiew of encloses-relation between product categories, products and process capabilities (Willmann 2016, p. 56)

case because there are also *process capabilities* which are specified for producing a category of *products* in general (e.g., it is assumed that p1.1.1 does not enclose pc1). One group of examples comprises capabilities of cooling processes or heating processes where probably the volume and density of the material are the only relevant constraints (e.g., pc1 comprises only *specifications* for *characteristics* A and B).

However, each *product* has its individually specified shape (e.g., p1.1.1 comprises *specifications* of *characteristics* A, B, C and D). Therefore, the *product's specifications* are a superset of the *specifications* of the *process capability* (e.g., p1.1.1 $\supseteq$ pc1). According to the definition of the relationship *encloses*, consequently the *product* does not enclose the *process capability*, and as further consequence the *product* would not require the *process capability*. But this decision is only half the story because this particular *process capability* is dedicated to a *product category* as mentioned before (e.g., pt1 encloses pc1).

How is it possible to determine, whether a categorized *product* requires this *process capability*? The solution is as follows. The *product* is categorized by a *product category* which may be a specialization of another *product category* (e.g., pt1 encloses pt1.1, p categorized by pt1.1 and therefore p is also categorized by pt1). The *product* p1.1.1 does not necessarily enclose pc1. However, both *specification sets* are for sure overlapping, and therefore, *process capabilities* of *product categories* can be possible candidates for their categorized *products*.

Based on these general relations between *products*, *product* categories and *process capabilities*, it is possible to determine existing *product-related information* of the target production system. However, there is still no way to recommend activities in order to reuse existing *device-dependent process-related information*. With respect to *process capabilities*, it is not only a question about enclosing their *specification*. It is also important to consider the subproducts which are used by the *process segment* in order to produce a subproduct on the next higher composition level.

A *process segment* setup uses an arbitrary number of *product categories*. It is the idea, that the same *process segment* can be used to satisfy different *process capabilities* just by variation of the used *product categories* or the adjustment of some setup parameters. As a consequence, *product categories* are used due to *process segment* setups in order to enable a certain *process capabilities*. So *product categories* do not only require *process capabilities* but they are also the input of *process* segments (through *process segment* setups) in order to satisfy *process capabilities*.

Due to this usage of *product categories*, *process segment* setups are also treated as *product category sets*. There is another concept in the information model which is treated as *product category set*—namely *product categories* themselves due to the fact that each *product category* is usually a composition of one or more subordinated *product categories*.

The composition of *product categories* through subordinated *product categories* requires afore mentioned *process capability*. The *process segment* setup which enables a *process capability* by utilizing a particular *process segment* uses these subordinated *product categories*. Reusing an existing *process capability* by a new *process capability* is therefore not only a question about enclosing or overlapping of *specification sets*. The *product categories* being used by the existing underlying *process segment* setup also need consideration.

The relations *encloses*, *overlaps* and *partially* of *specification sets* are already discussed above. In the context of *product category sets* the semantics of these relations are extended to sets of *specification sets*. However, the particular need for these semantics is limited to sets of *process categories* for the purpose of K-RAMP. A *product category set* is a set of *product categories* and provides relations ($encloses_{PTS}$, $overlaps_{PTS}$, $partially_{PTS}$ where PTS abbreviates *product category set*). *Product category sets* are used to match the immediate decomposition structure of *product categories* with the used *product categories* of *process segment* setups. For this purpose, *product categories* and *process segment* setups become specializations of *product category sets*. *Product category* is a specialization of *product category set* because every *product category* refers to a set of *product categories* which represent its immediate decomposition. *Process segment* setup is a specialization of *product category set* because every *process segment* setup refers to a set of used *product categories*.

What are the rules in order to assert those relations? A *product category set* pts1 $encloses_{PTS}$ a *product category set* pts2 if each member of pts1 *encloses*—mutual enclosing of *specification sets*—the equivalent member of pts2. This rule requires that each *product category* of pts1 encloses an equivalent counterpart in pts2. It is already highlighted as a premise, that each production system thoroughly specifies each *product category* with sufficient entirety, in order to distinct it from other entities in the knowledgebase. The latter ensures that for each member of the one set its counterpart can be determined in the other set.

A *product category set* pts1 *overlaps$_{PTS}$* with a *product category set* pts2 if at least one member of pts1 *overlaps*—in terms of *specification sets*—the equivalent member of pts2 instead of enclosing it. Again, each *product category* of pts1 has an equivalent counterpart in pts2. Each pair of members of both sets shares *specifications* with the same *characteristic* but either their *specification ranges* or their sets of *specification targets* are only overlapping.

A *product category set* pts1 is *partially$_{PTS}$* a *product category set* pts2 if a subset of members of pts1 overlaps or encloses equivalent members of pts2. In this case, there are members in each set which do not have equivalent counterparts in the other set.

After the introduction of *process category sets*, all information is available for asserting a series of possible recommendations about how newly introduced *products* and *product categories* can reuse existing *process capabilities* and thus the underlying *device-dependent process-related information* of the target production system.

Willmann (2016, pp. 65–70), determines 17 matchmaking scenarios. Possible efforts with respect to the adjustment of existing information before reuse lead to a categorization of these scenarios through penalty levels.

One scenario with the lowest possible penalty 0 (matchmaking scenario MS-01) (Fig. 15.9) represents a situation where the new *product category* (PT_i) encloses an existing *product category* (PT_x), which means that an existing *product category* has already the same of even tighter specifications than the new *product category* with respect to the relevant *characteristics*. Moreover, also the *product categories* on the next lower decomposition level of the introduced *product category* (PT_i-1 to PT_i-n) enclose (enclose$_{PTS}$) their counterparts of the existing *product category* (PT_x-1 to PT_x-n). In this case it is recommended to apply the existing *process capability* (PC_x), as well as the underlying *process segment* and its setup structure also for the production of the new *product category*.

The scenario with the highest possible penalty, 15, (MS-17) (Fig. 15.10) represents a situation where the new *product category* only partially overlaps an existing *product category*. Also the new introduced *process capability* (PC_i) only partially overlaps with the existing *process capability* (PC_x) related to the existing *product category*.

Moreover, also along the decomposition structure there is only a partial overlapping between the new *product category* and the existing *product category*. In such a situation it must be determined whether the overlapping subset of *specifications* on the level of *process capabilities* is a subset of the overlapping *specifications* on the *product category* level. For instance, there may be two *product categories* with a set of completely different *specifications* but one single common *specification* "body temperature" (e.g., "butter" and "chocolate coated cake") and

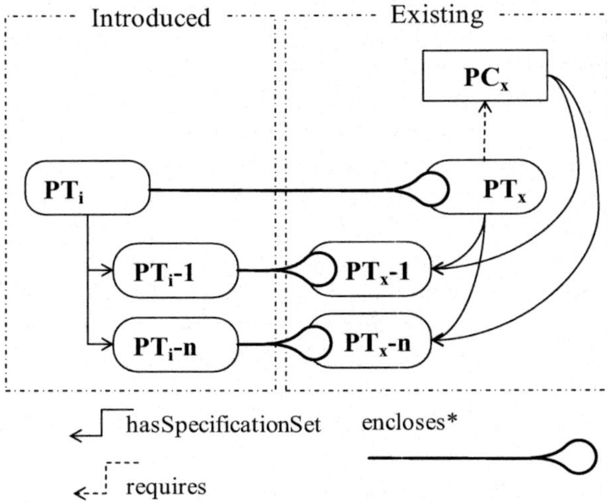

Fig. 15.9 Examplary matchmaking scenario MS-01 (Willmann 2016, p. 65)

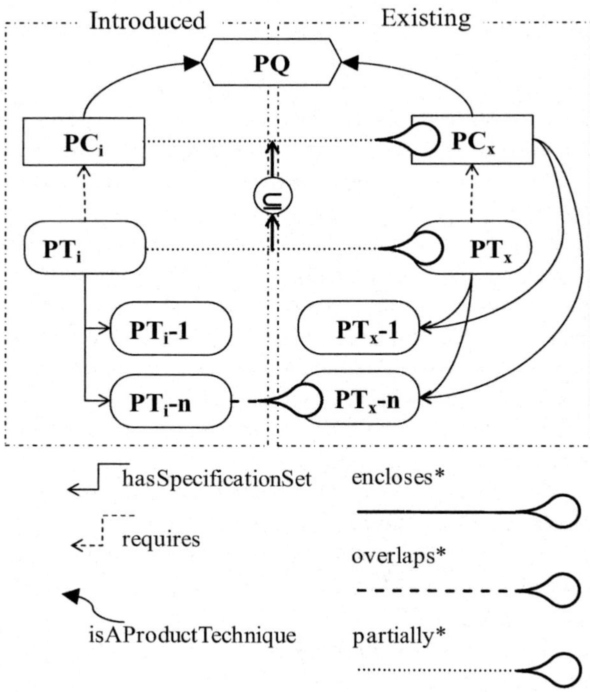

Fig. 15.10 Examplary matchmaking scenario MS-17 (Willmann 2016, p. 70)

both *product categories* require a *process capability* to ensure a certain "body temperature" (e.g. room temperature around 22 °C). In this scenario an existing *process segment* which brings cold butter after removing from the fridge to room temperature (e.g., "storing of material until room temperature achieved") may be also applicable for the new *product* to cool it down to room temperature after it was coated with hot chocolate couverture. However, there may be still additional *specifications* which need to be considered due to the new *process capability*. In order to consider those *specifications* through the potentially reusable *process segment* behind the existing *process capability*, new metrology operations need to be considered. This premise requires certain modification of the *process segment*. Moreover, it must be determined whether both *process capabilities* are implemented by *process segments* of the same *production technique*. For instance, two *process capabilities* may be specified by a *characteristic* "layer thickness". However, the existing *process capability* implemented by a *process segment* of the *production technique* "polishing" while the new introduced *process capability* is originally enabled by a *process segment* of the *production technique* "coating".

The summary of all matchmaking scenarios is comprehensively listed in Table 15.1. It provides an overview about the intensity of linkage between pairs of *product categories*, *process capabilities* or *product category sets* as well as change actions to be considered during ramp-up and the penalty of the recommendation. The higher the penalty, the higher are the duration and costs of the underlying process qualification. It is assumed that the replacement of the *product category* usages has the lowest costs. There are no modifications performed within the existing setup of the *process segment* but only the incoming material is modified. Parameter adjustment is more expensive because it requires at least the modification of some adjustable parameters of the *process segment*, which causes implicitly more expensive and time consuming pilot production runs than in the case of simple change of incoming material. If similarity of *process capabilities* is assumed due to the categorization by a common *production technique*, some penalty points are counted. The most expensive matchmaking scenarios are the ones where modifications of the *process segment's* sequences are encountered. Therefore, these recommendations are assigned with the highest penalty level – meaning the most expensive ramp-up activities. The need for activities is rated through binary digits for each scenario. The resulting binary number maps to the penalty level.

A recommendation about the adjustment of existing *device-dependent process-related information* holds all information which is needed in order to derive an appropriate *process segment* setup for enabling a *process capability* which can be required by the newly introduced *product category*. Based on the recommended actions in Table 15.1 which are flagged with 1, a recommendation sentence can be derived by the knowledge base.

The previous sections of this chapter discussed the underlying information model of the multi-disciplinary knowledge base and matchmaking scenarios for information which is exchanged between two production systems during a new product's ramp-up. Therefore, Q3 of Sect. 15.1 is answered.

Table 15.1 Summary of matchmaking scenarios, changes to be considered by recommendation and levels of severity (Willmann 2016, p. 71)

Match-making scenario	Levels of association			Changes to consider due to recommendation				Penalty
	Product category	Process capability	Product category set	Process segment	Production technique	Parameter adjustment	Product category usage	
MS-01	E	N./A	E	0	0	0	0	0
MS-02	O	E	E	0	0	0	0	0
MS-03	P	E	E	0	0	0	0	0
MS-04	*	E	O	0	0	0	1	1
MS-05	*	E	P	0	0	0	1	1
MS-06	O	O	E	0	0	1	0	2
MS-07	P	O	E	0	0	1	0	2
MS-08	O	O	O	0	0	1	1	3
MS-09	P	O	O	0	0	1	1	3
MS-10	O	O	P	0	0	1	1	3
MS-11	P	O	P	0	0	1	1	3
MS-12	O	P	E	0	1	1	0	6
MS-13	P	P	E	0	1	1	0	6
MS-14	O	P	O	1	1	1	1	15
MS-15	P	P	O	1	1	1	1	15
MS-16	O	P	P	1	1	1	1	15
MS-17	P	P	P	1	1	1	1	15

E Encloses, O Overlaps, P Partially

15.8 Needs for Standardization Across Enterprises

In Section 15.7 it turned out already that certain domains of the introduced information model are managed across the production systems which are involved in a product ramp-up scenario. These premises raise the need for standardization of the respective information domains if the scope is enhanced to the supply chain across enterprises. In the following sections such recommendations for standardization are introduced and reflected to existing initiatives thus answering Q4 of Sect. 6.1.

As highlighted in Sect. 15.7, automated recommendation concerning the reuse and the adjustment of reusable information, in case of a new product's ramp-up, comes with a set of premises which need to be fulfilled by a production system in order to maximize the benefit from the multi-disciplinary knowledge base.

- Design of *products* and *product categories* by consequent consideration of *modularity*, *scalability*, *compatibility* and *universality*.
- The structuring of *production processes* to *process segments* in accordance to the needs of designed subproducts and subproduct categories which consist of at least one metrology operation and an arbitrary number of material processing operations.
- All *process segments* deliver metrology data. Consequently, there are *process capabilities* specified by *characteristics* which are determined in this context.
- *Process capabilities* are managed within production systems, consequently decoupling *device-dependent process-related information* from *product-related information*. This aspect contributes to the vertical integration of *product-relation information* and *process-related information*.
- A commonly managed hierarchy of *product categories* which allows integration of *product-related information* horizontally along the supply chain.
- A categorization of *products* through *product categories*, consequently matching *products* and subproducts along the supply chain (horizontal integration of information) due to their membership in common *product categories*.
- *Products*, *product categories* and *process capabilities* are described uniquely as *specification sets*, thus enabling matchmaking between individuals of these concepts through the discussed relations *encloses* and *overlaps*. This matchmaking comprises vertical and horizontal integration of information.
- A taxonomy of *production techniques* across involved production systems enables the categorization or *process segments*, thus an approach to a common *process-related* taxonomy across production systems. Consequently, this aspect contributes to the horizontal integration of *process-related information*.
- A commonly managed set of *characteristics*—and consequently a commonly managed set of *engineering units*—across involved production systems is a further contribution to horizontal integration. Actually, this aspect is the core of all opportunities for matchmaking being listed previously.

Several of those premises need to be considered within individual enterprises as well as individual production systems. However, with respect to the enhanced scope of the supply chain across enterprises there are some needs for standardization.

A taxonomy or an ontology of *production techniques* across enterprises helps to categorize capabilities of production services for a more efficient lookup for service providers. Moreover, as described in Sect. 15.7, *production techniques* contribute if similarities between *process capabilities* need to be determined. Nowadays, generic or for specific industrial branches taxonomy models about *production techniques* are published (e.g., Koether and Rau 2008). However, there are no public standards available yet on this quality level and no *Semantic Web* based models which can be used automatically.

With respect to *product categories*, there is the initiative UNSPSC of the United Nations Organization (UNO) which also provides an ontology (United Nations Development Programme 2014). However, this ontology is a pure hierarchical model of unified names of *product categories* and *product service categories*. However, reviewing documentation and web-pages on the internet there are also a lot of specific standards and legal regulations for categories of products including detailed specifications. These detailed specifications are not yet considered by UNSPSC or another more enhanced ontology. Also in Chapter 6 there are several standards (e.g. STEP) introduced which are used to exchange product data between CA* systems. Eventually, it could be also useful to provide a platform, where product providers are able to provide the public specifications of their *products* thus supporting the exchange of detailed specifications of standard components.

Ontology models for *engineering units* are already very well researched through the QUDT ontology (Hodgson et al. 2014). The integration of QUDT could be a future extension of the K-RAMP concept. It can be also useful to unify the terminology on *characteristics*. Some industry (e.g., the semiconductor industry) has recently started to unify *characteristics* on the level of equipment data collection (Semiconductor Equipment and Material International 2013).

Willmann (2016) introduces a comprehensive ontology model which covers those information domains as part of K-RAMP. However, with respect to K-RAMP, currently asserted recommendations are only understood as the first step towards standardized ontology models for product ramp-up support on the supply chain level.

15.9 Outlook: Deterministic Product Ramp-up for Supply Chains

Achieving afore introduced needs for the reuse of existing production knowledge during a product ramp-up is not only a technical task but also an organizational challenge. As it could be highlighted, many techniques and methods are already known and best practices in many enterprises. Initiatives are driven today on several levels but following similar principles. We just need to compare the principles of *design for supply chain management* and the conversion enablers for manufacturing agility of production systems.

All those measures are crucial, and therefore they are premises in order to introduce an effective information model for knowledge reuse during a new product's ramp-up. K-RAMP provides an ontology model which is based on existing industry standards and with enhancements for facing this specific challenge. But its maximal contribution is also dependent on those premises.

There are technical concepts and best practices available how *Semantic Web* database are integrated with existing ICT-systems. K-RAMP provides ontology models which can be performed in the context of *Semantic Web* databases. These ontology models are used to perform automated recommendation about the reuse of production knowledge for new products during their ramp-up.

However, there is also a need for unified management of information across the involved production systems. This measure needs to be taken on the level of individual enterprises at least. A more effective approach is the unification of required information on a public level thus leading to more effective and more efficient product ramp-up on the level of global supply chains.

References

Abele, E. et al.: Global Production: A Handbook for Strategy and Implementation. Springer, Berlin (2008). ISBN:978-3-540-71652-5

Allemang, D., Hendler, J.: Semantic Web for the working ontologist—effective modelling in RDFS and OWL. s.l.:Elsevier (2011). ISBN:978-0-12-285965-5

Breyfogle III, F.W.: Implementing Six Sigma Second Edition. Austin, TX: Wiley (2003). ISBN:0-471-26572-1

Brickley, D., Guha, R.V., McBride, B.: RDF Schema (RDFS)—Release 1.1. [Online]. http://www.w3.org/TR/2014/REC-rdf-schema-20140225/ (2014). Accessed 30 Apr 2016

Cisek, R., Habicht, C., Neise, P.: Gestaltung wandlungsfähiger Produktionssysteme [eng. Design of agile production systems]. ZWF—Zeitschrift für wirtschaftlichen Fabrikbetrieb. **9**, 441–445 (2002)

Closs, D.J.: Design for supply chain management. Logist. Q. Mag. **13**(1), 9–10 (2007)

Deutsches Institut für Normung e. V: DIN 8580:2003-09: Manufacturing Processes—Terms and Definitions, Division. s.l.:Beuth (2003)

Dietrich, E., Schulze, A.: Statistische Verfahren zur Maschinen- und Prozessqualifikation [eng. Statistical methods for machine and process qualification]. Munich, Vienna (2003). ISBN:3-446-22077-1

Forschungsunion, National Academy of Science and Engineering: recommendations for implementing the strategic initiative INDUSTRIE 4.0—Final report of the Industrie 4.0 Working Group. [Online]. http://forschungsunion.de/pdf/industrie_4_0_final_report.pdf (2013). Accessed 8 Mar 2015

Fransoo, J.C., de Kok, A.G.: What Determines Product Ramp-up Performance? A Review of Characteristics Based on a Case Study at Nokia Mobile Phones. s.l.:Beta, Research School for Operations Management and Logistics (2007). ISSN:1386-9213

Harris, S., Seaborne, A., Prud'hommeaux, E.: SPARQL 1.1 query language. [Online]. http://www.w3.org/TR/2013/REC-sparql11-query-20130321/ (2013). Accessed 30 Apr 2016

Hodgson, R., Keller, P.J., Hodges, J., Spivak, J.: QUDT—Quantities, Units, Dimensions and Data Types ontologies. [Online]. http://qudt.org/ (2014). Accessed 1 May 2016

Horroks, I. et al.: SWRL: a Semantic Web Rule Language combining OWL and RuleML. [Online]. http://www.w3.org/Submission/SWRL/ (2004). Accessed 30 Apr 2016

Hüttenrauch, M., Baum, M.: Effiziente Vielfalt: Die dritte Revolution in der Automobilindustrie [engl. Efficient variety: the third revolution of the automotive industry]. Springer, Berlin (2008). ISBN:978-3-540-72115-4

ISA: ANSI/ISA-95.00.02-2001: enterprise control system integration—Part 2: Object model attributes. ISA-The Instrumentation, Systems, and Automation Society, Research Triangle Park, North Carolina, USA (2001). ISBN:1-55617-773-9

ISO—International Standards Organization: ISO 21747-2006—Statistical methods—Process performance and capability statistics for measured quality characteristics. [Online]. http://www.beuth.de/langanzeige/DIN-ISO-21747/de/93888258.html (2007). Accessed 17 July 2011

Kinkel, S., Maloca, S.: Modernisierung der Produktion—Flexibilitäts- und Stabilitätsstrategien in der deutschen Industrie [eng. Modern production—flexibility and stability strategies in the German industry]. [Online]. http://www.isi.fraunhofer.de/isi-wAssets/docs/i/de/pi-mitteilungen/pi54.pdf (2010). Accessed 6 May 2016

Koether, R., Rau, W.: Fertigungstechnik für Wirtschaftsingenieure [engl. Production engineering for industrial engineering and managemen], 3rd edn. Hansa, München (2008). ISBN 978-3-446-41274-3

Kurttila, P., Shaw, M., Helo, P.: Model factory concept—Enabler for quick manufacturing capacity ramp-up. s.l., s.n. (2010)

Macleod, P.: A review of flexible circuit technology and its applications. s.l.:PRIME Faraday Partnership (2002). ISBN:1-84402-023-1

Nenni, D., McLellan, P.: FABLESS: the transformation of the semiconductor industry. s.l.: SemiWiki.com (2013). ISBN:978-1-4675-9307-6

Nyhuis, P., Reinhart, G., Abele, E.: Wandlungsfähige Produktionssysteme—Heute die Industrie von morgen gestalten [eng. Agile production systems—shaping today the industry of tomorrow]. s.l.:978-3-939026-96-9 (n.d.)

Ong, S. K., Xu, Q. L., Nee, A. Y. C.: Design Reuse in Product Development Modeling, Analysis and Optimization. World Scientific Publishing, Singapore (2008). ISBN:13-978-981-283-262-7

Schmidt-Dilcher, J., Minssen, H.: Ewige Baustelle PPS—Strukturkonservative Rationalisierungsmuster im Maschinenbaubetrieb [en. Infinite construction site CAPP—structurally conservative rationalization patterns in systems engineering units]. In: Kooperationsnetze, flexible Fertigungsstrukturen und Gruppennetzwerke—Ein interdisziplinärer Ansatz [en: Cooperation networks, flexible manufacturing structures and group networks—an interdisciplinary approach], pp. 83–117. Leske + Budrich, Opladen (1996). ISBN:978-3-322-97326-9

SEMATECH, Inc.: Computer Integrated Manufacturing (CIM) Framework specification version 2.0. s.l.:Technology Transfer # 93061697J-ENG (1998)

Semiconductor Equipment and Material International: SEMI E164-0313—specification for EDA common metadata. s.l.:Semiconductor Equipment and Material International. (2013)

Slamanig, M., Winkler, H.: Management of product change projects: a supply chain perspective. Int. J. Serv. Oper. Manage. **11**(4), 481–500. ISSN 1744-2370 (Print), 1744-2389 (Online) (2012)

Terwiesch, C., Bohn, R. E., Chea, K. S.: International product transfer and production ramp-up: a case study from the data storage industry. In: R&D Management 31, 4, pp. 435–451. Blackwell, Oxford (2001)

Terwiesch, C., Xu, Y.: The copy exactly ramp-up strategy: trading-off learning with process-change, s.l.: s.n. (2003)

United Nations Development Programme: UNSPSC(R)—United Nations Standard Products ans Services Code. [Online]. https://www.unspsc.org/ (2014). Zugriff am 14-05-2016.

W3C—The World Wide Web Consortium: RDF 1.1 Primer—W3C Working Group note. [Online]. http://www.w3.org/TR/2014/NOTE-rdf11-primer-20140624/ (2014). Accessed Mar 2015

W3C—World Wide Web Consortium: RDF 1.1 Primer—W3C Working Group note. [Online]. http://www.w3.org/TR/2014/NOTE-rdf11-primer-20140624/ (2014). Accessed Mar 2015

Wappis, J., Jung, B., (2010): Taschenbuch—Null-Fehler-Management—Umsetzung von Six Sigma [engl. Zero-defects management—Implementation of six sigma], 3 edn. Munich: Carl Hanser Verlag. ISBN:978-3-446-42262-9.

Weber, A.: The Pros and Cons of cells. Assembly, 4 November. (2004)

Willmann, R.: Ontology Matchmaking of Product Ramp-up Knowledge in Manufacturing Industries. Vienna University of Technology, Vienna (2016)

Willmann, R., Biffl, S. & Serral Asensio, E.: Determining Qualified Production Processes for New Product Ramp-up Using Semantic Web.ACM, Graz (2014). ISBN:978-1-4503-2769-5

Willmann, R., Serral, E., Kastner, W., Biffl, S.: Shortening of product ramp-up by using a centralized knowledge base. s.l., IEEE, pp. 516–522 (2013). ISBN 978-1-4799-0752-6

World Wide Web Consortium (W3C): OWL 2 Web Ontology Language document overview, Second edn. http://www.w3.org/TR/owl2-overview/. s.n. (2012)

Chapter 16
Towards Model Quality Assurance for Multi-Disciplinary Engineering

Needs, Challenges and Solution Concept in an *AutomationML* Context

Dietmar Winkler, Manuel Wimmer, Luca Berardinelli, and Stefan Biffl

Abstract In multi-disciplinary engineering (MDE) projects, information models play an important role as inputs to and outputs of engineering processes. In MDE projects, engineers collaborate from various disciplines, such as mechanical, electrical, and software engineering. These disciplines use general-purpose and domain-specific models in their engineering context. Important challenges include model synchronization and model quality assurance (MQA) that are covered insufficiently in current MDE practices. This chapter focuses on the needs and approaches for MQA in MDE environments. We address the following two research questions (RQs): The first RQ focuses on investigating needs and expected capabilities that are required for a systematic review process that focuses on changes in MDE design models (RQ-MQA1). The second RQ focuses on how to extend a standard modeling language for MDE, such as the *AutomationML*, to address needs for storing process-relevant attributes in the context of quality assurance and review process support (RQ-MQA2). This chapter presents concepts and an initial evaluation of MQA approaches in the context of selected MDE processes, i.e., the addition, change, or removal of a component in an engineering discipline and an impact analysis on the integrated plant model. Main results are that (a) an adapted review process helps to systematically drive model reviews for MDE and (b) the standardized language

D. Winkler (✉)
SBA-Research gGmbH, Vienna, Austria

Technische Universität Wien, Institute of Software Technology and Interactive Systems, CDL-Flex, Wien, Austria
e-mail: dwinkler@sba-research.org; dietmar.winkler@tuwien.ac.at

M. Wimmer • L. Berardinelli
Technische Universität Wien, Institute of Software Technology and Interactive Systems, CDL-Flex, Wien, Austria

S. Biffl
Technische Universität Wien, Wien, Austria

© Springer International Publishing AG 2017

S. Biffl et al. (eds.), *Multi-Disciplinary Engineering for Cyber-Physical Production Systems*, DOI 10.1007/978-3-319-56345-9_16

description of *AutomationML* can be extended with process-related attributes that are useful for quality assurance and reviewing.

Keywords Model Quality Assurance • Multi-Disciplinary Engineering • Model Review • Defect Detection • *AutomationML*

16.1 Introduction

In *multi-disciplinary engineering* (MDE) projects, models play an important role as inputs to and outputs of engineering processes. Various engineering disciplines, such as mechanical, electrical, and software engineering are involved. Engineers coming from different disciplines typically use, create, and adapt generic and domain-specific models in their individual engineering context (Biffl et al. 2016a). Important challenges include *model synchronization* of often-heterogeneous inputs from various disciplines that need to be synchronized for efficient data exchange and *model quality assurance* (MQA) that is covered insufficiently in current MDE practices (Xiong et al. 2007; Whittle et al. 2016). In context of product lifecycle Management (PLM) (Chap. 4), engineering processes and tool chains (Chaps. 2 and 11) can help to support model synchronization in heterogeneous and multi-disciplinary engineering projects. Chapter 9 presents a set established tool chains from an industry perspective for different disciplines/phases in a typical production system engineering project. However, current approaches pay little attention to the quality of engineering models and the improvement of engineering processes.

Chapter 1 in this book explains in the context of *cyber-physical production systems* (CPPS), why additional complexity intensifies the need for model-driven systems and software engineering to address variability and flexibility of these production systems and the associated assessment of sufficient model quality (Lee 2008). Furthermore, quality assurance activities are insufficiently applied in context of model application in CPPS. Thus, there is a strong need for providing mechanisms for model quality assurance and for improving model quality in CPPS environments. In this context, this chapter focuses on the needs and approaches for MQA in MDE environments, where engineers from different disciplines have to collaborate. Based on needs for a standardized data exchange approach (Chap. 10), we focus on an integrated plant model of a production system, which can be represented as an *AutomationML* model.[1] *AutomationML* (Drath 2009) is an emerging standard for modeling artifacts derived from different engineering disciplines, such as the plant structure and interfaces between plant components, allowing to describe an integrated view on the plant model. Therefore, we investigate *AutomationML* models that aim at providing an integrated view on a draft of a production system. A major shortcoming of the current version of *AutomationML* is the lack of language means to express process states, e.g., whether a change is

[1]*AutomationML*: http://www.AutomationML.org

approved or still provisionary and under review by the engineering team (Göring and Fay 2012). Thus, the *AutomationML* standard does not provide process and tool support for MQA out-of-the-box. In addition, in MDE there is no established systematic review process for design models (Travassos et al. 1999; Winkler et al. 2016b), as a versioned integrated model of the draft production system is not always available. However, reviewing activities require domain experts (who are familiar with reviewing artifacts, associated tools and data models in related disciplines), high effort for review planning and execution, and cognitive skills. Blackwell et al. (2001) provide a framework of cognitive difficulties with notations. These difficulties make reviewing MDE models hard, such as the limited visibility of necessary information or hard mental operations to compare different parts of models, in particular, to find defects in different model views effectively and to assess the impact of changes between model versions.

Because engineering models can become quite large, an important review aspect is the definition of manageable scopes for reviewing tasks. In an engineering process with versioned engineering artifacts/models, the changes between versions allow defining a manageable scope for reviewing to assess the impact of recent changes on the quality and risks of the overall engineering model (Mordinyi et al. 2016a). Therefore, we propose a systematic review process focusing on changes in MDE design models, to reduce the cognitive load by limiting the review scope for finding defects in change sets and to assess and mitigate risks from the impact of changes to related engineering models.

In this kind of systematic review, the reviewers compare the review object, such as a selected model view, with a reference document, such as system requirements or usage scenarios, to guide the detection of important defects. Furthermore, to the best of our knowledge, there is only limited method and tool support for *model quality assurance* (MQA) in MDE teams along the lifecycle of engineering models. However, an important approach for early defect detection, e.g., in design models, is the systematic review of engineering models. In these reviews, domain experts investigate the overlaps of engineering models between disciplines to identify defects, such as syntactic and semantic mismatches, inconsistencies, and missing or unclear information. The identification of these kinds of defects can be a first step to design semi-automated support for efficient defect detection (Feldmann et al. 2015). Semantic data representations can support data integration and quality assurance in models of CPPS engineering projects (see chap. 12 for requirements and use cases) and how Semantic Web Technologies can help to improve defect detection in MDE projects (Winkler et al. 2017).

In the context of the research questions (RQs) of this book, two generic RQs are specifically relevant in context of this chapter. The book RQ M2 *"Modeling in CPPS lifecycle phases"* concerns how model-based methodologies support information creation and processing in different lifecycle phases of a CPPS, including addressing the quality needs for models of a CPPS. The book RQ I2 *"Quality assurance for information exchange"* concerns method and technology support assuring the required information quality for information exchange. From these generic RQs we derive the following research questions (RQs) for this chapter.

- **RQ MQA1: MDE Information Model Change Review Process.** Which requirements have to be addressed and which (tool) capabilities are needed to support systematic review processes with a focus on changes in MDE design models, expressed in *AutomationML*?
- **RQ MQA2: MDE Modeling Language with Process Concerns.** How can a standard modeling language for MDE, such as the *AutomationML* modeling language, be extended to address needs for storing process-relevant attributes to support quality assurance aspects?

Based on related work (Sect. 16.2) and identified research questions (Sect. 16.3), this chapter presents concepts of MQA approaches (Sect. 16.4) in the context of selected MDE processes, i.e., the addition, change, or removal of a component in an engineering discipline and impact analysis on the integrated plant model. In more detail, we present an effective, efficient, and usable systematic process approach *(MQA-Review)* that focuses on changes in MDE design models (Sect. 16.4.1). An *AutomationML* language metamodel extension will be introduced that allows storing process-relevant attributes to support a systematic review process (Sects. 16.4.2 and 16.4.3). Finally, this chapter presents a conceptual evaluation (Sect. 16.5) on the usability of MQA approaches in the context of selected MDE processes for round-trip-engineering for changes that focus on new, modified, or removed model elements or components in an engineering discipline and their impact on the integrated plant model, described in *AutomationML*. Finally, Sect. 16.6 summarizes, discusses limitations, and presents future work.

16.2 Background

This section summarizes background work on stakeholder needs, model-driven engineering, *AutomationML*, and quality assurance with focus on systematic quality assurance and review support.

16.2.1 Stakeholder Needs for Model Quality Assurance

In *MDE* projects for engineering production systems, such as a CPPS (Lee 2008), several stakeholder groups are critical for the success of MQA approaches for parallel engineering (Biffl et al. 2016b): *Engineers* create, update, and use engineering models in their discipline and exchange models with engineers from other disciplines. *System integrators* collect discipline-specific engineering views to create an integrated plant model, which allows checking for inter-disciplinary defects and understanding the impact of changes on other engineering disciplines. *Quality managers* need an integrated view on engineering plans to assess the overall progress, quality, and risks of the planned production system. However, in concurrent engineering projects, individual stakeholders work in parallel and need a stable support for model quality assurance. These parallel engineering activities result in the need for frequent synchronization of models, such as round-trip-engineering process approaches (Winkler et al. 2015) that represent an important sample engineering process in practice. Beyond model synchronization and process support (Biffl et al. 2016b), quality assurance (Winkler et al. 2016a) and model quality assurance are success-critical activities to identify defects early, effectively, and efficiently. Thus, managers, domain experts, and quality assurance personnel want effective, efficient, and usable methods and tools for the quality assessment and improvement of models that have significant impact on (automated) software engineering in the context of MDE teams. This chapter will provide practitioners with a process to effectively conduct MQA for design models, in particular, for integrated plant models expressed in *AutomationML*.

16.2.2 Model-Driven Engineering

Model-driven engineering promotes models as first-class citizens in engineering projects (Brambilla et al. 2012 and Chap. 11 of this book). Models are not only used for documentation purposes or early sketches of systems, but they are used in generative ways to produce parts of the systems (e.g., code generation) and in analytical ways to check for certain system properties (e.g., test case generation or system validation). Thus, models are used throughout the complete system lifecycle. Of course, this means that models must be of sufficiently high quality to realize and maintain systems effectively and efficiently.

Besides the traditional validation, verification, and testing approaches for models, additional support is needed to ensure and improve the quality of models by human inspection. Human-driven reviews and inspections are traditional approaches with a long tradition for formal requirements or design inspection of informal code walkthroughs (Aurum et al. 2002). Of course, these techniques are also required on the model level to ensure the quality of models.[2]

Unfortunately, there is still a lack of approaches and tools for performing model reviews. A pioneer on this is *EGerrit,*[3] an Eclipse plug-in that provides an integration of *Gerrit*[4] in Eclipse. *Gerrit* (Milanesio 2013) is a review tool that provides also dedicated support for modeling projects by combining *EMF*[5] *Compare*[6] and *Gerrit*. For instance, models and model differences may be directly commented and review tasks, such as voting, are supported. While *EGerrit* provides first tool support for performing model reviews, more research is needed to understand which review processes are beneficial for modeling projects.

[2]Model Reviews: http://agilemodeling.com/essays/modelReviews.htm

[3]EGerrit: http://eclipse.org/egerrit

[4]Gerrit: https://www.gerritcodereview.com/

[5]EMF: Eclipse Modeling Framework, https://eclipse.org/modeling/emf/

[6]EMF Compare: https://www.eclipse.org/emf/compare/

16.2.3 AutomationML

The representation of models is a success-critical issue to enable effective and efficient model management in related tool solutions. Model management includes model changes, such as adding, removing, or modifying model elements, the synchronization of different model views considering the involved disciplines, and model quality assurance to keep the overall model consistent (Brambilla et al. 2012).

AutomationML (AML) is an emerging standard in automation systems development for the data exchange between different engineering disciplines (Drath 2009). Based on a neutral, free, open, and XML-based standardized data exchange format, individual tools may exchange data without loss between related engineering disciplines within the MDE process. The ongoing standardization process covers the architecture and general requirements (IEC62714-1), role class libraries for modeling engineering information (IEC62714-2), and geometry and kinematics (IEC62714-3). Building blocks of the AML data exchange format (Berardinelli et al. 2015) focus on (a) the plant structure as a hierarchical representation of AML objects (Schleipen et al. 2008); (b) geometry and kinematics described in the COLLADA data description format for exchanging data between 3D modeling tools (ISO/PAS 17506); and (c) control logic based on *PLCOpen*[7] XML. CAEX (Drath 2009) enables the modeling of physical and logical system components encapsulating different aspects of the engineering project. CAEX data objects can facilitate the reuse of existing components to improve the engineering project based on a product line approach (Pohl et al. 2005) and therefore can improve engineering processes and projects. For example, the instantiation of classes by cloning existing prototype classes can support reuse of data objects. In addition, the hierarchical plant structure enables the definition of sub-elements, composition, or aggregation. Identified benefits of *AutomationML* can also support *Industrie 4.0* (Vogel-Heuser et al. 2016) initiatives and cyber-physical production systems (CPPS) (Lee 2008). However, the availability of data exchange formats for MDE projects requires method and tool support for making engineering and quality assurance processes more effective and efficient. To the best of our knowledge, there is only little work on MQA and review support for *AutomationML* models (Winkler et al. 2016b).

16.2.4 Quality Assurance and Model Review

Software Reviews and Inspections have been successfully applied in Software Engineering since more than 25 years (Aurum et al. 2002). The main purpose of a formal review process is to find defects in (software) engineering artifacts early in the development cycle, effectively and efficiently.

[7]PLCOpen: http://www.plcopen.org/

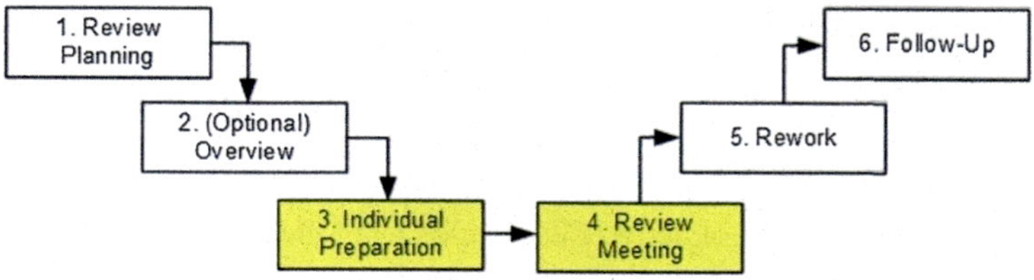

Fig. 16.1 Traditional software review process

Figure 16.1 presents the traditional review process consisting of six basic steps: (1) *Review Planning*, including team member selection, timing, preparation of the review material, and definition of entry criteria; (2) an optional *Overview* helps to get familiar with review objects and methods; (3) *Individual Preparation*, i.e., individual human-based review of artifacts based on provided guidelines and report of candidate defects; (4) *Review Meeting*, i.e., team discussion to agree on a team defect list; (5) *Rework* of review artifacts according to the team defect list; and (6) *Follow-Up*, e.g., planning of another review cycle, if important quality criteria have not been passed. Note that individual roles are defined for every task of the review process. See Laitenberger and DeBaud (2000) for a basic review framework, involved stakeholders and different review strategies.

In addition to systematic review processes, reading techniques (RTs) support reviewers in detecting defects in various engineering artifacts more effectively and efficiently, e.g., in requirements documents, specification documents, models, diagrams, and software code. However, reviews are not limited to software code documents but are applicable to various types of documents and models.

Reading guidelines and reading techniques aim at supporting reviewers or review teams by guiding them though the reading and defect detection process, e.g., based on checklists (checklist-based reading), use cases (usage-based reading), application scenarios (scenario-based reading), or perspectives (perspective-based reading). Benefits and limitations of reading technique approaches have been widely investigated in empirical studies in a variety of study contexts (Aurum et al. 2002; Kollanus and Koskinen 2007; Travassos et al. 1999). In MDE projects stakeholders come from different disciplines and typically have different perspectives on the projects, e.g., from mechanical, electrical, or software perspective. Following these different viewpoints, the Perspective-Based Reading (PBR) technique approach seems to be well-suited for model review in MDE contexts. Winkler and Biffl (2015) reported on a pilot study on an automation-supported review process based on perspectives in context of a hydro power plant systems development project. Main results were that different perspectives provide strong benefits for defect detection and analysis in a change management process approach. Although review processes and perspectives can help to increase defect detection performance, e.g., effectiveness and efficiency of defect detection processes, tool support is needed to facilitate reviewing of large models.

As already mentioned before, *EGerrit* or—more general—*Gerrit Code Review* is an established tool in software engineering that supports defect detection based on individual commits and code changes (Milanesio 2013). Main goal is to review every code change set (in the context of a single commit) for correctness prior to merging the change into a source code repository (e.g., GIT[8]). New software code (including the change) and old code (that comes from the repository) are presented and changes are highlighted in different colors. *Gerrit* uses color-coding for new, removed, and modified software parts. Thus, reviewers can easily inspect the code fragments, give comments and decide whether or not the code should be accepted (and committed to the repository) or rejected. It has been shown that *Gerrit* makes code reviews more efficient and effective (Milanesio 2013). Unfortunately, *Gerrit* is limited to software code and text documents and is not capable of supporting the review of changes related to models or images. Furthermore, the commenting feature of *Gerrit* allows remarks on individual aspects of the code but does not support annotating model elements. Nevertheless, *Gerrit* is a promising approach as part of a tool chain for model quality assurance in context of *AutomationML* models (Winkler et al. 2016b). For annotating model elements, *DefectRadar*[9] provides mechanisms to annotate elements in different types of documents, e.g., in PDF files, generated from engineering plans. *DefectRadar* was developed to support issue management for building automation. However, capabilities of *Gerrit* (as a code review tool) and *Defect Radar* (as issue tracking tool) are promising starting points to support reviews in MDE environments, e.g., based on *AutomationML* models.

16.3 Research Questions

Looking at the multi-disciplinary nature of the lifecycles of products, production systems, and production technologies described in Chap. 1, engineers require increasing support to make informed decisions and to ensure high quality work efficiently. An important requirement for this support is the effective quality assurance of information models along the lifecycles of products and production systems engineering. In this chapter, we investigate how model-based methodologies can support quality assurance of information models, which may come from heterogeneous data sources in MDE, in the different lifecycle phases of a CPPS (see RQ M2 and RQ I2 in Chap. 1). For this chapter, we derive the following research questions (RQs).

***RQ-MQA1*: MDE Information Model Change Review Process.** Which requirements have to be addressed and which (tool) capabilities are needed to support systematic review processes with a focus on changes in MDE design models, expressed in AutomationML? The size and complexity of large information

[8]GIT: git-scm.com/

[9]DefectRadar: www.defectradar.com

models in MDE may exceed human capabilities for a complete review within a reasonable period during the development process. Therefore, we investigate (a) needs and expected tool capabilities that for systematic review support in MDE contexts and (b) how a review process can be designed to focus the review scope on recent changes in an information model to guide reviewers to risky parts of the model and reduce the cognitive load for reviewers. We investigate the feasibility of the process with an industry use case in an AutomationML model of a production system part.

***RQ MQA2*: MDE Modeling Language with Process Concerns.** How can a standard modeling language for MDE, such as the AutomationML modeling language, be extended to address needs for storing process-relevant attributes to support model change review processes? The AutomationML modeling language allows expressing both discipline-specific views on a system and an integrated system view. This is a very useful capability for describing the design of a system and saving versions of these designs over the course of a project to represent the state of work over time. However, during the parallel engineering of a production system, the different model parts may represent finished system designs, early ideas that need refinement or explorative variants that need evaluation and may eventually be discarded. The current version of the AutomationML modeling language does not provide native means to express these process-relevant attributes as foundation for information model analysis from a management point of view. Therefore, we investigate how an extension of the AutomationML language can address these needs for storing process-relevant attributes with a change management use case (i.e., supporting the handling of new, modified, and removed components).

Our research approach is to collect requirements for MDE model quality assurance with a focus on the human review of *AutomationML* models of CPPS parts. As a study object, an important MDE process regarding the addition, change, or removal of a component in an engineering discipline and impact analysis on the integrated plant model has been selected. Related work has been revisited to build on and gaps in research that needs to be addressed. A systematic review process focuses on changes in MDE design models that is effective, efficient, and usable in a typical engineering team. Design extensions of the *AutomationML* language metamodel allow storing process-relevant attributes in general and in particular to support a systematic review process of information model changes. Section 16.5 presents an evaluation of the feasibility of the adapted review process focusing on changes in MDE design models and adequacy of the *AutomationML* language extensions with data from a use case on round-trip-engineering, i.e., the addition, change, or removal of a component in an engineering discipline and impact analysis on the integrated plant model described in automation.

16.4 Model Quality Assurance Concept

This section focuses on an adapted review process for model reviews in MDE context (Sect. 16.4.1), *AutomationML* extensions to support model annotation of model elements (Sect. 16.4.2), and the related prototype embedded in the *AutomationML* Editor[10] (Sect. 16.4.3).

16.4.1 Adapted Review Process for MDE and AutomationML MQA

Based on traditional review processes (see Fig. 16.1), MDE requires additional mechanisms to address (a) individual change types (e.g., new, removed, and modified model elements); (b) tool support for change highlighting; and (c) annotation and defect reporting. In context of this work we use *AutomationML* for efficient data exchange, model synchronization, and model quality assurance.

Figure 16.2 presents the adapted review process—*MQA-Review*—for MDE requirements based on the *AutomationML* data exchange format. Individual engineering artifacts from related stakeholders represent process inputs (Fig. 16.2, left hand side), are processed during individual process steps (Fig. 16.2, middle part) and create review process outputs (Fig. 16.2, right hand side). The process takes input data in *AutomationML* data format, e.g., the plant topology, systems requirements, and individual artifacts derived from engineering disciplines and prior review process steps. In context of traditional review processes, some artifacts are considered to be correct and represent reference documents, e.g., the requirements specification, and the plant topology. Reviewing objects, such as electrical or mechanical planning documents, are in focus of the review and are reviewed with the intent to identify defects and deviations. Therefore, fundamental results of the reviewing process are defects and deviations and intermediate results such as annotations, comments, or individual candidate defects. The process, depicted in Fig. 16.2, presents the basic process steps with concrete tasks and deliverables based on the *AutomationML*. Depending on the review purpose, examples in Fig. 16.2 focus on a process instance applicable in the design phase of a sequential engineering process. See Winkler et al. (2016b) for an example of a typical and sequential engineering process in MDE projects.

In detail, *MQA-Review* consists of three fundamental steps with sub-steps and inputs/outputs:

- *Step 1a. Review Planning.* The *Review Moderator* is responsible for planning, preparation, and coordination of the review process. Main tasks include defining the review scope, selecting reference documents, such as requirements

[10]*AutomationML* Editor: www.AutomationML.org

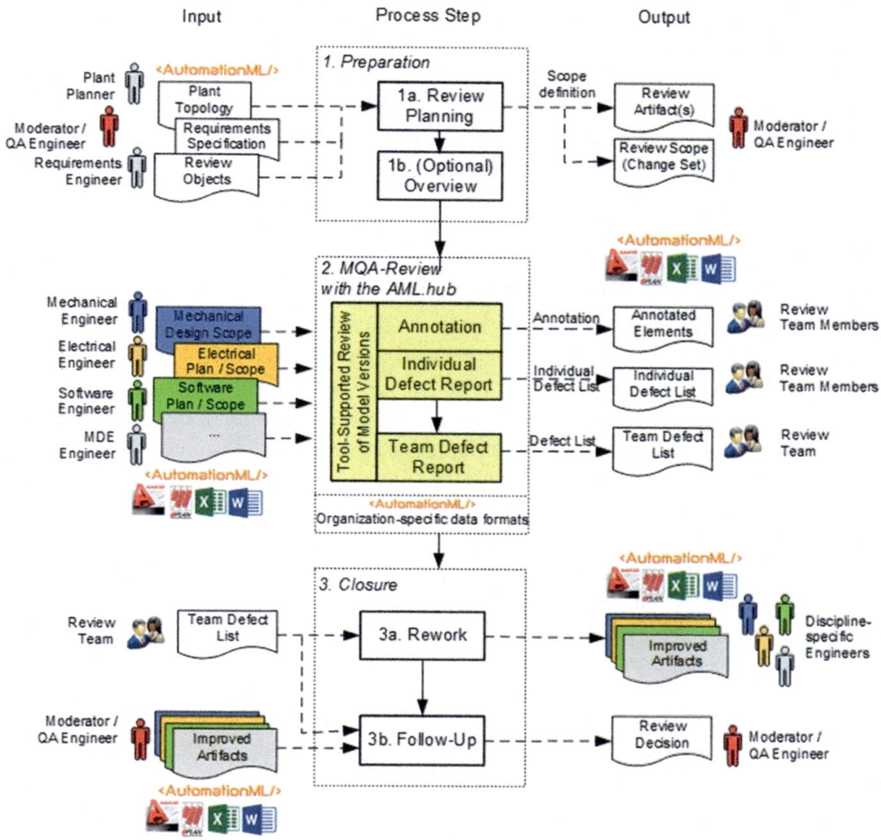

Fig. 16.2 Adapted Review Process (*MQA-Review*) based on an example for the Systems Design Engineering Phase (Winkler et al. 2016b)

specifications provided from *Requirements Engineers*, and appropriate guidelines, installing the review team, and scheduling the review process. In automation systems engineering projects, the moderator is typically supported by the *Plant Planner*. Output is the scope of the review and detailed plan for executing the review process and the review package including review objects, reference documents, and supporting material such as reading techniques for review. Note that review objects require a structured representation, such as provided by *AutomationML*. However, organization-specific data formats can be used in context of AML-Review.

- *Step 1b. (Optional) Overview*. As an introduction of the review package and the review process, an optional overview might be scheduled. Reasons for an optional overview can be high complexity of the engineering artifacts, novelty of the application domain, or limited experience of review team members with the method and/or the project context. Goal of this step is to introduce to participants the (sub-)system under review, reviewing and supporting material, and the review process. This process step has been skipped in Fig. 16.2 for simplicity reasons.

- *Step 2. MQA-Review*. This process step represents the core phase of the adapted review process. Note that it is required to have review artifacts available in machine-processing data formats, such as in *AutomationML* to enable automation-supported defect detection. However, organization-specific data formats can be used for (manual) defect detection. Basically, MQA-Review includes the following sub-process steps: (a) annotation and commenting; (b) individual defect detection and reporting; (c) generation of team defect report, i.e., an aggregated list of agreed defects. Annotations can help to set a current state on the model element, e.g., approved, rework, and declined, or add/assign defect attributes (such as defect type or severity). Comments can be used to better understand the identified defect or represent explanatory information on related model elements (e.g., relationships to other model elements). The list of annotations, identified candidate defects, and comments are the input for generating a team defect list (as a real team meeting or as nominal team meeting without communication and interaction). In real team meetings, individual candidate defects and comments are discussed to come to an agreed team defect list. In context of nominal teams, all individual candidate defects are scanned and merged by the moderator. However, the three process steps can be combined into one step supported by a tool solution or a tool chain. This tool chain can support different viewpoints, enables traceability (e.g., via the plant topology) and provides support for analyzing relationships within engineering plans and across engineering plans. The final output of this process is a team defect list that can be created during team meetings or generated by a tool solution.
- *Step 3a. Rework*. Based on the team defect list, annotations, and comments, individual and responsible engineers address assigned issues in their engineering plans, i.e., correction of defects or response to annotation and comments.
- *Step 3b. Follow-up*. Improved engineering artifacts and the team defect list represent the input for the last MQA process step. Typically, the review moderator (and the plant planner) check the modifications with respect to the team defect list and decide on accepting the work products or plan another review cycle. Outcome is a decision on the acceptance (or rejection) of the review artifacts and a review report summarizing reviewing results.

16.4.2 A Generic Reviewing Language

Based on *MQA-Review* and models that are available in a structured data format, such as in *AutomationML*, tool support can help to improve review processes and support engineers in making these processes more effective and efficient. The main question is how language extensions can help to annotate process elements in context of model quality assurance.

In this subsection, we introduce a generic reviewing language which may be reused for *AutomationML* and any modeling language. The design rationale behind this reviewing language is to attach the information about model reviews to the models and their model elements. By this, we follow the *"everything is a model"* principle of model-driven engineering (Brambilla et al. 2012), which allows to explicitly represent reviews as well as to apply the same techniques to reviews as for any other artifact. Furthermore, this allows to reason about review decisions directly in the context of the model, and provides an explicit model structure for such decisions, which enables automatic processing by model-driven engineering tools. For instance, also textual descriptions of model reviews may be automatically produced by *model-to-text* transformations.

Figure 16.3a shows an excerpt of a generic reviewing language (package *ReviewLanguage*) which is bound to a simple example structural modeling language just consisting of elements and connectors. In Fig. 16.3b, this language is used to annotate a base model consisting of two elements and one connector with reviewing decisions which are instantiations of the review decision meta-classes shown in Fig. 16.3a.

The goal of this reviewing language is to allow for annotating the results of a systematic examination of a model. For simplicity, we just introduce three reviewing decisions in the review language, namely *Approved* (i.e., the model element is correctly defined), *Rework* (i.e., the model element has to be modified by an engineer), and *Declined* (i.e., the proposed model element should be dropped from the model) which shall be applicable to every kind of element in every modeling language. Other example decision types which are not shown due reasons of brevity may include *Deprecate*, *Vote*, *Merge* to mention just a few. All of these concepts extend the abstract concept *review decisions*. By this, the applicability of these concepts is checked by reasoning about the base concept which is reviewed. A base concept may receive several review decisions and a review decision may relate to several base concepts.

In order to use the reviewing language for a concrete modeling language, the base concept has to be bound to the concepts of the modeling language which are valid targets to receive review decisions. In the example shown in Fig. 16.3a, the base concept is bound to the *NamedElement* concept of an exemplary base language, which provides the model elements and connections between them. In Fig. 16.3b, we show an example usage of the reviewing language. We have defined a simple model using the base modeling language, which consists of two elements, a robot and a working station, as well as a connection between the two elements, which means that the robot is attached to the working station. The model is simply shown in its abstract syntax, which is visualized in terms of a UML object diagram. In a reviewing process performed by *Harry* and *Sally*, two reviewing decisions are made: (a) the robot element is approved by *Harry* and (b) *Sally* performed a review concerning the working station element, which should be changed to represent not station *A* but station *B*. This rework task is assigned to *John*.

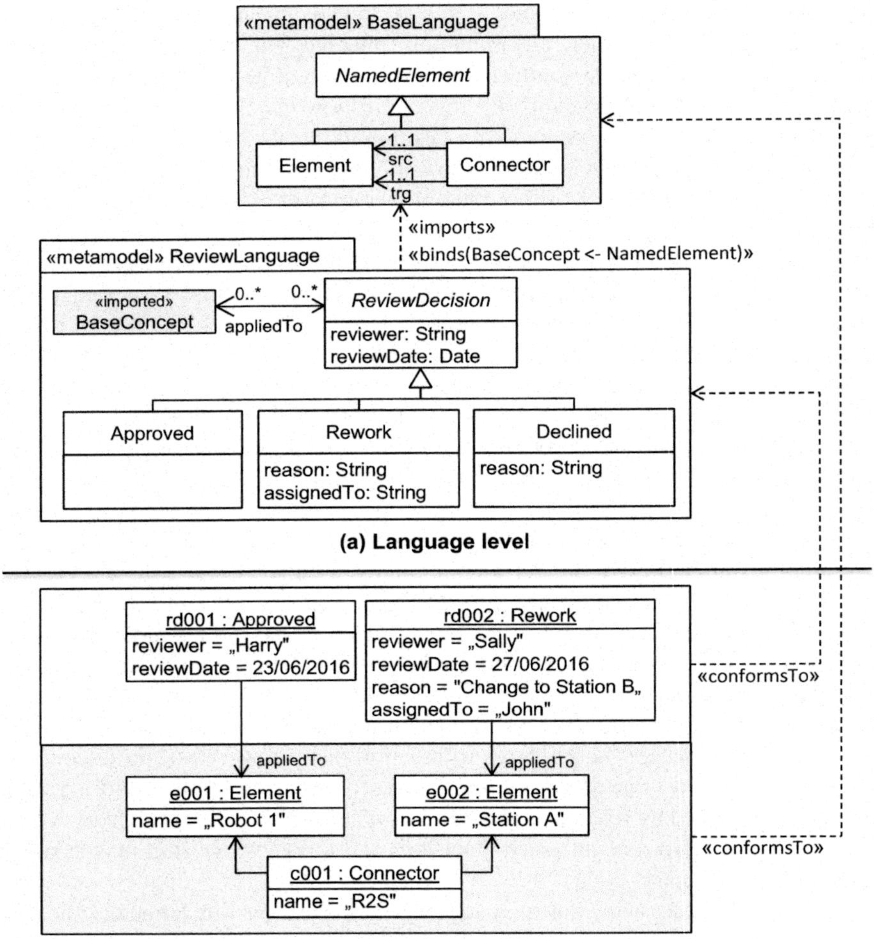

Fig. 16.3 Generic Reviewing Language: (**a**) Language definition and (**b**) examplary usage for an example model

16.4.3 Utilizing the Generic Reviewing Language for AutomationML

Reviewing decisions are important meta-information, which should be possible to be exchanged between tools, e.g., assume that John has to rework the model as requested by Sally in a different modeling tool. Therefore, we discuss in this subsection how to utilize the presented reviewing language for *AutomationML*. By this, not only the information about production systems should be exchanged, but

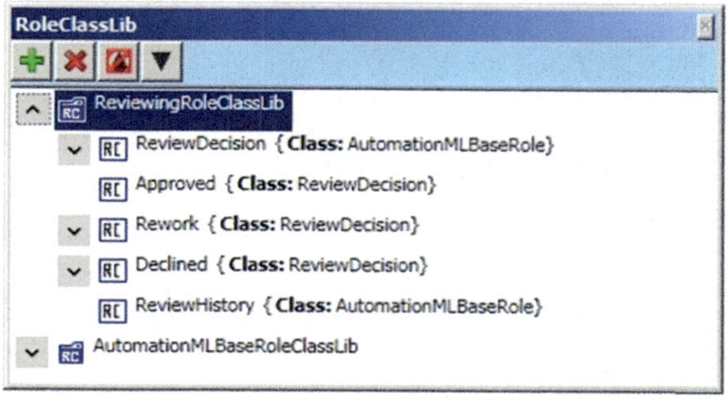

Fig. 16.4 Reviewing Language libraries in *AutomationML*: role class library

also current reviewing decisions which are attached to the model elements which are representing (parts of) the production systems.

The *AutomationML* realization of the reviewing language we have shown in Fig. 16.3 is realized in terms of *AutomationML* libraries (Fig. 16.4 presents role class library and Fig.16.5 the system unit class library) and applied on a simple example depicted in Fig. 16.6 (i.e., the instance hierarchy).

AutomationML allows in general two ways to integrate additional information resources going beyond CAEX, COLLADA, and *PLCopen* (Berardinelli et al. 2015). First, this additional information may be stored directly in the CAEX files by adding additional domain-specific layers to *AutomationML* based on dedicated libraries. Second, the additional information is stored outside and references by external data connectors from CAEX. As we aim to provide reviewing decisions in the context of the model, we aim for the first option. Based on the conceptual model of the reviewing language presented in Fig. 16.3a, we are able to derive a domain-specific layer for it consisting of two AML libraries, namely the *ReviewingRoleClassLib* (RClib) in Fig. 16.4 and the *ReviewingSystemUnitClassLib* (SUCLib) in Fig. 16.5. In particular, a new *RoleClass* and *SystemUnitClass* are introduced in the libraries for each review decision element shown in Fig. 16.3a.

The domain-specific semantics layering for reviews is then possible by assigning the reviewing-specific *Role Classes* (RCs) to the reviewing-specific *System Unit Classes* (SUCs), suitably mapping on SUCs the corresponding attributes (required for *MQA-Review*) from RCs through mapping objects.

Finally, these SUCs are instantiated to represent reviewing decisions in the instance hierarchy, which represents the system model under review (Fig. 16.6).

Review decisions can then be attached to internal elements (and subtype of internal elements). In particular, any ReviewedElement provides a ReviewHistory, which acts as a collection of ReviewDecisions of any kind for that internal element. Please note that this utilization of the reviewing language for *AutomationML* allows attaching review decisions to internal elements (and subtypes of internal elements) only. However, by this also review decisions on review decisions can be represented.

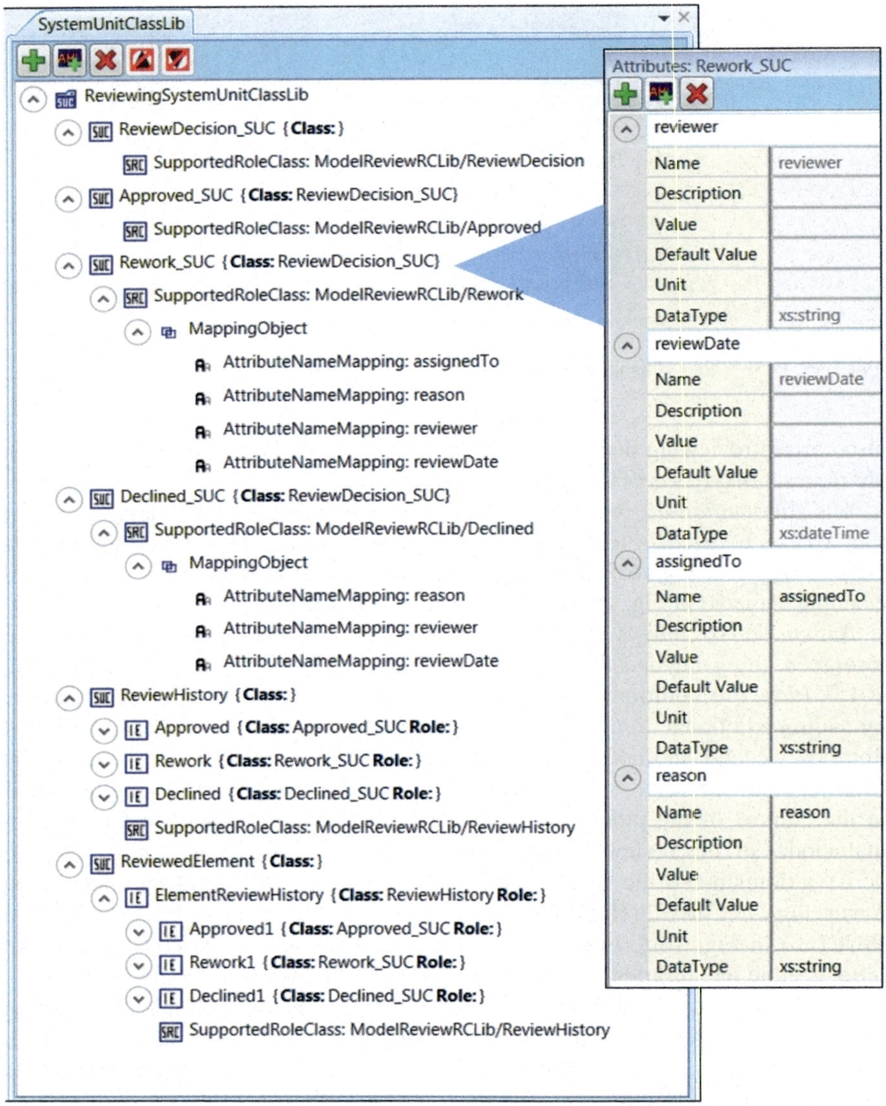

Fig. 16.5 Reviewing Language libraries in *AutomationML*: system unit class library

16.5 Conceptual Evaluation

This section focuses on the conceptual evaluation of the adapted reviewing process *(MQA-Review)*, the generic model language extension, and the prototype implementation by using the *AutomationML* language approach. We use the

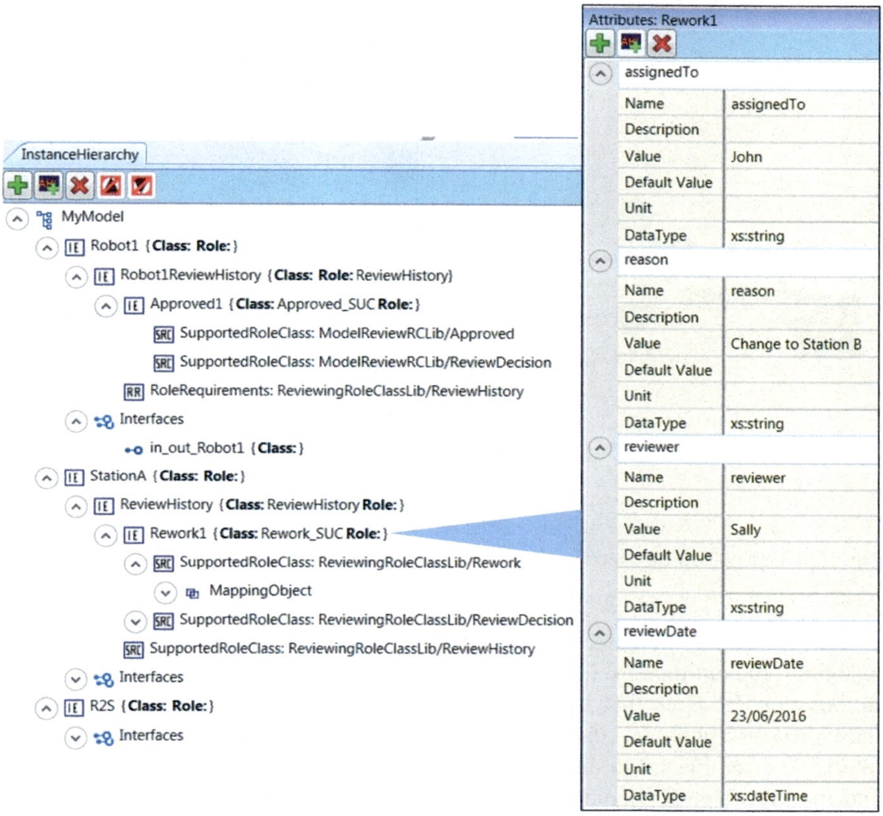

Fig. 16.6 Reviewing Language usage in *AutomationML*: instance hierarchy

round-trip-engineering process as an important and common engineering approach in MDE environments as illustrative use case.

16.5.1 Illustrative Use Case: Round-Trip-Engineering

In common industry projects several engineers, coming from different disciplines, have to collaborate and work in parallel. Important challenges arise from heterogeneous and distributed disciplines with limited interoperability and capabilities for data exchange and quality assurance (Winkler et al. 2016a). *AutomationML* (Drath 2009) enables efficient data exchange by providing a common data format for exchanging engineering data effective and efficient. However, tools are required that enable features to enable efficient data exchange based on the common *AutomationML* data format.

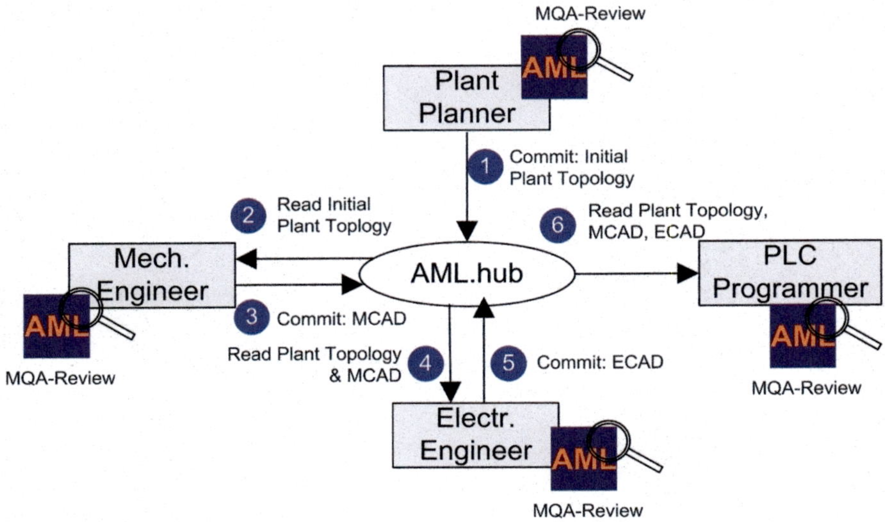

Fig. 16.7 Round-Trip-Engineering Process Approach (Sample Scenario)

The *AML.hub* (Winkler et al. 2015) provides the technical foundation for enabling efficient data exchange between tools and data models that are based on *AutomationML*. Note that the *AML.hub* is responsible for mapping and merging individual discipline-specific views and data elements and for propagating changes to related disciplines. Mordinyi et al. (2016a, b) describe the fundamental concepts and the prototype implementation of efficient data exchange for AutomationML and organization-specific artifacts. In context of this chapter, we will consider the *AML.hub* as *"black box"* for providing integrated data as foundation for *MQA-review*.

The round-trip-engineering process is a typical use case in MDE environments for parallel and distributed collaboration of related engineers from different disciplines. Figure 16.7 illustrates related stakeholders in a sample scenario based on the round-trip-engineering process and the contribution of *MQA-Review.*

The round-trip-engineering scenario includes four types of engineers (Plant Planner, Mechanical Engineer, Electrical Engineer, and Software Engineer, i.e., PLC programmers) and consists of six basic steps:

1. *Commit Initial Plant Topology*. The plant planner is responsible for providing the basic plant structure and the plant topology in *AutomationML* data format as architectural guideline for engineering.
2. *Read Plant Topology.* Mechanical Engineers receive/read the initial plant model usable in local engineering tools, e.g., mechanical design tools such as MCAD. Before adding/modifying model elements in the mechanical view they can review the received plans by using the extended *AutomationML Editor*. See Section

16.4.2 for a generic language extension and Sect. 16.4.3 for an *AutomationML* specific extension for review support.
3. *Commit MCAD.* After completing the mechanical design step (or even intermediate design versions) the mechanical engineer commits the current state of the work to the *AML.hub* for merging purposes.
4. *Read Plant Topology & MCAD.* In sequence of this round-trip engineering scenario the Electrical Engineer receives the plant topology and current MCAD version. After an optional *MQA-Review* process step he can add electrical plan data and commit these data to the *AML.hub*.
5. *Commit ECAD.* Committed electrical planning data are merged and changes are propagated by the *AML.hub* platform.
6. *Read Plant Topology, MCAD, ECAD.* The last step in this scenario focuses on the Software Engineer, who receives the software related view of the project, executes an optional *MQA-Review* and adds and commits modifications to the *AML.hub* for further processing.

16.5.2 MQA-Review Needs and Expected Tool Capabilities

The round-trip-engineering scenario follows a sequential engineering process derived from the "perfect engineering world". However, in practice, engineers work in parallel, identify defects and execute changes regularly, e.g., based on changed requirements from customers or related disciplines or based on reported defects. Thus, frequent changes, represented by various reads and commits from different perspectives arise in parallel along the project's course. Change types typically include added/modified/removed model elements or attributes. However, frequent interaction and data exchange may include risks of inconsistencies, defects in various engineering artifacts that need to be addressed efficiently and effective. *MQA-Review* can help to keep the project artifacts consistent on discipline level, where engineers can perform MQA-Reviews to assess the current state of the projects from their individual viewpoint. Furthermore, quality managers can initiate and drive formal review processes to identify defects according to the project plan.

As manual reviews become risky (if multiple disciplines are involved), time-consuming and expensive if experts from various disciplines are involved, tool-supported review can help to overcome these issues. Therefore needs and requirements have been elicited with research and industry partners to identify a set of needed capabilities for a tool solution that supports the *MQA-Review* process in *AutomationML* context. These results can be summarized as follows:

Process Capabilities

- *Traceable Review Process* focuses on the implemented review process that enables checking annotations and defect reports in various engineering models (across disciplines).

- *Defined Roles and Responsibilities.* Role definitions and responsibilities are required to organize reviewing activities.

Data Exchange Capabilities

- *AutomationML Support.* Efficient data exchange, e.g., by using the *AML.hub* requires model descriptions in machine-readable and machine-processable data format, such as *AutomationML*. However, depending on an agreed data exchange format, any structured data representation might be applicable but might need some work for data transformation, mapping, and merging of various tool data.
- *Language Extension for review process support.* Data exchange formats typically provide a description of pure engineering data without any additional functionality. However, for tool support of quality assurance and reviewing processes, these data formats should enable extensions for specific purposes, such as annotations for the review decision or for reporting and classifying defects.

Defect Detection Performance

- *Defect Detection Efficiency* refers to the number of defects identified over time, i.e., capturing as many defects as possible within a very short time interval. Thus, efficient tool-support should increase defect detection performance.
- *Defect Detection Effectiveness* refers to the capability of identifying most defects in the review objects. Thus, efficient tool-support should enable increasing the coverage of review artifacts (i.e., covering all relevant model parts in a measurable way).

Tool Capabilities

- *Model Element Annotations.* Annotation can help to provide additional comments or assessments results to model elements under review, i.e., review decisions or annotations of defects.
- *Browsing Capabilities.* The involvement of different disciplines (and discipline-specific views on the system) makes inter-disciplinary traceability difficult. However, browsing through the topology of a system can help to better identify relationships of relevant model elements.
- *Automation Supported Difference Checks.* Highlighting differences caused by changes (e.g., added, modified, or removed components) aims at accelerating the review process because reviewers are guided by changes leaving other components out of the review.
- *Reporting.* Finally, reporting capabilities have to be available to provide defect lists and lists of annotations for project and quality management.

16.5.3 Evaluation of MQA-Review with Tool Support

This subsection describes a conceptual evaluation based on expert estimations from research and industry with the focus on required capabilities of *MQA-Review*

Table 16.1 Comparison of Traditional Review and Tool-Supported Review Approaches (+ Good Support, O neutral Support, − Weak Support)

Required tool capabilities	Trad. Review	Gerrit Code Rev.	Defect Radar	MQA Review
Process capabilities				
Traceable Review Process	O	+	O	+
Defined roles and responsibilities	+	+	O	+
Data exchange format				
AutomationML Support	O	O	O	+
Language Extensions	−	−	−	+
Defect detection performance				
Efficiency	−	+	−	+
Effectiveness	−	+	−	+
Tool capabilities				
Model Element Annotations	O	+	+	+
Browsing	−	−	−	+
Difference Checks	−	+	−	+
Reporting	O	+	+	+

support (adapted review process including language extensions for review support). We compare the traditional (human-based) review process approach, *Gerrit* as representative for code review tools with focus on changes, *DefectRadar* as a representative tool for annotation and issue management in building automation, and *MQA-Review*. Table 16.1 summarizes the conceptual evaluation.

Process Capabilities All approaches are capable of supporting review process phases and *traceable review process* steps to some defined extent. However, it depends on the quality strategy and organizational constraints how well process phases are supported for the traditional review approach and for *DefectRadar*. *Gerrit* and *MQA-Review* include more formal process steps that enable process traceability. Similar comments are applicable for *defined roles and responsibilities*.

Data Exchange Format Models in *AutomationML* notation can be reviewed by all approaches, as traditional reviews also enable code reviews and *DefectRadar* supports informal reviews of documents, which might also be text documents and *AutomationML* models. However, we see strong benefits for *Gerrit* (focus on code changes) and *MQA-Review* (focus on *AutomationML* Models).

Defect Detection Performance Regarding *efficiency*, we observed benefits of *Gerrit* because of the focus on changes (and change types) that guide the review to most critical system parts. However, there is limited support for analyzing relationships and dependencies. For *MQA-Review,* there is limited support for change analysis but strong support for browsing capabilities that might increase defect detection efficiency. For *traditional review* and *DefectRadar* defect detection efficiency strongly depends on involved experts. No explicit tool-support is available. Similar comments apply for defect detection *effectiveness*. Language extensions of the

MQA-Review approach enable exploring the status of the review and provide measurable results of review coverage.

Tool Capabilities While *annotations* have to be executed manually in the traditional review approach by experts, *Gerrit* and *DefectRadar* provide commenting features and annotation of artifacts on a tool level. For the *MQA-Review approach* language extensions have been implemented to focus on specific annotation features required by the reviewing process. *Browsing capabilities* are not supported by traditional expert reviews, *Gerrit*, and *DefectRadar*. The topology representation and internal links (core components of *AutomationML*) enable efficient browsing. Tool-supported *Difference Checks* are not available for the *traditional review* and *DefectRadar*. Difference checks and visualization is a core feature of *Gerrit* and applicable in *MQA-Review*. Based on queries differences can be highlighted. Finally, paper-based *Reporting* is typically applied in the traditional approach, *Gerrit* and *Defect Radar* provide export functionality for reporting purposes. Finally MQA-Review supports reporting by querying capabilities provided by the *AML.hub*.

Summarizing the conceptual evaluation, the *traditional review approach* strongly relies on human experts and includes strong limitations for tool support. *Gerrit* focuses on analysis and review support for text documents and is applicable for *AutomationML* models. However, language extensions and browsing are not supported. *DefectRadar* enables the analysis of any types of documents by enabling annotations with limitations to *AutomationML* engineering plans. Finally, *MQA-Review* with *AutomationML* language extensions combines benefits of *Gerrit* and *DefectRadar*. The initial evaluation explored needs and required tool capabilities for MQA-Review. We observed limitations of individual tools to fully support MQA-Review. However, the introduction of a tool chain, incorporating features from individual tools, can help to support the MQA-Review process and improve defect detection performance and product quality.

16.6 Summary, Limitations, and Outlook

In Multi-Disciplinary Engineering, the synchronization and quality assurance of models is success critical for making informed decisions and ensuring high quality work products. An important requirement is the effective and efficient quality assurance of information models along the lifecycles of products and production systems engineering. In this chapter, we investigated how model-based methodologies can support quality assurance of information models, which may come from heterogeneous data sources in MDE, in the different lifecycle phases of a CPPS.

In the context of this chapter, we focused on two research questions: (a) what are the needs for systematic and tool-supported review processes and which review approaches support these processes with focus on changes in MDE Data models and (b) how can a standard modeling language for MDE, such as *AutomationML*, be extended to address needs for storing process-relevant attributes in context of quality assurance and review process support.

In context of systematic review process support (RQ-MQA1) we identified a set of needs and expected tool capabilities and adapted a traditional software review process for MDE projects based on the *AutomationML* data exchange format. Based on *AutomationML* and the *AutomationML* Editor we presented a generic reviewing language and an instance for *AutomationML* (RQ-MQA2) that is capable of supporting reviewing activities for *AutomationML* data.

For evaluation purposes, we focused on a selected scenario, a round-trip-engineering process, an important engineering process that includes the synchronization of models that typically change along the engineering lifecycle. Changes include added, modified, or removed components or model attributes. Based on this scenario, we derived a set of required tool capabilities for *MQA-Review* process support. These capabilities include process capabilities to support review process steps, data exchange capabilities that are needed for efficient and effective model synchronization and language extension capabilities to address needs for MQA, defect detection performance as measure for reviewing capabilities, and tool capabilities that are required to support MQA-Review processes. Based on the required capabilities, we conducted a conceptual evaluation of four approaches, i.e., traditional review, *Gerrit* as an established tool for code reviews, *DefectRadar*, an issue management for different types of documents usually used in context of issue management for building automation, and *MQA-Review*. Major results of the conceptual evaluation showed benefits of *Gerrit* for analyzing and reviewing text document, e.g., software code or *AutomationML* data) and benefits for *DefectRadar* for efficient annotation and the support of various engineering documents (beyond textual documents). However, *MQA-Review* combines the benefits of both approaches and includes additional features for effective and efficient model reviews, e.g., browsing capabilities.

Major findings are that (a) an adapted review process helps to systematically drive the review process and (b) *AutomationML* can be extended with process-related attributes that are useful for quality assurance and reviewing.

Limitations The presented approach for the *MQA-Review* process and tool support represents promising directions for supporting efficient and effective reviews in MDE environments. In this chapter, we focused on a conceptual evaluation of fundamental concepts (adapted process approach and *AutomationML* language extensions). An in-depth analysis of the evaluation results is required to fully understand these approaches in the context of review process support.

Outlook Future work will include an intensive evaluation of the conceptual prototypes, and the design and development of a tool chain that fully supports all parts of the review process (i.e., planning, execution, rework, and follow-up) based on *AutomationML* concepts. Finally, we will plan an empirical evaluation of the process with tool support in larger and more diverse MDE design models and scenarios from industry partners.

Acknowledgements Parts of this work were supported by the Christian Doppler Forschungsgesellschaft, the Federal Ministry of Economy, Family and Youth, and the National Foundation for Research, Technology and Development in Austria.

References

Aurum, A., Petersson, H., Wohlin, C.: State-of-the-art: software inspections after 25 years. Softw. Test. Verification. Reliab. **12**(3), 133–154 (2002)

Berardinelli, L., Biffl, S., Mätzler, E., Mayerhofer, T., Wimmer, M.: Model-based co-evolution of production systems and their libraries with *AutomationML*. In: Proceedings of the 20th IEEE Conference on Emerging Technologies & Factory Automation (ETFA). (2015)

Biffl, S., Lüder, A., Winkler, D.: Multi-disciplinary engineering for *Industrie 4.0*—semantic challenges, needs, and capabilities. In: Biffl, S., Sabou, M. (eds.) Semantic web technologies for intelligent engineering applications. Springer, Switzerland (2016a)

Biffl, S., Mordinyi, R., Steininger, H., Winkler, D.: Integrationsplattform für anlagenmodellorientiertes Engineering—Bedarfe und Lösungsansätze. In: Vogel-Heuser, B., Bauernhansl, T., ten Hompel, M. (eds.) Handbuch *Industrie 4.0*, p. 2. Springer, Auflage (2016b)

Blackwell, A.F., Britton, C., Cox, A., Green, T.R.G., Gurr, C., Kadoda, G., Kutar, M.S., Loomes, M., Nehaniv C.L., Petre, M., Roast, C., Wong, A., Young R.M.: Cognitive dimensions of notations: design tools for cognitive technology. In: Cognitive technology: instruments of mind, pp. 325–341. Springer, Berlin (2001)

Brambilla, M., Cabot, J., Wimmer, M.: Model-driven software engineering in practice. Morgan & Claypool Publishers, California (2012)

Drath, R. (ed.): Datenaustausch in der Anlagenplanung mit *AutomationML*: Integration von CAEX, PLCOpen XML und COLLADA. Springer, Berlin (2009)

Feldmann, S., Herzig, S.J.I., Kernschmidt, K., Wolfenstetter, T., Kammerl, D., Qamar, A., Lindemann, U., Krcmar, H., Paredis, C.J.J., Vogel-Heuser, B.: Towards effective management of inconsistencies in model-based engineering of automated production systems. In: Proceedings of 15th IFAC Symposium on Information Control in Manufacturing (INCOM), 48(3), pp. 916–923 (2015)

Göring, M., Fay, A.: Modeling change and structural dependencies of automation systems. In: Proceedings of 17th International Conference on Emerging Technologies and Factory Automation (ETFA). IEEE (2012)

IEC 62714-1: Engineering data exchange format for use in industrial automation systems engineering—Automation Markup Language—Part 1: Architecture and general requirement, International Standard, June 2014 (2014)

IEC 62714-2: Engineering data exchange format for use in industrial automation systems engineering—Automation Markup Language—Part 2: Role class libraries, International Standard, March 2015 (2015)

IEC 62714-3: Engineering data exchange format for use in industrial automation systems engineering—Automation Markup Language—Part 3: Geometry and Kinematics, International Standard, June 2017 (2017)

ISO/PAS 17506: Industrial automation systems and integration: COLLADA digital asset schema specification for 3D visualization of industrial data, International Standard (2012)

Kollanus, S., Koskinen, J.: Survey of Software Inspection Research: 1991-2005. Working papers WP-40, University of Jyväskylä (2007)

Laitenberger, O., DeBaud, J.-M.: An encompassing life cycle centric survey of software inspection. J. Syst. Softw. **50**(1), 5–31 (2000)

Lee, E.A.: Cyber physical systems: design and challenges, In: Proceedings of the 11th IEEE International Symposium on Project and Component-Oriented Real-Time Distributed Computing (ISORC), pp. 363–369 (2008)

Milanesio, L.: Learning Gerrit code review, 144 p. Packt Publishing (2013)

Mordinyi, R., Wimmer, M., Biffl, S.: Engineering model exchange in multi-disciplinary engineering with the AutomationML Hub. In: Proceedings of the 4th AutomationML User Conference, Esslingen, Germany (2016a)

Mordinyi, R., Winkler, D., Ekaputra, F.J., Wimmer, M., Biffl, S.: Investigating model slicing capabilities on integrated plant models with *AutomationML*. In: Proceedings of the 21st IEEE Conference on Emerging Technologies & Factory Automation (ETFA), IEEE (2016b)

Pohl, K., Böckle, G., van der Linden, F.J.: Software product line engineering: foundations, Principles and techniques. Springer, Berlin (2005)

Schleipen, M., Drath, R., Sauer, O.: The system-independent data exchange format CAEX for supporting an automatic configuration of a production monitoring and control system, In: Proceedings of ISIE, pp. 1786–1791 (2008)

Travassos, G., Shull, F., Fredericks, M., Basili, V.R.: Detecting defects in object-oriented designs—using reading techniques to increase software quality. In: ACM Sigplan, 34(10), pp. 47–56, ACM (1999)

Vogel-Heuser, B., Bauernhansl, T., ten Tompel, M. (eds.): Handbuch *Industrie 4.0*. Springer, Berlin (2016)

Whittle, J., Hutchinson, J., Rouncefield, M.: Model-driven development: a practical approach. Chapman & Hall/CRC, London, UK (2016)

Winkler, D., Biffl, S.: Focused inspections to support defect detection in multi-disciplinary engineering environments. In: Proceedings of the 16th International Conference on Product-Focused Software Process Improvement (PROFES), Research Preview. Springer (2015)

Winkler, D., Biffl, S., Steiniger, H.: Integration von heterogenen Engineering-Daten mit *AutomationML* und dem AML.hub: Konsistente Daten über Fachbereichsgrenzen hinweg. In: develop3, 3/2015 (2015)

Winkler, D., Ekaputra, F.J., and Biffl, S.: *AutomationML* Review Support in Multi-Disciplinary Engineering Environments. In: Proceedings of ETFA, IEEE (2016a)

Winkler, D., Mordinyi, R., Biffl, S.: Qualitätssicherung in heterogenen und verteilten Entwicklungsumgebungen für industrielle Produktionssysteme. In: Vogel-Heuser, B., Bauernhansl, T., ten Hompel, M. (eds.) Handbuch *Industrie 4.0*, p. 2. Springer, Auflage (2016b)

Winkler, D., Sabou, M., Biffl, S.: Improving quality assurance in multi-disciplinary engineering environments with semantic technologies. In: Kounis, L.D. (ed.) Quality control and assurance. INTEC Publishing (2017)

Xiong, Y., Liu, D., Hu, Z., Zhao, H., Takeichi, M., Hong, M.: Towards automatic model synchronization from model transformations, In: Proceedings of the 22nd International Conference on Automated Software Engineering (ASE), pp. 164–173. ACM (2007)

Chapter 17
Conclusions and Outlook on Research for Multi-Disciplinary Engineering for Cyber-Physical Production Systems

Stefan Biffl, Detlef Gerhard, and Arndt Lüder

This chapter summarizes and reflects on the material presented in this book regarding challenges and solutions for the required information processing and management in the context of the multi-disciplinary engineering of production systems.

The intention of this book has been to give insights into the requirements, technologies, and trends related to multi-* engineering, where * can stand for discipline, domain, and/or model. Therefore, the book discusses different views on the multi-disciplinary and multi-model nature of engineering processes looking on topics such as data and model integration, dependencies between the different objects to be engineered (especially products and their production systems), and the resulting needs with respect to information technology. All chapters in the book contribute to this discussion, each of them highlighting specific views and perceptions of the field of interests and each of them targeting different aspects of the selected research questions. The chapter authors describe the state of the art as well as research directions within their field and sketch the relevance of their findings towards the given research questions.

This chapter summarizes the discussions above. Related to the *research questions* (RQs), this chapter draws conclusions based on the chapter discussions in the three book parts. It is the intention of this chapter to consider which portions of

S. Biffl (✉) • D. Gerhard
Technische Universität Wien, Wien, Austria
e-mail: Stefan.Biffl@tuwien.ac.at; Detlef.Gerhard@tuwien.ac.at

A. Lüder
Otto-von-Guericke Universität Magdeburg, Magdeburg, Germany
e-mail: Arndt.Lueder@ovgu.de

© Springer International Publishing AG 2017

S. Biffl et al. (eds.), *Multi-Disciplinary Engineering for Cyber-Physical Production Systems*, DOI 10.1007/978-3-319-56345-9_17

the research questions are well addressed and to identify open issues for further investigation and research. Therefore, this chapter summarizes the contributions of the chapters towards the different research questions, and sum up the findings to discuss the state of the replies to research questions and open issues for future research.

The research questions M1 and M2 target advanced modelling capabilities applicable within the lifecycle of production systems. Thereby, the information sets created and applied within the cyber-physical production system (CPPS) lifecycle shall be representable reflecting both integration of the different lifecycle phases and different related views and disciplines as well as ensuring the consistency of modelled information sets. The applicable chapters contribute to the detailed consideration of these research questions in several ways.

RQ M1: Modelling the structure and behavior of CPPS. *How can model-based methodologies help address the specific multi-disciplinary requirements for the representation of the structure and behavior of CPPS?* Chapters 2, 5, 6, 7, and 11 address this research question.

Chapter 2 discusses the different procedural approaches within the design of technical systems. It compares the traditional multi-domain-oriented engineering approach of cooperating engineering domains with the systems-engineering-oriented approach. Following the comparison, the different impact factors on the engineering of a technical system are characterized and the importance of model-based engineering is highlighted. The chapter can be seen as plead for the application of modelling language sets, such as UML and SysML.

Chapters 5, 6, and 7 deal with the importance of information related to production systems components along the complete lifecycle of a production system. Together, these chapters identify a hierarchy of production system components and a raw version of a production system lifecycle (Chap. 5), depict the relevant sets of engineering-time-, run-time-, and deconstruction-time-related information for the different layers within this hierarchy (Chap. 6), and name relevant modelling means for the identified information sets (Chap. 7). Together the chapters provide a detailed consideration of all relevant information to be managed within the lifecycle of a cyber-physical system (CPS) on the different layers of a production system.

Finally, Chapter 11 examines the challenges of Model-Driven Systems Engineering especially against the background of applied modelling frameworks, such as SysML, MARTE, PMIF, and *AutomationML*. It sketches the different capabilities of these languages for requirement modelling, data exchange, or analysis, and discusses the possible combination of the different languages.

Altogether, the five chapters give a comprehensive overview on information sets relevant within the engineering of technical systems and on methodologies and technologies applicable to solve the challenges arising from the multi-domain nature of these information sets. The chapters discuss relevant requirements for model-based engineering and explore approaches for automating the multi-disciplinary engineering activities.

In addition, these chapters highlight an important, mostly open, challenge for the model-driven engineering of technical systems: methodologies transform models between modelling languages and methodologies to guarantee consistency between different models of the same modelling language or different modelling languages. In closed modelling frameworks, such as SysML, such transformations and consistency checks are possible, but in case of lifecycles of technical systems with discipline-crossing modelling technologies, there are several open issues, such as the identification of equal objects, the mutual mapping of model elements, the management of different levels of detail within and across the models, or the management of dependencies between characteristic parameters of modelled objects within and across the different applied models. Therefore, the chapters all together give important insights into the RQ without being able to solve the RQ challenges completely.

RQ M2: Modelling in CPPS lifecycle phases. *How can model-based methodologies support information creation and processing in the different lifecycle phases of a CPPS?* Chapters 2, 3, 4, 5, 6, 11, 12, 15, and 16 address this research question.

Chapter 2 provides to the discussion of RQ M2 the description of dependencies between engineering information sets and the sequential enrichment of information along the engineering process. Thereby, the chapter gives insights into the required tool support for multi-domain and systems engineering.

Chapter 3 considers the special case of CPPS enriched by *Product Service Systems* (PSS). This chapter discusses the importance of modelling means within requirement engineering and business cases of such enriched CPPS.

Chapter 4 deals with the universe of Product Lifecycle Management (PLM) systems and their use within the product (and production system) lifecycle. The chapter emphasizes the dependencies between product engineering, production system engineering, production, and product use, all of them considering product-related information with different focus points and related to different objects (classes of products versus instances of products). This chapter especially discusses information flows and the usability of IT technologies to enable a specific information flow.

Chapter 5 provides a short model of the lifecycle of a production system and its components. Based on this lifecycle, Chapter 6 discusses information reuse scenarios within and across the lifecycle phases *engineering*, *use*, and *end of life* of a production system. Chapter 6 describes and discusses requirements towards the information creation, distribution, and use related to production system components.

Chapter 11 describes model-based engineering technologies and their inherent information flows. Model exchange and model use are characterized to provide a set of standard modeling approaches that fit well to each other and can be iteratively extended to cover further modeling needs.

Chapter 12 discusses *Semantic Web* technologies, which assist the information use within multi-domain engineering by enabling the representation of information dependencies, including consistency rules. Thereby, automatic information use and reasoning about modelled information becomes possible enabling more comprehensive information use along the CPPS lifecycle.

Chapter 15 considers another special problem within the lifecycle of a production system, the production ramp-up process. The main emphasis of the chapter is on the required modelling of production processes and their relation to production resources and products. Thus, the chapter reflects on and models information emerging from and information used in different phases of the CPPS lifecycle to enable automatic reasoning, e.g., on production feasibility and ramp-up risks.

Chapter 16 studies the necessary quality assurance of engineering information in the CPPS lifecycle phases. The chapter describes a methodology to ensure engineering quality with model review based on a selected modelling methodology.

All chapters together present different views on the use of models to support information creation, exchange, and use. As this field is very broad, even the considerable number of chapters cannot consider all facets of model-based methodologies supporting information creation and processing. Nevertheless, the provided overview enables an evaluation of existing methodologies applicable for the automation of engineering, commissioning, and use of CPPS, the evaluation of capabilities to exploit services within the CPPS lifecycle, and to evaluate methodologies supporting quality assurance within engineering and run-time information of CPPS. Thus, these chapters enable the solution of RQ M2.

Beyond the evaluation, the chapters highlight needs of further research, such as the need for more detailed methodologies for consistency management within model systems as well as for a more detailed consideration of the different requirements along the lifecycle of CPPS, ranging from a component to a complete system.

Research questions I1, I2, and I3 consider the field of information representation within the lifecycle of production systems, focusing on CPPS. Here, modelling methodologies and model applications for different purposes are of interest.

RQ I1: Information integration in and across value chains. *Which methods and technologies support the integration of information within and across value chains of products, production systems, and production technologies? Are there benefits accessible from the exploitation of CPPS?* Chapters 2, 4, 6, 7, 9, 10, 11, 12, 14, and 15 contribute to addressing this research question.

Chapter 2 provides an overview on the engineering process structure for technical systems as well as on challenges emerging from the interdisciplinary views of model engineering, which point out links between product, production technology, and production systems engineering.

Chapter 4 introduces the concept of *Product Lifecycle Management* (PLM) for CPPSs and discusses data and information management issues, PLM activities, methods, and tools in a model-based, multidisciplinary context, therefore linking products to production technology and to production systems engineering. The chapter covers the phases engineering, production, and operation/use of smart products and CPPS, considering different product types and production concepts as well as forward and backward information flows between the phases, which provide needs and conceptual solutions for digital links between engineering and operation phases.

Chapter 6 identifies layers of CPPS within the production system. For each of these layers, Chapter 7 discusses the required information along all phases of the production system lifecycle as a foundation for analyzing needs and options for horizontal and vertical integration within production systems and production value chains.

Chapter 9 provides and discusses a selection of tool chains to support workflows in production system engineering. Further, relevant data formats for the horizontal integration in tool chains for production systems engineering are investigated, including the emerging standard *AutomationML*, which is illustrated with a case on virtual commissioning of a production system.

Chapter 10 discusses requirements for standardization in information exchange in production systems engineering and evaluates selected data exchange standards and formats, in particular the emerging standard *AutomationML*, as a foundation for better horizontal and vertical integration within production systems and production value chains.

Chapter 11 provides and discusses a selection of *Model-Driven System Engineering* (MDSE) standards to support better horizontal and vertical integration within the early phases of production systems engineering, in particular to facilitate the early verification of production system models regarding system structure, behavior, and non-functional system properties, such as performance, which is an important foundation for designing and evaluating CPPS.

Chapter 12 considers needs for semantic integration and identifies associated capabilities of *Semantic Web* technologies to support the horizontal and vertical integration of heterogeneous data generated by various tools and systems in multidisciplinary CPPS engineering.

Chapter 14 discusses on how to address issues of interoperability in CPPS, in particular heterogeneity, with *Service-Oriented Architecture* (SOA) approaches, SOA reference architectures, and derives a technology stack suitable for the operation of agile and flexible industrial plants.

Chapter 15 introduces a method for modeling, collecting, and analyzing production data in a multi-disciplinary knowledge base as foundation for the automated support of fact-based ramp-up project planning as a deterministic process.

Altogether, these chapters give insight into information modeling and information exchange along the lifecycle of a production system covering integration of engineering tools, engineering disciplines, and engineering roles. The chapters highlight the importance of information representation (modelling) and the application of common semantic concepts.

In addition, these chapters link to CPPS, mostly reflecting CPPS-based requirements to information modelling. Benefits possibly gained by exploiting CPPS based views have not been regarded. This seems to be an open issue for further investigation. Nevertheless, the discussed approaches have the potential to provide answers to RQ I1.

RQ I2: Quality assurance for information exchange. *Which methods and technologies support assuring the required information quality for information exchange?* Chapters 3, 6, 7, 9, 10, 11, 12, and 16 address this research question.

Chapter 3 introduces the concept of *Product Service Systems* (PSS), presents an interesting literature review on the challenges and benefits of PSS, and the study of required information in value chains related to CPPS as foundation for quality assurance on information provided for exchange along the production system lifecycle.

Chapter 6 defines three main lifecycle phases for production systems differing in information creation and use while Chap. 7 discusses the information required for different kinds of production system components along the phases of the production system lifecycle named in Chap. 6 as a foundation for quality assurance methods that can be applied within production systems and production value chains.

Chapter 9 reviews relevant data formats in software tool chains and their quality needs, motivating the need for the emerging standard *AutomationML*, which is illustrated with a case on virtual commissioning of a production system.

Chapter 10 discusses requirements for standardization in information exchange as foundation for studying the required information quality in different lifecycle phases of CPPS, and evaluates selected data exchange standards and formats, in particular the emerging standard *AutomationML*, regarding these criteria.

Chapter 11 provides and discusses a selection of *Model-Driven System Engineering* (MDSE) standards to facilitate the early verification of CPPS engineering models, a key enabler for many kinds of CPPS.

Chapter 12 explains how *Semantic Web* technologies support multi-disciplinary engineering processes by integration of heterogeneous data generated by various tools and systems as foundation for data quality assurance in multi-disciplinary knowledge.

Chapter 16 provides for *Model Quality Assurance* (MQA) a review process for *multi-disciplinary engineering* (MDE), which supports the detection of defects already in early phases of CPPS engineering.

All named chapters have one major commonality, the highlighting of the need of representation and application of dependencies and interrelations between information sets coming from several engineering disciplines, discipline-specific models, and lifecycle phases. In some cases, the chapters sketch initial ideas to solve these needs. However, altogether the chapters also show that there is still a wide field of research to be done to finally answer RQ I2. In particular, the chapters show that the identification and evaluation of the best-fitting or even relevant modelling means, e.g., semantic technologies, is still ongoing.

RQ I3: Description of plug-and-play capabilities and interfaces for engineering and run time. *Are there specific aspects of information exchange related to the lifecycle of CPPS?* The issues of plug-and-play capabilities of CPPS and interfaces for the linkage of engineering and run time are addressed, Chaps. 8, 9, and 15 intend to provide answers to this research question.

Chapter 8 on the engineering of next generation cyber-physical automation system architectures addresses the lack of implementation of control system architectures that have been developed and validated in the recent past by several research groups in industry. The chapter provides a summary of current alternative control system architectures that could be applied in industrial automation domain as well as a review of their commonalities. Further, the chapter points out the differences between the traditional centralized/hierarchical architectures and decentralized decision-making and control architectures. A vision of automation system integration over a common service-based infrastructure or integration layer that aims to ensure the transparent, secure, and reliable interconnection of shop-floor components and software systems in a plug-and-play fashion is proposed. However, the issue of linking the engineering and run time phases remains open.

Chapter 9 presents an overview on tool chains and the current workflows of mechanical design, electrical design, and software design in production system engineering process. The chapter highlights the necessity of a standardized data format to exchange engineering data along the entire production system engineering process, and presents *AutomationML* as example data format for the case of virtual commissioning of a production system. Nonetheless, several data formats are still required in the whole processes, in particular, if the interplay of product development and production system development needs to be addressed. Feedback of data from operation (run time) to production system engineering (e.g. virtual commissioning) in order to realize a digital twin is still an open issue not covered by any standard data format.

Chapter 15 introduces a method for modelling, collecting, and analyzing production data in a multi-disciplinary knowledge base on deterministic production ramp-up processes to support fact-based ramp-up project planning. So the interface of run time and engineering is directly addressed.

Altogether, these chapters discuss only a small part of the wide field of plug-and-play structures within the lifecycle of CPPS. Still, the chapters open up an essential view on the requirements within this area and discuss challenges to be tackled within the implementation of plug-and-play capabilities. Especially the topics of information modeling and the relation of information models to system architectures are discussed.

However, there are several open issues, such as the discussion of dependencies between engineering and run-time data, the mapping of required and available capabilities within a production system, or the identification of best-fitting data formats for information representation.

RQ C1: Modelling of CPPS flexibility and self-adaptation capabilities. *How can model-based approaches improve the flexibility and self-adaptation of production systems? What are the roles of product, production technology, and production system models in this context?* Chapters 4, 8, 13, and 15 address this research question.

Chapter 4 introduces the concept of *Product Lifecycle Management* (PLM) for CPPSs and discusses data and information management issues, PLM activities, methods, and tools in a model-based, multidisciplinary context, therefore linking products to production technology and production systems engineering. The chapter covers the phases engineering, production, and operation/use of smart products and CPPS, considering different product types and production concepts as well as forward and backward information flows between the phases, which provides needs and conceptual solutions for digital links between engineering and operation phases.

Chapter 8 gives a summary of non-hierarchical control system architectures that could be applied in industrial automation domain as well as a review of their commonalities. The chapter aims to point out the differences between the traditional centralized/hierarchical architecture and decentralized decision-making and control architectures. The chapter also explores the challenges and impacts that industries and engineers face in the process of adopting decentralized control architectures, analyzing obstacles for industrial acceptance and necessary new interdisciplinary engineering skills. In the end, the chapter gives an outlook on possible mitigation and migration activities required to implement decentralized control architectures.

Chapter 13 investigates existing studies of CPS with regard to advanced self-adaptation mechanisms, applied across the technology stack. From this investigation, the chapter authors derive recurring patterns, consolidating design knowledge on self-adaptation in CPS, in particular CPPS. The patterns and models can support future CPS designers with the realization and coordination of self-adaptation concerns. Finally, this chapter outlines a research agenda to advance self-adaptation and coordination in the domain of CPS.

Chapter 15 introduces a method for modeling, collecting, and analyzing production data in a multi-disciplinary knowledge base as foundation for the automated support of fact-based ramp-up project planning as a deterministic process. The chapter discusses the identification of capability fulfillment of product based production process requirements with resource-based production system capabilities and the exploitation of this identification.

Altogether, these chapters provide a detailed overview on the relations between production system architectures, product structures, and production technologies on the one hand and information representation on the other hand. The chapters give insight into modelling approaches that enable more flexibility in production systems by exploiting self-adaptation or even human-based adaptation.

Nevertheless, these chapters cannot provide a consistent approach. Rather, the chapters discuss building blocks that can be part of a solution to address RQ C1. Open issues related to the RQ include model management related to the different possibly involved modelling means or the necessary approaches for generating the information required within a self-adaptation process of CPPS.

RQ C2: Linking discipline-specific engineering views for flexible and self-adaptable CPPS. *How shall several disciplines in product and production system engineering be linked to support the engineering of flexible and self-adaptable CPPS?* Chapters 3 and 4 address this research question.

Chapter 3 introduces the concept of *Product Service Systems* (PSS), presents an interesting literature review on the challenges and benefits of PSS, and the study of required information in value chains related to CPPS.

Chapter 4 introduces the concept of *Product Lifecycle Management* (PLM) for CPPSs and discusses data and information management issues, PLM activities, methods, and tools in a model-based, multidisciplinary context, therefore linking products to production technology and to production systems engineering. The chapter covers the phases engineering, production, and operation/use of smart products and CPPSs, considering different product types and production concepts as well as forward and backward information flows between the phases, which provide needs and conceptual solutions for digital links between engineering and operation phases.

These two chapters give a view on the relations between engineering disciplines and their possible management. The chapters sketch approaches for dependency handling and for information propagation. Nevertheless, a consistent approach applicable in several cases is still lacking.

Beyond these intended relations between the book chapters and the research questions, all chapters within this book provide at least an implicit input to nearly all research questions. As examples, we point to Chaps. 4, 7, and 16.

Chapter 4 discusses with *engineering to order*, *build to order*, and *build to stock* different types of motivations for starting the product and/or the production system lifecycle. Thereby, the chapter discusses requirements for information creation, management, and use, having a strong impact on the modelling of structure and behavior of CPPS having an impact on the answers to RQ M1.

Chapter 7 discusses the information sets related to different lifecycle phases of production systems. As these information sets are partially interrelated, this chapter implicitly gives answers to the problem of linking engineering disciplines with each other and with runtime information. Thereby it gives input to RQ C2.

Finally, Chapter 16 discusses the quality management of engineering information. This quality management shall not be made only once in an engineering process but several times. Thereby, the fact becomes evident that the described approaches shall be applied within and between the different phases of the lifecycle of production system covering the different engineering steps and use phases (including redesign and maintenance). Thus, this chapter defines requirements to the information integration in and across value chains and, thereby, is related to RQ I1.

Summing up all chapters, this book discusses challenges and solutions for the required information representation, processing, and management capabilities within the context of multi-disciplinary engineering of products and production systems. The authors considered models, methods, architectures, and technologies applicable in CPPS use cases according to the viewpoints of product and production system engineering and their use. Thus, the main aim of this book is fulfilled.

As can be seen in this chapter, there are still several open issues for research identified within this book in relation to the research questions. The most relevant open issues are the following.

The information management within the complete lifecycle of products and production systems shall be based on commonly agreed and possibly standardized information representation means with more or less formal models as best choice. These information representation means shall be negotiated among all relevant stakeholders enabling a consistent and integrated information representation over all involved engineering and use activities, applied engineering tools and information processing entities of the use phase, as well as all involved disciplines during engineering and use.

Based on these information representation means, model integration and model consistency management shall be enabled. Different models views on the same object shall be transformable in to each other (clearly reflecting the different information sets they may cover). Dependencies between information sets within the different models must be ready for verification and appropriate evaluation.

Building on such a strong information representation foundation, an efficient information management shall be developed considering all relevant stakeholders along the complete lifecycle of a product and/or a production system. This approach shall enable a more efficient and high quality information generation, processing, and use.

Finally, the way to treat products as well as production systems within the combined engineering of products and production systems shall be reviewed. It can be discussed whether a system-engineering-oriented view on products and production systems is more likely to cover all requirements of the multi-disciplinary character of the information related to products and production systems compared to the current state of the practice.

Index

A
Action system, 51–53
Adaptability, 93
Adaptation, 335
Adapted review process, 442–444
Administration shell, 140
Agility of production systems, 410
AML.hub, 450
Analysis models, 33
Architecture, 332
Architecture-based adaptation, 335
Artifact class, 178
Artifacts, 175, 177–179
AS-IS scenario, 79
Assemble-to-Order (ATO), 92
Augmenting, 46
AutomationML (AML), 18, 245, 263, 274–276, 405, 434–455, 460
  standardization process, 253
AutomationML Editor, 442
Automation pyramid, 100

B
Behavior, 96
Big Data, 101
Bill-of-materials (BOM), 400
Bill of operations (BOO), 405
Business Model (BM), 65, 80
Business Model Canvas, 80
Business Process Model and Notation (BPMN), 79

C
CAEX, 438
Characteristics, 32, 400
Cloud, 192
Cloud manufacturing, 190
Collaboration, 95
Collaborative engineering, 97
COLLADA, 438
Collective Intelligence Systems (CIS), 333, 340, 357
Common concept ontology (CCO), 321
Compatibility, 411
Complexity, 92, 97
Component recovery, 133
Computer Aided Design (CAD), 90
Contextual factors, 46
Control activities, 131, 149
Conversion enablers, 411
Copy exactly, 409
Customization, 94, 97
Cyber-Physical Production System (CPPS), 2, 10, 12, 89, 261, 332, 402, 434, 460
  capabilities of, 13
  flexible and self-adaptable, 466
Cyber-Physical-Social Systems (CPSS), 65, 70, 357
Cyber-Physical Systems (CPS), 67, 89, 189, 332, 460
  capabilities, 199
  ecosystems, 200
  engineering, 199
  information systems, 200

© Springer International Publishing AG 2017

S. Biffl et al. (eds.), *Multi-Disciplinary Engineering for Cyber-Physical Production Systems*, DOI 10.1007/978-3-319-56345-9

infrastructures, 200
management, 199
Cyber Twin, 93

D
Data analytics, 101
Data and information flows, 29
Data exchange capabilities, 452
Data integration, 302, 313–314, 319
Defect detection performance, 452
DefectRadar, 440
Dependability, 96
Description means, 171, 178
Design for supply chain management, 412
Development process, 34
Digital factory, 93
Digital twin, 93
Distributed service-based integration layer, 193

E
Eclipse modeling framework (EMF), 437
EGerrit, 437, 440
Electrical design, 209
End-of-life (EoL) phase, 132–134, 159
artifacts, 160–163
scenarios, 132
Engineering artifacts, 142
Engineering chains, 237
Engineering models, 435
Engineering phase, 127–130
Engineer-to-Order (ETO), 5, 92
Enterprise Resource Planning (ERP), 90
Extended product, 69

F
Factory Social Media, 359
First pass yield (FPY), 406
Flexible and self-adaptable CPPS, 466
Flexible comparison, 318
Four Actions Framework, 82

G
Gerrit, 437
GIT, 440
Goal system, 49–50

H
Hierarchy layers, 134
Horizontal and vertical integration, 238
Horizontal data integration, 302
Hybrid ontology approach, 313
Hybrid ontology model, 318, 320

I
IDEF0, 79
IEC62264-3, 91
Individualization, 97
Industrie 4.0 component (I4.0 component), 64, 114, 140, 147, 159, 400, 438
Information driven production systems, 251
Information exchange technology, 244
Information integration, 462
Information representation means, 467
Information types, 173, 176, 178
Information utilization, 126
Inspections, 437
Instance hierarchy, 447
Internet of Things (IoT), 90
Interoperability, 96
ISA-95, 189
ISA-95/IEC 62264, 188
ISO 10303, 99
ISO 14306, 99
IT structures, 42, 44, 53

K
Knowledge integration, 316
Knowledge management, 103
Knowledge reuse, 103, 317

L
Life cycle integration, 98
Life cycle phases, 127
Linked data, 314
technologies, 311
Literature review, 118
Local ontologies, 321
Lot-size-one, 94

M
Machines-in-the-loop, 360
Machining tools, 95
Maintenance (service), repair, and overhaul (MRO), 91
Make-to-Order (MTO), 5, 92
Make-to-Stock (MTS), 92
Manufacturing execution systems (MES), 90
Matchmaking scenarios, 417–426
Material recovery, 133
Mechanical design, 209

Microservices, 105
Migration strategy, 194, 196–197
Mobility, 411
Model-based engineering (MBE), 54, 460
Model-based methodologies, 461
Model-based systems engineering (MBSE), 2, 42, 90
Model change review process, 436, 440
Model consistency management, 317–318
Model-driven engineering (MDE), 263, 265, 294, 437, 461
Model-driven system engineering (MDSE), 460, 463
Modeling and analysis of real-time embedded (MARTE) system, 263, 270–273
Modeling language, 436, 441
Model integration, 318
Model quality assurance (MQA), 434–455
Model review, 438–440
Model transformations, 263, 265–269, 277, 279, 288
Modularity, 411
Monitor, analyze, plan, execute and knowledge component (MAPE-K), 335
MQA-review, 442, 444, 448, 452–454
MQA-review needs, 451–452
Multi-agent systems (MAS), 189, 192, 336
Multi-disciplinary engineering (MDE), 434–455, 460, 464
Multi-disciplinary knowledge base, 463
Multi-* engineering, 15, 459
Multiple ontology approach, 313

N
Numerical Control (NC), 91

O
Object system, 50–51
Ontologies, 311, 314
Ontology matching, 313
Operation and maintenance phase, 130–132, 147
Operation phase, 93
Optimization, 98
Original equipment manufacturer (OEM), 209
Overall equipment effectiveness (OEE), 94

P
Patterns, 339, 351
Performance Model Interchange Format (PMIF), 263, 273–274
Plan Manufacture (PM), 209
Plant owner, 209
Plant topology, 450
PLCOpen7, 438
Plug-and-play capabilities, 464
Plug-and-produce technology, 190, 192
Predictability, 96
Problem-solving process, 31
Process capability, 404, 451
  index, 408
Process concerns, 436, 441
Process models, 29
Process operations, 404
Process segments, 405
Process system, 51
Process technology, 413
Product, 6
Product data management (PDM), 90
  systems, 43
Product design, 2
Production planning, 131
Production process, 7, 94
Production system, 6
  architectures, 118
  builder, 208, 209
  engineering, 2
  hierarchy layers, 119, 121
  life cycle, 7, 125, 236
  recovery, 133–134
Production technologies, 10
Productive use, 209
Product life cycle, 7, 92, 237
Product Lifecycle Management (PLM), 72, 90
Product manufacturing information (PMI), 99
Product models, 33
Product platforms, 414
Product Service Systems (PSS), 64, 67, 69
  lifecycle, 74
Programmable Logic Controller (PLC), 91, 208
Properties, 32

Q
Quality assurance, 438–440, 463
QUDT, 428
Queuing network (QN), 263, 269, 273, 277, 280, 282, 283, 294

R
Ramp-up phase, 93
Reading techniques, 439
Reasoning, 316–317

Reconfiguration, 96
Recycling/reuse, 209
Reference Architecture Model Industrie 4.0 (RAMI4.0), 200
Reliability, 96
Requirements engineering (RE), 65, 78
Resource Description Format (RDF(s)) Schema, 312
Reuse, 144, 156, 163
Reviewing language, 444–448
Review process, 439, 442–444
Reviews, 437
RoleClass, 447
Rolled throughput yield (RTY), 406
Round-trip-engineering, 449–451
  process, 450

S
Scalability, 411
Security, 96
Self-adaptation, 96, 332
  mechanisms, 466
Self-* capabilities, 192
Self-configuration, 96
Self-optimization, 96
Self-organization, 96, 336
Semantic mediation, 79
Semantic modelling, 314
Semantics, 251
Semantic Web, 402
Semantic Web knowledge representation languages, 312
Semantic Web technologies (SWTs), 302, 310, 435
Serious gaming, 79
Service Life-cycle Management (SLM), 72, 73
Service management, 91
Service-oriented architecture (SOA), 105, 189, 192
Servitization, 64, 69
Simulation, 98
Single ontology approach, 313
Six Paths Framework, 82
Smart factories, 101
Smart production, 100
Smart products, 94
Software design, 209
Spiral model, 37
Stakeholder needs, 436–437
Standard, 438
Standardization, 251
  in information exchange, 463
STEEP-Analysis, 80
Storytelling method, 78
Strategy Canvas, 80
Sustainability, 96
Syntax, 251
Systematic mapping study, 339
Systematic review, 435, 436
System development, 209
Systems engineering, 40
Systems Modeling Language (SysML), 79, 263, 269–271, 275, 277–286, 288–290, 460, 461
System unit, 448
SystemUnitClass, 447

T
Tailoring, 46
Technical functionality, 121
TO-BE scenario, 79
Tool capabilities, 451–452, 453
Tool support, 452–454
Traditional automation control systems, 188

U
Uncertainty, 332, 333, 345
Unified Modeling Language (UML), 460
Universality, 411
UNSPSC, 428

V
VDI 2206, 91
  V-model, 98
VDI 2219, 91
VDI 2221, 91
VDI 5600, 100
Virtual Commissioning (VC), 209
Virtual engineering (VE), 209
Virtualization, 93
Virtual representation, 140
V-model, 36

W
Waterfall model, 36
Web ontology language (OWL), 312
Web services, 190, 192

Printed by Printforce, the Netherlands